Marine Composites

Related Titles

Advanced Composite Materials for Aerospace Engineering
(ISBN: 9780081009390)

SHM in Aerospace Structures
(ISBN: 9780081001486)

Lightweight Composite Structures in Transport Applications
(ISBN: 9781782423256)

Marine Applications of Advanced Fiber-reinforced Composites
(ISBN: 9781782422501)

Woodhead Publishing Series in Composites Science and Engineering

Marine Composites

Design and Performance

Edited by

Richard Pemberton

John Summerscales

Jasper Graham-Jones

Woodhead Publishing is an imprint of Elsevier
The Officers' Mess Business Centre, Royston Road, Duxford, CB22 4QH, United Kingdom
50 Hampshire Street, 5th Floor, Cambridge, MA 02139, United States
The Boulevard, Langford Lane, Kidlington, OX5 1GB, United Kingdom

Copyright © 2019 Elsevier Ltd. All rights reserved.

No part of this publication may be reproduced or transmitted in any form or by any means, electronic or mechanical, including photocopying, recording, or any information storage and retrieval system, without permission in writing from the publisher. Details on how to seek permission, further information about the Publisher's permissions policies and our arrangements with organizations such as the Copyright Clearance Center and the Copyright Licensing Agency, can be found at our website: www.elsevier.com/permissions.

This book and the individual contributions contained in it are protected under copyright by the Publisher (other than as may be noted herein).

Notices
Knowledge and best practice in this field are constantly changing. As new research and experience broaden our understanding, changes in research methods, professional practices, or medical treatment may become necessary.

Practitioners and researchers must always rely on their own experience and knowledge in evaluating and using any information, methods, compounds, or experiments described herein. In using such information or methods they should be mindful of their own safety and the safety of others, including parties for whom they have a professional responsibility.

To the fullest extent of the law, neither the Publisher nor the authors, contributors, or editors, assume any liability for any injury and/or damage to persons or property as a matter of products liability, negligence or otherwise, or from any use or operation of any methods, products, instructions, or ideas contained in the material herein.

Library of Congress Cataloging-in-Publication Data
A catalog record for this book is available from the Library of Congress

British Library Cataloguing-in-Publication Data
A catalogue record for this book is available from the British Library

ISBN: 978-0-08-102264-1 (print)
ISBN: 978-0-08-101913-9 (online)

For information on all Woodhead publications visit our website at https://www.elsevier.com/books-and-journals

Publisher: Matthew Deans
Acquisition Editor: Gwen Jones
Editorial Project Manager: Charlotte Rowley
Production Project Manager: Swapna Srinivasan
Cover Designer: Christian Bilbow

Typeset by SPi Global, India

Dedication

The editors dedicate this book to all those who seek to prevent loss of life at sea, especially the Royal National Lifeboat Institution, HM Coastguard and the SOLAS Convention.

Contents

Contributors	xiii
Preface	xvii
Acknowledgments	xxiii

Part One	**Materials and process engineering**	**1**
Section A	**Materials selection, characterization and performance**	**2**

1 Materials selection for marine composites   3
John Summerscales
1.1	Introduction	4
1.2	The matrix	4
1.3	The reinforcement	14
1.4	The fiber-matrix interface	16
1.5	Reinforcement forms	18
1.6	Sandwich structures	21
1.7	Degradation of marine composites	23
1.8	Life cycle considerations	24
1.9	Conclusions	26
	Acknowledgments	26
	References	26

2 Thermoplastic matrix composites for marine applications   31
Mael Arhant, Peter Davies
2.1	Introduction	31
2.2	Material options	32
2.3	Manufacturing options	33
2.4	Influence of the marine environment on thermoplastic composites	39
2.5	Underwater structures	44
2.6	Repair	45
2.7	Recycling and environmental impact	46
2.8	Conclusion	48
	References	49

3	**Experimental and theoretical damage assessment in advanced marine composites**	**55**
	Phuong Tran, Abdallah Ghazlan, Tu Phan Nguyen, Rebecca Gravina	
	3.1 Introduction	55
	3.2 Damage to marine structures	59
	3.3 Nondestructive damage detection for maritime composites	64
	3.4 Numerical and theoretical modeling of composite damages	70
	3.5 Conclusions	79
	References	79
4	**Durability testing and evaluation of marine composites**	**85**
	Oliver Parks, Paul Harper	
	4.1 Introduction	86
	4.2 Loading and durability requirements	87
	4.3 Material selection	90
	4.4 Current sea water conditioning techniques	93
	4.5 Mechanical testing of saturated specimens	98
	4.6 Defining the limits of accelerated aging techniques	103
	4.7 Modelling of accelerated moisture absorption	106
	4.8 Constituent-level predictive methods	111
	4.9 Summary and future work	111
	References	112
5	**Fire performance of maritime composites**	**115**
	Quynh Thuy Nguyen, Phuong Tran, Xin Ren, Guomin Zhang, Priyan Mendis	
	5.1 Introduction	115
	5.2 Advanced polymer composites and design for maritime fire	117
	5.3 Test methods and requirements for fire safety of maritime composites	126
	5.4 Fire reaction of maritime composites	132
	5.5 Structural performance of maritime composite during fire and postfire mechanical performance	141
	5.6 Numerical analysis of naval composite structure performance in fire	146
	5.7 Enhancement of maritime composite structures subjected to fire	152
	5.8 Conclusions	156
	References	156
	Further reading	159
6	**Effective use of composite marine structures: Reducing weight and acquisition cost**	**161**
	Luis F. Sánchez-Heres, Jonas W. Ringsberg, Erland Johnson	
	6.1 Introduction	161
	6.2 General objective and methodology	162
	6.3 Material safety factors	163

6.4	Material characterization	170
6.5	Structural design exploration	174
6.6	Conclusions	181
References		182

Section B   Sandwich structures   185

7   Core materials for marine sandwich structures   187
*Nikhil Gupta, Steven Eric Zeltmann, Dung D. Luong,
Mrityunjay Doddamani*

7.1	Introduction	187
7.2	PVC foams	190
7.3	Syntactic foams	203
7.4	Summary	219
Acknowledgments		219
References		219

Section C   Manufacture   225

8   Resin infusion for the manufacture of large composite structures   227
Ned Popham

8.1	Introduction	228
8.2	Physics of resin infusion	231
8.3	Materials selection and characterization	237
8.4	Tooling	244
8.5	Plant equipment, setup, and redundancy	246
8.6	Infusion prediction, strategy, and setup	250
8.7	Resin delivery and management	258
8.8	Manufacturing process	261
8.9	Process control and preinfusion checks	263
8.10	Postinfusion management	266
8.11	Conclusion/summary	267
References		267

Section D   Advanced concepts and special systems   269

9   Smart composite propeller for marine applications   271
H.N. Das, S. Kapuria

9.1	Introduction	271
9.2	Flow solution	273
9.3	Deformation of composite propeller	274
9.4	Modeling of shape memory alloy	277
9.5	Fluid-structure interaction	279
9.6	Material failure	279

9.7	Analysis of different propellers	279
9.8	Conclusions	296
References		296
Further reading		297

Part Two Naval architecture and design considerations 299

10 A structural composite for marine boat constructions 301
Alexandre Wahrhaftig, Henrique Ribeiro, Ademar Nogueira

10.1	Introduction	301
10.2	Basic core materials	304
10.3	Composite structure concepts	305
10.4	Economic viability	307
10.5	Case study: A vessel structural computational design	309
10.6	Conclusions	313
References		313

Part Three Applications 315

11 Offshore wind turbines 317
Puyang Zhang

11.1	Introduction	317
11.2	The load-bearing characteristics of composite bucket foundation	321
11.3	Model tests on the bearing capacity of composite bucket foundation	329
11.4	Model tests on the installation of composite bucket foundation	333
11.5	Conclusions	342
References		343

12 Marine renewable energy 345
Ramona B. Barber, Michael R. Motley

12.1	Introduction	345
12.2	Bend-twist deformation coupling	347
12.3	General turbine design parameters	350
12.4	Composite-specific design considerations	354
12.5	Potential performance benefits of composites	358
12.6	Conclusions	359
References		360
Further reading		362

13 Propulsion and propellers 363
Y. Hong, X.D. He, G.F. Qiao, R.G. Wang

13.1	Introduction	363
13.2	The characteristics of composite propeller	364
13.3	The calculation and evaluation method of composite propeller	369

13.4	Performances of composite propeller	381
13.5	Conclusions and future trends	386
References		387

14 Offloading marine hoses: Computational and experimental analyses — 389
Maikson L.P. Tonatto, Pedro Barrionuevo Roese, Volnei Tita, Maria M.C. Forte, Sandro C. Amico

14.1	Introduction	391
14.2	Types of models	400
14.3	Offloading hoses: Computational and experimental analyses	406
14.4	Concluding remarks	413
Acknowledgments		413
References		414

15 Modern yacht rig design — 417
Hasso Hoffmeister

15.1	Introduction	418
15.2	"Why" is a rig?	419
15.3	Modern rig configurations	421
15.4	Selected design considerations	422
15.5	Why weight savings?	432
15.6	Material selection	432
15.7	Rig analysis technologies	437
15.8	Statics and dynamics	438
15.9	Rig loads	439
15.10	Design criteria; safety margins, reserve factors	442
15.11	Future trends	445
References		448
Further reading		448

16 Composite materials for mooring applications: Manufacturing, material characterization, and design — 451
Eduardo A.W. de Menezes, Laís V. da Silva, Filipe P. Geiger, Rogério J. Marczak, Sandro C. Amico

16.1	Introduction	452
16.2	Design of composite cables	456
16.3	Mathematical modeling of cables with linearized kinematics	459
16.4	Manufacturing of composite cables	463
16.5	Mechanical characterization and aging of composite cables	466
16.6	Finite element modeling of composite cables	476
16.7	Concluding remarks	485
Acknowledgments		486
References		486

Index — 491

Contributors

Sandro C. Amico Post-Graduation Program in Mining, Metallurgical and Materials Engineering, Federal University of Rio Grande do Sul (UFRGS), Porto Alegre, Brazil

Mael Arhant Marine Structures group, IFREMER (The French Ocean Research Institute), Plouzané, France

Ramona B. Barber Department of Civil and Environmental Engineering, University of Washington, Seattle, WA, United States

Laís V. da Silva Post-Graduation Program in Mining, Metallurgical and Materials Engineering, Federal University of Rio Grande do Sul (UFRGS), Porto Alegre, Brazil

H.N. Das Naval Science & Technological Laboratory, DRDO, Visakhapatnam, India

Peter Davies Marine Structures group, IFREMER (The French Ocean Research Institute), Plouzané, France

Eduardo A.W. de Menezes Post-Graduation Program in Mechanical Engineering, UFRGS, Porto Alegre, Brazil

Mrityunjay Doddamani Advanced Manufacturing Laboratory, Department of Mechanical Engineering, National Institute of Technology Karanataka, Surathkal, India

Maria M.C. Forte Post-Graduation Program in Mining, Metallurgical and Materials Engineering, Federal University of Rio Grande do Sul (UFRGS), Porto Alegre, Brazil

Filipe P. Geiger Post-Graduation Program in Mechanical Engineering, UFRGS, Porto Alegre, Brazil

Abdallah Ghazlan Department of Infrastructure Engineering, The University of Melbourne, Melbourne, VIC, Australia

Rebecca Gravina Department of Civil and Infrastructure Engineering, RMIT University, Melbourne, VIC, Australia

Nikhil Gupta Composite Materials and Mechanics Laboratory, Department of Mechanical and Aerospace Engineering, New York University Tandon School of Engineering, Brooklyn, NY, United States

Paul Harper Aerospace Engineering, University of Bristol, Bristol, United Kingdom

X.D. He Science and Technology on Advanced Composites in Special Environment Laboratory, Harbin Institute of Technology, Harbin, China

Hasso Hoffmeister DNV GL, Hamburg, Germany

Y. Hong Science and Technology on Advanced Composites in Special Environment Laboratory, Harbin Institute of Technology, Harbin, China

Erland Johnson RISE Research Institutes of Sweden, Department of Safety—Mechanics Research, Borås, Sweden

S. Kapuria CSIR–Structural Engineering Research Centre, Taramani, Chennai; Department of Applied Mechanics, Indian Institute of Technology Delhi, New Delhi, India

Dung D. Luong Composite Materials and Mechanics Laboratory, Department of Mechanical and Aerospace Engineering, New York University Tandon School of Engineering, Brooklyn, NY, United States

Rogério J. Marczak Post-Graduation Program in Mechanical Engineering, UFRGS, Porto Alegre, Brazil

Priyan Mendis The University of Melbourne, Parkville, VIC, Australia

Michael R. Motley Department of Civil and Environmental Engineering, University of Washington, Seattle, WA, United States

Tu Phan Nguyen Department of Infrastructure Engineering, The University of Melbourne, Melbourne, VIC, Australia

Quynh Thuy Nguyen The University of Melbourne, Parkville, VIC, Australia

Ademar Nogueira Department of Mechanical Engineering, Polytechnic School, Federal University of Bahia, Salvador, Brazil

Oliver Parks Aerospace Engineering, University of Bristol, Bristol; AEL Airborne, Hungerford, United Kingdom

Ned Popham Sunseeker International Limited, Poole, Dorset, United Kingdom

G.F. Qiao School of Civil Engineering, Harbin Institute of Technology, Harbin, China

Xin Ren Nanjing Tech University, Jiangsu, PR China

Henrique Ribeiro Bahia Federal Institute of Education, Salvador, Brazil

Jonas W. Ringsberg Chalmers University of Technology, Department of Mechanics and Maritime Sciences, Division of Marine Technology, Gothenburg, Sweden

Pedro Barrionuevo Roese Petrobras E&P, Rio de Janeiro, Brasil

Luis F. Sánchez-Heres Chalmers University of Technology, Department of Mechanics and Maritime Sciences, Division of Marine Technology, Gothenburg, Sweden

John Summerscales University of Plymouth, Plymouth, United Kingdom

Volnei Tita Department of Aeronautical Engineering, São Carlos School of Engineering, University of São Paulo, São Carlos, Brazil

Maikson L.P. Tonatto Post-Graduation Program in Mining, Metallurgical and Materials Engineering, Federal University of Rio Grande do Sul (UFRGS), Porto Alegre, Brazil; Centre for Innovation and Technology in Composite Materials (CIT^{e}C), University of São João Del Rei, São João Del Rei, Brazil

Phuong Tran Department of Civil and Infrastructure Engineering, RMIT University, Melbourne, VIC, Australia

Alexandre Wahrhaftig Department of Construction and Structures, Polytechnic School, Federal University of Bahia, Salvador, Brazil

R.G. Wang Science and Technology on Advanced Composites in Special Environment Laboratory, Harbin Institute of Technology, Harbin, China

Steven Eric Zeltmann Composite Materials and Mechanics Laboratory, Department of Mechanical and Aerospace Engineering, New York University Tandon School of Engineering, Brooklyn, NY, United States

Guomin Zhang Department of Civil & Infrastructure Engineering, RMIT University, Melbourne, VIC, Australia

Puyang Zhang State Key Laboratory of Hydraulic Engineering Simulation and Safety, Tianjin University,Tianjin, China

Preface

Introduction

The marine environment is challenging for traditional engineering materials due to the corrosion of metals or the bio-deterioration of natural materials. Consequently, the use of fiber-reinforced polymer matrix composites in the seas and oceans has grown in diversity of components, the size of structures and production numbers. Composites consistently demonstrate good performance, with many technologies and developments only realized because of their use. This chapter introduces the book, signposts to the topics covered, and briefly considers some of the recent innovations not otherwise included in the text.

The incorporation of fibres (e.g., aramid, carbon, glass) into a polymeric matrix produces a composite material. With well-selected constituents, these fiber-reinforced plastics (FRP) can provide excellent performance in the marine environment while being resistant to the biological and chemical attack that can compromise other materials. For readers new to the area of composites, we recommend the following starter texts (Scott Bader, 2005; Gurit, 2013; Hull and Clyne, 1996; Åström, 1997).

The use of composites in marine structures has been the subject of a number of earlier books (Smith, 1990; Eric Greene Associates, 1999; Shenoi and Wellicome, 2008a, b). This book seeks to provide a summary of some recent developments as a complement to Marine Applications of Advanced Fiber-Reinforced Composites (MAAFRC) (Graham-Jones and Summerscales, 2016), but is inevitably limited in its coverage as a comprehensive treatment would require a complete encyclopædia with the consequent costs. Over the past 5 years, there have been a number of review papers pertinent to the topic of this book, which complement the chapters of this text. Of especial importance are those addressing impact (Sutherland, 2018a, b, c, d), durability (Davies, 2014), marine fouling (Myan et al., 2013. Yang et al., 2014) and the impact of plastics on the marine environment (Wright et al., 2013. Gall and Thompson, 2015).

For the purpose of this book, we define composites as fiber-reinforced polymer (FRP) systems which use continuous fiber reinforcements. Further, we define four categories of composite:

- monolithic composite material: all layers aligned parallel,
- laminated composite structure: orientation changes between layers,
- hybrid structures: more than one type of fiber (e.g., carbon/glass),
- sandwich structures: composite skins and lightweight core.

One additional chapter addresses a different form of composites as an essential foundation for effective implementation of offshore renewable energy generation.

What is in the book ...

This book comprises two distinct sections and many of the topics build on those covered within MAAFRC. In Part one, the focus is on the materials and the processes associated with them. Chapter 1 describes materials selection for marine composites, which leads to Chapter 2, detailing thermoplastic matrices for composites. The ability of composites to withstand hostile environmental conditions are covered in Chapters 3–5. Chapter 6 considers using composites effectively for marine structures, both in terms of acquisition cost and savings due to weight reduction. Whilst manufacturing sandwich structures was described in Chapter 3 of MAAFRC, this book contains a review of core materials for sandwich structures in Chapter 7. Composite manufacturing methods were described in Chapter 2 of MAAFRC and this is developed further in Chapter 8, dealing in particular with the techniques required to infuse large scale structures. Novel materials are discussed, be they smart materials (Chapter 9) or an innovative variation on existing, well known materials (Chapter 10).

Part two focuses on specific applications of composites. The marine renewables sector is dealt with specifically in Chapter 11 (foundations for offshore wind turbines) and Chapter 12 (marine renewable devices), complementing Chapter 9 of MAAFRC. Chapter 13 considers the application and modeling of composite propellers, whilst Chapter 14 describes composite marine hoses. Chapter 15 describes large yacht masts, as a complement to MAAFRC Chapter 12 on the use of composites within the yacht rigging market. Finally, the use of composite materials for mooring applications is covered in Chapter 16.

What is not in the book ...

The RAMSSES (Realization and Demonstration of Advanced Material Solutions for Sustainable and Efficient Ships: https://www.ramsses-project.eu/) project aims to produce a 70 m long glass fiber reinforced vinyl ester composite hull, then test the structure under real-life conditions on the high seas.

The new EU Horizon 2020 FIBRESHIP (http://www.fibreship.eu/) research project aims to revolutionize shipbuilding by replacing steel with composite materials for the construction of light commercial vessels, passenger and leisure transport and oceanographic vessels over 5 m long.

A partnership between the Oak Ridge National Laboratory (ORNL) Manufacturing Demonstration Facility (MDF) and the Navy's Disruptive Technology Lab has produced the first 3D-printed submarine hull.

Innovative and high-performance craft

A variety of innovative new vessels have emerged since the previous book, including:

- the 42 m carbon fiber composite hybrid-electric sightseeing vessel "Vision of the Fjords", designed by Brødrene Aa,

- the two near-identical 53 m Latitude trimaran super-yachts "Galaxy" and "Galaxy of Happiness", and
- the 50 knot CFRP Princess Yachts/Ben Ainslie Racing (PY/BAR) R35 prototype super-boat with active foiling that promises to be one of the "most exciting and revolutionary products" the company has ever produced.

In the context of high-performance:

- Multiplast used North Thin Ply Technology prepregs, (as described in Chapter 14 of MAAFRC (Graham-Jones and Summerscales, 2016)), lightweight glue films and Automated Tape Laying to build the Groupama Team France yachts, allowing optimized lay-ups and fiber orientations in the aft wing flaps and removing critical weight high up in the yacht's wing rig.
- In June 2017, the New Zealand NZ America's Cup boat flew on foils for 100% of race.
- In July 2017, the Transpac monohull course record was smashed by "Comanche" with an elapsed time of 5 days, 1 hour, 55 minutes and 26 seconds: over half a day off the previous record.
- In December 2017, François Gabart in the 30m trimaran MACIF finished his solo circum-navigation of the globe in a record time of 42 days, 16 hours, 40 minutes and 35 seconds.

Novel applications

Further novel applications have also been published. Acciona (Spain) and Huntsman Advanced Materials (United States) have created an all-composite (CFRP, GFRP and hybrids) lighthouse, completed in less than 20 days with only 6 hours for installation using a lightweight crane. Yu et al. (2017) have reviewed the design and analysis of reinforced thermoplastic pipes for offshore applications. Weller et al. (2015) have recently reviewed the use of synthetic mooring ropes for marine renewable energy applications. The US Naval Station Mayport took delivery of the first set of fiber-reinforced polymer (FRP) composite "camels" (floating structures designed to separate a large vessel and the mooring wharf) for berthing nuclear powered aircraft carriers

Marine renewable energy devices

An important potential market for composites is marine renewable energy where potential devices have been reviewed by Chen et al. (2013) and Chen and Lam (2015). Ocean Renewable Power has deployed a second next-generation, commercial-scale ocean tidal energy power turbine generator unit (TGU) in the Bay of Fundy on the Canada-US border with helical composite foils and a hybrid carbon/glass fiber-reinforced composite direct driveshaft. Three composite rotor blades, manufactured by AC Marine and Composites, have been installed on the 1.5 MW AR1500 tidal turbine located in the Pentland Firth between the Orkney Islands and Caithness.

End-of-life

The market for marine leisure has expanded rapidly throughout the period since composites became the material of choice for such products. A recent estimate suggested there are about one million boats in France and 600,000 in the United Kingdom which will reach their end-of-life in the next decades. APER (https://www.aper.asso.fr/) reported that only 20% of end-of-life craft are currently partly or totally recycled, with the other 80% buried or incinerated. ECONAV (http://www.econav.co/) suggested that the construction stage of recreational boat represents only 20% of their carbon footprint. There is scope for the development of a circular economy to enhance the ecological standing of end-of-life boats (ELB) with potential for complementary disposal routes for other large composite structures (e.g., wind turbine blades).

A European Commission paper on nautical tourism (European Union, 2016), identifies that there is only a limited scale recycling and dismantling for ELB, reflecting the unfavourable economics of the business (high costs and few revenue opportunities). In turn, this discourages operators from providing facilities and boat owners from seeking appropriate means of disposal. A lack of boat owner registration systems makes effective monitoring, control and enforcement of ELB rules difficult. All stakeholders must become engaged to consider systems to collect and deconstruct these vessels and equipment, to process the waste streams, and to develop markets for the recycled materials.

A European Commission report (Directorate-General for Environment Financial Incentive for Ship Recycling, 2016) has recommended that any ship over 500 gross tonnage (gt) calling at an EU port would need to pay for a ship recycling licence. The ultimate ship-owner would be entitled to a proportion of the accrued Ship Recycling Fund if the vessel were recycled at an EU approved facility, but would forfeit the rights to this payment if the ship were scrapped at a nonapproved facility. However, this is a consideration for the long term future given the current limited number of vessels (e.g., Royal Navy Hunt class Mine Counter Measures Vessels: 725 tons, Tripartite minehunters: 605 tons, Swedish Navy Visby stealth corvette: 600 tons) above the specified displacement, although composite components of larger vessels (e.g., superstructures on cruise liners and the 900 tons composite deckhouse of the US Navy DDG 1000 destroyer) would be implicated.

The technologies for ELB were reviewed in MAAFRC Chapter 8.

Enjoy the book!

Jasper, John and Richard.

Richard Pemberton
John Summerscales
Jasper Graham-Jones

References

Åström, B.T., 1997. Manufacturing of Polymer Composites. CRC Press, Boca Raton, FL, ISBN-13: 978-0-7487-7076-2.

Chen, L., Lam, W.-H., 2015. A review of survivability and remedial actions of tidal current turbines. Renew. Sustain. Energy Rev. 43, 891–900.

Chen, Z., H, Y., M, H., Meng, G., Wen, C., 2013. A review of offshore wave energy extraction system. Adv. Mech. Eng. J.. 5623020.

Davies, P. (Ed.), 2014. Durability Of Composites In A Marine Environment. Springer, Dordrecht, ISBN 978-94-007-7416-2.

Directorate-General for Environment Financial Incentive for Ship Recycling, 2016. Financial Instrument to Facilitate Safe and Sound Ship Recycling—Final Report. In: Directorate-General for Environment Financial Incentive for Ship Recycling, European Union, Luxembourg, ISBN 978-92-79-59773-2.

European Union, 2016. Assessment Of The Impact Of Business Development Improvements Around Nautical Tourism—Final Report. European Union, Luxembourg, ISBN 978-92-79-67732-8.

Gall, S.C., Thompson, R.C., 2015. The impact of debris on marine life. Marine Pollut. Bull. 92 (1–2), 170–179.

Graham-Jones, J., Summerscales, J. (Eds.), 2016. Marine Applications of Advanced Fibre-Reinforced Composites. Elsevier/Woodead, Cambridge, ISBN 978-1-78242-250-1.

Gurit, 2013. Guide_to_Composites. Gurit, Newport ∼ Isle of Wight.

Hull, D., Clyne, T.W., 1996. An Introduction to Composite Materials, second ed. Cambridge University Press, Cambridge ISBN-10: 0-521-38855-4.

Eric Greene Associates, 1999. Marine Composites, second ed. Eric Greene Associates, Annapolis, MD, ISBN 0-9673692-0-7.

Myan, F.W.Y., Walker, J., Paramor, O., 2013. The interaction of marine fouling organisms with topography of varied scale and geometry: a review. Biointerphases 8, 30.

Scott Bader, 2005. Crystic Composites Handbook. Scott Bader, Wollaston.

Shenoi, R.A., Wellicome, J.F. (Eds.), 2008a. Composite Materials in Maritime Structures. Fundamental Aspects, vol. 1. Cambridge University Press, Cambridge, ISBN 978-0-521-08993-7.

Shenoi, R.A., Wellicome, J.F. (Eds.), 2008b. Composite Materials in Maritime Structures. Practical Considerations, vol. 2. Cambridge University Press, Cambridge, ISBN 978-0-521-08994-4.

Smith, C.S., 1990. Design of Marine Structures in Composite Materials. Elsevier Applied Science Publishers, London, ISBN 1-85166-416-5 ISBN.

Sutherland, L.S., 2018a. A review of impact testing on marine composite materials: part I—marine impacts on marine composites. Comp. Struct. 188, 197–208.

Sutherland, L.S., 2018b. A review of impact testing on marine composite materials: part II—impact event and material parameters. Comp. Struct. 188, 503–511.

Sutherland, L.S., 2018c. A review of impact testing on marine composite materials: part III—damage tolerance and durability. Comp. Struct. 188, 512–518.

Sutherland, L.S., 2018d. A review of impact testing on marine composite materials: part IV—scaling, strain rate and marine-type laminates. Comp. Struct. 200, 929–938.

Weller, S.D., Johanning, L., Davies, P., Banfield, S.J., 2015. Synthetic mooring ropes for marine renewable energy applications. Renew. Energy 83, 1268–1278.

Wright, S.L., Thompson, R.C., Galloway, T.S., 2013. The physical impacts of microplastics on marine organisms: a review. Environ. Pollut. 178, 483–492.

Yang, W.J., Neoh, K.-G., Kang, E.-T., Teo, S.L.-M., Rittschof, D., 2014. Polymer brush coatings for combating marine biofouling. Prog. Polym. Sci. 39 (5), 1017–1042.

Yu, K., Morozov, E.V., Ashraf, M.A., Shankar, K., 2017. A review of the design and analysis of reinforced thermoplastic pipes for offshore applications. JRPC 36 (20), 1514–1530.

Acknowledgments

We are most grateful to Gwen Jones, Charlotte Cockle, and Charlotte Rowley at Elsevier/Woodhead for initiating this project and driving us to completion in a timely manner. Further, we commend the excellent work of all at SPI Publishing Services for their accurate typesetting and picking up fine detail we missed, and especially Swapna Srinivasan at Elsevier Global Book Production for coordinating the final corrections to the proofs.

Our sincere thanks go to the many colleagues, friends, and family, who have provided inspiration, suggestions and ideas, especially Lucy Pemberton, Dr Elke Graham-Jones and Carolyn Thomas.

Part One

Materials and process engineering

Section A

Materials selection, characterization and performance

Materials selection for marine composites

John Summerscales
University of Plymouth, Plymouth, United Kingdom

Chapter Outline

- 1.1 Introduction   4
- 1.2 The matrix   4
    - 1.2.1 Thermosetting resins   5
    - 1.2.2 Thermoplastic polymers   8
- 1.3 The reinforcement   14
    - 1.3.1 Aramid fibers   14
    - 1.3.2 Carbon fibers   14
    - 1.3.3 Glass fibers   16
    - 1.3.4 Other reinforcement fibers   16
- 1.4 The fiber-matrix interface   16
- 1.5 Reinforcement forms   18
    - 1.5.1 Woven fabrics   19
    - 1.5.2 Stitched or knitted fabrics   19
    - 1.5.3 Braids   20
    - 1.5.4 Three-dimensional woven fabrics   20
    - 1.5.5 Preforms   21
    - 1.5.6 Preimpregnated reinforcements (prepregs)   21
- 1.6 Sandwich structures   21
- 1.7 Degradation of marine composites   23
    - 1.7.1 Diffusion   23
    - 1.7.2 Osmosis and blistering   23
    - 1.7.3 Environmental stress concentration   24
    - 1.7.4 Marine fouling   24
    - 1.7.5 Cavitation erosion   24
    - 1.7.6 Galvanic corrosion   24
- 1.8 Life cycle considerations   24
    - 1.8.1 Microplastics   25
    - 1.8.2 Marine-sourced materials   25
- 1.9 Conclusions   26
- Acknowledgments   26
- References   26

1.1 Introduction

The salt-water environment of the seas and oceans is corrosive to most engineering metals and, in combination with marine animals such as the naval shipworm (*Teredo navalis*) and gribble (*Limnoriidae*), causes rapid deterioration of wood. In consequence, the excellent properties of fiber-reinforced polymer-matrix composites, using the continuous fibers that became available during the middle of the 20th century, have come to dominate the materials of choice for marine structures.

The majority of marine structural composites employ *E*-glass fibers in an unsaturated polyester resin matrix. Where higher stiffness is required, and cost allows, the structures may be manufactured using carbon-fiber-reinforced epoxy resin. Different reinforcement and matrix materials may be selected to sensibly meet the design requirements and end-of-life considerations for a specific application.

1.2 The matrix

The matrix is the medium that transfers load from the external environment into the reinforcement fibers. The term resin will be used exclusively for thermosetting systems below, while thermoplastics will be termed polymers. This terminology is specific to the following text and the reader may find these terms used in either context in other texts. The thermosetting resins are generally of single use (i.e., not amenable to easy recycling), whereas thermoplastic-matrix systems can be recycled with relative ease.

A polymer will normally have more than one characteristic temperature, including (in the normal ascending order):

- T_g: the glass transition temperature.
- T_v: the topology freezing transition temperature (for vitrimers: may be above or below Tg)
- T_c: the peak crystallization temperature
- T_m: the crystalline melting point (not applicable to amorphous polymers).
- T_p: the processing temperature (for thermoplastics).
- T_d: the degradation temperature.

As the temperature rises through the *glass transition temperature*, short segments of the polymer backbone with insufficient energy for movement other than atomic vibration start to move as a group of atoms. On cooling through this temperature, it is normal to refer to segmental motion being frozen out. The mechanical properties of the polymer are then:

- below Tg: normally elastic and brittle (with good resistance to creep deformation)
- above Tg: normally viscoelastic and tough (however, creep deformation can be a problem)

Tm may be a narrow range of temperatures rather than a single point. Wholly amorphous and the amorphous part of partially crystalline polymers do not have a crystalline melting point.

1.2.1 Thermosetting resins

The principal commercial groups of thermosetting resins are (a) phenolic resins, (b) epoxy, (c) unsaturated polyester, and (d) vinyl ester. These materials are normally supplied as a liquid resin that can be solidified using chemicals and heat. The reaction results in a solid which will degrade rather than melt, so end-of-life issues are primarily about disposal rather than recycling. For thermosetting resins, the glass transition temperature generally follows the maximum temperature experienced during the cure cycle.

1.2.1.1 Phenol-formaldehyde resin (PF)

Phenol-formaldehyde resins are among the earliest polymer systems to be exploited commercially. The base resin is manufactured by a condensation reaction between phenol (C_6H_5OH: CAS 108-95-2) and formaldehyde (HCHO: CAS 50-00-0). The system is cured (cross-linked) by continuation of the condensation reaction.

1.2.1.2 Epoxide resin (Ep)

The epoxide ring can be considered as a di-alcohol (abbreviated to "diol": two –OH alcohol groups) on adjacent carbon atoms with water removed to create a highly strained (~60 degree bond angles instead of 109° 28″ normally associated with sp^3 bonding) two-carbon-and-one-oxygen three-membered ring. Epoxy resins are normally manufactured by a condensation reaction between a phenolic compound and epichlorohydrin ($CH_2(O)CHCH_2Cl$: CAS 106-89-8). The system is cured by ring-opening of the epoxide (a.k.a. oxirane) group initiated by hardeners (normally acids, anhydrides, amides, or amines). The cross-linking (curing) reaction occurs without release of the water molecules (eliminated on oxirane ring formation), reducing the formation of voids in the resin/composite.

The principal base resins for epoxy systems are (a) diglycidyl ether of bisphenol A (DGEBA) and (b) tetraglycidyl 4,4′-diaminodiphenylmethane (TGDDM). The glycidyl chemical entity is the epoxy (oxirane) ring bonded to a methylene group ($CH_2(O)CH-CH_2^-$). Table 1.1 shows how the Tg of DGEBA epoxy resins changes with the chosen hardener (curing agent).

There have been recent developments towards recyclable epoxy resins using cleavable amine hardeners, including (a) Recyclamine (Connora Technologies—USA), which enables the recovery of the thermoset epoxy resin as its thermoplastic counterpart, and (b) Cleavamine® (Adesso Advanced Materials—China and UK), aliphatic and aromatic curing agents in Recycloset® rAFL-1001 epoxy resin systems.

1.2.1.3 Unsaturated polyester resin (UPE)

Unsaturated polyester resins are normally manufactured by a condensation reaction between a di-acid and a diol where some of the monomers contain double bonds (unsaturation). The resin is normally supplied diluted by styrene ($C_6H_5CHCH_2$:

Table 1.1 Glass transition temperatures for DGEBA resins cured with aromatic diamines at the stoichiometric mix ratio

Resin	Hardener	Tg(°C)	Test	Source
Epon 828/DDM	DDM	170	DMA	Galy et al. (1986)
Pure DGEBA/DDM	DDM	176	DSC	Bellenger et al. (1987a), Bellenger et al. (1987b), and Bellenger et al. (1988)
Pure DGEBA/DDS	DDS	184	n/a	Zukas (1994)
EPN834/DDS	DDS	186	n/a	Zukas (1994)
DER332/DDS	DDS	190	DSC	Galy et al. (1986)
EPN828/DDS	DDS	211	n/a	Zukas (1994)
EPN825/DDS	DDS	222	n/a	Zukas (1994)

Relative reactivity is ranked 4,4' DDS* < 3,3' DDS < 4,4' DDM ≈ m-PDA, where DDM = diaminodiphenylmethane or DDS = diaminodiphenylsulphone, (Varma and Gupta, 2000).

CAS 100-42-5, a reactive diluent which also has unsaturation) and cure takes place by an addition-reaction initiated by a peroxide. During cure, the initiator converts one double bond into two free radicals (FR), then a chain reaction follows as pairs of FR combine to cross-link the resin network (Fig. 1.1: the reaction has been simplified showing all double bonds opened simultaneously).

The chemicals chosen for the manufacture of unsaturated polyester resin affect the relative properties. The straight chain (aliphatic) precursors allow local flexibility, while conjugated cyclic (aromatic) precursors confer stiffness, strength, and thermal stability to the molecule. For a mono-substituted benzene ring (C_6H_5X, which is phenol when X=OH), further substitution can take place at the first (ortho-), second (meta-), or third (para-/tere-) positions away from the substituted molecule. If a link in the polymer chain passes through the benzene ring, the ortho-position will impose a 60 degrees turn in the steric conformation of the molecule, while meta- will change the direction by 120 degrees and para- will maintain the line of the molecule (180 degrees). Orthophthalic UPE follows a convoluted path, while (meta- or para-) Isophthalic UPE is a straighter molecule which can pack more densely and hence has higher stiffness, higher strength, and improved environmental resistance. Isophthalic polyester resins are the preferred option (vs. orthophthalic) for marine applications as they generally have better resistance to water permeation, better chemical/weather resistance, better mechanical strength and strain to failure, and higher heat distortion temperatures. However, the good performance does come at a marginally higher cost. Premium polyester resins often have neopentyl glycol (NPG: 2,2-dimethylpropane-1,3-diol, CAS 126-30-7) as the alcohol component to reduce weight loss when heated and confer higher degradation temperatures.

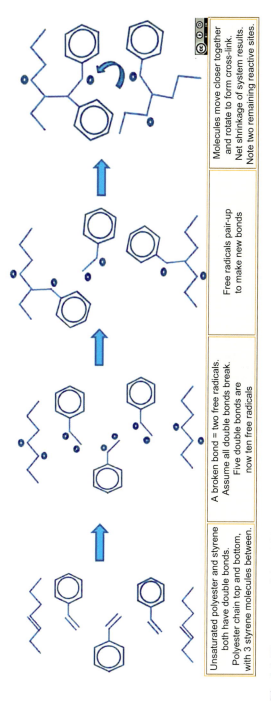

Fig. 1.1 The progress of the addition curing reaction in unsaturated polyester.
The Figure by John Summerscales is licensed under a Creative Commons Attribution-ShareAlike 4.0 International License.

1.2.1.4 Vinyl ester resin (VE)

Vinyl ester resins have a similar backbone chemistry to epoxide resins, but have unsaturated reactive sites positioned at the extreme ends of the molecule. The lower ester group content reduces their susceptibility to hydrolysis. They can be produced by reacting bisphenol A with monocarboxylic acids to produce difunctional molecules with terminal vinyl groups. The cross-linking sites on unsaturated polyester resins are within the backbone of the polymer and the consequent pendant groups at chain ends might reduce Tg, whereas the chain-end cross-linking sites in vinyl ester resins produce a different network structure without pendant groups and hence potentially higher glass transition temperatures. They are cured in a similar manner to unsaturated polyesters, but the chemical structure produces more resilient/tougher cured polymers.

Methacrylic resins are similar to vinyl ester or unsaturated polyester, but with different reactive diluents, for example, acetoacetoxyethyl methacrylate (AAEM, $CH_3COCH_2CO_2CH_2CH_2O_2CC(CH_3)=CH_2$, CAS 021282-97-3), butyl methacrylate ($CH_2=C(CH_3)CO_2(CH_2)_3CH_3$, CAS 97-88-10), glycidyl methacrylate ($CH_2=C(CH_3)CO_2CH(O)CH_2$, CAS 106-91-2), or methylmethacrylate ($CH_2=C(CH_3)CO_2CH_3$, CAS 80-62-6), either singly or in mixtures with styrene (Malik et al., 2000; Kumar et al., 2010).

1.2.1.5 Bismaleimide (BMI) resins

The BMI resins are addition polymers characterized by the -(CO)-NX-(CO)- group, that is, a tertiary amine between two carbonyl groups. When X is an aromatic group, the conjugation of alternating double bonds will spread across the aromatic ring, the amine, and both carbonyls to produce a relatively large mechanically and thermally stable planar structural entity. A very wide range of structurally different BMI resins is commercially available for use as adhesives or as the matrix for composites. BMI resins can be cured using standard processing equipment for high-temperature epoxies (i.e., $\sim$180°C) followed by free-standing postcure in an oven at $\sim$250°C to complete the polymerization and thus achieve even higher properties (Table 1.2).

1.2.2 Thermoplastic polymers

The principal thermoplastic polymers to find use as the matrix for composites are (a) polypropylene, (b) polyamide, (c) polyester, and (d) PEEK. For crystalline polymers, the melting point value is normally $\sim$200 ($\pm$50)°C above the glass transition temperature ($T_m \approx T_g + 200$°C).

1.2.2.1 Poly(propylene) (PP)

Poly(propylene) (PP) is produced by the homopolymerization of propylene ($CH_3-CH=CH_2$). If the glass transition occurs below ambient temperature, then the polymer should not be used in highly stressed components as it will creep.

Table 1.2 Properties of thermosetting resins

Resin	Density (liquid kg/m²)	Density (cured kg/m²)	Initial viscosity (mPa.s (cP) at 25°C, mixed system)	Shrinkage[a] (%: L = linear, V = volumetric)	Water absorption[b] (%)	Thermal (°C, T_g or HDT)	Modulus (MPa)	Strength (MPa)	Elongation at break (%)	Description	References
Unsaturated polyester resin											
Crystic® 474PA	1100	1220	530	L: 2.7 V: 8.2	"28 mg"	HDT: 112	3700	57	1.8	"Thixotropic, preaccelerated orthophthalic polyester resin"	June 2014 data sheet
Crystic® 491PA	1100	1200	650	?	"17 mg"	HDT: 77	3500	77	4.0	"Thixotropic isophthalic polyester resin with good water and chemical resistance", "approved by Lloyd's Register of Shipping"	February 2013 data sheet
Crystic® 701PAX	1080	1190	160	?	"10 mg"	HDT: 75	3580	66	2.5	"Preaccelerated isophthalic polyester resin for vacuum injection"	January 2017 data sheet
Crystic® 703PA	?	?	160	?	"11 mg"	HDT: 59–64	2758–3162	38–49	1.3–2.1	"Preaccelerated DCPD-based polyester resin", "approved by Lloyd's Register of Shipping"	February 2013 data sheet

Continued

Table 1.2 Continued

Resin	Density (liquid kg/m²)	Density (cured kg/m²)	Initial viscosity (mPa.s (cP) at 25°C, mixed system)	Shrinkage[a] (%: L = linear, V = volumetric)	Water absorption[b] (%)	Thermal (°C, T_g or HDT)	Modulus (MPa)	Strength (MPa)	Elongation at break (%)	Description	References
Crystic®785PA (LR)	1100	~	220	~	"23 mg"	HDT: 56	2900	49	1.82	"Preaccelerated, DCPD-based, polyester resin designed for RTM"	February 2013 datasheet
Vinyl ester resin											
Derakane®8084	1020	1140	360	L: 2.7 V: 8.2	~	Tg: 115, HDT: 82	2900	76	8–10	"[Elastomer modified] epoxy vinyl ester resin"	30 August 2011 Ashland data sheet
Dion®9100 series	~	1050–1130	400–500	~	~	HDT: 104	3172	80	5.2	"Bisphenol-epoxy vinyl ester resins"	July 2012 Reichhold data sheet
Epoxy resins											
AMPREG 21	1100	1140–1150	432–1194	L: 1.3–1.6 V: 3.9–4.8	~	Tg: 67–115, HDT: 66–77	2200–3250*	42–63*	3.0–4.3	"Epoxy wet laminating system"	Ampreg 21-8-0515 data sheet
AMPREG 26	~	1110–1130	292–1150	L: 1.6 V: 4.8	0.9–1.3	Tg: 55–109, HDT: 48–89	3450–3910	58–85	1.9–5.1	"Epoxy laminating system"	PDS-Ampreg 26-12-0515 data sheet

CYCOM®977-2	?	1310	?	?	Tg: 212	3520	81	?	"Toughened epoxy resin for autoclave or press molding"	Cytec AECM-00007 Rev 01 2012 data sheet
PRIME™27	1080–1090	1130–1140	115–590	L: 1.05–1.8 & tba V: 3.2–5.4	0.36–0.97	Tg: 69–75, HDT: 60–68	3430–3600	70–78	3.3–5.0	"Lowest viscosity PRIME™ infusion resin"	Gurit PDS-PRIME27-7-1215 data sheet
Bismaleimide (BMI) resin										
Cycom®5250-4	?	1250	?	?	3.9–4.2	Tg(dry): 254–271 Tg(wet) 199–207	4600	103–104	4.8–4.9	"One-part homogeneous BMI resin" for the RTM process	Cytec AECM-00019 Rev 01 2012 data sheet

[a] Italic data calculated from quoted value using the approximation linear shrinkage is one third of volumetric shrinkage.
[b] Water absorption in mg after "24 h at 23°C."
[c] Resin modulus and strength reduced by factor 10 assuming error on datasheet!

1.2.2.2 Polyamide (PA)

An amide (—CONH—) is the product of the reaction between a carboxylic acid (—CO$_2$H) and an amine (—NH$_2$) with water as a by-product. Linear polyamide molecules can be produced using either (i) an ω-aminoacid or by ring-opening of a cyclic amide (lactam) or (ii) by the reaction of a difunctional acid with a difunctional amine. The former group (i) polyamides are normally denoted by PAx, while the two monomer systems (ii) are denoted by PA$x,y$ where x is the number of carbon atoms in the amine monomer and y is the number of carbon atoms in the acid monomer (Table 1.3). Polymers with shorter CH$_2$ sections are more likely to crystallize and to have higher levels of hydrogen bonding between molecules, but they are also more likely to be hydrophilic.

1.2.2.3 Polyesters

An ester (—CO$_2$—) is the product of the reaction between a carboxylic acid (—CO$_2$H) and an alcohol (—OH) with water as a by-product. The most common linear polyester molecules are polyethylene terephthalate (PET, terephthalic acid and ethylene glycol) and polybutylene terephthalate (PBT, from ring-opening of cyclic butylene terephthalate (CBT)). PET is the only condensation polymer marked for recycling (#1 PET or PETE). In-process polymerization of CBT is finding increased use in liquid composite molding (LCM) process, especially resin transfer molding and resin infusion under flexible tooling, although process temperatures are normally around 60°C above ambient temperature.

1.2.2.4 Poly aryl ether ketones (PAEK) (Cogswell, 1992)

The highest performance available from thermoplastic polymers is achieved when a number of aryl (benzene ring) groups are sufficiently close on the backbone of the polymer for electron delocalization to extend the electron cloud across more than one aryl group. These polymers are known generically as poly aryl ether ketones (PAEK), where ether is an oxygen atom in the backbone conferring molecular flexibility and where ketone is the double-bonded carbon-oxygen group. The principal polymers are poly ether ether ketone (PEEK) and poly ether ketone ketone (PEKK).

1.2.2.5 Biopolymers

There is increasing interest in bio-based polymers, including poly(butylene succinate) (PBS), poly(hydroxyalkanoates) (PHA), and poly(lactic acid) (PLA). These materials may have comparable mechanical properties to synthetic polymers combined with enhanced sustainability (subject to confirmation by an appropriate lifecycle assessment), biodegradability, and/or biocompatibility.

Table 1.3 Performance characteristics of commercially important thermoplastic-matrix systems for composites

Monomer(s)	Polymer	Density (kg/m^3)	T_g (°C)	T_m (°C)	M_∞ (%) (Colin and Verdu, 2014)	E (GPa)	σ' (MPa)
Propylene (CH$_3$—CH=CH$_2$)	PP	900	−5	175	<0.1	1.1–1.3	29–34
Polyamides							
1,4-diaminobutane adipic acid HOOC—(CH$_2$)$_4$—COOH	PA4,6	1180		295	<10 (high)	3	80
Caprolactam ⌊(CH$_2$)$_5$—CO—NH⌋ Hexamethylene diamine H$_2$N—(CH$_2$)$_6$—NH$_2$ adipic acid HOOC—(CH$_2$)$_4$—COOH	PA6 PA6,6	1130 1140	56 (dry) 76 (dry)	215–221 254–264	<10 (high) <10 (high)	2.8 3	76 80
Hexamethylene diamine H$_2$N—(CH$_2$)$_6$—NH$_2$ sebacic acid HOOC—(CH$_2$)$_8$—COOH	PA6,10	1090	215		<10	2.1	55
Laurolactam ⌊(CH$_2$)$_{10}$—CO—NH⌋	PA11	1040	185	190–200	<10	1.4	38
Dodecanelactam ⌊(CH$_2$)$_{11}$—CO—NH⌋	PA12	1020	175	180–210	<10	1.4	45
Polyesters							
Terephthalic acid and ethylene glycol	PET	1390	67–75°C	252–265°C	0.55–2.0 (24 h) Brydson (1999)		117–173
Cyclic butylene terephthalate	PBT	1310–1340	25–52°C	225–228°C	<3.0	2.34	56
Poly aryl ether ketones							
Poly ether ether ketone (Cogswell, 1992)	PEEK	1264–1400	143°C	340°C	0.5	3.6	92

Where ⌊—⌋ indicates a cyclic molecule (Brydson, 1999; Anon, 2000a; Colin and Verdu, 2014), T_g = glass transition temperature, T_m = crystalline melting point, M_∞ = equilibrium moisture content, E = elastic modulus, and σ' = strength.

1.3 The reinforcement

1.3.1 Aramid fibers

Aramid is a contraction of **ar**omatic **amid**e (the final "e" remains in the French name!). The chemical make-up of the synthetic organic fibers alternates the (aromatic) benzene ring ($-C_6H_4-$) with the amide ($-CONH-$) group found in polyamide ("nylon"). The fibers are produced by spinning the rigid-rod-like polymer from a liquid crystal solution in concentrated sulfuric acid. The covalent bonds in the polymer molecule backbone are oriented along the fiber principal axis, while the molecules are bound by hydrogen bonds in the radial direction.

Aramid fibers have a unique combination of high modulus, high strength, toughness, thermal stability, and chemical resistance. However, due to the molecular arrangement, the fibers have poor transverse strength and poor compressive properties (compressive strength is generally about 20% of tensile strength). Further, the amide group is hydrophilic, so the fibers absorb moisture. The fibers are degraded by ultraviolet light. Table 1.4 summarizes the principal forms of aramid fibers.

Special tools and techniques are appropriate for machining aramid composites (Pinzelli, 1990, 1991) and these may be more relevant than traditional routes for machining natural fiber composites. For example (Pinzelli, 1990, 1991), the band saw should have a fine tooth blade (550–866 teeth/m) with straight-set or raker-set teeth and operate at high speed to stretch and shear the material. To minimize the production of fuzz and to keep the teeth from snagging fibers run the blade in reverse (teeth pointing upwards) (Pinzelli, 1990, 1991). "As shipped, Kevlar® aramid fiber products do not pose a hazard. Kevlar® staple and pulp contain a small amount of respirable fibers which may become airborne during opening, mixing, carding, or regrinding waste products containing Kevlar®. When mechanically working Kevlar® fiber or materials containing Kevlar® in operations such as cutting, machining, grinding, crushing or sanding, airborne respirable fibers may be formed. Repeated or prolonged inhalation of excessive concentration of respirable fibers may cause permanent lung injury" (Bryant, 2002).

1.3.2 Carbon fibers

Carbon fibers (Table 1.5) are normally produced from one of three precursor fibers: poly(acrylonitrile) (PAN), pitch, or reconstituted cellulose (rayon). The process for PAN will typically involve oxidation under tension at 200°C–250°C, carbonization in a nonoxidizing atmosphere at 1000°C, then graphitization in a nonoxidizing atmosphere at 2500–3000°C. The resulting fibers have a similar structure to graphite, but with a turbostratic layer structure and consequent increased interlayer spacing. The layers become closer to each other, with increased axial elastic modulus, during graphitization, especially at higher temperatures.

Table 1.4 Chemical nature, trade names, and uses of aramid fibers (Lovell, 1981; Kevlar aramid fibre Technical Guide, 2000; Anon, 2012a)

Chemical nature	Trade names	Uses	Density (kg/m³)	Modulus (GPa)	Strength (MPa)	Elongation at break (%)
Poly para-phenylene terephthalamide PPPT, PPTA	Kevlar® (DuPont) Twaron (Teijin)	K: rubber tyres, hoses, conveyor belts K29: cables, ropes, ballistics K49: reinforcement fibers	K: ~ K29: 1440 K49: 1440 T: 1440–1450	K: ~ K29: 70.3 K49: 112 T: 60–120	K: ~ K29: 2.92 K49: 3.00 T: 2.4–3.6	K: ~ K29: 3.6 K49: 2.4 T: 2.2–4.4
Poly meta-phenylene isophthalamide PMTA	Nomex® (DuPont)	paper for honeycomb cores	~	~	~	~
Para aramid copolymer: para-phenylenediamine/diaminodiphenylether/terephthaloyl chloride (Ozawa and Matsuda, 1989)	Technora (Teijin)		1390	74	3.4	4.5

Table 1.5 **Characteristics and composition of carbon fibers**

Precursor	Tensile modulus (GPa)	Tensile strength (GPa)	Strain to failure (%)	Density (kg/m^3)	Carbon assay (%)
Rayon	41	1.1	2.5	1600	99
Pitch	161	1.4	0.9	1900	97
PAN	231	3.4	1.4	1800	94
Pitch	385	1.8	0.4	2000	99
PAN	392	2.5	0.6	1900	100
Pitch (K13D2U)	931	3.7	0.4	2200	>99

1.3.3 Glass fibers

Glass fibers are formed by melt spinning specific combinations of oxides dependent on the fiber type (Table 1.6). *E*-glass dominates the market.

1.3.4 Other reinforcement fibers

A variety of other fibers have been proposed as the reinforcements for composites, including Ultra-High Molecular Weight Poly Ethylene (UHMWPE, e.g., Dyneema and Spectra), the polybenz[*x*]azole (PBX) rigid-rod polymers (e.g., PBI: poly(benzimidazole), PBO: poly(benzoxazole), or PBT: poly(benzthiazole where [*x*] is nitrogen, oxygen, or sulfur, respectively).

Sustainability issues have revived interest in natural fibers derived from animal (feather, hair, silk, or wool), mineral (basalt), or vegetable resources. Plant fibers may be extracted from the stem (bast), leaf, fruit/seed, root, grass, or wood. The most extensively explored bast fibers are flax, hemp, or nettle from temperate climates or jute, kenaf, or ramie from tropical climates.

1.4 The fiber-matrix interface

The rules-of-mixture for the estimation of the mechanical properties of a composite assume that the fiber-to-matrix interface is completely bonded. Interfacial bonding has been reviewed by several authors (Metcalfe, 1974; Plueddemann, 1974; Caldwell, 1990; Hughes, 1991; Kim and Mai, 1991; Andrews et al., 1996; Drzal and Herrera-Franco, 2000; Thomason, 2012). The reinforcement fibers are normally supplied with a surface coating ("size" or "sizing") which promotes good bonding between the two adherends. Coupling agents are often small molecules where the silane end bonds to the fibers and the other end is specific to the matrix system.

Table 1.6 Characteristics and composition (excludes constituents at <0.1%) of glass fibers

Type	Uses	Mechanical properties					Composition														References
		Density (kg/m³)	Modulus (GPa)	Strength (GPa)	Elongation at break (%)	SiO₂	Li₂O	Na₂O	K₂O	BeO	MgO	CaO	B₂O₃	Al₂O₃	TiO₂	Fe₂O₃	ZrO₂	CeO₂	PbO	F₂	
A-glass	High alkali content for chemical resistance	2460	73	3.1	3.6	72%	?	12.5%	1.5%	?	0.9%	9%	0.5%	2.5%	?	0.5%	?	?	?	?	Lovell (1981)
C-glass	Chemical corrosion resistance	2460	74	3.1	?	65%	?	8.5%	?	?	3.0%	14%	5%	4%	?	0.5%	?	?	?	?	Lovell (1981)
D-glass	Low dielectric quartz glass: good transparency to radar	2140	55	2.5	?	73%	?	✓	✓	?	✓	✓	23%	✓	?	?	?	?	?	?	Lovell (1981)
E-glass	Electrical insulator. High strength.	2550	71	3.4	3.37	55.2%	?	0.3%	0.2%	?	3.3%	18.7%	7.3%	14.8%	?	0.3%	?	?	?	0.3%	Lovell (1981)
L-glass	High lead content for radiation protection	?	?	?	?	62.9%	?	?	?	?	10.3%	?	13.6%	2.6%	?	?	2.1%	?	8.5%	?	Fluegel (n.d.)
M-glass	High modulus (rare)	?	?	?	?	53.7%	3.0%	?	?	8.0%	9.0%	12.9%	?	?	8.0%	0.5%	2.0%	3.0%	?	?	Hausrath and Longobardo (2010)
R-glass	Reinforcement grade	2550	86	4.4	5.2	60%	?	?	?	?	6%	9%	?	25%	?	?	?	?	?	?	Lovell (1981)
S-glass	High strength, high modulus, and high-temperature resistance grade: ballistics	2500	85	4.58	4.6	65%	?	?	?	?	10%	?	?	25%	1.4%	?	?	?	?	?	Lovell (1981)

Table 1.7 **Reinforcement achievable fiber volume fractions and reinforcing efficiencies**

Configuration	Volume fraction range	Fiber orientation distribution factor
Random	10%–30%	0.375 (All directions in plane)
2-D fabric	30%–60%	0.5 (Ignoring out-of-plane crimp)
Unidirectional	50%–80%	1.0

For example, vinylsilane is used with unsaturated polyester or vinylester resins, while epoxysilane is used with epoxy resins. The chemical activity of these two systems is very different. A quick *rule-of-thumb* would be that glass fibers are sized for vinyl curing systems (unsaturated polyester or vinyl ester) and "advanced" (e.g., aramid or carbon) fibers are sized for use with epoxy. When unsure what system is in use, it is appropriate to check the nature of the fiber surface coating with the supplier before committing to manufacture safety critical structures.

1.5 Reinforcement forms

The reinforcement fibers can be arranged in a variety of different ways. The configuration of the reinforcement sets ranges for both the achievable fiber volume fraction and the reinforcing efficiency (Table 1.7).

Such et al. (2014) present an overview of the history of aligned discontinuous fiber composites. They focus on the process and application of highly aligned advanced composites with properties that approach those of continuous fiber composites. Aligned discontinuous fiber composites may be created by aligning short fibers or introducing discontinuities into aligned fibers in order to produce components with complex topologies.

There is increasing interest in using very thin layers within laminates. In 1984, Richards et al. described an aircomb technique for spreading reinforcement tows. North TPT (Thin Ply Technology) (Pearson, 2015; Anon, 2015) now has produced lightweight unidirectional carbon fiber/epoxy prepreg tapes as light as 15 g^{-2} (gsm). Borg (2015)) and Ohlsson (2015) have recently reviewed the use of spread tow reinforcements.

Planar reinforcements (fabrics) are produced by textile processes including weaving, knitting, and stitching. The terminology for the yarns in a cloth is given in Table 1.8. Fabrics are characterized by the mode of construction (see below) and their areal weight (measured in grams per square meter: gm^{-2} or gsm). Continuous fibers for the reinforcement of advanced composites are normally supplied to textile processes on spools. The spools are mounted onto a creel frame to control the unwinding, tensioning, and guiding of the fibers, yarns, or tapes. Glass fibers may be pulled from the inside of the spool when there is no precise requirement for tension control. Tension controllers (Whiteside, 2010) (known as tensioners or dancers) may use a

Table 1.8 The terminology for fibers within a fabric

	Direction	Woven cloth	Individual threads	Knitted fabric
	Along the roll length	warp	end	wale
	Across the roll width	weft/woof/fill	pick/shot	course

(Diagram shows: Weft/woof/fill (pick/shot) course — across; Warp (end) wale — along)

servo-controlled torque motor, a pneumatic cylinder, a magnetic particle brake, or a magnetic particle clutch.

1.5.1 Woven fabrics

Woven fabrics normally consist of two sets of interlaced orthogonal fibers produced on a loom. For a balanced fabric (equal numbers of tows/meter with the same linear density in each direction), the in-plane fiber orientation distribution is 0.5 when loaded on one set of fibers or 0.25 when loaded on the bias (stress at ±45 degrees to the fibers). The interlacing causes undulation of the fibers out-of-plane. Crimp is defined (Farnfield and Alvey, 1975) as "the waviness of a fiber" and is normally expressed numerically as either "the number of waves or crimps per unit length" or "the difference in distance between points on the fiber as it lies in a crimped condition and the same two points when the fiber is straightened under suitable tension". The fiber orientation distribution factor will be reduced depending on the degree of out-of-plane crimp.

When used for the reinforcement of composites, textile fabrics are normally woven in one of three styles: plain, twill, or satin (Fig. 1.2).

While the vast majority of woven reinforcements are in the form of two-dimensional planar fabrics with orthogonal fibers, it is also possible to produce *triaxial weaves* where the fibers are at 0 degree and ± 60 degrees.

1.5.2 Stitched or knitted fabrics

Stitched fabrics are bound together by a lightweight fiber loop sewn or knitted around the reinforcement tow. The fabrics may be just a single unidirectional layer, or multiple layers. A multilayer fabric may be biaxial (0°/90° or ± 45° cross-plied for two layers, or 0°/90°/0° for three layer with two parallel layers) or multi-axial (e.g., triaxial as 0°/+45°/−45° or quadriaxial as 0°/+45°/−45°/90°). The reinforcement tow remains aligned in the plane without crimp, so these reinforcements are commonly referred to as noncrimp fabrics (NCF). NCF have the potential for a higher fiber orientation distribution factor than for a woven fabric and each unidirectional layer can pack with the

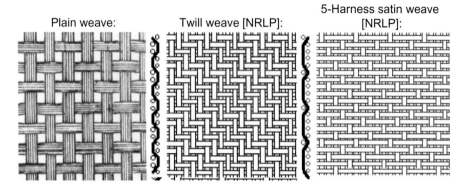

Fig. 1.2 Plain, twill, and satin weave structures [NRLP] indicate. Figures drawn by Neil Pearce.

possibility of higher fiber volume fractions than is possible in woven fabrics. Full consideration must be given to the selection of the stitching fiber: poor resin penetration or a poor bond to this thread could be a precursor to laminate failure.

Modern knitting machines permit the insertion of unidirectional reinforcement tows in a "matrix" of knitted loops. Where fibers are inserted across the width of the machine, a weft-insertion warp-knit fabric is produced but the reinforcement length is limited to the roll width. The alternative is warp-insertion weft-knit (WIWK) fabric where the reinforcement length is only limited by the size of the roll.

1.5.3 Braids

Braiding is defined by the Textile Institute (Farnfield and Alvey, 1975) as "the process of interlacing three or more threads in such a way that they cross one another and are laid in a diagonal formation" and is the subject of a book by Kyosev (2014). Flat tubular or solid constructions may be formed in this way. A typical braiding machine operates using a "maypole" action, whereby one-half of the yarn carriers rotate on a clockwise sinusoidal path, while the remaining half of the yarn carriers follow a concentric counter-clockwise sinusoidal path. In the tubular braided reinforcement, each fiber follows a helical path around the principal axis of the braid. Ayranci and Carey (2008) have reviewed the use of 2-D braided composites for stiffness critical applications. The modeling of the mechanical properties of textile braids has been considered by a number of authors (e.g., Cox et al., 1994; Xu et al., 1995; McGlockton et al., 2003; Naik, 1995; Qingda and Cox, 2003; Ayranci and Carey, 2008; Rana and Fangueiro, 2015).

1.5.4 Three-dimensional woven fabrics

There is increasing interest in 3-D textile reinforcements (Brandt et al., 1996; Mouritz et al., 1999; Kamiya et al., 2000; Bogdanovich and Mohamed, 2009; Ansar et al., 2011) as preforms for LCM processes, including resin transfer molding (RTM) and

Table 1.9 Prepreg classification

System	Normal cure temperature
Low temperature	60°C ± 30°C
Medium temperature	120°C ± 30°C
High temperature	180°C ± 30°C

resin infusion under flexible tooling (RIFT). Bogdanovich and Mohamed (2009) make a clear distinction between *3D interlock weave* (either layer-by-layer with the interlock only running through a limited number of layers, or angle-interlock with binder fibers running through the full fabric thickness) and *3D orthogonal noncrimp weave* (with binder fibers running at 90 degrees to the two other sets of tows).

1.5.5 Preforms

A preform is an assembly of reinforcement that closely reproduces the component shape. Preforms are normally prepared outside the mold tool and loaded as a single entity to permit reduced overall cycle times. A preform must normally have sufficient rigidity and integrity to permit transfer into the tool without losing net-shape.

1.5.6 *Preimpregnated reinforcements (prepregs)* (HexPly® Prepreg Technology, 2013)

For vacuum-bagging (and autoclave cure) processes, control of the resin mixing and quality can be transferred to the material supplier by having the reinforcement preimpregnated with the resin system. Systems with an unsaturated polyester resin matrix are normally supplied just-in-time (JIT) and are known as "molding compounds". For higher performance structures, epoxy prepregs are generally classified according to their cure temperature (Table 1.9).

1.6 Sandwich structures

Sandwich structures consist of stiff, strong skins on either side of a lightweight core. This configuration moves the reinforcement materials away from the neutral axis of a beam or plate, conferring higher stiffness without a significant increase in weight. A sandwich structure can be considered as an I-beam where the skins replace the flanges and the core replaces the web. The selection of skin materials follows the approach described above for monolithic composites. The core may be a natural material, polymeric foam, corrugated structure, or honeycomb (Table 1.10).

Table 1.10 Typical properties of common core materials (the combination of top values from every range is probably not available, mechanical properties in honeycombs will vary between lengthways and widthways directions)

System	Density (kg/m³)	Elastic modulus (MPa)	Shear modulus (MPa)	Compression strength (MPa)	Shear strength (MPa)	Elongation at break (%)	Thermal conductivity (mW/m K)	Thermal expansion (µm/m K)	References
End-grain balsa	130	3000	230	7		1.2	100	6.5	Compare balsa to cork (n.d.)
Compressed cork	150	20	10	1		15	40	200	Compare balsa to cork (n.d.)
Polyurethane (PUF)	60	20	4.1	0.42	0.41	30	30		Anon (2000b)
Polyvinyl chloride (linear PVC)	50–80	37–56	15–21	0.4–0.9	0.5–1.2	80	33–35		Anon (2000b)
Polyvinyl chloride (cross-linked PVC)	40–200	26–223	12–77	0.5–4.6	0.4–3.5	10–31	29–42		Anon (2000b)
Styrene-acrylonitrile (SAN)	200–400	155–350	60–240	4–13	3–8	6–7	48–55		Anon (2000b)
Polymethacrylimide (PMI)	32–110	35–180	13–70	0.80–3.6	0.8–2.4	3.0–7	29–31	3–37	Anon (n.d.-a)
Nomex® honeycomb	29–48	60–140	17–40	0.9–2.4	0.35–1.2				Anon (n.d.-b)

1.7 Degradation of marine composites

In addition to the "normal" degradation of composite materials in air, there are a number of additional considerations for materials exposed to moist atmospheres, or submerged in (sea)water, that are briefly introduced below and discussed in detail in (Searle and Summerscales, 1999; Davies et al., 2003, 2012; Davies and Choqueuse, 2008; Choqueuse and Davies, 2008; Davies and Rajapakse, 2014).

1.7.1 Diffusion

Polymer molecules can adopt many different conformations (distinguishable spatial arrangements of the molecular segments), especially in their amorphous volumes. The porosity and permeability of the material are determined by the fractional free volume (FFV). At the macroscopic scale, the diffusion of liquids, vapors, or gasses is normally modelled by Fick's law. The fluid content gradually rises and then approaches a saturation level depending on the polarity of the polymer. Saturation levels can be divided into the following ranges (Colin and Verdu, 2014):

- <0.1% water: non-hydrophilic polymers (e.g., polyethylene, polypropylene, silicones)
- <3% water: moderately hydrophilic groups (e.g., polyesters, polycarbonate)
- <7% water: strongly hydrophilic-cured network polymers (e.g., amine-cured epoxies)
- <10% water: strongly hydrophilic crystalline thermoplastics (e.g., polyamides)
- water-soluble strongly hydrophilic amorphous thermoplastics (e.g., polyvinyl alcohol, polyacrylamide).

When the material has a non-Fickian response, the diffusion coefficient may be derived from sorption isotherms such as Henry's or Langmuir's laws (Colin and Verdu, 2014).

Wright (1981) plotted the fall in glass transition temperature for data from epoxy resins (from five separate published papers) as a function of moisture content and found "as a rough rule-of-thumb" that there was a drop of 20°C for each 1% of water pick-up (data available up to 7% moisture content).

1.7.2 Osmosis and blistering

Osmosis is a process whereby solution strengths are equalized by the passage of a solvent (normally water) through a semi-permeable membrane. Any soluble material in a composite may dissolve (in diffusing fluids) and create a local strong solution. Further solvent will then diffuse through the matrix to dilute the solution until the concentration gradient is zero. As the volume of material in solution increases with dilution, it exerts pressure on the surrounding matrix or interface until stresses exceed the material strength. Delamination then occurs leading to degraded fiber-matrix interfaces and blisters on the laminate surface. The factors implicated in osmosis and blistering, and measures to reduce or eliminate the problem, have been reviewed in (Searle and Summerscales, 1999).

1.7.3 Environmental stress concentration

During extended exposure of marine composites to seawater, especially when under sustained or cyclic stress, the combination of stress and chemicals may accelerate, or retard, degradation mechanisms.

1.7.4 Marine fouling

Any system operating in a marine environment should be protected against fouling organisms. The commercial technologies available to deter or displace fouling by marine organisms include toxic formulations (e.g., cuprous oxide paints), self-polishing (exfoliating) surfaces, or low-surface energy coatings (e.g., silicone or PTFE). Each technique imposes environmental burdens, that is, poisoning beneficial species, plastics microdebris, or aggressive chemicals during coating, respectively. Alternative methods have been proposed including pulsed electric fields, ultrasonic vibration, or biomimetic topography.

1.7.5 Cavitation erosion

When a bubble collapses in a fluid (cavitation), it will create a high-power jet directed towards the closest solid surface. In turbulent flow (especially around stern-gear or hydrofoils), millions of such bubble collapse events may occur, each of which may generate a stress pulse up to 1000 MPa (comparable to the strength of many materials) with a duration of $\sim$2–3 μs. Each pulse can lead to local erosion of the material surface, with their repetition leading to severe loss of material from the structure.

1.7.6 Galvanic corrosion

Corrosion involves the flow of an electrical current within a material. As the majority of the constituents of composite structures are insulating materials, corrosion is not normally a problem for these materials (but there are other forms of degradation as introduced above). However, carbon lies between platinum and titanium in the galvanic series, and hence, acts as a noble metal. Carbon fibers should not be used in conjunction with (especially light alloy) metals when a circuit could be completed by the presence of a conducting liquid (e.g., seawater): the metal will corrode at a rate increasing with the separation of the two materials in the galvanic series. Polymer liners around bolt-holes, or a thin-insulating layer on the composite surface, may be sufficient to prevent formation of a galvanic corrosion cell.

1.8 Life cycle considerations

The selection of materials for a marine composite must recognize both the economic and environmental implications. The cost of producing a component or system has traditionally been constrained by relative pricing of competitor products. However, corporate social responsibility (CSR) now requires consideration of whole life costs as reflected in the new discipline of life cycle costing (LCC) (Ilg et al., 2016; Bernardo et al., 2016; Hueber et al., 2016). Further, increasing regulation of environmental issues has thrown the focus towards demonstrating minimal environmental burdens. The appropriate methodology is life cycle assessment (LCA) as defined by the ISO14040 series of international standards (Anon, 2006, 2012b; ISO, 2006) and reviewed by a number of authors (Islam et al., 2016; Pizzol et al., 2015; Chang et al., 2014; Ekvall and Finnveden, 2001). Two key issues, which have recently come to prominence, are dealt with below.

1.8.1 Microplastics

There is growing concern about polymer microparticles entering the marine environment and subsequently marine animals (Thompson et al., 2004, 2005, 2009) (Thompson et al., 2004, 2005, 2009). A wide range of organisms including commercially important species of fish and crustaceans are known to ingest microplastic (e.g., Lusher et al., 2013) and laboratory studies indicate that plastic microparticles ingested by mussels pass into their circulatory system rather than simply being excreted (Browne et al., 2008). Further evidence from laboratory studies suggests that ingestion of plastic can compromise the ability of marine invertebrates to store energy (Wright et al., 2013). Teuten et al. (2007) report that plastic particles may concentrate and transport hydrophobic contaminants such as phenanthrene from the surface of the sea. If these particles are ingested, chemicals will be released and there is some evidence of associated toxicological consequences for biota (Browne et al., 2013; Rochman et al., 2013). A recent review by Cole et al. (2011) identified that these microplastics can enter the marine food web.

Plastic Oceans (www.plasticoceans.org) is a global network of independent not-for-profit and charitable organizations, united in their aims to change the world's attitude towards plastic within a generation. They aim to prevent plastic waste from entering the environment. Their three key pillars are education, business and sustainability, and science. Composite materials are an important subset of plastics, but their use may be impacted by such pressure groups, unless we all take a responsible attitude to the systems we place in the marine environment.

1.8.2 Marine-sourced materials

In the light of the foregoing discussion of marine microplastic debris, is there scope for deeper consideration of the use of marine-sourced materials in the context of marine composites? Kim (2017) has produced useful summary of potential marine biomaterials, albeit focused on biological and biomedical applications. Potential sources include algae (especially seaweed alginate, carrageenan, and fibers), exoskeletons (e.g., chitin from lobster/crab/shrimp), and collagen (vertebrate protein in bones, skin, muscle, and cartilage). If such materials can be used for structural applications in the marine environment, the abrasion and degradation products should be no more harmful than the waste generated from the dead organisms.

1.9 Conclusions

Well-informed materials selection is essential for the creation of composite products that will survive the harsh marine environment. A conflict exists between design for manufacturability and design for durability against design for minimum environmental impact (especially for end-of-life composites).

Acknowledgments

The author is grateful to Elke Graham-Jones for checking the chemical accuracy of the text and to Aitor Hernandez Michelena (doctoral student) for drawing attention to references that may otherwise have been omitted from the text. Indicative data for materials may not fully represent the range of possibilities. Trade Marks and similar designations are included for traceability: the author does not intend to endorse a specific product by any such inclusion.

References

Andrews, M.C., Bannister, D.J., Young, R.J., 1996. The interfacial properties of aramid epoxy model composites. J. Mater. Sci. 31 (15), 3893–3913.

Anon, 2000a. DSC Transitions of Common Thermoplastics. Perkin Elmer Datasheet PETech-44.

Anon, 2000b. Cores—properties. In: Composite Materials Handbook Cores. SP Systems, Newport IOW. Intro-1-1098-9.

Anon, 2006. ISO 14040:2006 (confirmed 2016), In: Environmental management—life cycle assessment—principles and frameworks. International Organization for Standardization, Geneva.

Anon Twaron—A Versatile High-Performance Fiber, Teijin Aramid report 40-00-001, 2012a, Accessed 17 June 2015.

Anon, 2012b. ISO/TR 14047:2012, In: Environmental management—life cycle assessment—illustrative examples on how to apply ISO 14044 to impact assessment situations. International Organization for Standardization, Geneva.

Anon, 2015. World's thinnest prepreg gets even thinner, Materials Today online. Reinf. Plast. 59 (5), 224.

Anon MK Plastics Technical Data, Accessed 22 July 2017.

Anon Nomex Aramid Honeycomb, Accessed 22 July 2017.

Ansar, M., Xinwei, W., Chouwei, Z., 2011. Modeling strategies of 3D woven composites: a review. Compos. Struct. 93 (8), 1947–1963.

Ayranci, C., Carey, J., 2008. 2-D braided composites: a review for stiffness critical applications. Compos. Struct. 85 (1), 43–58.

Bellenger, V., Dhaoui, W., Verdu, J., 1987a. Packing density of nonstoichiometric epoxide-amine networks. J. Appl. Polym. Sci. 33 (7), 2647–2650.

Bellenger, V., Verdu, J., Morel, E., 1987b. Effect of structure on glass transition temperature of amine crosslinked epoxies. J. Polym. Sci. Part B: Polym. Phys. 25 (6), 1219–1234.

Bellenger, V., Dhaoui, W., Morel, E., Verdu, J., 1988. Packing density of the amine-crosslinked stoichiometric epoxy networks. J. Appl. Polym. Sci. 35 (3), 563–571.

Bernardo, C.A., Simões, C.L., Costa Pinto, L.M., 2016. Environmental and economic life cycle assessment of polymers and polymer matrix composites: a review. Cienc. Tecnol. Mater. 28 (1), 55–59.

Bogdanovich, A.E., Mohamed, M.H., 2009. Three-dimensional reinforcements for composites. SAMPE J. 45 (6), 8–28.
Borg, C., 2015. An introduction to spread tow reinforcements, Part 1: manufacture and properties. Reinforced Plastics 59 (4), 194–198.
Brandt, J., Drechsler, K., Arendts, F.-J., 1996. Mechanical performance of composites based on various three-dimensional woven-fibre preforms. Compos. Sci. Technol. 56 (3), 381–386.
Browne, M.A., Dissanayake, A., Galloway, T.S., Lowe, D.M., Thompson, R.C., 2008. Ingested microscopic plastic translocates to the circulatory system of the mussel, *Mytilus edulis* (L.). Environ. Sci. Technol. 42, 5026–5031.
Browne, M.A., Niven, S.J., Galloway, T.S., Rowland, S.J., Thompson, R.C., 2013. Microplastic moves pollutants and additives to worms, reducing functions linked to health and biodiversity. Curr. Biol. 23 (23), 2388–2392.
Brydson, J., 1999. Plastics Materials, Seventh ed. Butterworth-Heinemann, Oxford. ISBN 0-7506-4132-0.
Caldwell, D.L., 1990. Interfacial analysis. In: Lee, S.M. (Ed.), Encyclopedia of Composites. vol. 2. VCH Publishers, New York. ISBN 0-89573-732-9, pp. 361–377. PU CSH Library.
Chang, D., Lee, C.K.M., Chen, C.-H., 2014. Review of life cycle assessment towards sustainable product development. J. Clean. Prod. 83, 48–60.
Choqueuse, D., Davies, P., 2008. Ageing of composites in underwater applications. In: Martin, R. (Ed.), Ageing of Composites. Woodhead Publishing, Cambridge. ISBN 1-84569-352-3, pp. 467–517 (Chapter 18).
Cogswell, F.N., 1992. Theroplastic Aromatic Polymer Composites. Butterworth-Heinemann, Oxford. ISBN 0-7506-1086-7.
Cole, M., Lindeque, P., Halsband, C., Galloway, T.S., 2011. Microplastics as contaminants in the marine environment: a review. Mar. Pollut. Bull. 62 (12), 2588–2597.
Colin, X., Verdu, J., 2014. Humid ageing of organic matrix composites. In: Davies, P., Rajapakse, Y.D.S. (Eds.), Durability of Composites in a Marine Environment. Springer, Dordrecht. ISBN 978-94-007-7416-2, pp. 47–114 (Chapter 3).
Compare balsa to cork, MakeItFrom.com, Accessed 22 July 2017.
Cox, B.N., Carter, W.C., Fleck, N.A., 1994. A binary model of textile composites I: formulation. Acta Metall. Mater. 42 (10), 3463–3479.
Davies, P., Choqueuse, D., 2008. Ageing of composites in marine vessels. In: Martin, R. (Ed.), Ageing of Composites. Woodhead Publishing, Cambridge. ISBN 1-84569-352-3, pp. 326–353 (Chapter 12).
Davies, P., Rajapakse, Y.D.S., 2014. Durability of Marine Composites. Springer, Dordrecht NL. ISBN 978-94-007-7416-2. https://doi.org/10.1007/978-94-007-7417-9.
Davies, P., Choqueuse, D., Roy, A., 2003. Fatigue and durability of marine composites. In: Harris, B. (Ed.), Fatigue in Composites: Science and Technology of the Fatigue Response of Fibre-Reinforced Plastics. Woodhead Publishing, Cambridge. ISBN 978-1-85573-608-5, pp. 709–729 (Chapter 27).
Davies, P., Choqueuse, D., Devaux, H., 2012. Failure of polymer matrix composites in marine and off-shore applications. In: Robinson, P., Greenhalgh, E., Pinho, S. (Eds.), Failure Mechanisms in Polymer Matrix Composites: Criteria, Testing and Industrial Applications. Woodhead Publishing, Cambridge, pp. 300–336. ISBN: 987-1-84569-750-1 (Chapter 10).
Drzal, L.T., Herrera-Franco, P.J., Henjen, H.O., 2000. Kelly, A., Zweben, C. (Eds.), Fibre-Matrix Interface Tests. In: Comprehensive Composite Materials, vol. 5. Pergamon, Oxford. ISBN 0-08-043723-0, pp. 71–111 (Chapter 5).
Ekvall, T., Finnveden, G., 2001. Allocation in ISO14041—a critical review. J. Clean. Prod. 9 (3), 197–208.

Farnfield, C.A., Alvey, P.J., 1975. Textile Terms and Definitions, Seventh ed. The Textile Institute, Manchester. ISBN 0-900739-17-7.

A Fluegel, Glass Composition, Glass Types, Accessed 17 June 2015.

Galy, J., Sabra, A., Pascault, J.P., 1986. Characterization of epoxy thermosetting systems by differential scanning calorimetry. Polym. Eng. Sci. 26, 1514–1523.

Hausrath, R.L., Longobardo, A.V., 2010. High-strength glass fibers and markets. In: Wallenberger, F.T., Bingham, P.A. (Eds.), Fiberglass and Glass Technology: Energy Friendly Compositions and Applications. Springer. ISBN 978-1-4419-0736-3, pp. 197–225 (Chapter 5).

HexPly® Prepreg Technology, Hexcel Publication FGU 017c, 2013, Accessed 18 June 2015.

Hueber, C., Horejsi, K., Schledjewski, R., 2016. Review of cost estimation: methods and models for aerospace composite manufacturing. Adv. Manuf.: Polym. Compos. Sci. 2 (1), 1–13.

Hughes, J.D.H., 1991. The carbon fibre/epoxy interface: a review. Compos. Sci. Technol. 41 (1), 13–45.

Ilg, P., Hoehne, C., Guenther, E., 2016. High-performance materials in infrastructure: a review of applied life cycle costing and its drivers ∼ the case of fiber-reinforced composites. J. Cleaner Prod. 112 (1), 926–945.

Islam, S., Ponnambalam, S.G., Lamb, H.L., 2016. Review on life cycle inventory: methods, examples and applications. J. Cleaner Prod. 136 (B), 266–278.

ISO 14044: 2006 (Confirmed 2016), Environmental management—life cycle assessment—requirements and guidelines, International Organization for Standardization, Geneva, 2006.

Kamiya, R., Cheeseman, B.A., Popper, P., Chou, T.-W., 2000. Some recent advances in the fabrication and design of three-dimensional textile preforms: a review. Compos. Sci. Technol. 60 (1), 33–47.

Kevlar aramid fibre Technical Guide, DuPont report H-77848, 2000, Accessed 17 June 2015.

Kim, S.-K. (Ed.), 2017. Marine Biomaterials: Characterization, Isolation and Applications. CRC Press, Boca Raton FL. ISBN 978-1-138-07638-9.

Kim, J.-K., Mai, Y.-w., 1991. High strength, high fracture toughness fibre composites with interface control: a review. Compos. Sci. Technol. 41 (4), 333–378.

Kumar, M., Verma, V., Kumar, V., Nayak, S.K., Yadav, S.B., Verma, P., 2010. Studies on effect of acrylates diluents on the properties of glass vinyl ester resin composite. J. Chem. Pharm. Res. 2 (5), 181–192.

Kyosev, Y., 2014. Braiding Technology for Textiles: principles, design and processes. Woodhead Publishing, Cambridge. ISBN 978-0-85709-135-2.

Lovell, D.R., 1981. Reinforcements. In: Hancox, N.L. (Ed.), Fibre Composite Hybrid Materials. Applied Science Publishers, Barking. ISBN 0-85334-928-2, pp. 23–35 (Chapter 2A).

Lusher, A.L., McHugh, M., Thompson, R.C., 2013. Occurrence of microplastics in the gastrointestinal tract of pelagic and demersal fish from the English Channel. Mar. Pollut. Bull. 67 (1–2), 94–99.

Malik, M., Choudhary, V., Varma, I.K., 2000. Effect of reactive diluents on the curing and thermal behaviour of vinyl ester resins. In: Mathur, G.N., Kandpul, L.D., Sen, A.K. (Eds.), Recent Advances in Polymers and Composites. Allied Publishers, Mumbai - India, pp. 599–602. ISBN 81-7764-066-6.

McGlockton, M.A., Cox, B.N., McMeeking, R.M., 2003. A binary model of textile composites III: high failure strain and work of fracture in 3D weaves. J. Mech. Phys. Solids 51 (8), 1573–1600.

Metcalfe, A.C., 1974. Broutman, L.J., Crock, R.H. (Eds.), Interfaces in Metal Matrix Composites. In: Composite Materials Series, vol. 1. Academic Press, New York and London. ISBN 0-12-136501-8.

Mike Bryant (BFG Industries Inc), Material Safety Data Sheet, 2002, Accessed 10 December 2004 12:13.

Mouritz, A.P., Bannister, M.K., Falzon, P.J., Leong, K.H., 1999. Review of applications for advanced three-dimensional fibre textile composites. Compos. A: Appl. Sci. Manuf. A30 (12), 1445–1461.

Naik, R.A., 1995. Failure analysis of woven and braided fabric reinforced composites. J. Compos. Mater. 29 (17), 2334–2363.

Ohlsson, F., 2015. An introduction to spread tow reinforcements, Part 2: design and applications. Reinf. Plast. 59 (5), 228–232.

Ozawa, S., Matsuda, K., 1989. Aramid coplymer fibres. In: Lewin, M., Preston, J. (Eds.), High Technology Fibers Part B (Handbook of Fiber Science and Technology III). Marcel Dekker, New York. ISBN 0-8247-8066-3 (Chapter 1).

Pearson, W.E., 2015. Textiles to composites: 3D moulding and automated fibre placement for flexible membranes. In: Graham-Jones, J., Summerscales, J. (Eds.), Marine Applications of Advanced Fibre-Reinforced Composites. Elsevier/Woodhead, Cambridge. ISBN 978-1-78242-250-1 (Chapter 14).

Pinzelli, R., 1990. Cutting and machining of composites based on aramid fibres. Composites Plastiques Renforces Fibres de Verre 0754-087630 (4), 17–23. (in French) and 23–25 (in English).

Pinzelli, R., 1991. Cutting and machining of composite materials based on aramid fibres. DuPont report H-23157(1M), January.

Pizzol, M., Weidema, B., Brandão, M., Osset, P., 2015. Monetary valuation in life cycle assessment: a review. J. Clean. Prod. 86, 170–179.

Plueddemann, E.P., 1974. Broutman, L.J., Crock, R.H. (Eds.), Interfaces in Polymer Matrix Composites. In: Composite Materials Series, vol. 6. Academic Press, New York and London. ISBN 0-12-136506-9.

Qingda, Y., Cox, B., 2003. Spatially averaged local strains in textile composites via the binary model formulation. J. Eng. Mater. Technol. 125 (4), 418–425.

Rana, S., Fangueiro, R.M.E.S., 2015. Braided Structures and Composites: Production, Properties, Mechanics, and Technical Applications. CRC Press. ISBN 978-1-4822-4500-4.

Richards, T., Short, D., Summerscales, J., 1984. In: An aircomb technique to produce thin layers of continuous unidirectional mixed-fibre reinforcement. 14th British Plastics Federation Reinforced Plastics Congress, Brighton, 5–7 November 1984, Paper 18, pp. 77–78.

Rochman, C.M., Hoh, E., Kurobe, T., Teh, S.J., 2013. Ingested plastic transfers hazardous chemicals to fish and induces hepatic stress. Nat. Sci. Rep. 3. article 3263.

Searle, T.J., Summerscales, J., 1999. Review of the durability of marine laminates. In: Pritchard, G. (Ed.), Reinforced Plastics Durability. Woodhead Publishing, Cambridge. ISBN 1-85573-320-X, pp. 219–266 (Chapter 7).

Such, M., Ward, C., Potter, K., 2014. Aligned discontinuous fibre composites: a short history. J. Multifunct. Compos. 2(3).

Teuten, E.L., Rowland, S.J., Galloway, T.S., Thompson, R.C., 2007. Potential for plastics to transport hydrophobic contaminants. Environ. Sci. Technol. 41 (22), 7759–7764.

Thomason, J., 2012. Glass fibre sizings: a review of the scientific literature. James Thomason, Glasgow.

Thompson, R.C., Olsen, Y., Mitchell, R.P., Davis, A., Rowland, S.J., AWG, J., McGonigle, D., Russell, A.E., 2004. Lost at sea: where is all the plastic? Science 304 (5672), 838.

Thompson, R., Moore, C., Andrady, A., Gregory, M., Takada, H., Weisberg, S., 2005. New directions in plastic debris (letter). Science 310 (5751), 1117.

Thompson, R.C., Moore, C.J., vom Saal, F.S., Swan, S. (Eds.), 2009. Plastics, the environment and human health. Philos. Trans. R. Soc. 364(1526).

IK Varma and VB Gupta, Thermosetting resin—properties, (Chapter 2.01) Comprehensive Composite Materials, Elsevier, 2000, v2, 1–56. ISBN (set): 0-08-042993-9 (v2): ISBN: 0-08-043720-6.

Whiteside, D., 2010. Understanding dancer tension control systems.PLACE (Polymers, Laminations, Adhesives, Coatings, and Extrusions) Conference, TAPPI, Albuquerque NM, 18–21 April.

Wright, W.W., July 1981. The effect of diffusion of water into epoxy resins and their carbon fibre reinforced composites. Composites 12 (3), 201–205.

Wright, S.L., Rowe, D., Thompson, R.C., Galloway, T.S., 2013. Microplastic ingestion decreases energy reserves in marine worms. Curr. Biol. 23 (23), R1031–R1033.

Xu, J., Cox, B.N., McGlockton, M.A., Carter, W.C., 1995. A binary model of textile composites II: the elastic regime. Acta Metall. Mater. 43 (9), 3511–3524.

Zukas, W.X., 1994. Torsional braid analysis of the aromatic amine cure of epoxy resins. J. Appl. Polym. Sci. 53 (4), 429–440.

Thermoplastic matrix composites for marine applications

2

Mael Arhant, Peter Davies
Marine Structures group, IFREMER (The French Ocean Research Institute), Plouzané, France

Chapter Outline

2.1 Introduction 31
2.2 Material options 32
2.3 Manufacturing options 33
   2.3.1 Product forms 33
   2.3.2 Manufacturing processes 34
   2.3.3 Particular morphology of thermoplastic composites induced by processing 39
2.4 Influence of the marine environment on thermoplastic composites 39
   2.4.1 Carbon/peek 40
   2.4.2 Carbon/polyamide 6 41
   2.4.3 Glass and carbon reinforced acrylic 43
2.5 Underwater structures 44
2.6 Repair 45
2.7 Recycling and environmental impact 46
2.8 Conclusion 48
References 49

2.1 Introduction

Thermoplastic matrix composites have been around for many years, with studies for aerospace applications being performed in the 1970s (Hoggatt, 1975; Maximovich, 1977). However, those early studies focused on amorphous thermoplastics which were susceptible to solvent attack. It was the introduction of semicrystalline polyetheretherketone (PEEK), in prepreg form, in the early 1980s which allowed thermoplastic composites to be considered for aircraft applications and this resulted in many detailed test programs in the 1990s. The aim here is not to give a history of high-performance thermoplastic composites, the reader can find this elsewhere (Cogswell, 1992a; Carlsson, 1991; Kausch, 1993), but rather to focus on the specific advantages of thermoplastic composites for marine applications.

First, an overview of available material options will be provided, followed by a discussion of manufacturing possibilities. Then the influence of the marine environment on the behavior of some thermoplastic composites will be described. A key area for future developments is underwater structures, and offshore applications will be discussed. Results showing how these materials behave under hydrostatic pressure loading will be given. The chapter will conclude with discussion of two related aspects, repair and recycling. These are specific to thermoplastics and could provide significant benefits for marine structures in the future.

2.2 Material options

There is an increasing number of thermoplastic matrix polymers available on the market. Even if these materials have been studied for many years, they are only starting to find applications today. Two main families of materials can be distinguished. The low cost thermoplastic polymers which include Polypropylene (PP), Polyethylene (PE), and the family of Polyamides (PA 6, PA 6.6, PA 11, PA 12) and other high-performance polymers such as PEEK, PolyEtherKetoneKetone (PEKK), Polyphenylene Sulfide (PPS), Polyetherimide, and Polycarbonates (PC). The application and the environment (temperature, depth) in which the material will be used determine its choice. These polymers are mostly associated with glass or carbon fibers, again depending on the chosen application and requirements. Also, a new thermoplastic acrylic resin called Elium™ has been added to the market recently and was developed by Arkema. Contrary to other thermoplastic matrix, this resin has the main advantage to be infusible in a similar way to thermoset composite parts. It is also worth citing the Anionic Polyamide 6 (A-PA6) resin and the CBT (butylene terephthalate) from Cyclics™ that can also be infused.

Fundamental differences exist between thermoset and thermoplastic polymers. Concerning the former, these are produced through a chemical reaction between a monomer, an accelerator, and a catalyst or between a prepolymer and a hardener that will start the reaction. The liquid resin then passes into the solid state through a crosslinking process and this reaction is irreversible. Once this reaction has occurred, it is no longer possible to reshape the thermoset resin. Exposures to high temperature will degrade the material rather than melting it, unlike thermoplastic polymers. Thermoplastic polymers are fundamentally different from their thermoset counterparts because the matrix has already undergone polymerization prior to processing, except for infusible thermoplastics. Semicrystalline polymers are mostly characterized by three main properties, the glass transition temperature T_g, the melting temperature T_m, and another parameter that is the degree of crystallinity X_c. The melting temperature is characteristic of semicrystalline polymers as thermosets are not able to crystallize.

Thermoplastic composites offer considerable advantages compared to their thermoset counterparts such as increased toughness, recyclability, reparability, and an excellent environmental resistance, essential for marine applications. They also have a near infinite shelf life, which is a significant advantage over thermoset prepregs which have to be refrigerated and stored for a maximum duration of 6 months generally. They allow shorter processing times than thermosets since they do not require any postcuring stage after manufacture. These are also particularly interesting for underwater structures which have to withstand high hydrostatic pressures. It is indeed possible to manufacture thick composite parts with very few defects, unlike thick thermoset composites where a significant exotherm can add internal porosities. Finally, they also have the possibility to be welded, which can greatly reduce the weight of composite structures that are often assembled using rivets.

The advantages and drawbacks of thermoset and thermoplastic composites are summarized in Table 2.1.

2.3 Manufacturing options

The melting temperatures of semicrystalline polymers are usually high (above 200°C) and their high viscosity in the melt state does not make them ideal when using conventional processes such as resin infusion or RTM (resin transfer moulding). Therefore, it became necessary to develop new manufacturing processes able to withstand these high temperatures and to develop new product forms (preimpregnated plies, powder-coated fibers, commingled yarns, etc.) adapted for these materials. However, the implementation of these new processes and products in the industry is not an easy task and is one of the reasons why thermoplastic composites are only starting to be used today, as the processes are starting to reach maturity.

Several manufacturing techniques can be used to manufacture thermoplastic composite parts, depending on the shape of the final product (sheets, tubes) and the initial product form of the constituents (matrix and fibers). First, the product forms that can be used will be presented, then the manufacturing processes will be described.

2.3.1 Product forms

Different types of products or semiproducts can be used to manufacture thermoplastic composites, as shown in Fig. 2.1. Here, only the semiproducts in which the matrix has already undergone polymerization are presented. Depending on the semiproduct, the

Table 2.1 Advantages and drawbacks of thermoset and thermoplastic composites

	Thermoset composites	Thermoplastic composites
Advantages	Low viscosity	Infinite shelf life
	Suitable for high temperatures	Recyclable/reparable
	Low processing temperatures	Impact resistance
	Well-established properties	Chemical resistance
	Excellent bonding with fibers	No emissions
Drawbacks	Limited shelf life	High viscosity
	Difficult to manufacture thick composite parts	High manufacturing temperatures
	Nonrecyclable	Generally more expensive

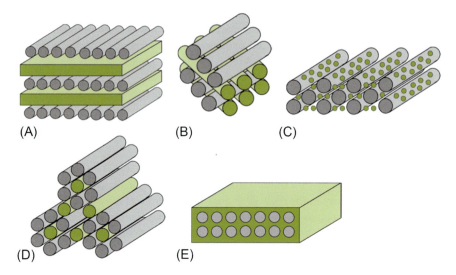

Fig. 2.1 Types of products, semiproducts (A), Film Stacking (B), Cowoven fabrics/hybrid weave (C), powdered fabrics (D), Commingled yarns (E), Preimpregnated plies.

quality of the final composite part can differ drastically. Five different types of products are presented here, starting in Fig. 2.1A from the less expensive and lower quality ones to the more expensive ones in Fig. 2.1E.

Due to the high viscosity of thermoplastics compared to thermosets, it is quite difficult to impregnate the fibers, more especially because the bonding of fibers to thermoplastics is difficult in general. Therefore, preimpregnated products such as the ones observed in Fig. 2.1E usually lead to higher quality parts in terms of compaction and voids when simple parts are produced than the film-stacking technique in Fig. 2.1A, which requires high pressures to ensure a good quality part. However, the drapability of thermoplastic prepregs is not always adequate when manufacturing complex composite parts because the prepreg has already undergone polymerization, unlike thermoset prepregs. Therefore, depending on the shape of the composite part, a compromise must be made.

2.3.2 Manufacturing processes

Two different types of manufacturing processes will be differentiated here, the ones that use semiproducts such as the ones shown in Fig. 2.1, that is, those which have already undergone polymerization prior to processing, and the ones where the matrix is infused and polymerizes during the process.

First, for processes using semiproducts, the processing cycle can be divided into three main stages shown in Fig. 2.2.

- Heating: during the first stage, the composite ply/plies is/are heated up above the melting temperature of the thermoplastic matrix. No pressure or low pressure is used at this stage.

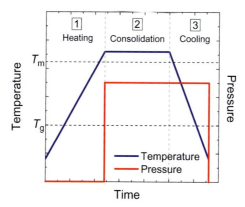

Fig. 2.2 Typical process cycle for thermoplastic composites.

– Consolidation: when the composite plies have reached the proper temperature, that is, above the melting temperature, pressure is added (the level of pressure depends on the manufacturing process) while the temperature is being stabilized. This stage allows the diffusion of the long polymer chains in order to consolidate the composite part within the mould and obtain high quality parts.
– Cooling: finally, the cooling stage starts and the cooling rate at which the composite part is being cooled down determines many parameters such as the microstructure (crystallinity) and the residual stress level. The composite part encounters two main transitions during the cooling stage. First, the crystallization temperature, after which crystallization and solidification occur. Then, when the glass transition temperature has been reached, it is possible to release the pressure and demould the composite part from the mould.

Hereafter are presented the main processes that use semiproducts.

2.3.2.1 Autoclave

Autoclave processes are used to produce high-performance composites and are widely used in the industry, especially for thermosets. In this process, both heat and pressure are used to produce the given composite part. First, prepreg plies are stacked together at the chosen orientation. These are then covered with a release film, a breather fabric, a bleeder, and a vacuum bag (Fig. 2.3). The autoclave is then heated up to the desired temperature and pressure is applied (from 1 to 10 bar typically). The main advantage

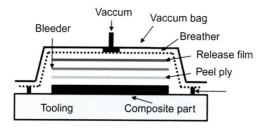

Fig. 2.3 Autoclave process.

of this process is that it is highly reproducible and that, due to the isostatic pressure that is applied using the vacuum bag, complex parts can be produced. In the case of thermoplastic composites, its main drawback is the processing time that is relatively long because of the convective heating; it takes time to heat the entire system.

2.3.2.2 Compression moulding

The compression moulding process is quite similar to the autoclave one. However, it has the main advantage to be constituted of two solid parts (upper and bottom mould, Fig. 2.4) and uses much higher pressures (above 10 bar) that can facilitate the adhesion of the fibers and matrix when using nonimpregnated semiproducts. Also, the processing cycles are much shorter than those using the autoclave because of the conductive heating used during the process.

2.3.2.3 Automated tape placement

This process uses narrow bands (tapes) of prepregs (typically 1 in., ½ inch, or ¼ inch wide) that are laid ply-by-ply on a mandrel or surface (flat or curved). These are consolidated upon deposition using a compaction roller. Intimate contact between the plies is ensured by a local increase of the tape temperature, above the melting temperature of the thermoplastic matrix (Fig. 2.5).

This process was first developed in the early 70s for both thermosets and thermoplastics (Lukaszewicz et al., 2012). Over the years, several changes have been made, more especially concerning the heating source, which was initially based on convection heating strategies, whereas today radiation is mostly used. These improvements are the main reason why thermoplastic composites are being used more extensively today.

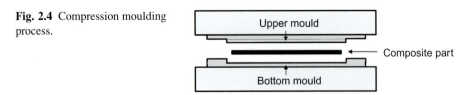

Fig. 2.4 Compression moulding process.

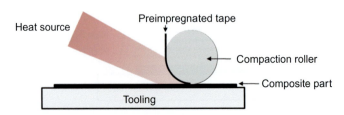

Fig. 2.5 Automated tape placement (ATP) process.

Concerning convection heating, considerable work is found in the literature, with heating systems using hot gas torches (Mantell and Springer, 1992; Mazumdar and Hoa, 1996; Tierney and Gillespie, 2003). However, the main problem is that an inert gas has to be used to prevent oxidation of the tapes and it is much harder to regulate the temperature at the nip point. To address these issues, developments were made by using radiation heating with the help of lasers, first CO_2 lasers in the late 90s (Pistor et al., 1999; Agarwal et al., 1992, 1996), and nowadays the near infrared diode lasers (Stokes-Griffin, 2015; Stokes-Griffin and Compston, 2015; Comer et al., 2015) (Fig. 2.6).

Several difficulties can be highlighted with this process. Contrary to other processes, a considerable number of processing parameters have to be mastered such as the laser angle, the tape back tension, the manufacturing speed, the mandrel temperature, compaction pressure, etc. Additionally, parameters related to the hardware itself such as the roller type (metallic or silicone), material aspects (fiber distribution through the tape thickness, tape surface roughness, void content, Lamontia et al., 2009; Khan et al., 2010), and the composite structure (edge connections between tapes: gaps, overlaps Lan et al., 2015) have also proved to have a significant influence. For each material of interest, optimal parameters have to be determined and the time needed to obtain these depends on the operator's experience with ATP (Mazumdar, 1994). Extensive studies have been performed to investigate the effects of all these parameters. It may be noted that most of the results reported in the literature focused

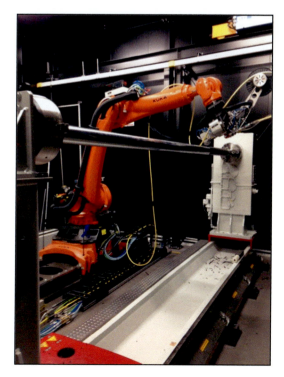

Fig. 2.6 Manufacturing of thermoplastic composite cylinders for underwater applications using assisted tape placement.

on C/PEEK laminates. For instance, optimum heating temperatures (Khan et al., 2010; Mazumdar, 1994) are to be found to reach the maximal mechanical properties as well as optimum mandrel temperatures that have been shown to reduce residual stresses (Qureshi et al., 2014). It has also been shown that slower laying speeds usually lead to improved mechanical properties (Grouve et al., 2010; Stokes-Griffin and Compston, 2015) as the increased healing time leads to a better intimate contact (Grouve et al., 2010).

2.3.2.4 Infusion

The infusion process is similar in some ways to the autoclave one shown in Fig. 2.3. The main differences are that the infusion of marine structures involves resin flow over longer distances, usually takes place at room temperature, and the pressure used to impregnate the fibers is the atmospheric pressure. The pressure differential between the resin container and the vacuum creates a resin flow that impregnates the fibers. The speed at which the resin flows within the fibers can be calculated using Darcy's law, which takes into consideration three main parameters which are: the viscosity of the matrix, the permeability of the fibers (fabrics), and the pressure differential. This process is being used extensively in the boat-building industry, more especially for pleasure boats, because of its low cost compared to other processes, improved working conditions compared to hand layup, and its ease of manufacturing large composite parts.

As stated in the Material Options section, the three main thermoplastic resins that are available for this process are the A-PA6 (Anionic Polyamide 6), the newly developed Elium™ resin from Arkema, which was used to successfully manufacture a racing yacht hull (Fig. 2.7), and cyclic butylene terephthalate (CBT) from Cyclics™, a matrix system based on commercial PBT which could take on particular importance for liquid impregnation processes.

Fig. 2.7 Internal structure of 6.5 m racing yacht manufactured from carbon/acrylic composite. Photo copyright Arkema used with permission.

2.3.3 Particular morphology of thermoplastic composites induced by processing

Due to their semicrystalline nature, the manufacturing parameters have a strong influence on the microstructure and the mechanical performance of many thermoplastics. The understanding and the control of these relationships are essential to optimize structures. First, it is important to make sure that the thermoplastic matrix does not degrade during manufacturing, since the manufacturing temperatures are high. Significant work can be found in the literature concerning this aspect (Cogswell, 1992b; Patel et al., 2010). Second, the degree of crystallinity of the thermoplastic resin within the composite part after manufacture depends on the cooling rate used during the process. The higher the cooling rate, the lower is the degree of crystallinity. When designing a thermoplastic composite part, it is essential to take this parameter into account, as it can have a major influence on the mechanical properties (Davies and Cantwell, 1994; Gao and Kim, 2001). It also has a significant effect on the residual stress level (Manson and Seferis, 1992; Deshpande and Seferis, 1996; Sonmez and Akbulut, 2007), because amorphous and semicrystalline regions do not have the same specific volumes (Nairn and Zoller, 1987) and, as a consequence, they shrink by different amounts during cooling. Another factor responsible for the high level of residual stresses in thermoplastic composites comes from the high melting temperature of thermoplastics and the large temperature differential during cooling. Therefore, when manufacturing thermoplastic composite parts, these aspects have to be taken into account carefully. Finally, the degree of crystallinity within a composite part has an effect on the diffusion of water (and other fluids) and the quantity that the thermoplastic polymer will absorb, since only the amorphous regions are able to absorb fluids. It will be shown in the next section that water absorption can have a significant effect on the mechanical properties.

2.4 Influence of the marine environment on thermoplastic composites

In parallel with defining the manufacturing method, it is essential to ensure that the thermoplastic matrix composites selected for a marine structure will provide adequate long-term behavior in a seawater environment. Many thermoplastics take up very little water, but this gives no guarantee that the composites properties will not degrade; it is not easy to bond thermoplastics to glass or carbon fibers, and water ingress can weaken an already weak interface region. Juska showed for a range of high-performance thermoplastics that while the carbon-reinforced materials retained adhesion after immersion, the glass fiber thermoplastics showed significantly lower mechanical properties after immersion (Juska, 1993).

It is, therefore, essential to understand how quickly seawater enters these composites and what the kinetics of the degradation mechanisms are. One of the difficulties in using polymer matrix composites rather than metals is that the influence of the presence of water will affect different properties to different extents, so generalizations about the effects of aging may be misleading. It requires considerable effort to obtain data which can be used for design.

Three examples will be examined in more detail below. The first is carbon/PEEK, a high-performance composite which is starting to find aerospace applications; then a lower performance matrix, polyamide 6, will be discussed; and finally some recent results from tests on a new infusible thermoplastic acrylic will be shown.

2.4.1 Carbon/PEEK

PEEK matrix is remarkably insensitive to solvents, as has been shown by many authors (e.g., Stober et al., 1984). If the microstructure is amorphous, PEEK will be more sensitive to water, but under usual manufacturing conditions of composites, a semicrystalline matrix is obtained. Any influence of water will then depend mainly on the interfaces with fibers (Meyer et al., 1994).

The influence of water on carbon/PEEK composites has been studied by many authors. For example, Selzer showed how immersion in distilled water at 23°C, 70°C, and 100°C affected a range of properties of carbon/PEEK and carbon/Epoxy composites (Selzer and Friedrich, 1997). The former were hardly affected, while the epoxy matrix composites showed significant property losses.

Choqueuse et al. (1997) showed results from specimens tested in distilled water for 2 years at 5°C, 20°C, 40°C, and 60°C, with and without applied pressure of 10 MPa. The diffusion coefficients for C/PEEK were low and property losses at saturation (flexural modulus and strength, ILSS) were less than 15%.

In recent tests at IFREMER, compression moulded carbon/PEEK samples were immersed in natural sea water at 80°C until saturation (Fig. 2.8). This plot shows a Fickian behavior and a very low weight gain in the saturated state. Table 2.2 shows compression test results for this material before and after aging.

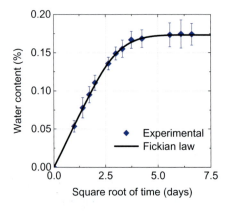

Fig. 2.8 Weight gain of unidirectional C/PEEK at 80°C in natural sea water.

Thermoplastic matrix composites for marine applications 41

Table 2.2 Influence of immersion in seawater at 80°C on compression properties of compression moulded unidirectional carbon/PEEK

Material	σ_{1f} (MPa)	ε_{1f} (%)	E_1 (GPa)
C/PEEK	1645 ± 29	1.37 ± 0.03	118 ± 4
Aged at 80°C in sea water for 2 months (saturated)	1783 ± 245	1.41 ± 0.20	121 ± 3
Aged at 80°C in sea water for 20 months	1648 ± 436	1.38 ± 0.51	121 ± 9

There is an increase in variability of results after aging; further work is underway to explain this, but even after 20 months in seawater at 80°C, compression strength still exceeds 1200 MPa.

In summary, the durability of carbon/PEEK composites in seawater environments is generally excellent, and they have become the reference material for critical marine applications.

2.4.2 Carbon/polyamide 6

While carbon/PEEK shows very good durability, it is also expensive, and cheaper, lower performance polymers such as polyamides might provide adequate properties for some marine applications. Indeed, there has been some interest in polyamides for offshore wind turbine blades (Van Rijswijk et al., 2005) and liquid infusion is possible for anionic polyamide 6 (Rijswijk et al., 2006). There are many different polyamide grades available (e.g., PA6, PA66, PA11, PA12); PA6 is one of the cheapest, but is known to be very sensitive to moisture (Silva et al., 2013; Reimschuessel, 1978). The first step is, therefore, to study the diffusion kinetics for PA6 in seawater. Fig. 2.9 shows the weight gain of an unreinforced PA6 immersed at 15°C. This

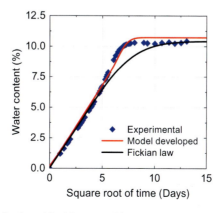

Fig. 2.9 Weight gain of polyamide 6 immersed in sea water at 15°C.

indicates that the polymer can pick up 10% by weight of water. This water has a significant effect on properties, but after drying the initial properties are completely recovered (Le Gac et al., 2017).

Diffusion studies on this material pose problems, as the dry PA6 glass transition temperature is quite low, around 60°C. During service, when water enters the material, the T_g drops significantly, and so the composite matrix will change from the glassy to the rubbery state. Passing through the glass transition results in an acceleration of water ingress as seen in Fig. 2.9 and the need for a specific diffusion model accounting for this transition. Such a model has been developed, and more details of this work are given in (Arhant et al., 2016a).

Underwater pressure vessels need to be quite thick in order to resist hydrostatic pressure, so it is also important to know their composite diffusion kinetics, as in service water ingress may be confined to a small external region of the wall thickness. Immersion tests have, therefore, been performed on carbon fiber reinforced composites. Fig. 2.10.

The presence of fibers, about 48% by volume here, limits the amount of absorbed water at saturation to around 3%. This is comparable to weight gains in many epoxy matrix composites.

Once the diffusion kinetics have been established, the influence of water content on composite properties can be examined. Fig. 2.11 shows how various properties change with seawater aging. Tests were performed on specimens which had been saturated in water at different saturation levels, in order to avoid the presence of a water gradient. These data are very useful as they can be introduced into an aging model for any geometry once the local water profile has been established, in order to examine how structural behavior is affected with aging time.

There is, indeed, a drop in strength with water ingress for all the matrix-dependent properties. However, the changes are predictable and, even under compression loading, the residual strength values remain around 600 MPa at complete saturation. These

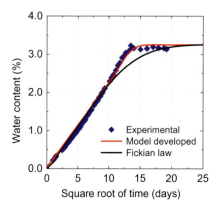

Fig. 2.10 Carbon fiber reinforced PA6 composite weight gain, immersion sea water 15°C.

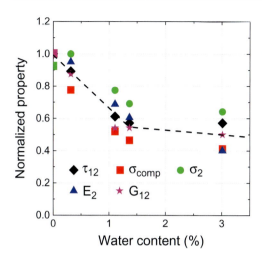

Fig. 2.11 Influence of water content on various properties, carbon/PA6 (Arhant et al., 2016b).

properties have also been shown to be largely recoverable after drying, indicating that matrix plasticization is the main physical mechanism involved.

Based on the water diffusion kinetics and the associated loss in compression properties, it is possible to estimate the response of a composite tube in deep sea. For example, for a 12 mm thick carbon/polyamide 6 pressure vessel of 120 mm inner diameter at 15°C, it would take around 75 years to saturate the wall thickness, so the use of these materials for deep sea applications such as drifting profilers (2 years' immersion at 2500 m depth) appears quite feasible. More details are available in (Arhant et al., 2015, 2016b).

2.4.3 Glass and carbon reinforced acrylic

A third type of thermoplastic composite, developed in the last few years (2014), is based on acrylic polymer (Anon, n.d.-a) and sold under the trade name of Elium™. The advantage of this material is that it can be infused as a liquid at room temperature, then polymerizes to become a thermoplastic which is potentially recyclable. This is a major innovation as it would allow boatyards with traditional infusion technology to produce thermoplastic composites. One obvious question is whether this material is stable in a marine environment. The very recent commercialization of Elium™ acrylic has limited the amount of long-term data available. However, some results have been published from a study in which acrylic resin and glass and carbon fiber reinforced acrylic composites were aged in seawater for up to 18 months (Davies et al., 2017a). Fig. 2.12 shows two examples of results, which indicate that the resin is less sensitive to water than a traditional marine grade epoxy, and that the main water-induced mechanism in wet aged carbon/acrylic samples is reversible plasticization.

So far there are few published results on the recycling of composites based on this resin, but should that prove to be economically viable, these composites could become very popular for marine applications.

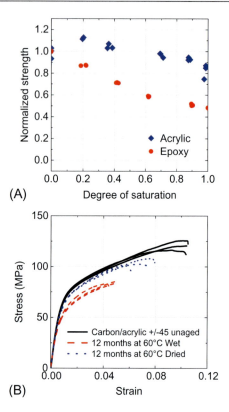

Fig. 2.12 Influence of seawater immersion on (A) resin tensile properties, compared to a marine epoxy, and (B) stress-strain behavior of a ± 45° carbon/acrylic composite.

2.5 Underwater structures

Today, there are few underwater applications of thermoplastic composites. In spite of promising results from laboratory studies, various factors have limited their use in the past. However, this is changing dramatically, with the recent adoption by the Offshore Oil & Gas industry of thermoplastic composite flexible risers (Picard et al., 2007; Osborne, 2013; Echtermeyer and Steuten, 2013; DNVGL-RP-F119, n.d.). There is considerable interest in a range of pipe materials, from glass fiber reinforced polyethylene, for low internal pressures, through to carbon fiber reinforced PEEK (Melot, 2017). There are currently two main producers of bonded flexible thermoplastic composite risers; Airborne Oil & Gas, who pioneered the bonded pipe application and offer a range of materials (Anon, n.d.-b), and Magma, who are focusing on carbon/PEEK (Anon, n.d.-c).

One initial concern about thermoplastic composites for underwater applications was their compression strength, as early test results on carbon reinforced polysulfone suggested significantly lower values than for epoxy matrix equivalents. However, subsequent tests showed that this drop was related to fiber waviness (Adams and Hyer, 1993), and compression strengths for unidirectional carbon/PEEK above 1400 MPa have been reported (Lee and Trevett, 1987; Zhang et al., 1996). The influence of cooling rate on matrix microstructure was also a concern when these materials were first proposed, but this was thoroughly examined by various authors (e.g., Davies et al., 1991a).

A second factor is the cost of high-performance thermoplastics, both the raw materials and the manufacturing costs, which remain high. Whether these costs are justified will depend on the application, but in critical applications such as subsea equipment with a lifetime of 20 years or more, operating cost reductions may significantly outweigh a higher initial capital expenditure.

From a technical point of view, it has been demonstrated that carbon/PEEK cylinders can withstand the high hydrostatic pressures associated with deep sea applications (Gruber et al., 1995; Davies et al., 2005). There are two specific questions to be answered for this environment, the first is whether pressure will affect the water ingress, and the second is whether resistance to hydrostatic pressure (biaxial compression) can be accurately predicted.

Aging tests performed in pressure vessels at pressures up to 500 bar (>5000 m depth) have shown that, for most composites, hydrostatic pressure will only affect water ingress if voids are present (Humeau et al., 2016). There is competition between the effect of pressure which will try to push water into the polymer, and the free volume available which reduces as pressure increases. Tests on the carbon/PA6 material described above showed no acceleration by a hydrostatic pressure of 500 bar (Davies et al., 2017b).

More details of the behavior of composites in deep sea can be found in (Davies, 2016; Davies et al., 2016), which include results from implosion tests. An example of a test recording from such a test on a carbon/PEEK cylinder is shown in Fig. 2.13 below. The insert photo shows the tube after the test.

Prediction of implosion behavior of composite cylinders is not simple as was clearly illustrated by the World Wide Failure Exercise (Soden et al., 2004). Failure criteria under biaxial compression remain approximate and emphasis has been placed on testing to determine failure pressures. This is an area which requires more work, particularly if large subsea structures such as composite separators are to be developed (Bigourdan et al., 2003).

2.6 Repair

One of the frequently cited advantages of thermoplastic composite materials is the possibility to repair them, either directly by local heating and crack healing or through fusion bonding of a repair patch. There was considerable interest in this area in the 1990s, and Stokes examined various options including resistance welding and ultrasonic welding (Stokes, 1989).

Several authors have provided overviews of composite fusion bonding processes (Davies et al., 1991b; Davies and Cantwell, 1993; Xiao et al., 1994; Ageorges et al., 2001; Yousefpour et al., 2004; Stavrov and Bersee, 2005).

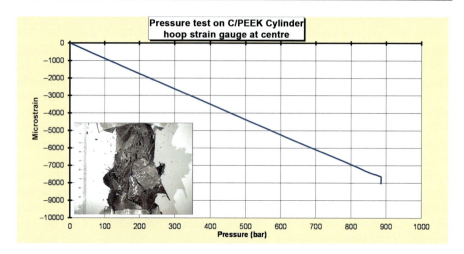

Fig. 2.13 Results from hydrostatic pressure test on carbon/PEEK cylinder, 6 mm thick, inner diameter 55 mm, length 110 mm, failure at a pressure of 88 MPa (equivalent to >8 km immersion depth).

From a practical point of view, Otheguy et al. (2009) describe a study of the repair of an experimental RIB (rigid inflatable boat) manufactured in 2004 from woven polypropylene-glass commingled fabric. The boat hull was impacted by allowing a 500 kg steel striker to fall 6.5 m onto the hull. This test condition was chosen to provide the same kinetic energy as a grounding incident. Damage was extensive, with perforation and multiple delaminations in the 12 mm thick hull. A fusion bonded patch repair was performed, and tensile tests indicated excellent strength recovery.

Glass/polypropylene repair was also examined by Reyes and Sharma (Reyes and Sharma, 2010), who showed that significant recovery in the flexural strength and modulus could be achieved after repair. They applied a one-step compression moulding process following the original manufacturing process parameters (180°C).

High-performance composites can also be repaired; for example, Wang and Shin (2002) showed that reconsolidation after fatigue damage in carbon/PEEK composites could recover a large proportion of initial strength. Cantwell et al. showed similar results for impact damaged specimens (Cantwell et al., 1991). Tarpani et al. (2014) also showed results for carbon/PPS composites reconsolidated after impact and fatigue and, for impacts of 10 and 20 J, mechanical strength was fully recovered.

2.7 Recycling and environmental impact

Another potential advantage of thermoplastic matrix composites is the possibility for recycling. This is of increasing importance as the use of composite materials extends, and two areas where large amounts of composite are generated are in pleasure boats, already a current problem, and in the near future from wind turbine blades (Cherrington et al., 2012), particularly offshore. For the former, the average lifetime of composite pleasure boats is around 30 years, and very little of the material used is recycled. Efforts have been made to install material separation plants for boats at their end of life.

The expected quantities of composites in future offshore wind farms are huge; the largest current blades are around 80 m long and weigh more than 70 tons. Various projects have examined the use of thermoplastics for wind blades; one example is the European FP7 project WALiD (Wind Blade Using Cost-Effective Advanced Lightweight Design), (Anon, n.d.-d) in which thermoplastic composites and foams and processes such as tape laying have been studied for the replacement of various turbine components.

There are several aspects which need to be considered with respect to thermoplastic composites; the first is the possibility to remelt the thermoplastic matrix and reform it, thus reusing all the material. This is of interest irrespective of the type of application. A second aspect is the reuse of the carbon fiber reinforcement by removal of the matrix. If high-performance applications become more widespread, it will be essential to reuse the fibers even if complete recycling is not possible. A third avenue is the development of bio-sourced and/or biodegradable thermoplastics, and this offers new possibilities to reduce environmental impact. In particular, natural fiber reinforced bio-sourced plastics such as PLA (poly(lactic acid)) are being considered as a replacement for low-performance glass fiber reinforced structures and offer potential for bio-sourced, biodegradable composites. A final and essential point is the evaluation of the overall environmental impact of such processes with respect to the existing composite solutions.

First, remelting of thermoplastic composites during a recycling operation is possible for all semicrystalline matrix materials (e.g., PE, PP, PA, PEEK), provided the melting window is sufficient to allow reforming without degradation. Schinner et al. and Stewart provide several examples (Schinner et al., 1996; Stewart, 2011).

For the second case, recovery of carbon fibers from thermoplastic composites, it is necessary to develop specific procedures. For example, Dandy et al. proposed a solvent process to decompose PEEK matrix and recover the fibers (Dandy et al., 2015). This is based on applying a cosolvent system of ethanol and water and can be performed at 350 °C within 30 min.

Bio-sourced thermoplastics, such as poly (lactic acid) (PLA) from corn plants, have been extensively developed for packaging applications. Recycling trials on flax reinforced PLA (Le Duigou et al., 2008) using injection moulding indicated that tensile modulus remained stable even after six recycling operations, while strength and failure strain started to drop after three.

PLA is also biodegradable, which may be an advantage at the end of life. Smith provides an overview of biodegradable polymers (Smith, 2005), which include polymers from both natural and petrochemical sources. These materials can be composted; so long-term durability in marine applications will clearly be a challenge. Le Duigou et al. (2009, 2014) examined the wet aging behavior of PLA and its flax reinforced composites for up to 2 years at sea and showed a linear relationship between water uptake and property loss, so protection of the fibers is essential.

Fig. 2.14 Flax/PLA kayak demonstrator.
Photo University Betagne Sud, used with permission.

Fig. 2.14 shows a flax fiber reinforced PLA kayak, manufactured as a demonstrator prototype (Anon, 2011).

It has been shown recently that flax/PP composites can be produced by ATP, and this may extend the applications of these materials (Baley et al., 2016).

Finally, the aim with all these solutions is to evaluate their contributions to environmental impact with respect to existing solutions under comparable conditions. Life cycle analysis (LCA) provides a tool to do this, but requires reliable input data on raw materials, transport, and processing. Unfortunately, these are rarely available. For example, while thermoset polymers are difficult to recycle, they can be used to construct large composite structures at relatively low temperatures, whereas melting thermoplastics consumes more energy to achieve the required higher temperatures with potential consequences for climate change. Further, the enhanced toughness of thermoplastics increases the energy required to reduce the waste material to appropriate sizes for reuse. To the best of the authors' knowledge, a comparative LCA for these situations has not been performed.

2.8 Conclusion

Thermoplastic matrix composites offer distinct advantages for certain marine structures. However, the need for elevated temperature requires a change in manufacturing technology which is difficult to justify for small boatyards. As a result, existing applications are limited to high-performance structures such as flexible pipes where significant benefits can be obtained.

Development of new materials such as infusible thermoplastics, acrylics or anionic polyamides, may widen the range of applications considerably and allow the marine industry to respond to a pressing need for more material recycling.

References

Adams, D.O.H., Hyer, M.W., 1993. Effects of layer waviness on the compression strength of thermoplastic composite laminates. J. Reinf. Plast. Compos. 12 (4), 414–429.

Agarwal, V., Guçeri, S.I., Mccullough, R.L., Schultz, J.M., 1992. Thermal characterization of the laser-assisted consolidation process. J. Thermoplast. Compos. Mater. 5, 115–135.

Agarwal, V., Mccullough, R.L., Schultz, J.M., 1996. The thermoplastic laser-assisted consolidation process-mechanical and microstructure characterization. J. Thermoplast. Compos. Mater. 9, 365–380.

Ageorges, C., Ye, L., Hou, M., 2001. Advances in fusion bonding techniques for joining thermoplastic matrix composites: a review. Composites Part A 32, 839–857.

Anon Environmentally friendly composites for marine applications, the NAVECOMAT project, JEC Magazine, August–September 2011.

Anon, Arkema website: http://www.arkema.com/en/products/product-finder/range-viewer/Elium-resins-for-composites/.

Anon, Airborne Oil & Gas website: http://airborne-oilandgas.com/.

Anon Magma website: https://www.magmaglobal.com/.

Anon EU WALiD (Wind Blade Using Cost-Effective Advanced Lightweight Design) project website: http://www.eu-walid.com/.

Arhant, M., Davies, P., Burtin, C., Briançon, C., 2015. Thermoplastic matrix composites for underwater applications.Proceedings 20th International Conference on Composite Materials, (ICCM20), Copenhagen.

Arhant, M., Le Gac, P.Y., Le all, M., Burtin, C., Briancon, C., Davies, P., 2016a. Modelling the non Fickian water absorption in polyamide 6. Polym. Degrad. Stab. 133, 404–412.

Arhant, M., Le Gac, P.Y., LeGall, M., Burtin, C., Briancon, C., Davies, P., 2016b. Effect of seawater and humidity on the tensile and compressive properties of carbon-polyamide 6 laminates. Compos. A: Appl. Sci. Manuf. 91 (1), 250–261.

Baley, C., Kervoelen, A., Lan, M., Cartié, D., Bourmaud, A., LeDuigou, A., Davies, P., 2016. Flax/PP manufacture by automated fibre placement (AFP). Mater. Des. 94, 207–213.

Bigourdan, B., Chauchot, P., Astrugue, J.C., Micheaux, D., Alary, V., Dilosquer, S., Falcimaigne, J., Rigaill, C., Gérard, P., 2003. Composite Materials for Subsea Oil Separation, DOT. Marseille, France.

Cantwell, W.J., Davies, P., Kausch, H.H., 1991. Repair of impact damaged carbon fiber PEEK laminates. SAMPE J. 27 (6), 30–35.

Carlsson, L.A. (Ed.), 1991. In: Thermoplastic Composite materials, Composite materials Series, vol. 7. Elsevier Publishers.

Cherrington, R., Goodship, V., Meredith, J., Wood, B.M., Coles, S.R., Vuillaume, A., Feito-Boirac, A., Spee, F., Kirwan, K., 2012. Producer responsibility: Defining the incentive for recycling composite wind turbine blades in Europe. Energy Policy 47, 13–21.

Choqueuse, D., Davies, P., Mazéas, F., Baizeau, R., 1997. Aging of composites in water: Comparison of five materials in terms of absorption kinetics and evolution of mechanical properties. ASTM STP 1302, 73–96.

Cogswell, F.N., 1992a. Thermoplastic Aromatic Polymer Composites, first ed. Butterworth-Heinemann.

Cogswell, F.N., 1992b. Thermoplastic Aromatic Polymer Composites: A Study of the Structure, Processing, and Properties of Carbon Fibre Reinforced Polyetheretherketone and Related Materials. Butterworth-Heinemann, Oxford.

Comer, A.J., Ray, D., Obande, W.O., Jones, D., Lyons, J., Rosca, I., Higgins, R.M.O.'., McCarthy, M.A., 2015. Mechanical characterization of carbon fibre–PEEK manufactured by laser-assisted automated-tape-placement and autoclave. Compos. Part Appl. Sci. Manuf. 69, 10–20.

Dandy, L.O., Oliveux, G., Wood, J., Jenkins, M.J., Leeke, G.A., 2015. Accelerated degradation of Polyetheretherketone (PEEK) composite materials for recycling applications. Polym. Degrad. Stab. 112, 52–62.

Davies, P., 2016. Behavior of marine composite materials under deep submergence. (Chapter 6). In: Graham-Jones, J., Summerscales, J. (Eds.), Marine Applications of Advanced Fibre-Reinforced Composites. Woodhead Publishing, pp. 125–145.

Davies, P., Cantwell, W.J., 1993. Bonding and Repair of Thermoplastic Composites. Chapter 11. In: Kausch, H.H. (Ed.), Advanced Thermoplastic Composites. Hanser Publishers, pp. 337–366.

Davies, P., Cantwell, W.J., 1994. Fracture of glass/polypropylene laminates: Influence of cooling rate after moulding. Composites 25 (9), 869–877.

Davies, P., Cantwell, W.J., Jar, P.Y., Richard, H., Neville, D.J., Kausch, H.H., 1991a. Cooling rate effects in carbon fibre/PEEK composites. ASTP STP 1110, 70–88.

Davies, P., Cantwell, W.J., Jar, P.Y., Bourban, P.E., Zysman, V., Kausch, H.H., 1991b. Joining and repair of a carbon fibre-reinforced thermoplastic. Composites 22 (6), 425–431.

Davies, P., Riou, L., Mazeas, F., Warnier, P., 2005. Thermoplastic composite cylinders for underwater applications. J. Thermoplast. Compos. Mater. 18 (5), 417–431.

Davies, P., Choqueuse, D., Bigourdan, B., Chauchot, P., 2016. Composite cylinders for deep sea applications: an overview. ASME J. Pressure Vessel Technol.. 138(6).

Davies, P., Le Gac, P.-Y., Le Gall, M., 2017a. Influence of sea water aging on the mechanical behaviour of acrylic matrix composites. Appl. Compos. Mater. 24 (1), 97–111.

Davies, P., Le Gac, P.Y., Le Gall, M., Arhant, M., 2017b. Marine aging behavior of new environmentally-friendly composites. In: Davies, P. (Ed.), Durability of Composites in a Marine Environment 2. Springer, Rajapakse YSD.

Deshpande, A.P., Seferis, J.C., 1996. Processing characteristics in different semi-crystalline thermoplastic composites using process simulated laminate (PSL) methodology. J. Thermoplast. Compos. Mater. 9 (2), 183–198.

DNVGL-RP-F119 Thermoplastic Composite Pipes.

Echtermeyer, A., Steuten, B., 2013. Thermoplastic composite riser guidance note.Proceedings Offshore Technology Conference. OTC-24095.

Gao, S.-L., Kim, J.-K., 2001. Cooling rate influences in carbon fibre/PEEK composites. Part II: Interlaminar fracture toughness. Compos. Part Appl. Sci. Manuf. 32 (6), 763–774.

Grouve, W.J.B., Warnet, L., Akkerman, R., Wijskamp, S., Kok, J.S.M., 2010. Weld strength assessment for tape placement. Int. J. Mater. Form. 3 (1), 707–710.

Gruber, M.B., Lamontia, M.A., Smoot, M.A., Peros, V., 1995. Buckling performance of hydrostatic compression loaded 7-inch diameter thermoplastic composite monocoque cylinders. J. Thermoplast. Compos. Mater. 8, 94–108.

Hoggatt, J.T., 1975. Thermoplastic resin composites.20th National SAMPE conference, May. p. 606.

Humeau, C., Davies, P., Jacquemin, F., 2016. Moisture diffusion under hydrostatic pressure in composites. Mater. Des. 96, 90–98.

Juska, T., 1993. Effect of water immersion on fibre/matrix adhesion in thermoplastic composites. J. Thermoplast. Compos. Mater. 6 (4), 256–274.

Kausch, H.H. (Ed.), 1993. Advanced Thermoplastic Composites. Hanser Publishers.

Khan, M.A., Mitschang, P., Schledjewski, R., 2010. Identification of some optimal parameters to achieve higher laminate quality through tape placement process. Adv. Polym. Technol. 29 (2), 98–111.

Lamontia, M.A., Gruber, M.B., Tierney, J.J., Gillespie Jr., J.W., Jensen, B.J., Cano, R.J., 2009. In situ thermoplastic ATP needs flat tapes and tows with few voids.*30th International SAMPE Europe Conference,* Paris.

Lan, M., Cartié, D., Davies, P., Baley, C., 2015. Microstructure and tensile properties of carbon–epoxy laminates produced by automated fibre placement: Influence of a caul plate on the effects of gap and overlap embedded defects. Compos. Part Appl. Sci. Manuf. 78, 124–134.

Le Duigou, A., Pillin, I., Bourmaud, A., Davies, P., Baley, C., 2008. Effect of recycling on mechanical behaviour of biocompostable flax/poly(l-lactide) composites. Compos. A: Appl. Sci. Manuf. 39 (9), 1471–1478.

Le Duigou, A., Davies, P., Baley, C., 2009. Seawater ageing of flax/poly(lactic acid) biocomposites. Polym. Degrad. Stab. 94 (7), 1151–1162.

Le Duigou, A., Bourmaud, A., Davies, P., Baley, C., 2014. Long term immersion in natural seawater of flax/PLA biocomposite. Ocean Eng. 90 (1), 140–148.

Le Gac, P.Y., Arhant, M., Le Gall, M., Davies, P., 2017. Yield stress changes induced by water in polyamide 6: Characterization and modeling. Polym. Degrad. Stab. 137, 272–280.

Lee, R.J., Trevett, A.S., 1987. Compression strength of aligned carbon fibre reinforced thermoplastic laminates. Proceedings of ICCM-V ECCM-2. Elsevier Applied Science Publishers, London. pp. 1.278–1.287.

Lukaszewicz, D.H.-J.A., Ward, C., Potter, K.D., 2012. The engineering aspects of automated prepreg layup: History, present and future. Compos. Part B Eng. 43 (3), 997–1009.

Manson, J.-A.E., Seferis, J.C., 1992. Process simulated laminate (PSL): a methodology to internal stress characterization in advanced composite materials. J. Compos. Mater. 26 (3), 405–431.

Mantell, S.C., Springer, G.S., 1992. Manufacturing process models for thermoplastic composites. J. Compos. Mater. 26 (16), 2348–2377.

Maximovich, M.G., 1977. Evaluation of selected high-temperature thermoplastic polymers for advanced composite and adhesive applications. ASTM STP. 617, p. 123.

Mazumdar, S.K., 1994. "Automated Manufacturing of Composite Components by Thermoplastic Tape Winding and Filament Winding," Concordia University. PhD Thesis, Montreal, Canada.

Mazumdar, S.K., Hoa, S.V., 1996. Determination of manufacturing conditions for hot-gas-aided thermoplastic tape winding. J. Thermoplast. Compos. Mater. 9 (1), 35–53.

Melot, D., 2017. Present and future composite requirements for the offshore oil and gas industry. In: Davies, P. (Ed.), Durability of Composites in a Marine Environment 2. Springer, Rajapakse YSD.

Meyer, M.R., Latour, R.A., Shutte, H.D., 1994. Long-term durability of Fiber/matrix interfacial bonding in hygrothermal environments. J. Thermoplast. Comp. 7 (3), 180–191.

Nairn, J.A., Zoller, P., 1987. The development of residual thermal stresses in amorphous and semicrystalline thermoplastic matrix composites. ASTM STP 937, 328–341.

Osborne, J., 2013. Thermoplastic pipes—lighter, more flexible solutions for oil and gas extraction. Reinforced Plastics. 26.

Otheguy, M.E., Gibson, A.G., Findon, E., Cripps, R.M., 2009. Repair technology for thermoplastic composite boats.Proc ICCM17, Edinburgh.

Patel, P., Hull, T.R., Mccabe, R.W., Flath, D., Grasmeder, J., Percy, M., 2010. Mechanism of thermal decomposition of poly(ether ether ketone) (PEEK) from a review of decomposition studies. Polym. Degrad. Stabil. 95 (5), 709–718.

Picard, D., Hudson, W., Bouquier, L., Dupupet, G., Zivanovic, I., 2007. Composite Carbon Thermoplastic Tubes for Deepwater Applications.Offshore Technology Conference. OTC19111.

Pistor, C.M., Yardimci, M.A., Güçeri, S.I., 1999. On-line consolidation of thermoplastic composites using laser scanning. Composites, Part A 30, 1149–1157.

Qureshi, Z., Swait, T., Scaife, R., El-Dessouky, H.M., 2014. In situ consolidation of thermoplastic prepreg tape using automated tape placement technology: potential and possibilities. Compos. Part B Eng. 66, 255–267.

Reimschuessel, H.K., 1978. Relationships on the effect of water on glass transition temperature and Young's modulus of nylon 6. J Polym. Sci. Polym. Chem. Ed. 16 (6), 1229–1236.

Reyes, G., Sharma, U., 2010. Modeling and damage repair of woven thermoplastic composites subjected to low velocity impact. Compos. Struct. 92, 523–531.

Van Rijswijk, K., Bersee, H.E.N., Jager, W.F., Picken, S.J., 2006. Optimization of anionic polyamide-6 for vacuum infusion of thermoplastic composites: choice of activator and initiator. Compos. A: Appl. Sci. Manuf. 37 (6), 949–956.

Schinner, G., Brandt, J., Richter, H., 1996. Recycling carbon-fiber-reinforced thermoplastic composites. J. Thermoplast. Compos. Mater. 9, 239–245.

Selzer, R., Friedrich, K., 1997. Mechanical properties and failure behaviour of carbon fibre-reinforced polymer composites under the influence of moisture. Composites A 28, 595–604.

Silva, L., Tognana, S., Salgueiro, W., 2013. Study of the water absorption and its influence on the Young's modulus in a commercial polyamide. Polym. Test. 32 (1), 158–164.

Smith, R. (Ed.), 2005. Biodegradable Polymers for Industrial Applications. Woodhead Publishing.

Soden, P.D., Kaddour, A.S., Hinton, M.J., 2004. Recommendations for designers and researchers resulting from the world-wide failure exercise. Compos. Sci. Technol. 64 (3-4), 589–604.

Sonmez, F.O., Akbulut, M., Sep. 2007. Process optimization of tape placement for thermoplastic composites. Compos. A: Appl. Sci. Manuf. 38 (9), 2013–2023.

Stavrov, D., Bersee, H.E.N., 2005. Resistance welding of thermoplastic composites-an overview. Composites Part A 35, 39–54.

Stewart, R., 2011. Thermoplastic composites - recyclable and fast to process. Reinf. Plast. 55 (3), 22–28.

Stober, E.J., Seferis, J.C., Keenan, J.D., 1984. Characterization and exposure of polyetheretherketone (PEEK) to fluid environments. Polymer 25, 1845–1852.

Stokes, V.K., 1989. Joining methods for plastics and plastic composites: An overview. Polym. Eng. Sci. 29 (19), 1310–1324.

Stokes-Griffin, C.M., 2015. A Combined Optical-Thermal Model for Laser-Assisted Fibre Placement of Thermoplastic Composite Materials. The Australian National University. PhD thesis, Canberra.

Stokes-Griffin, C.M., Compston, P., 2015. The effect of processing temperature and placement rate on the short beam strength of carbon fibre–PEEK manufactured using a laser tape placement process. Compos. Part Appl. Sci. Manuf. 78, 274–283.

Tarpani, J.R., Canto, R.B., Saracura, R.G.M., Ibarra-Castanedo, C., Maldague, X.P.V., 2014. Compression after impact and fatigue of reconsolidated fiber reinforced thermoplastic matrix solid composite laminate. Proc. Mater. Sci. 3, 485–492.

Tierney, J., Gillespie, J.W., 2003. Modeling of heat transfer and void dynamics for the thermoplastic composite tow-placement process. J. Compos. Mater. 37, 1745–1768.

Van Rijswijk, K., Joncas, S., Bersee, H.E., Bergsma, O.K., Beukers, A., 2005. Sustainable vacuum-infused thermoplastic composites for MW-size wind turbine blades-preliminary design and manufacturing issues. J. Sol. Energy Eng. 127 (4), 570–580.

Wang, C.M., Shin, C.S., 2002. Residual properties of notched [0/90]$_{4S}$ AS4/PEEK composite laminates after fatigue and re-consolidation. Compos. Part B 33, 67.

Xiao, X.R., Hoa, S.V., Street, K.N., 1994. Repair of thermoplastic resin composites by fusion bonding. Composites Bonding. ASTM STP 1227, 30–44.

Yousefpour, A., Hojjati, M., Immarigeon, J.-P., 2004. Fusion bonding/welding of thermoplastic composites. J. Thermoplast. Compos. Mater. 17, 303–341.

Zhang, G., Latour, R.A., Kennedy, J.M., Schutte, H.D., Friedman, R.J., 1996. Long-term compressive property durability of carbon fibre-reinforced polyetheretherketone composite in physiological saline. Biomaterials 17 (8), 781–789.

Experimental and theoretical damage assessment in advanced marine composites

Phuong Tran*, Abdallah Ghazlan[†], Tu Phan Nguyen[†], Rebecca Gravina*
*Department of Civil and Infrastructure Engineering, RMIT University, Melbourne, VIC, Australia, [†]Department of Infrastructure Engineering, The University of Melbourne, Melbourne, VIC, Australia

Chapter Outline

3.1 Introduction 55
3.2 Damage to marine structures 59
    3.2.1 Damage due to impulsive loading 59
    3.2.2 Damage due to impact loading 60
    3.2.3 Damage due to environmental effects 62
3.3 Nondestructive damage detection for maritime composites 64
    3.3.1 Experimental methods 64
    3.3.2 In situ damage detection 69
3.4 Numerical and theoretical modeling of composite damages 70
    3.4.1 Finite element modeling 71
    3.4.2 Hashin damage model for fiber/matrix failure in composite layups 72
    3.4.3 Cohesive damage model for interlaminar fracture 74
    3.4.4 Artificial neural networks for damage detection in laminated composites 75
    3.4.5 Modeling cyclic fatigue damage to composites 78
3.5 Conclusions 79
References 79

3.1 Introduction

Marine structures are susceptible to damage throughout their design life, which can arise from accidental impacts or explosions (Anon, 2013, 2016, 2017a,b), fire (Anon, 2012a, b), wear and tear due to aggressive environments, and so on. An important event that occurred recently was the USS Fitzgerald warship's collision with a merchant vessel carrying shipping containers. Several casualties occurred due to flooding of the accommodation compartments due to severe damage to the hull of the naval vessel (Fig. 3.1), whereas the merchant container ship did not suffer any casualties (CNN, 2017).

Marine Composites. https://doi.org/10.1016/B978-0-08-102264-1.00003-0
Copyright © 2019 Elsevier Ltd. All rights reserved.

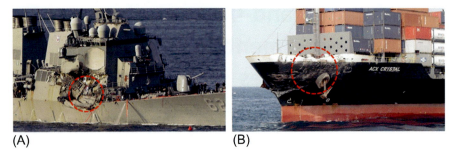

Fig. 3.1 (A) Impact damage to the USS Fitzgerald naval vessel; (B) Impact damage to the hull of the merchant vessel carrying shipping containers (CNN, 2017).

In this regard, composite structures are attractive because they offer significant advantages compared to traditional steel structures, namely high strength to weight ratio, good impact properties, low infrared, magnetic and radar signatures, excellent durability, and high resilience to extreme loads. Composites have been considered for a diverse range of marine applications since the early 1950s, including laminated glass plastic construction in naval boats (Bushby et al., 1952), reinforced plastic piping (McKenzie and Stark, 1953), and glass fiber-reinforced plastics (GFRP) for marine structures and ships (Buermann and DellaRocca, 1960; Alfers, 1966; Lindsay, 1966). The use of fiber-reinforced polymers (FRP) for marine applications has found steady growth since the 1950s (Shenoi and Wellicome, 1993). Although the applications of FRP were initially limited to small crafts such as lifeboats, they have found applications in large-scale structures such as ship hulls and superstructures, submarines, and offshore structures. Given the advances in materials, production methods, and design procedures, it is now possible to produce light-weight, high strength, and cost-effective composite structures that outperform their metal counterparts.

For two decades before the late 1960s, varieties of FRP marine vessels were built, but their size restricted to 20 m. A mine counter measures vessel (MCMV) combines the roles of seeking (hunter) and eliminating (sweeper) naval mines. From 1968, when a section of an MCMV was tested, it became possible to construct ships with 60 m long hulls. This was facilitated by single skin, woven E-glass polyester composite hulls for replacing wooden hulls. On the other hand, double-skinned (or sandwich type) hulls have not shared the same monotonic progress as single-skinned hulls due to bonding issues between the shear carrying core materials with the external skins for withstanding explosive shock loads. However, there are now many examples of double skin composite configurations in high-speed boats and even nonmaritime applications.

Many marine structures and vehicles such as sailboards, dinghies, lifeboats, power boats, fishing boats, workboats, patrol boats, and MCMVs, as well as sonar domes and the structure of many submarines outside the pressure hull, are now almost universally made from (GFRP (Shenoi and Wellicome, 1993). The widespread use of carbon fiber-reinforced plastic (CFRP) in aircraft, namely 70% of the air-frame, has incurred significant weight savings compared to traditional engineering materials, which resulted in significant improvements in performance. FRPs have found such widespread marine applications due to their resistance to corrosion and rot, thereby reducing maintenance costs considerably, which outweighs the initial cost of the materials.

Although the savings in weight due to composites are evident, they can be problematic for long naval vessels, such that deflections of the hull girder have been estimated to be about 240% higher than steel (Alm, 1983). This can affect the alignment in the propeller shaft-line, as well as lead to fatigue cracking around joints and connections. The Skjold class corvette vessel operated by the Royal Norwegian Navy is one of the largest naval patrol boats build purely from composite materials (Storman, 1999; Foxwell, 1999). The hull is 47 m long and 14 m wide, with 270 tons full-load displacement. The vessel is built from a sandwich composite, namely glass- and carbon-fiber laminate skins with a poly(vinyl chloride) (PVC)-foam core. The sandwich composite structure was chosen over steel or aluminium alloy because it simplified the construction of the hull and superstructure. However, conducting materials, need to be employed to provide electromagnetic shielding to sophisticated electronic equipment used on the boat when GFRP is employed as the sole composite material, which contributes a higher construction cost.

Since World War 2, high-quality timber from which MCMVs were originally built is not readily available. Timber was used because its nonmagnetic properties enabled the ships to operate in waters protected by magnetic sea mines. Wood structures have the ongoing maintenance disadvantage. To replace wood, the US Navy attempted to build a 16 m long minesweeper (XMSB-23) from a honeycomb sandwich composite (Spaulding, 1966). Due to the poor fabrication quality, mechanical performance and water resistance, seawater seeped into the hull. Therefore, the vessel failed to be utilized for mine countermeasure operations. The design and development of mine hunting ships continued during the 1960s and 1970s in the USA and UK (Lankford and Angerer, 1971; Dixon et al., 1972). The HMS Wilton was the first minehunter built using composites in 1973, which was 47 m long and 450 tonnes full-load displacement. This was the largest ship made purely from GFRP (Chalmers et al., 1984). This highly successful composite construction paved the path for the construction of over 200 all-composite minehunters since the early 1980s. The use of composites in mine countermeasure vessels has driven innovative ship hull designs, which can resist local buckling, facilitate higher hull girder stiffness, and have superior resistance to underwater shocks. Other criteria that naval operators consider in selecting hull types are acquisition and through-life maintenance costs, magnetic signature, acoustic damping, and fire performance (Schutz, 1984).

Composite materials and structures for marine applications offer many advantages over traditional engineering materials, namely light-weight, corrosion resistance, and excellent behavior under extreme and cyclic fatigue loading, which leads to a significant reduction in maintenance costs. Specifically, CFRP has high strength-to-weight and stiffness-to-weight ratios compared to steel and aluminium. This significantly improves the performance of marine structures without sacrificing their mobility. Although the effectiveness of GFRP was historically challenged due to fears of its vulnerability under fire, it has been found that GFRP offers an excellent fire barrier because its low thermal conductivity prevents fire propagation by avoiding flashover into adjacent cabins (Shenoi and Wellicome, 1993).

Sandwich composites made from carbon, glass, and/or aramid fiber-vinylester skins with a PVC-foam core have been employed by the Royal Swedish Navy for constructing large patrol craft (including the 73 m CFRP sandwich Visby class stealth corvette). Using this sandwich composite material, they built a 30 m long surface effect ship, known as the Smyge MPC2000 (Makinen et al., 1988). These composite materials exhibit light-weight, superior corrosion resistance, and good resistance to underwater shocks, in addition to stealth properties such as low thermal and magnetic signatures and good noise dampening properties. Despite the superior properties of the Skjold and Smyge MPC2000, large patrol boats continue to be built from steel and aluminium alloys due to the higher costs associated with composite materials.

FRPs are still a large part of a wide range of naval structures being developed. Increased range, stealth, stability, and payload are the operational performance indicators that drive the need for the development warships and submarines, as well as the reduction in ownership costs arising from reduced maintenance and fuel consumption. Some of the different material systems used for naval composite structures are explored herein. There has been growing interest in increasing the size of patrol boats up to 55 m, even though patrol boats are rarely built longer than 20 m because their hull girder suffers from low stiffness. To this effect, feasibility studies have been conducted for comparing the cost, weight, and structural performance of large patrol boats made from steel, aluminium, and sandwich composites. These studies have shown that patrol boats made from GFRP can exhibit a reduction of weight of up to 10% than that of an aluminium boat and up to 36% than that of a steel boat of similar size (Goubalt and Mayes, 1996).

The disadvantages with FRPs typically stem from the lack of user confidence in plastics outperforming metals in terms of structural integrity. However, in-service monitoring, which embeds sensors in the composite to assess its performance during its service life may convince sceptics, as well as regulatory authorities of the superiority of FRPs (Marsh, 2010). Another disadvantage with composite structures is their variability between the same types of structural applications due to production methods, which affect their quality, mechanical properties, and additional costs incurred due to the resulting variations in the development of design and manufacturing procedures. This issue has been addressed by suggesting closer collaborations between designers and fabricators to build productivity from the beginning of the manufacturing process. In terms of recyclability, thermoset plastics differ from thermoset resins that dominate today in terms of recyclability, but thermoplastics have disadvantages of high energy requirements due to the elevated manufacturing process temperatures, less powerful interfacial resin/fiber bonding, and poor adhesion when secondarily bonded.

This chapter reviews experimental and theoretical damage assessment methods that have been employed by researchers over the past few decades. Firstly, the modes of damage due to extreme events such as blast and impact, cyclic loads from environmental effects such as water slamming, moisture issues, and humidity are explored in Section 3.2. Subsequently, damage detection techniques are reviewed in Section 3.3 with a focus on the different failure modes of maritime composites. Destructive and nondestructive techniques for assessing residual damage to marine composites subjected to extreme loads are also discussed in Section 3.3. Finally, numerical and experimental methods for predicting damage to composites are discussed in Section 3.4.

3.2 Damage to marine structures

Marine structures have a high susceptibility to extreme loads, particularly impact and blast-induced damage. This can significantly reduce their structural performance or cause instantaneous catastrophic failure. Furthermore, environmental effects can cause cyclic fatigue loading on marine structures, as well as damage due to seawater moisture, humidity, and so on. The damage modes that composites undergo are quite complex compared to traditional materials such as isotropic metals, namely matrix cracking, fiber fracture, interface debonding, layer delamination, fiber pull-out, and compound failures including kink bands (buckling delamination due to kinking of fibers under compression) and barely visible impact damage (BVID) cone of fracture. Advancements in damage detection and computational modeling techniques have facilitated accurate predictions of the behavior of marine composite structures under extreme loads and aggressive environments.

3.2.1 Damage due to impulsive loading

The physics of blast loading is well-established whereby a blast shockwave is generated by a rapid increase in pressure driven by the rapid combustion of detonation products, followed by an exponential decay due to the blast wind that trails the shockwave (Ngo et al., 2015, 2007; Tran et al., 2016; Ding et al., 2015; Ghazlan et al., 2016). A typical blast pressure-time history plot is shown in Fig. 3.2A. Arora et al. (2014) studied the residual compressive strength of sandwich composite materials after they were exposed to blast loading. The sandwich panel consisted of CFRP composite skins on a styrene–acrylonitrile polymer closed-cell foam core. Residual cracks are clearly visible on the GFRP sample postblast (Fig. 3.2B), which indicate flexure and shear

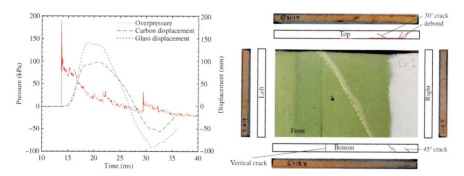

Fig. 3.2 (A) Blast pressure profile and deflection of the CFRP and GFRP panels; (B) Residual damage on the GFRP panel after blast loading (Arora et al., 2014).

Fig. 3.3 Damage zones on the laminated composite sandwich panel after underwater blast loading (Wei et al., 2013a).

cracking, as well as delamination modes of damage in the laminated composite skins. An interesting observation obtained from digital image correlation (DIC) also showed that wrinkling occurred in the front skin, which indicates that the face-sheets maintained their bonded connectivity with the foam core. Subsequent compression tests on the damaged panels also showed significant residual compressive strength.

Wei et al. (2013a) investigated the behavior of sandwich composite panels subjected to underwater blast loading, which consisted of E-glass fiber plies reinforcing a vinylester matrix. They simulated the underwater blast experiment by launching a projectile from a gas gun onto the inlet of an anvil tube, which causes a shock to propagate through a water channel and subsequently impact the specimen. Fig. 3.3 illustrates the damage patterns to the composite face-sheets and the core of the sandwich panel. Significant damage can be observed at the clamped edge and midspan of the panel due to shearing and crushing of the fibers. The delamination patterns were compared to those of a monolithic panel of equal areal mass, which experienced catastrophic failure near the clamped edges via interlaminar delamination.

3.2.2 Damage due to impact loading

Many studies on the impact behavior of laminated composites, which are typically used in marine applications, have been conducted compared to blast (Nasirzadeh and Sabet, 2014; Perillo and Jorgensen, 2016; Castilho et al., 2015; Muscat-Fenech et al., 2014a, b; Feng and Aymerich, 2013; Petrucci et al., 2015; Moriniere et al., 2014; Lopez-Puente et al., 2007), particularly to assess the structural integrity of composite sandwich panels by investigating their postimpact residual strength. Bull et al. (Bull and Edgren, 2004) employed foil strain gauges and damage speckle photography to investigate the residual compressive strength of CFRP-foam core sandwich panels for marine applications after low velocity impact loading from a drop weight. Microbuckling is a major mode of failure that can be observed from the kink bands (up to 22 mm in length) in the composite laminates shown in Fig. 3.4A, which are growing perpendicular to the loading direction. Fig. 3.4B illustrates the damage propagation in the laminated plate prior to failure, which is highly localized. The postimpact residual compressive strength of the composite laminates was found to be approximately 20% of the initial compressive strength.

Feng and Aymerich (2013) investigated the damage mechanisms of composite sandwich panels consisting of carbon/epoxy faces bonded to a PVC-foam core under out of plane low velocity impact loading. Fig. 3.5 illustrates the onset and growth of several damage mechanisms in the composite skin of a typical face-sheet from the impact zone, namely fiber fracture and matrix cracking and delamination. Both

Experimental and theoretical damage assessment in advanced marine composites 61

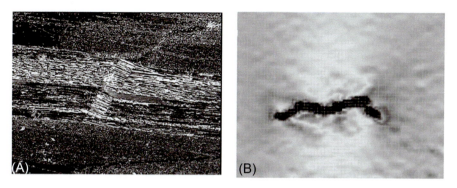

Fig. 3.4 (A) Micro-buckling observed in a composite laminate with kink bands in the fibers growing perpendicular to the loading direction; (B) Damage progression prior to failure (Bull and Edgren, 2004).

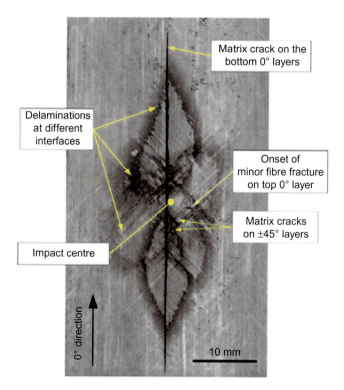

Fig. 3.5 Damage modes in a typical face-sheet of the sandwich composite panel under out of plane low velocity impact loading (Feng and Aymerich 2013).

intralaminar and interlaminar damage were observed, which indicates that the laminated composite panel is capable of dissipating a high amount of energy under impact loading via mixed modes of failure. In particular, it can be seen that delamination occurred across several interfaces near the impact zone and the damage was distributed quite well in the vicinity of the impactor. Such complex damage mechanisms can effectively be simulated via traction-separation cohesive zone models (CZM) and the Hashin damage model for laminated composites, which are detailed in the modeling section below.

3.2.3 Damage due to environmental effects

Environmental effects, including fatigue due to cyclic loads such as water slamming, moisture effects, and saline corrosion, are also important to the design of marine composites due to their aggressive environments. FRP composites are typically used in the construction of composite-based civil and military marine crafts. They are typically exposed to aggressive marine environments which induce moisture, saline corrosion, ultraviolet (UV) radiation, and cyclic loading that cause significant degradation to the mechanical properties of FRP composites (Afshar et al., 2015; Altunsaray et al., 2012; Siriruk and Penumadu, 2014; Arhant et al., 2016; Veazie et al., 2004; Segovia et al., 2007; Riaz et al., 2014). Afshar et al. (2015) investigated the effects of fatigue coupled with ultraviolet UV radiation and moisture effects from the marine environment on the damage degradation of carbon fiber/vinylester composite laminates. They employed an environmental chamber that is capable of exposing the laminates to UV radiation and moisture/condensation as found in the marine environment. They found that aging the specimens in marine environments causes a 10%–20% reduction in stiffness (Fig. 3.6A), but a significant reduction in tensile strength. Furthermore, scanning electron micrographs (SEM) show significant flaws induced by the environmental aging process after 1000 h (Fig. 3.6B). It can be seen that the damage is confined to the 30–50μm thick surface layers of the 1.4 mm thick specimens, which means that at

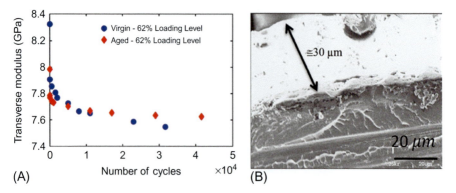

Fig. 3.6 (A) Stiffness comparison between aged laminated composite panels; (B) SEM of damage induced due to UV radiation and moisture in environmental chamber (Afshar et al., 2015).

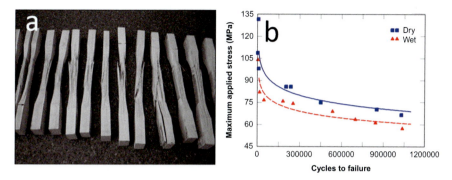

Fig. 3.7 (A) Damage to dry and wet samples induced by cyclic fatigue loading, which initiated in the gage region; (B) The relationship between the maximum applied stress and number of fatigue cycles to induce failure in both atmospheric and seawater environments (Altunsaray et al., 2012).

least 90% of each specimen is unaffected by the environmental aging process. Therefore, this damage localization due to the aging process was attributed to the amplified effects of fatigue damage on the laminate stiffness.

Altunsaray et al. (2012) investigated the behavior of GFRP consisting of E-glass noncrimp fabrics (NCF) reinforcing a polyester matrix, which is widely used for small marine crafts, under environmental fatigue loading. The specimens (Fig. 3.7A) were cut out from 5 and 10 mm thick panels in three different directions (0, 45 and 90 degrees). The panels were manufactured by hand layup in workshop of a boat builder, which is widely employed in the boat building industry because of its lower cost and ease of obtaining uniform thickness throughout the lamination process. These composites were tested under atmospheric fatigue and fatigue induced by simulated seawater environments. It is seen from Fig. 3.7 that failure occurred in the gauge region for all the specimens, regardless of whether they were dry or wet. However, the effects of the seawater environment can clearly be seen via the propagation of damage throughout the wet samples, which significantly reduced the fatigue lifetime. Also shown in Fig. 3.7B is the relationship between the maximum applied stress and the number cycles required to induce fatigue failure on the 5 mm thick dry and wet specimens. It can be seen that the maximum applied stress in both the dry and wet samples reaches a steady state at around 80 MPa and the effects of the wet environment do not pose much of an effect on the fatigue behavior of the composite. Similar results were obtained for 10 mm thick samples with different fiber orientations, which show that the fatigue behavior is not dependent on the thickness of the specimen or orientation of the fibers. However, the stress amplitudes to achieve failure in the seawater environment are lower than those at atmospheric conditions.

Segovia et al. (2007) investigated the corrosive effects of saline mediums on an E-glass composite material. The composite was immersed in a saline medium for up to 36,000 h and loaded under continuous bending to simulate the concurrent working stress that composites undergo while immersed in aggressive environments. The

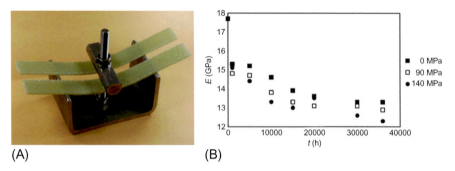

Fig. 3.8 (A) Bending device for applying working stresses to a composite material in a saline medium; (B) Degradation to the flexural modulus under different bending stresses (Altunsaray et al., 2012).

device used to apply the working stresses is shown in Fig. 3.8A. The saline medium had a salt concentration of 5%, which mimicked seawater conditions. Fig. 3.8B illustrates the degradation in the flexural modulus under different bending stresses. It is seen that the flexural modulus degraded by up to 31%. Furthermore, the flexural strength degraded by up to 35%, while the specific fracture energy was reduced by up to 51%. This was attributed to diffusion of salt water into the matrix and fiber junctions, which induced the corrosive effects of the chemical ions in the saltwater medium.

3.3 Nondestructive damage detection for maritime composites

Advances in damage detection techniques have overcome difficulties in detecting damage to laminated composite structures, which may not be immediately visible, such as shear failure in the core, or subsurface intralaminar cracking. Nondestructive techniques include optically excited lock-in/infrared thermography (OLT/IRT) for the assessment of subsurface defects in thick composite laminates, structural health monitoring (SHM), vibration-based ultrasonic (US) methods, DIC described in more detail in Sánchez-Heres et al. elsewhere in this book, artificial neural networks (ANN), and fiber optic damage assessment and detection systems (Montanini and Freni, 2012; Montesano et al., 2015; Silva-Munoz and Lopez-Anido, 2008; Zhang et al., 2013, 2014; Ibrahim, 2014; Herman et al., 2013; Zhou and Sim, 2002).

3.3.1 Experimental methods

Montanini and Freni (2012) employed OLT/IRT to quantitatively assess the simulated subsurface defects in thick GFRP, which are used for manufacturing yachts. OLT/IRT involves detecting subsurface defects through an infrared camera, schematically shown in Fig. 3.9A, by exploiting a harmonically modulated excitation of the optical

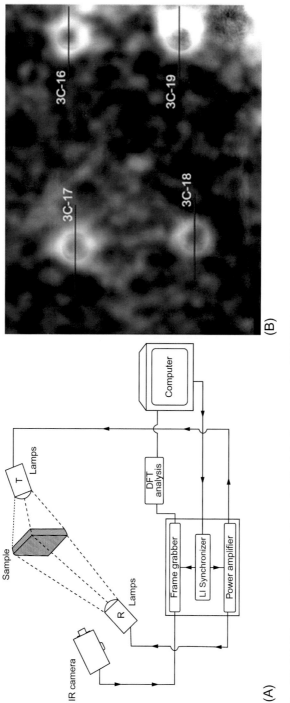

Fig. 3.9 (A) Schematic of the optically excited lock-in thermography (OLT/IRT) setup; (B) Subsurface detects detected in composite specimen using OLT/IRT (Montanini and Freni, 2012).

source to induce a heat flux into the sample material. Fig. 3.9B illustrates the processed image, showing subsurface flaws that were detected by the OLT/IRT method. Although OLT/IRT has some limitations when applied for the inspection of relatively thick GFRP marine structures, which are characterized by substantial inhomogeneity, a correlation was obtained between the flaw depth assessment and flaw geometry.

Montesano et al. (2015) employed infrared thermography to assess damage evolution in woven carbon fiber/epoxy composites under tensile static and fatigue loading. An infrared camera is able to detect the evolution of weft yarn cracking during the initial stage, as well as the initiation and growth of interply delamination cracking during the final stage of cyclic damage evolution. The temperature profiles captured during displacement controlled quasi-static tests have shown three stages of evolution where the thermoelastic limit was accurately captured, namely damage initiation at the beginning of the second stage. Matrix cracking is the main damage mechanism detected within the weft yarns under quasi-static loading, and the cracks are detected as localized dissipated heat sources via infrared thermography. This is confirmed by subsequent scanning electron microscope (SEM) observations (Fig. 3.10).

Zhang et al. (2013) employed vibration-based technique for detecting delamination in composites in graphite/epoxy panels with an eight-ply fiber orientation sequence [0/−45/45/90]. Delamination between the layers was simulated using pairs of rectangular Teflon release films covering the full width of the beams between the layer interfaces. Excitation was induced with sound from a loud speaker behind the specimen

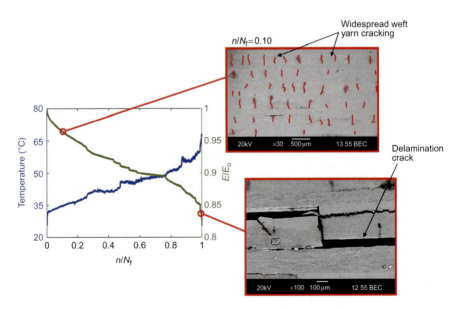

Fig. 3.10 Cracking delamination in the matrix and distributed cracking in the yarns of a woven fiber composite, which were detected by an infrared camera using infrared thermography and confirmed in subsequent SEM observations (Montesano et al., 2015).

Experimental and theoretical damage assessment in advanced marine composites 67

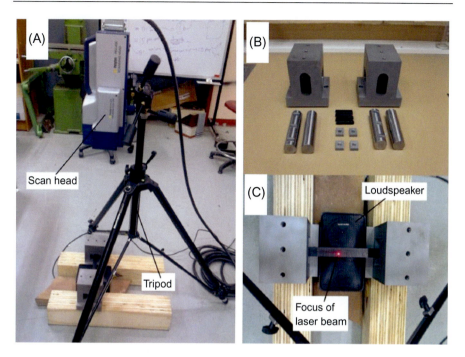

Fig. 3.11 Experimental setup for a simply supported beam with: (A) the scan head; (B) the simple support fixture; and (C) the loud speaker for inducing excitation, and a scanning laser beam (Zhang et al., 2013).

from which a set of frequencies was obtained to assess damage in the composite. This data was used to build a surrogate-assisted genetic algorithm, which uses artificial neural networks (ANN) as the surrogate model for approximation, which was found to predict delamination parameters accurately using numerical data as well as experimental data with measurement errors (Fig. 3.11).

Li et al. (2015) employed electromagnetic coupled spiral inductors (CSI) for the nondestructive evaluation of CFRP plates. Through this approach, they were able to evaluate three types of representative damage in CFRP composites, namely BVID, subsurface defects, and internal microcracks. Through signal data acquisition, the size and location of the subsurface defects were quantitatively assessed. Basically, a CSI sensor is used to scan the top of a CFRP strip with embedded grooves as illustrated in Fig. 3.12A. This detection method is shown to be accurate in Fig. 3.12B, where the peaks of the transmitted signal represent the variability in height of the grooves from Fig. 3.12A. The width at the base of the signal is also representative of the width of these grooves. Subsequently, a three-point bend test was conducted to introduce microcracks in a CFRP plate. The length of the crack extracted from the CSI sensor data was within 10% agreement with the experiment.

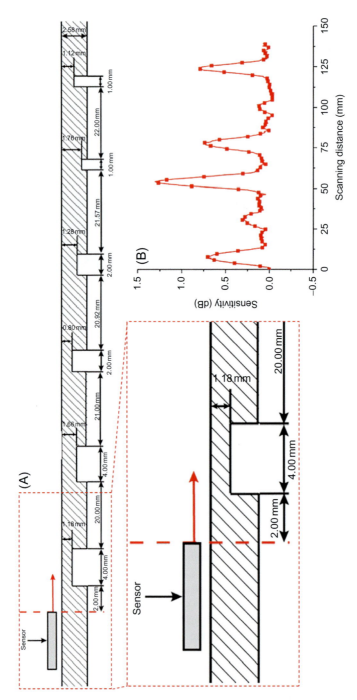

Fig. 3.12 (A) Cross section of CFRP strip with grooves used for nondestructive damage evaluation; (B) Detection of subsurface grooves, with each peak showing accurate detection of the groove width and height (Li et al., 2015).

3.3.2 In situ damage detection

Silva-Munoz and Lopez-Anido (2008) employed fiber Bragg grating (FBG) strain sensors for structural health monitoring (SHM) of composite joints. They subjected secondary bonded woven E-glass/vinylester composite doubler plate joints to fatigue tension loading to induce stable crack propagation. The three locations of the sensors were capable of detecting the changes in strain due to crack extension during the fatigue loading. They concluded that the optimum location for detecting crack extension is the interior base plate sensor, since this location is close enough to capture the changes in strain when crack growth occurs. At the same time, the location is unaffected by large strain gradients near the crack tip (Fig. 3.13).

Genest et al. (2009) utilized the pulsed thermography (PT) technique to detect debonding in a composite patch joint. Fig. 3.14A illustrates a patch specimen, which was made from graphite epoxy plies with a thickness of 1.52 mm. The patch was used to repair composite plates with a size of $400 \times 50 \times 6$ mm. PT experiments were

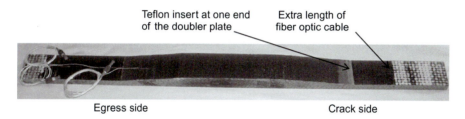

Fig. 3.13 Doubler plate with three embedded strain sensors for structural health monitoring (SHM) (Silva-Munoz and Lopez-Anido, 2008).

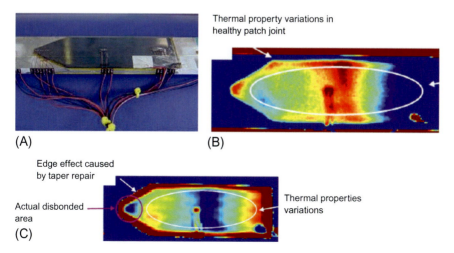

Fig. 3.14 (A) Patch specimen used to repair a composite plate; thermal images obtained from (B) a healthy patch specimen and (C) a partially debonded specimen (Genest et al., 2009).

carried out using a commercial PT system. The surface temperature of the specimen was monitored by an infrared (IR) camera, which is sensitive to long wavelengths of 8–12 μm. The thermal images between a healthy patch joint and a debonded joint are compared in Fig. 3.14B and C. It is clear that there are variations at the edge of the debonded specimen compared to the healthy specimen, particularly due to the discontinuity present at the tapered edge of the patch indicated in Fig. 3.14C. Hence, with an experienced operator, the PT system is an effective nondestructive method that can be employed in the field for detecting damage in composite joints.

Herszberg et al. (2005) present another application of embedded Bragg grating sensors, which complement the aforementioned discussion, for assessing the effects of artificially induced debonding in composite ship joints. The basic idea behind a Bragg grating is that it reflects only a narrow band of wavelengths propagating in the optical fiber. Hence, the Bragg grating spacing, and reflected wavelength, changes as the fiber is strained. A typical composite ship joint is illustrated in Fig. 3.15A, in which a T-joint is used to connect the bulkhead to the hull. Fig. 3.15B illustrates the reflection profiles in a tensile specimen at various strains. At large strains (greater than 3000 $\mu\varepsilon$), there are significant distortions in the reflective spectra, which make it difficult to distinguish the peaks to determine the linear relationship between the applied load and strain. This type of distortion is believed to be characteristic of nonuniform strain distributions at the grating location, which may be attributed to the close proximity of the sensor to resin-rich volumes. This is typical of woven fabric composites, especially in heavy fabrics. Hence, the fiber optic Bragg sensors are effective as a nondestructive technique, provided that they are embedded in locations of composite ship joints where their performance and durability are not affected.

3.4 Numerical and theoretical modeling of composite damages

Finite element modeling (FEM) of laminated composite panels is well-established and has been employed extensively in studies on novel composite structures, which focused on predicting fiber and matrix damage, as well as delamination between

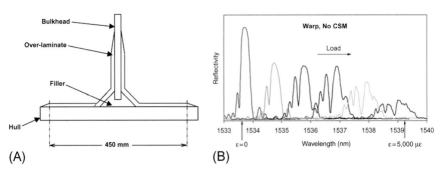

Fig. 3.15 (A) Schematic of a composite ship joint; (B) Reflection profiles obtained at different strains from a Bragg grating sensor (Herszberg et al., 2005).

composite layups in FRP composites under extreme loads such as impact and blast (Tran et al., 2016; Ghazlan et al., 2016; Wei et al., 2013a; Meng et al., 2016; Nimje and Panigrahi, 2015). Other studies have employed a similar approach in multiscale modeling of carbon nanotubes and woven fibers, where a specific unit cell was extracted from the laminates (Lua et al., 2006; Wicks et al., 2014). Specifically, the Hashin damage model (Hashin, 1980) and CZM (Camanho and Davila, 2002) can predict interlaminar and intralaminar damage quite accurately, as demonstrated in the aforementioned experimental/numerical studies on the behavior of FRPs under extreme loading. Artificial neural networks have also been shown to predict delamination in FRP composites quite well and have shown good comparison with nondestructive thermography-based damage detection techniques (Zhang et al., 2013). Furthermore, cumulative damage modeling for capturing the fatigue behavior due to effects of the environmental aging process on the degradation of carbon vinylester laminates has also been performed (Altunsaray et al., 2012).

3.4.1 Finite element modeling

A typical finite element model of a laminated composite is shown in Fig. 3.16 to give a general overview of the modeling approach. The schematic description of a typical multilayer E-glass fiber/vinylester composite panel with a composite layup similar

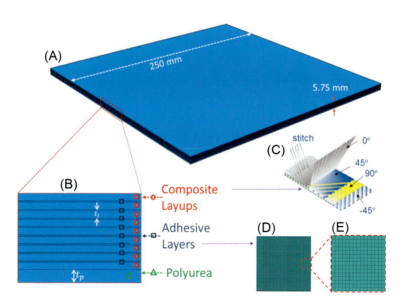

Fig. 3.16 (A) Numerical model of E-glass fiber/vinylester composite; (B) elevation view showing individual composite layups bonded together by a vinylester adhesive; (C) fiber orientations in individual composite laminates forming each composite layup (laminates 1–4 and 5–9 have fiber orientations of 0°/45°/90°/−45° and 45°/90°/−45°/0°, respectively); (D) finite element mesh of the adhesive layer; and (E) close-up view of the finite element mesh of the adhesive layer (Tran et al., 2016).

to that reported by Latourte et al. (2011) is given hereafter. The 250 × 250 × 5.75 mm elastomer/composite panel is composed of nine FRP composite laminates, which are modeled using eight-node quadrilateral in-plane continuum shell elements with an average size of 2 × 2 × 0.55 mm³. Each lamina is designed with a thickness of $t_1 = 0.55$ mm, with four plies (having different fiber orientations) uniformly distributed through its thickness (Fig. 3.16C). The top four laminates, which are closest to the blast source, are designed to have identical fiber orientations of (0°/45°/90°/−45°), while the remaining (inverted) five laminates are assigned fiber orientations of (45°/90°/−45°/0°) (Fig. 3.16B and C). The FRP composite structure is then bonded/tied together with the vinylester adhesive layers, which are modeled with zero thickness, eight-node cohesive elements having an average size of 1 × 1 mm.

3.4.2 Hashin damage model for fiber/matrix failure in composite layups

Given that the behavior of laminated composites depends on the direction of fiber orientation, a linear orthotropic constitutive law is normally assigned to each ply forming the composite laminas. Hashin's model is typically applied to investigate the damage initiation and material failure of each unidirectional composite ply, which incorporates four material damage mechanisms: fiber damage in axial tension and compression, and matrix damage in transverse tension and compression. The damage initiation equations are:

Fiber tension ($\hat{\sigma}_{11} \geq 0$) (3) and fiber compression ($\hat{\sigma}_{11} < 0$):

$$F_f^t = \left(\frac{\hat{\sigma}_{11}}{X^T}\right)^2 + \alpha \left(\frac{\hat{\tau}_{12}}{S^L}\right)^2 \tag{3.1}$$

$$F_f^C = \left(\frac{\hat{\sigma}_{11}}{X^C}\right)^2 \tag{3.2}$$

Matrix tension ($\hat{\sigma}_{22} \geq 0$) (5) and matrix compression ($\hat{\sigma}_{22} < 0$):

$$F_m^t = \left(\frac{\hat{\sigma}_{22}}{Y^T}\right)^2 + \left(\frac{\hat{\tau}_{12}}{S^L}\right)^2 \tag{3.3}$$

$$F_m^C = \left(\frac{\hat{\sigma}_{22}}{2S^T}\right)^2 + \left[\left(\frac{Y^C}{2S^T}\right)^2 - 1\right]\frac{\hat{\sigma}_{22}}{Y^C} + \left(\frac{\hat{\tau}_{12}}{S^L}\right)^2 \tag{3.4}$$

In the above equations, a value of 1.0 or greater indicates that damage initiation has occurred. X^T and X^C denote the longitudinal tensile and compressive strength, respectively. Y^T and Y^C denote the transverse tensile and compressive strength, respectively. S^L and S^T denote the longitudinal and transverse shear strength, respectively. α is a coefficient that determines the contribution of the shear stress to the fiber tensile initiation criterion, which is taken as zero. $\hat{\sigma}_{11}, \hat{\sigma}_{22}, \hat{\tau}_{12}$ are components of the effective stress tensor, $\hat{\sigma} = M\sigma$, in which σ is the nominal stress tensor and M is the damage operator (Wei et al., 2013a):

$$M = \begin{bmatrix} \dfrac{1}{1-d_f} & 0 & 0 \\ 0 & \dfrac{1}{1-d_m} & 0 \\ 0 & 0 & \dfrac{1}{1-d_s} \end{bmatrix} \quad (3.5)$$

where d_f, d_m, d_s are internal damage variables that characterize fiber, matrix, and shear damage. These variables are derived from the damage variables, which correspond to the four modes discussed previously:

$$d_f = \begin{cases} d_f^T & \text{if } \hat{\sigma}_{11} \geq 0 \\ d_f^C & \text{if } \hat{\sigma}_{11} < 0 \end{cases}$$

$$d_m = \begin{cases} d_m^T & \text{if } \hat{\sigma}_{22} \geq 0 \\ d_m^C & \text{if } \hat{\sigma}_{22} < 0 \end{cases} \quad (3.6)$$

$$d_s = 1 - \left(1 - d_f^T\right)\left(1 - d_f^C\right)\left(1 - d_m^T\right)\left(1 - d_m^C\right)$$

The fracture toughness for fiber tension and compression and matrix tension and compression modes ($G_{ft}^C, G_{fC}^C, G_{mt}^C, G_{mC}^C$) are input into the damage evolution law, and a bilinear softening model is assumed. Typical properties for unidirectional E-glass/vinylester composite laminas are given in Table 3.1. Furthermore, a visual of the failure surfaces for the fiber and matrix, as well as the traction-separation law describing damage in this type of laminated composite, is given in Fig. 3.17.

Table 3.1 E-glass/vinylester unidirectional composite lamina properties (Wei et al., 2013a)

Material property	Value	Material property	Value
Density, ρ	1850 kg/m³	X^T (Tensile)	1.2 GPa
E_{11}	39 GPa	X^T (Compressive)	620 MPa
E_{22}	11.5 GPa	Y^T (Tensile)	50 MPa
G_{12}	3.5 GPa	Y^T (Compressive)	128 MPa
G_{13}	3 GPa	S^L	89 MPa
In-plane Poisson's ratio, ν_{12}	0.28	S^T	60 MPa
		G_{ft}^C, G_{fc}^C	35 MPa mm
		G_{mt}^C, G_{mc}^C	2 MPa mm

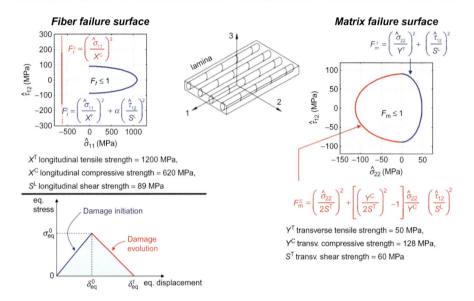

Fig. 3.17 Failure surfaces and traction-separation law describing damage in E-glass/vinylester laminated composite given by the Hashin model.

3.4.3 Cohesive damage model for interlaminar fracture

Delamination is an important failure mode, as well as one of the means of energy dissipation in composite materials when subjected to transverse loadings. A weak interface, however, can cause significant damage and degradation to the structural integrity of a composite panel. A stronger interface, on the other hand, could prevent the effective energy absorption capabilities of the composite panel. The interlaminar debonding behavior is typically simulated by the CZM (Tran et al., 2008, 2010, 2011; Wei et al., 2013b).

The bilinear cohesive model, as illustrated in Fig. 3.18B, is extensively utilized for modelling delamination in adhesive bonds. The cohesive elements with finite thickness connect two volumetric elements, as shown in Fig. 3.18A, with traction-

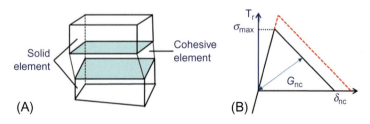

Fig. 3.18 (A) Schematic of a three-dimensional eight-node cohesive element sandwiched between two volumetric finite elements; (B) rate-independent bilinear softening model.

separation laws that relate the cohesive traction vector $T = \{t_n, t_s, t_t\}$ and the displacement jump $\delta = \{\delta_n, \delta_s, \delta_t\}$, where the subscripts n, s, and t, respectively, denote one normal and two tangential components. A simple bilinear quadratic cohesive law is adopted for damage initiation:

$$\left\{\frac{t_n}{t_n^0}\right\}^2 + \left\{\frac{t_s}{t_s^0}\right\}^2 + \left\{\frac{t_t}{t_t^0}\right\}^2 = 1 \tag{3.7}$$

where t_{n0}, t_{s0}, t_{t0} represent the peak values of the nominal stress when the deformation is purely normal to the interface, or purely tangential in the first or the second direction, respectively. The power law form was adopted to describe the rate of stiffness degradation:

$$\left\{\frac{G_n}{G_n^C}\right\}^2 + \left\{\frac{G_s}{G_s^C}\right\}^2 + \left\{\frac{G_t}{G_t^C}\right\}^2 = 1, \tag{3.8}$$

where G_n^C, G_s^C, G_t^C refer to the fracture energy required to cause failure in the normal and shear directions, respectively. The cohesive damage model adopted for a well-known vinylester adhesive has the following properties (Table 3.2).

3.4.4 Artificial neural networks for damage detection in laminated composites

Artificial neural networks (ANN) are well-established as a series of simple processing elements, also referred to as units or nodes, connected together to mimic neurons found in the human brain (Priddy and Keller, 2005; Skabar, 2007). The human brain consists of billions of information-processing units (neruons) with trillions of connections between them (synapses). The neuron is composed of dendrites (or minicomputers), which divert information to the cell body (input) and axons, which take information from the cell body (output). ANN mimic this behavior by employing activation functions in each neuron, which take in a weighted sum of inputs, apply a mathematical function, and produce a set of outputs as shown schematically in Fig. 3.19. This representation of a neural network is set up for detecting interfacial delamination in composites, which is explored further below (Zhang et al., 2013). In this case, the inputs are the first seven bending frequencies of a beam and the outputs are the

Table 3.2 **Vinylester cohesive material model**

t_n^0, t_s^0, t_t^0	80 MPa
G_n^C, G_s^C, G_t^C	1 mJ/mm^2
ρ	1850 kg/m^3
E_{nn}	4 GPa
E_{ss}, E_{tt}	1.5 GPa

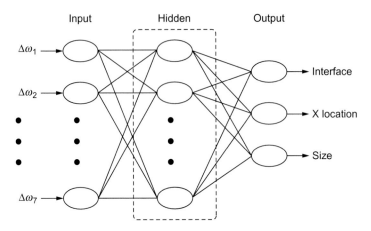

Fig. 3.19 Schematic of an artificial neural network (ANN) for detecting delamination in composites (Zhang et al., 2013).

interface at which delamination occurs, the crack position, and the size of the crack. The frequency changes are obtained by inverse algorithms, which compute the frequency shift between delaminated and nondelaminated composites, when subjected to an excitation (see Fig. 3.11 for example). The main disadvantage of ANN is that they require large amount of data for model training, that is, in this example, the database of frequency changes was obtained from the FE model of delaminated beams. Nevertheless, ANN have found a broad range of applications such as facial recognition, predicting damage in composites, stock market predictions, blast window hazard predictions, structural health monitoring and geometry optimization of aircraft composites, predicting fatigue crack propagation in composite patch repairs for aircraft, predicting internet search history patterns, fault detection in gearbox bearings, and so on (Bangalore and Tjernberg, 2015; Dworakowski et al., 2015; Gajewski et al., 2017; Okafor et al., 2017; Sharif-Khodaei et al., 2015).

Artificial neural networks (ANN) have been applied extensively in damage detection, but have relied on human experience due to lack of available training data, a large consumption of computational resources, and requirement of validation datasets to assess their efficacy (Dworakowski et al., 2015; Ramasamy and Sampathkumar, 2014; Nazarko and Ziemianski, 2016; de Oliveira and Inman, 2017; De Fenza et al., 2015). However, they have been employed extensively for different purposes in composites. Crivelli et al. (2015) employed ANN to predict matrix cracking and delamination in carbon fiber-reinforced composites. The application of ANN successfully separated the two typical damage mechanisms in the matrix, namely cracking and delamination. De Fenza et al. (2015) used ANN to identify the location and entity of damage in aluminium and fabric composite materials. They employed finite element modeling (FEM) to obtain the training data. ANN was successfully able to predict the location of damage with up to 99% accuracy. Farhana et al. (2016) used ANN to predict the fiber-matrix volume fraction in GFRP composites.

Kitahara (1992) used ANN to determine the crack depth in steel plates from ultrasonic backscattering data. McCrory et al. (2015) used ANN to classify damage in CFRP composites. Mohanty et al. (2009) employed ANN to predict fatigue crack propagation in aluminium alloys. Nazarko and Ziemianski (2016) employed ANN for damage detection in aluminium and composite elements. O'Brien et al. (2017) employed ANN as pattern recognition systems to detect and classify the level of damage on composite beams. Ramasamy and Sampathkumar (2014) employed ANN to predict impact damage tolerance of drop impacted woven GFRP composites. Vo-Duy et al. (2016) used ANN to detect damage in laminated composites.

Generally, many ANN applications found in the literature make their predictions by using a unidirectional flow of information whereby the neurons are represented as nodes. Every neuron in the input layer is connected to a hidden neuron and every hidden neuron has a link to an output neuron. This configuration is known as forward propagation, that is, the information propagates forward through ANN to make their predictions. The behavior of the neurons in the hidden layer is influenced by a weight and bias. In other words, the hidden neuron applies an activation function (e.g., sigmoid, hyperbolic tangent, step, and so on) to the weighted input data (and the bias) to produce its output, that is, the results are transmitted over its outgoing connections to other neurons (Zhang et al., 2013; Gurney, 1997). Note that the bias shifts the activation function into a certain range as the activation function normally places the outputs in the range between zero and one. The number of hidden layers required depends on the complexity of the problem and the choice of activation function, of which the significance will become apparent further below. The hidden layers are normally referred to as a black box because the user can generally observe and interpret the inputs and outputs.

To be successful in making valid predictions, ANN first needs to be trained using an extensive dataset. For example, in the case above (see Fig. 3.19), the bending frequencies of the beam will produce the interface where delamination occurs and the x location and size of the crack. As such, if ANN is not run with many such cases, it would produce unusable outputs. One of the most popular approaches is to use a Gaussian (or normal) distribution of the training data to initiate the training process, where the reason will be understood better in the description hereafter, i.e., speeding up convergence. It is first worth noting that there are different training methods that can speed up convergence of the training phase with very large datasets depending on the application (e.g., Bayesian methods in probabilistic neural networks), where each has its own set of criteria for organizing the training data efficiently (Masters, 1993a). The training method consists of a combination of forward propagation and back propagation steps, whereby a data fitting approach is employed to scale the weights (or influence of the input) using an error function (Yadav and Sahu, 2017). Basically, the real (or actual) inputs are fed through ANN and the predicted outputs are acquired. They are then compared to the actual outputs through an error function (e.g., mean squared error (MSE)), which is also known as an optimization or objective function. The training procedure repeatedly attempts to minimize the gradient of this function to find a local minimum by adjusting the weights and biases mentioned above, which is referred to as the gradient descent method (Masters, 1993b). Hence, one can deduce

3.4.5 Modeling cyclic fatigue damage to composites

Cumulative damage indices have been proposed as a simplified approach for quantifying the effects of fatigue on composites mathematically (Afshar et al., 2015). This simplification is employed because composite laminates undergo complex modes of damage, including fiber-matrix delamination, matrix cracking, and fiber failure. As such, these will affect the macroscopic properties of the laminates differently and there needs to be a method to quantify their cumulative effects on these properties, including the strength and stiffness of the laminate. The normalized cumulative damage index is represented mathematically as:

$$D = 1 - \frac{E(N) - E(N_f)}{E_0 - E(N_f)} \tag{3.9}$$

where the damage index D is in the range $[0, 1]$, E_0 is the initial stiffness of the composite, $E(N)$ is the stiffness of the composite after N fatigue cycles, and $E(N_f)$ is the modulus at the last cycle. The relationship between stiffness and number of fatigue cycles (Fig. 3.6A) can be plotted in terms of the damage index as shown in Fig. 3.20. Hence, using an exponential fitting approach, the stiffness-based damage index can be represented as:

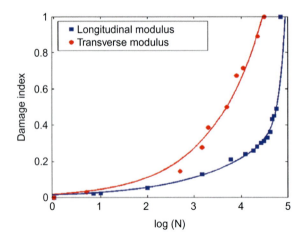

Fig. 3.20 Relationship between the damage index and the number of fatigue cycles (Afshar et al., 2015).

$$\frac{dD}{d\log N} = \overline{C}e^{nD} \tag{3.10}$$

where n and $\overline{C}$ are fitting constants and $d/d \log (N)$ is the slope of the curve in Fig. 3.20, which represents the rate of change in damage with the number of fatigue cycles. Equation (3.10) is a function of the maximum stress, amplitude of stress, frequency of loading, temperature, and so on (Tang et al., 2000).

3.5 Conclusions

This chapter reviewed the advances in experimental and theoretical approaches currently employed to predict damage to composites. Fiber-reinforced composites were found to be popular in marine applications, such as ship hulls and patrol boats, owing to their light-weight and high resilience to damage. Several damage modes were identified, namely, fiber pull-out, fiber bridging and micro-buckling, and matrix cracking and delamination. These were observed through destructive and nondestruction techniques, which have the ability to detect damage to composites without any contact. Finally, finite element modeling approaches that can accurately simulate damage in laminated composites were explored, as well as neural networks as a tool that can be trained for automatically predicting key damage parameters, such as the size and location of the damage interface.

References

Afshar, A., Alkhader, M., Korach, C.S., Chiang, F.P., 2015. Synergistic effects of fatigue and marine environments on carbon fiber vinyl-ester composites. J. Eng. Mater. Technol. 137, 1–8.

Alfers, J.B., 1966. Plastics in ships. Nav. Ship Syst. Command Tech. News 15, 5–8.

Alm, F., 1983. GRP versus steel in ship construction. Nav. Forces 4 (5), 82–86.

Altunsaray, E., Neser, G., Erbil, C., Gursel, K.T., Unsalan, D., 2012. Environmental fatigue behavior of non-crimp. *E-glass fiber reinforced polyester composites for marine applications.* Mat.-wiss. u. Werkstofftech. 43 (10), 1053–1058.

Anon. news.com.au. *Engine fire aboard Collins-class submarine.* 2012a; Available from: http://www.news.com.au/national/engine-fire-aboard-collins-class-submarine/news-story/17edf130795b81cfe4587efe0fd74c80.

Anon. DailyMail. *Vacuum cleaner started $400m U.S. Navy nuclear submarine fire that wrecked ship.* 2012b; Available from: http://www.dailymail.co.uk/news/article-2155659/Nuclear-submarine-linked-vacuum-cleaner-causing-400m-damages.html.

Anon Theguardian. *18 trapped after Indian submarine explodes and sinks in Mumbai.* 2013; Available from: https://www.theguardian.com/world/2013/aug/14/18-trapped-indian-submarine-explodes-sinks-mumbai.

Anon. TheNationalInterest. *Top 5 Worst Submarine Disasters.* 2016; Available from: http://nationalinterest.org/blog/the-buzz/top-5-worst-submarine-disasters-15496.

Anon. Wikipedia. *Russian submarine Kursk (K-141).* 2017a; Available from: https://en.wikipedia.org/wiki/Russian_submarine_Kursk_(K-141).

Anon. MaritimeHerald. *Offshore oil platform caught fire in Gulf of Mexico.* 2017b; Available from: http://www.maritimeherald.com/2017/offshore-oil-platform-caught-fire-in-gulf-of-mexico/.

Arhant, M., Le Gac, P.Y., Le Gall, M., Burtin, C., Briancon, C., 2016. Effect of sea water and humidity on the tensile and compressive properties of carbon-polyamide 6 laminates. Compos. Part A 91, 250–261.

Arora, H., Kelly, M., Worley, A., Del Linz, P., Fergusson, A., Hooper, P.A., Dear, J.P., 2014. Compressive strength after blast of sandwich composite materials. Phil. Trans. R. Soc. A 372, 1–27.

Bangalore, P., Tjernberg, L., 2015. An artificial neural network approach for early fault detection of gearbox bearings. IEEE Trans. Smart Grid 6 (2), 980–987.

Buermann, T.M., DellaRocca, R.J., 1960. Fiberglass-reinforced plastics for marine structures. Proceedings of the Spring Meeting of the Society of Naval Architects and Marine Engineers.

Bull, P.H., Edgren, F., 2004. Compressive strength after impact of CFRP-foam core sandwich panels in marine applications. Compos. Part B 35, 535–541.

Bushby, A.C., Fyfe, R.A., Corkum, F.G., Frost, W.P., Alfers, J.B., 1952. Laminated glass plastic construction with special reference to boats. In: Proceedings of the Society of Naval Architects and Marine Engineers.

Camanho, P.P., Davila, C.G., 2002. Mixed-mode decohesion finite elements for the delamination in composite materials. NASA/TM-2002-211737, pp. 1–37.

Castilho, T., Sutherland, L.S., Guedes Soares, C., 2015. Impact resistance of marine sandwich composites. In: Santos, G.S. (Ed.), Maritime Technology and Engineering. Taylor & Francis Group, London.

Chalmers, D.W., Osburn, R.J., Bunney, A., 1984. Hull construction of CMVs in the United Kingdom.*Proceedings of the International Symposium on Mine Warfare Vessels and Systems*, London.

CNN. *Missing sailors found dead in flooded compartments on US Navy destroyer.* 2017; Available from: http://edition.cnn.com/2017/06/17/us/missing-sailors-found/index.html.

Crivelli, D., Guagliano, M., Eaton, M., Pearson, M., Al-Jumaili, S., Holford, K., Pullin, R., 2015. Localisation and identification of fatigue matrix cracking and delamination in a carbon fibre panel by acoustic emission. Compos. Part B 74, 1–12.

De Fenza, A., Sorrentino, A., Vitiello, P., 2015. Application of artificial neural networks and probability ellipse methods for damage detection using lamb waves. Compos. Struct. 133, 390–403.

Ding, C., Ngo, T., Ghazlan, A., Lumantarna, R., Mendis, P., 2015. Numerical simulation of structural responses to a far-field explosion. Aust. J. Struct. Eng. 16 (3), 226–236.

Dixon, R.H., Ramsey, B.W., Usher, P.J., 1972. Design and build of the GRP hull of HMS Wilton.*RINA Symposium on GRP Ship Construction*, London.

Dworakowski, Z., Ambrozinski, L., Packo, P., Dragan, K., Stepinski, T., 2015. Application of artificial neural networks for compounding multiple damage indices in lamb-wave-based damage detection. Struct. Control Health Monit. 22, 50–61.

Farhana, N.I.E., Abdul Majid, M.S., Paulraj, M.P., Ahmadhilmi, E., Fakhzan, M.N., Gibson, A.G., 2016. A novel vibration based non-destructive testing for predicting glass fibre/matrix volume fraction in composites using a neural network model. Compos. Struct. 144, 96–107.

Feng, D., Aymerich, F., 2013. Damage prediction in composite sandwich panels subjected to low-velocity impact. Compos. Part A 52, 12–22.

Foxwell, D., 1999. Skjold class comes in from the cold. Jane's Navy Int. 104 (6), 14–20.

Gajewski, J., Golewski, P., Sadowski, T., 2017. Geometry optimization of a thin-walled element for an air structure using hybrid system integrating artificial neural network and finite element method. Compos. Struct. 159, 589–599.

Genest, M., Martinez, M., Mrad, N., Renaud, G., Fahr, A., 2009. Pulsed thermography for nondestructive evaluation and damage growth monitoring of bonded repairs. Compos. Struct. 88, 112–120.

Ghazlan, A., Ngo, T., Tran, P., 2016. Three-dimensional voronoi model of a nacre-mimetic composite structure under impulsive loading. Compos. Struct. 153, 278–296.

Goubalt, P., Mayes, S., 1996. Comparative analysis of metal and composite materials for the primary structure of a patrol boat. Nav. Eng. J. 108 (3), 387–397.

Gurney, K., 1997. Neural Networks - an Overview. In: An introduction to neural networks. Taylor and Francis, London, pp. 2–6.

Hashin, Z., 1980. Failure criteria for unidirectional fiber composites. J. Appl. Mech. 47, 329–334.

Herman, A.P., Orifici, A.C., Mouritz, A.P., 2013. Vibration modal analysis of defects in composite T-stiffened panels. Compos. Struct. 104, 34–42.

Herszberg, I., Li, H.C.H., Dharmawan, F., Mouritz, A.P., Nguyen, M., Bayandor, J., 2005. Damage assessment and monitoring of composite ship joints. Compos. Struct. 67, 205–216.

Ibrahim, M.E., 2014. Nondestructive evaluation of thick-section composites and sandwich structures: A review. Compos. Part A 64, 36–48.

Kitahara, M., 1992. Neural network for crack-depth determinatioon from ultrasonic backscattering data. Rev. Prog. Quant. Nondestr. Eval. 11, 701–708.

Lankford, B.W., Angerer, J.F., 1971. Glass reinforced plastic developments for application to minesweeper construction. Nav. Eng. J. 83 (5), 13–26.

Latourte, F., Gregoire, D., Zenkert, D., Wei, X.D., Espinosa, H.D., 2011. Failure mechanisms in composite panels subjected to underwater impulsive loads. J. Mech. Phys. Solids 59 (8), 1623–1646.

Li, Z., Soutis, C., Gibson, A.A.P., Sloan, R., Karimian, N., 2015. Damage evaluation of carbonfibre reinforced polymer composites using electromagnetic coupled spiral inductors. Adv. Compos. Lett. 24 (3), 44–47.

Lindsay, E.M., 1966. Glass fiber-reinforced plastic as a marine structural material. *Proceedings of the Second Marine Systems and ASW Conference*, California.

Lopez-Puente, J., Zaera, R., Navarro, C., 2007. An analytical model for high velocity impacts on thin CFRPs woven laminated plates. Int. J. Solids Struct. 44, 2837–2851.

Lua, J., Gregory, W., Sankar, J., 2006. Multi-scale dynamic failure prediction tool for marine composite structures. J. Mater. Sci. 41, 6673–6692.

Makinen, K., Hellbratt, S.E., Olsson, K.A., 1988, Springer. The development of sandwich structures for naval vessels during 25 years. In: Mechanics of Sandwich Structures.pp. 13–28. link.springer.com/chapter/10.1007/978-94-015-9091-4_2.

Marsh, G., 2010. Marine composites - drawbacks and successes. Reinf. Plast. 18–22.

Masters, T., 1993a. Probabilistic neural networks. In: Practical Neural Network Recipes in C++. Academic Press/Morgan Kaufman Publishers, California, pp. 201–276.

Masters, T., 1993b. Multilayer Feedforward Networks. In: Practical Neural Network Recipes in C++. Academic Press/Morgan Kaufman Publishers, California, pp. 77–116.

McCrory, J., Al-Jumaili, S., Crivelli, D., Pearson, M.R., Eaton, M.J., Featherston, C.A., et al., 2015. Damage classification in carbon fibre composites using acoustic emission: A comparison of three techniques. Compos. Part B 68, 424–430.

McKenzie, A.M., Stark, H.J., 1953. Progress on naval use of reinforced plastic piping. J. Am. Soc. Nav. Eng. 65 (1), 57–70.

Meng, M., Rizvi, M.J., Le, H.R., Grove, S.M., 2016. Multi-scale modelling of moisture diffusion coupled with stress distribution in CFRP laminated composites. Compos. Struct. 138, 295–304.

Mohanty, J.R., Verma, B.B., Parhi, D.R.K., Ray, P.K., 2009. Application of artificial neural network for predicting fatigue crack propagation life of aluminum alloys. Arch. Comput. Mater. Sci. Surf. Eng. 1 (3), 133–138.

Montanini, R., Freni, F., 2012. Non-destructive evaluation of thick glass fiber-reinforced composites by means of optically excited lock-in thermography. Compos. Part A 43, 2075–2082.

Montesano, J., Fawaz, Z., Bougherara, H., 2015. Non-destructive assessment of the fatigue strength and damage progression of satin woven fiber reinforced polymer matrix composites. Compos. Part B 71, 122–130.

Moriniere, F.D., Alderliesten, R.C., Benedictus, R., 2014. Modelling of impact damage and dynamics in fibre-metal laminates—a review. Int. J. Impact Eng. 67, 27–38.

Muscat-Fenech, C.D.M., Cortis, J., Cassar, C., 2014a. Impact damage testing on composite marinesandwich panels. Part 2: Instrumented drop weight. J. Sandw. Struct. Mater. 16 (5), 443–480.

Muscat-Fenech, C.D.M., Cortis, J., Cassar, C., 2014b. Impact damage testing on composite marine sandwich panels, part 1: Quasi-static indentation. J. Sandw. Struct. Mater. 16 (4), 341–376.

Nasirzadeh, R., Sabet, A.R., 2014. Study of foam density variations in composite sandwich panels under high velocity impact loading. Int. J. Impact Eng. 63, 129–139.

Nazarko, P., Ziemianski, L., 2016. Damage detection in aluminum and composite elements using neural networks for lamb waves signal processing. Eng. Fail. Anal. 69, 97–107.

Ngo, T., Mendis, P., Gupta, A., Ramsay, J., 2007. Blast loading and blast effects on structures—an overview. EJSE: Special Issue 76–91.

Ngo, T., Lumantarna, R., Whittaker, A.S., Mendis, P., 2015. Quantification of the blast-loading parameters of large-scale explosions. J. Struct. Eng. 141 (10), 1–11.

Nimje, S.V., Panigrahi, S.K., 2015. Interfacial failure analysis of functionally graded adhesively bonded double supported tee joint of laminated FRP composite plates. Int. J. Adhes. Adhes. 58, 70–79.

O'Brien, R.J., Fontana, J.M., Ponso, N., Molisani, L., 2017. A pattern recognition system based on acoustic signals for fault detection on composite materials. Eur. J. Mech. A/Solids 64, 1–10.

Okafor, A.C., Singh, N., Singh, N., Oguejiofor, B.N., 2017. Acoustic emission detection and prediction of fatigue crack propagation in composite patch repairs using neural network. J. Thermoplast. Compos. Mater. 30 (1), 3–29.

de Oliveira, M.A., Inman, D.J., 2017. Performance analysis of simplified fuzzy ARTMAP and ProbabilisticNeural networks for identifying structural damage growth. Appl. Soft Comput. 52, 53–63.

Perillo, G., Jorgensen, J.K., 2016. Numerical/experimental study of the impact and compression after impact on GFRP composite for wind/marine applications. Proc. Eng. 167, 129–137.

Petrucci, R., Santulli, C., Puglia, D., Nisini, E., Sarasini, F., Tirillo, J., et al., 2015. Impact and post-impact damage characterisation of hybrid composite laminates based on basalt fibres in combination with flax, hemp and glass fibres manufactured by vacuum infusion. Compos. Part B 69, 507–515.

Priddy, K.L., Keller, P.E., 2005. Introduction. In: Artificial Neural Networks—An Introduction. The International Society of Optical Engineering, Washington, pp. 1–11.

Ramasamy, P., Sampathkumar, S., 2014. Prediction of impact damage tolerance of drop impacted WGFRP composite by artificial neural network using acoustic emission parameters. Compos. Part B 60, 457–462.

Riaz, U., Nwaoha, C., Ashraf, S.M., 2014. Recent advances in corrosion protective composite coatings based on conducting polymers and natural resource derived polymers. Prog. Org. Coat. 77, 743–756.

Schutz, H., 1984. Apects of materials selection for MCMV hulls.*Proceedings of the International Symposium on Mine Warfare Vessels and Systems*, London.

Segovia, F., Salvador, M.D., Sahuquillo, O., Vicente, A., 2007. Effects of long-term exposure on e-glass composite material subjected to stress corrosion in a saline medium. J. Thermoplast. Compos. Mater. 41 (17), 2119–2128.

Sharif-Khodaei, Z., Ghajari, M., Aliabadi, M.H., 2015. Impact damage detection in composite plates using a self-diagnostic electro-mechanical impedance-based structural health monitoring system. J. Multiscale Modell. 6 (4), 1–23.

Shenoi, R.A., Wellicome, J.F., 1993. Composite Materials in Maritime Structures: Fundamental Aspects. vol. 1. Cambridge University Press, New York.

Silva-Munoz, R.A., Lopez-Anido, R.A., 2008. Structural health monitoring of marine composite structural joints using embedded fiber Bragg grating strain sensors. Compos. Struct. 89 (2), 224–234.

Siriruk, A., Penumadu, D., 2014. Degradation in fatigue behavior of carbon fiber–vinyl ester based composites due to sea environment. Compos. Part B 61, 94–98.

Skabar, A., 2007. Modeling the spatial distribution of mineral deposits using neural networks. Nat. Resour. Model. 20 (3), 435–450.

Spaulding, K.B., 1966. Fibreglass boats in naval service. Nav. Eng. J. 78, 333–340.

Storman, K.H., 1999. The Skjold class fast patrol boat. Nav. Forces 5, 38–43.

Tang, H.C., Nguyen, T., Chuang, T., Chin, J., Lesko, J., Wu, H.F., 2000. Fatigue model for fiber-reinforced polymeric composites. J. Mater. Civ. Eng. 12 (2), 97–104.

Tran, P., Kandula, S.S.V., Geubelle, P.H., Sottos, N.R., 2008. Hybrid spectral/finite element analysis of dynamic delamination of patterned thin films. Eng. Fract. Mech. 75 (14), 4217–4233.

Tran, P., Kandula, S.S.V., Geubelle, P.H., Sottos, N.R., 2010. Dynamic delamination of patterned thin films: A numerical study. Int. J. Fract. 162 (1–2), 77–90.

Tran, P., Kandula, S.S., Geubelle, P.H., Sottos, N.R., 2011. Comparison of dynamic and quasi-static measurements of thin film adhesion. J. Phys. D-Appl. Phys. 44(3).

Tran, P., Ngo, T., Ghazlan, A., 2016. Numerical modelling of hybrid elastomeric composite panels subjected to blast loadings. Compos. Struct. 153, 108–122.

Veazie, D., Robinson, K.R., Shivakumar, K., 2004. Effects of the marine environment on the interfacial fracture toughness of PVC core sandwich composites. Compos. Part B 35 (6–8), 461–466.

Vo-Duy, T., Ho-Huu, V., Dang-Trung, H., Nguyen-Thoi, T., 2016. A two-step approach for damage detection in laminated composite structures using modal strain energy method and an improved differential evolution algorithm. Compos. Struct. 147, 42–53.

Wei, X., Tran, P., de Vaucorbeil, A., Ramaswamy, R.B., Latourte, F., Espinosa, H.D., 2013a. Three-dimensional numerical modeling of composite panels subjected to underwater blast. J. Mech. Phys. Solids 61, 1319–1336.

Wei, X., Tran, P., de Vaucorbeil, A., Ramaswamy, R.B., Latourte, F., Espinosa, H.D., 2013b. Three-dimensional numerical modeling of composite panels subjected to underwater blast. J. Mech. Phys. Solids 61 (6), 1319–1336.

Wicks, S.S., Wang, W., Williams, M.R., Wardle, B.L., 2014. Multi-scale interlaminar fracture mechanisms in woven composite laminates reinforced with aligned carbon nanotubes. Compos. Sci. Technol. 100, 128–135.

Yadav, A., Sahu, K., 2017. Wind forecasting using artificial neural networks: A survey and taxonomy. Int. J. Res. Sci. Eng. 3 (2), 148–155.

Zhang, Z., Shankar, K., Ray, T., Morozov, E.V., Tahtali, M., 2013. Vibration-based inverse algorithms for detection of delamination in composites. Compos. Struct. 102, 226–236.

Zhang, Z., Shankar, K., Morozov, E.V., Tahtali, M., 2014. Vibration-based delamination detection in composite beams through frequency changes. J. Vib. Control. 22 (2), 496–512.

Zhou, G., Sim, L.M., 2002. Damage detection and assessment in fibre-reinforced composite structures with embedded fibre optic sensors—review. Smart Mater. Struct. 11, 925–939.

Durability testing and evaluation of marine composites

Oliver Parks[*,†], Paul Harper[*]
[*]Aerospace Engineering, University of Bristol, Bristol, United Kingdom, [†]AEL Airborne, Hungerford, United Kingdom

Chapter Outline

4.1 Introduction 86
4.2 Loading and durability requirements 87
    4.2.1 Fire requirements 88
    4.2.2 Temperature 88
    4.2.3 Moisture absorption and degradation 88
4.3 Material selection 90
    4.3.1 Resin selection 90
    4.3.2 Fiber selection 91
    4.3.3 Sizing selection 92
    4.3.4 Manufacturing processes 92
    4.3.5 Challenges facing the research community 92
4.4 Current sea water conditioning techniques 93
    4.4.1 Effect of temperature 94
    4.4.2 Effect of pressure 94
    4.4.3 Effect of water composition 95
    4.4.4 Effect of testing environment 95
    4.4.5 Effect of pH 96
    4.4.6 Effect of specimen dimensions 96
    4.4.7 Effects of the manufacturing process 97
    4.4.8 Current aging procedures 97
4.5 Mechanical testing of saturated specimens 98
    4.5.1 Testing methodology 98
    4.5.2 Static testing of conditioned specimens 99
    4.5.3 Impact testing of conditioned specimens 101
    4.5.4 Fatigue testing of conditioned specimens 102
    4.5.5 Further testing of conditioned specimens 102
4.6 Defining the limits of accelerated aging techniques 103
    4.6.1 Effect on moisture absorption of increased temperature 103
    4.6.2 Effect on mechanical properties of increased temperature 104
4.7 Modelling of accelerated moisture absorption 106
    4.7.1 Fickian diffusion 106
    4.7.2 Effect of temperature on diffusivity 107
    4.7.3 Effect of temperature on saturation level 108

Marine Composites. https://doi.org/10.1016/B978-0-08-102264-1.00004-2
Copyright © 2019 Elsevier Ltd. All rights reserved.

4.7.4 Application of the model 108
4.7.5 Limitations of the model and potential improvements 110
4.8 Constituent-level predictive methods 111
4.9 Summary and future work 111
References 112

4.1 Introduction

Traditional materials such as wood and steel have been used for many years as construction materials for marine structures. Advances in materials science have led to the development of high-grade metal alloys and fiber-reinforced polymer–matrix composite materials. The latter are generally preferred for marine applications due to their greater static and fatigue mechanical properties, increased resistance to corrosion, and ability to be formed into continuous, complex geometries.

Composite materials have seen a wide and varied history of use in marine structures, ranging from boats and yachts to more recent developments such as renewable energy devices (OpenHydro, 2017) and composite passenger ferries (Brødrene, 2017). The acceptance of composites as construction materials has grown in many global industries; including the marine sector. This has been attributed to general improvements in resin and fiber performance, a greater understanding of composite laminate behavior, improvements in manufacturing capabilities, and the emergence of new commercial applications.

Current trends indicate that the use of composite materials in critical marine structures will continue to expand, in many cases replacing steels as key structural materials. For example, the European network for Lightweight Applications at Sea (E-LASS) is leading research in light-weighting commercial ship structures using composites and other advanced materials. The global shipping industry is a substantial market, thought to account for over 90% of global trade (IMO, 2012), but estimated to produce only 2.4% of global greenhouse gas emissions per year (IMO, 2014). While more efficient lightweight ship structures will help reduce emissions, the lower fuel consumption will also result in lower operating costs, a clear incentive for ship operators. However, financial risk is a major limitation preventing the use of composite materials in such applications due to uncertainties in both the manufacturing variability and long-term durability of composite materials. It is not surprising, therefore, that ship builders, owners, and operators are hesitant to adopt such drastic structural modifications considering the net worth of their vessels and cargo. The "Realisation and Demonstration of Advanced Material Solutions for Sustainable and Efficient Ships" (RAMSSES) and FIBERSHIP projects are specific examples of large European research efforts aiming to tackle this issue. Commercial shipbuilders are active in these projects, clearly indicating that there is an industrial, as well as academic, interest in this field.

Commercial shipbuilding is an example of the potential future adoption of composites in a large, established industry. In contrast, marine renewable energy is an example of the current application of composites in a small, but growing industry. It has been predicted that, by 2030, there will be 22,000 wave and 50 tidal stream devices operating globally (QinetiQ, 2013). However, these figures may be optimistic when one considers industrial trends and recent events in the years since the QinetiQ document was published. The major factor preventing tidal steam growth is the relatively high cost of energy compared to wind, solar, and fossil fuel energy sources. This is largely a result of the high installation and maintenance costs associated with positioning energy harvesting devices in high-strength tidal flows in remote locations. An improved understanding of long-term composite durability in marine environments will allow for reductions in cost through more efficient designs and reduced commercial risk. This is true for most marine applications and highlights the importance of the ability to accurately predict long-term saturated properties of these materials.

This chapter will begin by briefly exploring the loading and durability requirements for typical marine structures, and how these can influence material selection. The effect of moisture absorption on laminate mechanical properties is discussed, and test data is presented to show these effects. The importance of physical testing is highlighted alongside a description of current accelerated conditioning techniques and durability testing. A summary of recent testing is also presented to investigate the extent to which aging can be accelerated further in a manner representative of the in-service environment. Finally, the reader is introduced to a number of numerical methods that could be used, alongside or potentially in place of current testing procedures, to reduce the future cost and duration of high-performance composite commercial projects.

4.2 Loading and durability requirements

Selection of suitable materials is a critical step towards a successful structural design. To do this effectively, the designer must understand the operational loads and environment. For high-performance marine structures, the combination of high loads and hostile environment demands materials that offer exceptional mechanical performance and long-term durability.

This is especially the case for tidal turbine blades, where fatigue loading is a key design driver due to highly turbulent flows and the high density of water. Therefore, the selected materials must provide exceptional static and fatigue properties, while being able to be moulded into smooth slender designs. Alternatively, critical military vessel structures such as the hull must be designed to resist significant blast and impact loads, while also providing high strength and stiffness to weight ratios.

The common environmental concern for all submerged marine structures is moisture absorption and its negative impact on mechanical properties. Other environmental issues such as fire and elevated service temperatures may also be of concern to the designer.

4.2.1 Fire requirements

Fire safety is a key design consideration for commercial ship structures and has been an issue of great interest for many years as organic resins are naturally flammable and may release toxic fumes during combustion. The risk of fire poses serious implications for commercial and civilian vessel designs. Much research has been done to investigate and improve the fire resistance of composite materials for these applications. Two options are open to the designer; one can either aim to achieve sufficient fire resistance through suitable resin and fiber selection, which can negatively impact cost and in-plane laminate performance. Alternatively, one can apply additional fire insulation, traditionally comprised of mineral wool, although new ceramic blankets are available offering reduced weight (CSC, 2009). Additional insulation can add to the weight and cost of designs, reducing the benefit of using composites over steels.

4.2.2 Temperature

Ocean temperatures are typically within the range of 0–35°C. This is a fairly narrow range of moderate temperatures, indicating that thermal loading is not a design issue for submerged applications (although nonsubmerged structures exposed to direct sunlight may experience higher temperatures, and should be considered on an individual basis). Furthermore, sea temperatures become even less variable with depth, so for deep-sea applications, temperature is even less of a concern.

The glass transition temperature (Tg) of typical marine resins lies around 80–100°C. Moisture absorption acts to reduce the Tg of a composite laminate, which will be discussed in a later section. Even so, the wet Tg of a typical marine laminate is normally sufficient to allow operation in most marine environments without additional post-cure cycles or special consideration for temperature.

Fire safety requirements also dictate that certain ship structures must be able to withstand high temperatures. Fortunately, composites are naturally good insulators, especially in the form of foam-cored sandwich panels. Research has shown that it is possible to meet fire safety requirements with such panels (CSC, 2009).

4.2.3 Moisture absorption and degradation

Moisture absorption of sea water is common among all applications and will, therefore, be the main focus of this chapter. Most materials are affected to varying degrees by submersion in sea water. For example, steels typically corrode, and in the offshore oil and gas industry, a simple approach to tackling this issue in steel risers is to allow corrosion to occur in additional sacrificial layers of material. One can then predict with reasonable accuracy the rate of corrosion and therefore estimated life of the component.

Composite materials absorb sea water, and while there may be no visible degradation, this typically results in a reduction in mechanical properties. There are three major mechanisms by which water can enter a fiber-reinforced composite laminate. The first and primary mode of moisture transport through a composite material is diffusion within the resin matrix, in particular at the fiber/resin interface. Moisture also moves through the composite by two other modes; capillary transport between the fiber and matrix, and movement through any microcracks in the matrix. These two processes are considered to occur at a much slower rate than diffusion for most composite materials. There are many mathematical models of varying complexity that can be used to predict the overall moisture absorption behavior in composite materials. Fick's law is one of the simplest tools and is widely accepted as being sufficiently accurate to describe the diffusion process, and hence, overall moisture absorption process in composite laminates for most industrial applications.

For Fickian absorption, the quantity of water in a laminate increases linearly with the square root of time. This is followed by a plateau, at which point the laminate has reached saturation. An example of typical experimental data showing this trend is highlighted in Fig. 4.1 and relevant modelling techniques are discussed in more detail in section 4.7.

Multiple processes occur during moisture absorption to alter the mechanical properties of the composite material. These can be characterized as reversible and nonreversible effects. Reversible effects include plasticization and swelling of the matrix, the latter of which can lead to irreversible effects such as fiber-matrix interfacial failure (Schutte, 1994) and microcracking due to the internal buildup of stresses. Composite materials may also suffer from irreversible chemical degradation known as hydrolysis. During this process, polymer chains are broken due to the presence of H^+ and OH^- ions, negatively affecting the mechanical properties of the resin matrix and composite laminate. Esters, amides, and imides are generally more susceptible to this (Choqueuse and Davies, 2008). The severity of these effects depends largely on the selection of resin, fiber, and sizing, together with environmental factors such as conditioning media and temperature.

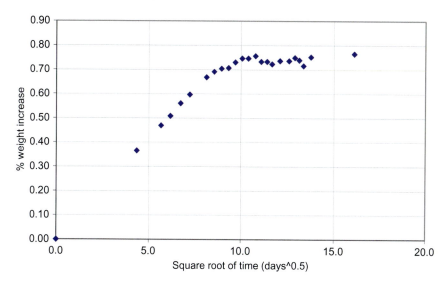

Fig. 4.1 Typical weight gain plot due to moisture absorption for an epoxy/glass laminate.

4.3 Material selection

4.3.1 Resin selection

Different resin types exhibit varying degrees of sensitivity to moisture absorption and degradation. Moisture absorption not only affects mechanical properties, but also causes the resin to swell. Generally, higher levels of moisture absorption are an indication of greater degradation, so it is advantageous to select resins that absorb low levels of moisture. Manufacturers' technical data sheets can provide an insight into the expected moisture absorption of neat resins; however, this data is often presented in different ways, meaning comparison can be difficult. Epoxy, vinyl ester, and polyester resins are the three most common choices for composite marine structures.

Polyester resins are generally not considered for high-performance applications due to their inferior mechanical properties and greater levels of moisture degradation when compared with epoxies and vinyl esters. Polyester resins are commonly used for less critical applications such as the production of smaller leisure craft, where their lower cost makes them an attractive choice. In such applications, resin-rich areas of more durable resins, gel coats, or other protective barriers between the composite and sea water can be used to shield the internal glass/polyester laminate. This may not prevent eventual degradation, and defects such as osmotic blistering may still occur (Clegg, 2006). Searle and Summerscales (1999) present a detailed discussion of osmosis and blistering in composite laminates.

Both polyester and vinyl ester resins contain styrene, resulting in harmful vapor diffusing from the laminate during cure. This potential health risk to employees requires expensive fume extraction and protective safety equipment. Even with these precautions, the smell of styrene is often clearly apparent on the factory floor. Low styrene content resins are, therefore, preferred.

Vinyl esters typically sit between epoxies and polyesters in terms of mechanical performance and cost. These resins exhibit relatively low moisture absorption values, supporting their traditional use in boatbuilding and other marine structures. Processing of these resins is also more tolerant to variations in catalyst/resin mixture ratios. Furthermore, their improved resistance to harsh chemicals and environmental degradation over polyesters make them suitable for a wide range of marine applications. Ashland (2016) provides a guide detailing the comparative resistance of their vinyl ester resin grades to various chemical environments.

Epoxy resins offer greater mechanical properties than either vinyl ester or polyester resins and are more regularly used for high-performance structures due to their higher cost. Processing epoxy resins is slightly more challenging due to the tight tolerances required for hardener/resin mixture ratios. Even so, epoxy resins are widely used as they can also offer greater resistance to environmental degradation.

More recent developments in polymer chemistry have produced resins that process like vinyl ester resins, but offer similar mechanical performances to epoxies. To reflect this, these resins are marketed as epoxy-vinyl ester resins. Two examples of such marine-grade resins are Dion 9100 and Derakane 8084. Both resins have been approved by DNV-GL for production of ship structures.

4.3.2 Fiber selection

Carbon and glass are the two most common fiber types for high-performance marine applications. Glass fibers are generally preferred for less critical applications where cost is a key driver, while carbon fibers can provide laminates with superior strength and stiffness properties, resulting in more efficient structures at a higher cost.

To achieve the necessary strength to weight ratios required in high-performance applications, laminates with continuous aligned fibers are generally used over alternatives such as chopped strand mats. Continuous aligned fibers allow for the construction of optimized laminates, the orthotropic properties of which provide a significant advantage over steels, justifying the often greater manufacturing costs.

Fibers are the major contributor to the laminate mechanical properties, and therefore, fiber degradation is an important consideration when selecting suitable materials for marine applications. The vulnerability to moisture degradation varies among the range of available fibers. While carbon fibers are susceptible to chemical attack, they are generally inert in sea water at moderate temperatures (Echtermeyer et al., 2018). Unprotected glass fibers, on the other hand, suffer a reduction in strength when exposed to sea water. This effect is reduced if the fibers are protected by a resin matrix, which is true for most applications. Nevertheless, designers often select glass fibers, accepting the greater vulnerability to moisture degradation. In some applications, it is simply not commercially viable to select carbon fibers, and in others, a combination (hybrid) of glass and carbon fibers may be used as the best compromise between cost and performance.

Glass fibers are available in various different grades, selected based upon their properties and the specific application. E-glass is the most affordable and widely used grade of glass fiber. S-glass fibers offer greater tensile strength, but come at a higher cost. E-CR glass, a boron-free glass variant, offers improved chemical and electrical resistance. Renaud and Greenwood (2005) demonstrated that the advantages of utilizing E-CR glass are more apparent when materials are exposed to either acidic/alkaline solutions or tap/deionized water. Nevertheless, this study still indicates that E-CR glass is more corrosion-resistant than E-glass in sea water environments. Conversely, Kennedy et al. (2016) found that both E-glass and E-CR glass/vinyl ester laminates experienced similar reductions in tensile strength after a 30-month conditioning period in sea water at 40°C. Thus, the true cost benefit of using E-CR glass over E-glass for marine applications is unclear.

Recent concerns regarding the environmental impact of composite structures have fuelled a growing interest in more sustainable materials. Natural fibers can be sourced from many different plant species, and Sanjay et al. (2018) provide an overview of the typical properties of each type. Hemp and flax are common choices, and multiple papers have investigated the effects of water submersion on these materials (Yan and Chouw, 2015; Cheour et al., 2016). A major disadvantage with natural fibers is their susceptibility to moisture degradation, which has restricted their use in many structures requiring long-term durability. However, it has been demonstrated that fiber treatment can reduce this effect (Liu et al., 2017). As of today, the application of natural fibers in the marine industry is limited to a few small case studies; examples include the all-flax composite trimaran (Flaxcomposites, 2014) and the jute-reinforced biocomposite sailboat (De Mony-Pajol and Cail, 2014).

4.3.3 Sizing selection

The choice of sizing (interface coupling agent) can have considerable impact on not only the immediate static performance, but also the long-term durability of a composite laminate. As water enters a composite laminate, it acts to degrade the interface between the resin and fiber. It is, therefore, crucial that a good bond is formed between resin and fiber, which can be achieved through both well-controlled manufacturing processes and selection of a compatible sizing. It has been demonstrated that degradation of the sizing on rolls of dry fibers can also occur over time depending on storage conditions (Peters, 2016). Sizing degradation can result in a poor fiber/resin interface, impacting laminate mechanical performance. A number of studies at AEL Airborne have shown that different resins may be more susceptible to this than others, and that the degradation effect may not become apparent until after submersion in sea water. As this is predominantly an issue concerning the chemical formulation of the constituents, it is extremely difficult for engineers to predict the effect of sizing selection without extensive coupon testing.

4.3.4 Manufacturing processes

It is also important to consider how the manufacturing process may affect the mechanical performance and durability of a composite structure. The manufacturing process is generally limited by other factors such as cost, size of the structure, available equipment and capabilities, material availability, and existing material durability data. For example, the higher cost of prepreg materials often prevents them being used in large marine structures with high material volume. Furthermore, the high temperature curing requirements of prepreg materials mean that structures must be able to fit inside an oven or autoclave. For larger structures, this can be difficult. Infused resins generally have lower temperature curing requirements, allowing large structures to be produced within heated tents or even at room temperature. Vacuum bag infusion also offers lower material and manufacturing costs, provided the process is repeatable (dependent on the knowledge and experience of the manufacturer). For these reasons, vacuum bag infusion is a common manufacturing process for high-performance marine structures.

Alternatively, hand layup, a process more commonly used in boatbuilding, is arguably the simplest method for manufacturing a composite structure. However, the resulting laminate properties of a hand layup structure are significantly inferior to vacuum bag infusion and prepreg, both of which can fully exploit the advantages of higher fiber volume fractions in composite materials. Also, increasing concerns for workers' health and safety have led to a push away from open layups towards closed infusions to limit human exposure to hazardous chemicals and vapors.

4.3.5 Challenges facing the research community

A significant amount of academic research is still being conducted to investigate the moisture degradation of various composite materials. However, it is generally accepted that this process is being complicated by the wide range of commercially available fibers, resins, and sizings, combined with potential variations in laminate mechanical properties due to manufacturing processes. As resin and fiber suppliers continue to improve their products, older versions become obsolete and more difficult, or even impossible to obtain. Design engineers can also be pressured to utilize the most advanced and/or widely available materials to enhance product performance or reduce costs, gaining an edge over competitors. These considerations, combined with the wide range of marine applications, mean that it is difficult for academia and industry to select a "universal laminate" on which to focus research efforts.

Because of the aforementioned variations in marine laminates and their properties, it is unlikely that laminates produced in an industrial setting will have identical mechanical properties and moisture degradation behavior to those quoted in literature. Furthermore, laminates produced in industry are expected to vary between companies, even for the same applications. Therefore, for high-performance applications, companies often reduce risk by conditioning and testing their own laminates. These tests also often account for batch variation in constituent materials and internal manufacturing variability of the laminates.

4.4 Current sea water conditioning techniques

The general purpose of seawater conditioning is to identify the knockdown in mechanical properties due to long-term submersion in sea water for a specified composite material, which can then be incorporated into the product design. For example, in tidal turbine designs, these degraded material properties are used in place of dry material properties when performing design analysis. This is based upon the conservative assumption that, in service, the entire structural laminate will be fully saturated. In practice, it is difficult to predict whether this is the case after 25 years of service. Due to the combination of protective coatings, low temperatures, and thick laminates, it is entirely possible that large sections of the laminate will remain dry. However, it is difficult to verify this due to the lack of real data (inspectable sections of blades that have seen a full service life). Where the designer is faced with uncertainty, it is better to build-in reasonable factors of safety into the design. A better understanding and method of predicting the level of saturation seen in service would allow for more efficient designs.

Current sea water conditioning techniques involve the continued submersion of test specimens in water at a specified constant temperature. Moisture absorption is measured by periodic weighing of the submerged specimens. Saturation is considered to occur when the specimen weight reaches a point of constant mass. Many improvements to this basic setup have been made to increase the accuracy of the overall conditioning process, including a constant supply of fresh water, pH control, controlled diffusion through laminates in single directions, and defined minimum specimen weights to reduce weighing error.

The most accurate approach to investigate the long-term sea water effects on composite materials would be to condition samples in an environment identical to the predicted in-service environment. However, a typical service life would consist of

20–30 years submergence in sea water between 0°C and 35°C. In an industrial setting, where companies need to conduct their own tests, it is clearly not feasible to wait this long for design data. Instead, companies use accelerated conditioning procedures to obtain this data faster.

4.4.1 Effect of temperature

The rate of moisture absorption into a polymer can be increased by raising the conditioning temperature. For investigations concerning the mechanical properties of saturated laminates, the conditioning temperature is limited by the Tg of the constituents. Wright (1981) showed, as a rough rule-of-thumb, the Tg of epoxy resins reduces by 20°C for each 1% weight gain due to moisture absorption. This is true for weights gains between 0% and 7%. Based upon previous experience of typical saturated weight gains of marine composites at AEL Airborne, 45°C was determined to be the maximum "safe" conditioning temperature.

Conditioning near the wet Tg can promote further degradation, affecting weight gain measurements and mechanical properties of the laminate. Previous research has demonstrated that conditioning at elevated temperatures (~80°C) can result in weight loss of the specimen after an initial linear weight gain (Guermazi et al., 2016). Non-Fickian effects due to elevated conditioning temperatures have also been demonstrated in Boisseau et al. (2011). Similar results have been found during an internal investigation at AEL Airborne. It is thought that, in this case, conditioning at temperatures near the wet Tg of the polyester stitching (50–60°C) results in a secondary weight gain after initial saturation.

Merdas et al. (2001) present a formula for predicting the wet Tg of a composite matrix. Using this method, one can predict a knockdown in Tg for a typical epoxy marine resin (dry Tg ~ 350 K) to be 20.1 K, assuming 3% weight gain at saturation (of the resin only, assumed to be equivalent to ~1% weight gain in a laminate) (Davies and Choqueuse, 2008). This matches fairly closely to the above rough rule-of-thumb.

The effect of high temperature conditioning (50°C+) on specific composite materials has been investigated. It is apparent that some materials can be conditioned at temperatures of 60–80°C (Tual et al., 2015), while others clearly cannot (Dawson et al., 2016). For industrial applications, conditioning procedures must be robust and applicable to a multitude of relevant materials.

4.4.2 Effect of pressure

Humeau et al. (2016) have investigated the effect of hydrostatic pressure on moisture uptake. It was found that the effect of pressure varies depending on the manufacturing process, with hand layup laminates experiencing greater moisture uptake under high pressure compared to infused laminates. Additionally, no effect of pressure on moisture uptake was observed with prepreg laminates. The authors suggest that porosity in the hand layup and infused panels is the major cause. The study also found that this increase in moisture uptake did not significantly affect the mechanical performance of the laminates. A previous investigation also found similar trends; with carbon fiber filament wound laminates experiencing greater saturation levels under high pressure, with little effect on mechanical performance (Davies et al., 1997). This study also demonstrated how increased pressure can result in a greater initial rate of moisture uptake.

These findings are interesting for deep-sea applications, as they suggest that it may be possible to avoid pressurized conditioning chambers when testing certain monolithic laminates. They also suggest that pressurizing sea water chambers is not an effective method for accelerating conditioning procedures.

4.4.3 Effect of water composition

Deionized water can be used in place of sea water for conditioning procedures. This has been shown to have little effect on the initial rate of moisture uptake; however, there is a significant increase in the saturation level when compared with sea water (Dawson et al., 2016). Therefore, conditioning processes using deionized water are thought to be more conservative than sea water. However, the use of deionized water allows better comparison of data between companies and researchers, as its composition is fairly well-controlled. The composition of sea water can vary based upon when and where it was collected. Therefore, the designer must take care to source sea water from suitable locations that represent the in-service environment.

It is important to note that the sensitivity to conditioning media varies among different materials. A study by Renaud and Greenwood (2005) suggests that conditioning laminates in deionized water rather than sea water has a greater effect on the mechanical properties of E-glass compared to ECR-glass. Therefore, using deionized water as a conditioning medium to investigate the long-term durability of composites for sea water applications may be less suitable for laminates composed of E-glass compared to ECR-glass. It is difficult to predict these effects without conducting physical testing of specimens. For the purposes of testing multiple types of material, it is therefore much safer to condition with a medium that closely represents in-service conditions.

4.4.4 Effect of testing environment

As previously discussed, composites experience reversible and nonreversible effects when undergoing sea water conditioning. For fully submerged applications, it is important to capture all of these effects during material testing.

Generally, the testing environment only significantly affects results when tests are of long duration, as moisture diffusion out of a laminate is not instantaneous. For short duration static tests, it is acceptable to test the coupons in dry conditions, provided they are tested within a reasonable time of removal from the conditioning unit. For tests of longer duration, it is generally thought that maintaining the conditioning environment during testing will increase the accuracy of results. Smith and Weitsman (1996) demonstrated that fatigue coupons tested in air exhibit a longer fatigue life than those tested in sea water. The authors propose that water trapped within cracks in the composite are the cause of this observation. However, recent studies on a glass/epoxy laminate have shown that testing fatigue properties while the test specimen is submerged in water produces the same results as for a standard, dry testing environment (Dawson et al., 2016). These two findings highlight the wide variation and complexities of composite materials testing.

4.4.5 Effect of pH

Harsh chemicals are known to cause significant degradation of composite laminates, with the effects varying between different chemicals and composite materials. Surendra Kumar et al. (2007) found that increasing the concentration of HCl in water resulted in a further knockdown in interlaminar shear strength (ILSS) of E-glass/epoxy conditioned specimens. A 16%–18% reduction in ILSS was observed for specimens conditioned in a 5% concentrated solution of HCl. This study also showed that an increase in concentration of HCl acts to reduce the moisture uptake at saturation.

Interestingly, Stamenović et al. (2011) found that glass/polyester specimens conditioned in a low pH (1–2) solution experienced a slight (0%–13%) increase in tensile strength. The same study also shows that the same specimens conditioned in a higher pH (12–14) solution experienced a reduction in tensile strength of up to 27%.

It is, therefore, imperative that the conditioning environment seen by the composite specimens accurately represents the expected in-service environment. To ensure this, pH monitoring and control methods should be used. For typical marine applications, a pH of 7.5–8.4, representative of fresh sea water, is desired. In closed conditioning tanks with no continuous circulation of fresh sea water, the pH can vary over time as materials such as metals corrode in the sea water. In such cases, active pH control is required.

4.4.6 Effect of specimen dimensions

The specimen dimensions can heavily influence the rate of diffusion in specific directions (Beringhier et al., 2016). For example, a flat, thin specimen will experience a greater diffusion rate through thickness due to the larger area of the top and bottom exposed surfaces. Panels of ∼3 mm thick are common for weight gain measurements. In practice, it is unlikely that laminate edges will be in direct contact with sea water, and further improvements can be made to the conditioning process to reflect this. The edges of the laminate can be covered with an impermeable material, so only diffusion through the faces of the laminate is possible. Alternatively, conditioning chambers can be designed to hold specimens in such a way that only one face is in direct contact with water. This is the most representative conditioning test for most marine structures.

The time for a typical composite specimen to saturate is proportional to the specimen thickness. This is due to the bias in diffusion through thickness and means that conditioning thicker laminates will take significantly longer. This can be an issue when investigating the impact response of conditioned specimens as these test laminates are typically 10–15 mm thick, representing real world structures.

4.4.7 Effects of the manufacturing process

Specimen manufacturing quality can have a significant impact on both laminate mechanical properties and moisture absorption characteristics. Rough handling of materials, especially lightly stitched dry fabrics, can result in poor laminate quality and incorrect fiber orientations. For material characterization and qualification, it is imperative that fiber orientation within test coupons is tightly controlled. Tolerances are specified in the relevant test standards; for example, ASTM D2344 defines a ply orientation tolerance of +/−0.5° with respect to a datum.

Ply orientation can also affect the moisture absorption characteristics of a laminate. As discussed previously, one mechanism for moisture transport through a laminate is capillary action between the fibers and matrix. While the fiber orientation may have an initial effect on moisture absorption, it is not thought to have a significant effect once the specimen has been saturated. Furthermore, as most conditioned test specimens are thin, the primary path of diffusion is through thickness.

Thomason (1995) demonstrated that porosity and voids can significantly increase the initial rate of absorption and the saturation level of a composite laminate. This is of particular importance for industrial applications where manufacturing variability and batch variation can occur. Indeed, Carlsson and Du (2016) present data showing that two similar composite panels manufactured at different locations exhibit very different moisture uptakes. Dawson et al. (2016) also directly compared the moisture uptake of infused and prepreg specimens. Interestingly, these results indicate that while the infused laminates seem to saturate faster, the prepreg specimens experience greater weight gains at saturation.

4.4.8 Current aging procedures

One must, therefore, make a compromise between simulating a realistic service environment and achieving a sufficiently fast absorption rate when designing a suitable conditioning procedure. A common approach is to condition samples in sea water at 45°C. This temperature has been shown in practice to be sufficiently high to significantly accelerate the conditioning process while being far enough below the wet Tg as to not cause additional degradation. In this scenario, the engineer places the importance of risk and accuracy of conditioning data over conditioning speed.

The major limitation with current methods is the elapsed time required to fully saturate specimens. It can take between 4 and 6 months to fully saturate a laminate of ∼3 mm thickness, which is typical for many mechanical tests. This conditioning time is added to the overall project duration, as the design process cannot be finalized without this data. By accelerating this process, the overall project duration and cost can be reduced. This in turn will improve the competitiveness of composite products against alternative materials such as steel.

It should be noted that these procedures are valid for predicting the long-term performance of composite structures exposed to relatively mild temperatures; for example, marine current turbines and ship structures. These procedures are of limited use when investigating high temperature applications such as those found in the offshore oil and gas industry, where the materials are often operating near their maximum service temperature (Echtermeyer et al., 2018).

4.5 Mechanical testing of saturated specimens

4.5.1 Testing methodology

In order to investigate the mechanical degradation of composite laminates in a chosen conditioning medium, specimens are tested to failure after conditioning and compared against identical, nonconditioned dry specimens.

The extent of testing required is dependent on the purpose of the tests and use of the data. To compare the moisture degradation of different laminates for the purpose of material selection, it is generally considered acceptable to investigate only two mechanical properties;

Interlaminar shear stress (ILSS): This value is highly dependent on resin properties. It is a good indication of resin and resin/fiber interface strength, and hence, resin degradation when comparing dry and conditioned results. There are multiple test methods available for obtaining this property, the simplest being the three-point bend short beam shear test. The simplicity of coupon geometry and test procedure is particularly advantageous when conducting a large number of tests in an industrial environment. ASTM D2334 and ISO 14130 provide detailed accounts of the testing procedure. Other tests are available such as double-notched shear; however, coupon preparation is more complex. A Double Beam Shear (DBS) test using five-point loading has been the subject of an international ISO round-robin.

Flexural strength: This property is an indication of fiber and fiber/matrix interface strength, as well as fiber and sizing degradation. ASTM D7264 and ISO 14125 both define procedures for obtaining this value from three- and four-point bending tests. Once again, the test coupons are quite simple and therefore preferred for comparative investigations.

These two properties give a quick and fairly reliable indication of general laminate performance and allow for a quick comparison between different laminates and materials. It is important to understand the assumptions and limitations of both tests, even when following the test standards. The combination of specimen geometry and constituent properties will determine the proportion of directional loading within the specimen, which determines the failure mode; an essential consideration.

For detailed structural design, more extensive testing is required. Generally, this will entail static and fatigue testing of various coupons and load conditions; including tension, compression, and in-plane and interlaminar shear. Larger specimens can be tested separately with intentional defects such as holes and ply drops to be more representative of real-world structures. These tests are often conducted on dry specimens as thicker coupons can take significantly longer to saturate. The separate effects of local defects and moisture degradation are normally combined during product design. Designs incorporating thick monolithic laminates may also require knowledge of out-of-plane properties.

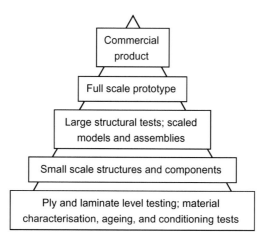

Fig. 4.2 Typical testing pyramid for high-performance marine structures.

The test pyramid describes a standard approach used in many industrial design projects. Fig. 4.2 displays a test pyramid for a typical high-performance marine structure. A larger number of more affordable ply and laminate-level tests support fewer large scale structural tests. The entire structural design may be validated with a full scale prototype. The goal of a test pyramid is to help the designer achieve a suitable compromise between cost and structural validation. For example, the certification standard for tidal and wave energy converters (DNV, 2008) suggests supporting analytical design with material and prototype testing, among others.

A greater understanding of the physical processes by which materials degrade over time in a marine environment will enable designers to replace a significant portion of the lengthy, costly test programs with faster, cheaper simulations and numerical predictions. Numerical models are being developed; however, they are at early stages and require extensive test data for validation. It has been suggested that, to help accelerate this process, as well as improve the applicability of such models for both academic and industrial purposes, a limited number of resin, sizing, and fiber combinations should be selected for investigation and commercial use.

4.5.2 Static testing of conditioned specimens

Recent studies at AEL Airborne have investigated the reduction in ILSS and flexural strength of a number of typical marine laminates conditioned using currently accepted methods for accelerating the aging process. Figs. 4.3 and 4.4 show the degradation in ILSS and flexural strength, respectively, for these laminates.

Fig. 4.3 indicates that marine composite laminates typically experience a 5%–30% reduction in ILSS after sea water saturation, and that both the wet and dry strength vary across the range of materials. In Fig. 4.4, flexural strength appears to vary more across wet specimens, with an overall variation of 43% compared to 12% for dry

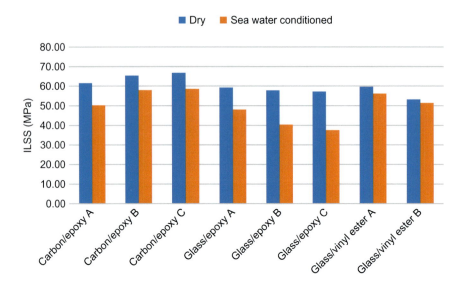

Fig. 4.3 Comparison of dry and sea water-conditioned (45°C) ILSS of typical marine laminates.

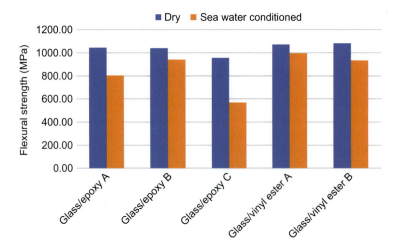

Fig. 4.4 Comparison of dry and sea water-conditioned (45°C) flexural strength of typical marine laminates.

specimens. It is clear that moisture degradation varies significantly among composite laminates constructed from resins and fibers that suppliers claim are all suitable for marine use. This variation in material behavior highlights the importance of material testing and selection at an early stage in the design process.

The relationship between moisture content and reduction in mechanical properties has been studied previously. For example, Shen and Springer have demonstrated that

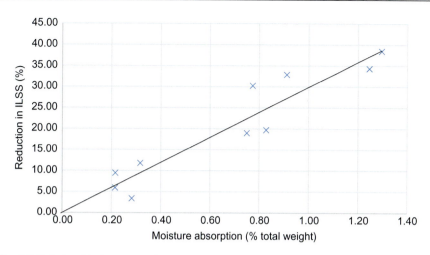

Fig. 4.5 Relationship between ILSS knockdown and moisture absorption of typical marine composite laminates.

the tensile strength and elastic modulus of a specific composite laminate reduce with increased moisture content (Shen and Springer, 1976, 1977). Fig. 4.5 displays data for various epoxy and vinyl ester glass fiber laminates conditioned at 45°C and 55°C at AEL Airborne. The data indicates that the reduction in ILSS of these materials is proportional to the level of moisture absorbed. This trend agrees with findings from previous studies of similar materials (Dawson et al., 2016).

This is a significant trend, as it potentially allows selection of suitable constituent materials based solely on moisture absorption at saturation. Therefore, if this trend holds true for a wider range of materials, there may no longer be a need to initially conduct mechanical testing on such a wide range of different laminates for the purpose of material selection.

4.5.3 *Impact testing of conditioned specimens*

It is important to test laminates with thicknesses representative of real structures, as impact damage is dependent on laminate thickness (Sutherland and Guedes Soares, 2005). Thicker laminates have been shown to suffer from delamination at lower impact loads than thinner laminates. Some thick laminates can take years to fully saturate in conditioning tanks, meaning it can be difficult, and often unfeasible, to fit such test programs into a typical design project.

Moisture uptake can result in a softening of the composite matrix over time, predominantly due to plasticization. This effect has been shown to affect the impact response of composite laminates (Imielińska and Guillaumat, 2004).

It is also important to consider the postimpact strength of a laminate. Local impact damage in the form of fiber breakage and delamination can significantly reduce the in-plane properties of a composite laminate, especially when combined with other environmental factors. Compression after impact tests is common for investigating such effects. ASTM D7137 provides a detailed explanation of this method. Sala (2000) and Imielińska and Guillaumat (2004) both present results for compression after impact tests for wet and dry laminates. Their results clearly indicate a reduction in compressive strength due to moisture uptake in addition to impact damage.

4.5.4 Fatigue testing of conditioned specimens

Marine structures such as tidal turbine blades must be designed to withstand harsh fatigue loading. However, complete fatigue testing of conditioned laminates is both time-consuming and costly, and so is often limited in industry to a few select laminates. It is, therefore, difficult for companies to compare the fatigue performance of a wide range of composite laminates, and for this reason, cheaper, quicker static tests are often used instead for the purposes of material selection.

The cost to conduct fatigue tests is proportional to the number of cycles performed. Careful consideration is, therefore, required when planning fatigue testing. In some cases, it may be sufficient to conduct fatigue tests with a low number of cycles and estimate strength for higher cycles via extrapolation (Echtermeyer et al., 2018). However, experimental proof of the fatigue performance of a material will help to reduce project risk and is often crucial for fatigue-driven designs.

As with static mechanical properties, moisture absorption into the laminate acts to reduce the fatigue strength of the composite structure. This can be shown graphically as a drop in the S–N curve compared to dry coupons. The degradation of tensile fatigue strength has been investigated for an epoxy/E-glass laminate (Vesna Jaksic et al., 2018). This work has demonstrated that the fatigue degradation of saturated laminates, ranging from 8% to 25% compared to dry, is dependent on the number of loading cycles. Therefore, the gradient of the S–N curve can also vary depending on material. Sun and Dawson (2013) also found this to be true for other glass and carbon epoxy laminates. This is an important consideration for the designer, as knockdown in fatigue strength can vary with both application and material. This highlights the need for full fatigue testing of both dry and saturated coupons for high-performance design projects. Such test programs are both costly and time-consuming as previously discussed.

4.5.5 Further testing of conditioned specimens

The ability of a material to resist constant load over an extended duration is also of significant interest to designers. Creep and stress rupture tests can be used to investigate long-term damage and potential failure due to the application of a constant load significantly lower than the ultimate tensile strength.

It may also be desirable to combine an applied stress with environmental effects to further investigate the limits of the chosen material, generally referred to as stress-aging. For example, one may wish to investigate the effects of a constant tensile stress applied to a specimen submerged in sea water, or for deep-sea applications to condition specimens under a constant, elevated pressure. Alternatively, a worst-case scenario may be investigated where a hot/wet-conditioned specimen is subject to compression-after-impact or creep/fatigue loading. These tests are thought to be more accurate as they can account for coupling effects between multiple conditions that may or may not be apparent to the designer and can be difficult to predict analytically. Stress corrosion cracking of glass fibers is one such effect that is commonly investigated and heavily dependent on the external environment (Maxwell et al., 2005). Kennedy et al. (2016) describe a test procedure in which vinyl ester/E-glass specimens are exposed to 10,000 fatigue cycles between 0.8 and 8 kN prior to wet and dry stress-aging for 21 months. These loads were selected to generate cracking in the 90° plies, simulating expected in-service damage of a tidal turbine blade.

Larger, thicker coupons are more representative of real-world structures and give a better indication of performance for designs incorporating thick laminates. Research has suggested that a scaling effect is present whereby the mechanical properties are dependent on specimen dimensions (Wisnom et al., 2009). This is, therefore, another factor that should be taken into account when designing durability tests.

4.6 Defining the limits of accelerated aging techniques

As previously discussed, a major limitation with current durability testing of composite laminates is the extensive test duration. This section presents results for the first stage of an investigation into the feasibility of further increasing the conditioning temperature beyond current practice. Such improvements may lead to reduced time and cost of the design process for high-performance marine structures such as tidal turbine blades.

The wet Tg of typical marine composite laminates is estimated to lie between 60°C and 80°C. It is, therefore, suggested that increasing the conditioning temperature from 45°C to 55°C could be a suitable way to accelerate the conditioning process further. However, it is important to investigate the potential impact this could have on laminate moisture absorption and mechanical properties.

4.6.1 Effect on moisture absorption of increased temperature

This effect has been investigated for unidirectional specimens constructed from glass and carbon fibers and epoxy and vinyl ester resins. In all cases, conditioning at 55°C has resulted in an increase in the initial rate of moisture absorption and final saturation level.

Fig. 4.6 indicates that raising the conditioning temperature accelerates the saturation time from 73 days at 45°C, to 52 days at 55°C for these particular specimens. The higher temperature raises the saturation level by approximately 0.1% total weight gain. It is evident that the weight gain at 55°C follows a similar trend to that at 45°C. Both sets of data exhibit close agreement with Fickian moisture absorption.

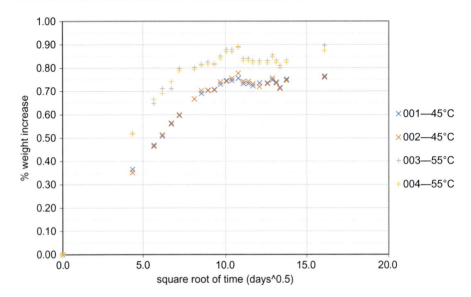

Fig. 4.6 Comparison of weight gain due to moisture absorption of glass/epoxy laminates conditioned in sea water at 45°C and 55°C.

4.6.2 Effect on mechanical properties of increased temperature

Figs. 4.7 and 4.8 show the knockdowns (wet/dry) in ILSS and flexural strength due to conditioning at 45°C and 55°C compared to unaged specimens. It is evident that the slight increase in saturation level observed for specimens conditioned at 55°C results

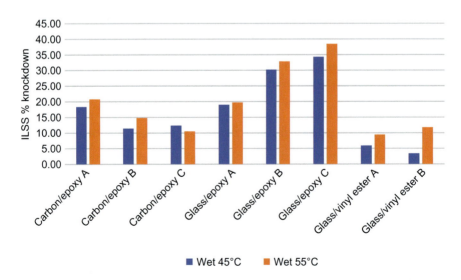

Fig. 4.7 Comparison of ILSS of typical marine laminates conditioned at 45°C and 55°C.

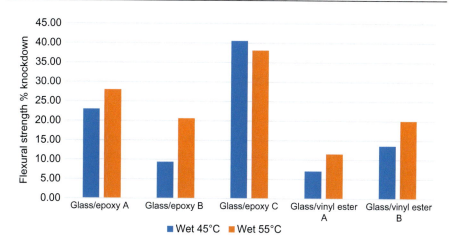

Fig. 4.8 Comparison of flexural strength of typical marine laminates conditioned at 45°C and 55°C.

in a further reduction in mechanical properties of between 1 and 10% depending on the material.

The results in Figs. 4.7 and 4.8 suggest that conditioning composite laminates in sea water at 55°C produces generally conservative results compared to lower temperatures. The effect of higher temperatures varies among the materials investigated. For most specimens, the difference in strength between the two temperatures is small; around 5%. However some specimens experience a reduction in strength of approximately 10% due to the higher conditioning temperature. Furthermore, there are two examples where conditioning at 45°C results in a greater strength knockdown compared to 55°C. By raising the conditioning temperature to 55°C, the results indicate, on average, lower material strengths, but the two outliers suggest that the resultant impact on laminate strength is variable. In some cases, this variation is unacceptable despite the benefit of faster conditioning. Further work is suggested to identify the materials for which this procedure is suitable.

Previous studies have shown that increasing the conditioning temperature from 20°C to 50°C has a detrimental effect on shear modulus and strength of polyester, vinyl ester, and epoxy laminates (Davies and Choqueuse, 2008). In this study, the specimens were conditioned for approximately 14 months, and the results indicate that the effect of raised temperature is dependent on resin selection.

It is also interesting to consider the potential difference in strength knockdowns between specimens saturated in sea water at 45°C and ambient temperature. Future investigations into the potential conservative nature of current conditioning practices are planned, which will indicate the suitability of further acceleration via increased temperatures.

A possible solution to account for the conservative nature of further accelerated conditioning procedures may be to use theoretical models to predict the effect of temperature on mechanical performance, thereby allowing the correction of data to account for conservatism in experimental results. Such models may also be used to

4.7 Modelling of accelerated moisture absorption

4.7.1 Fickian diffusion

It was previously discussed that for flat composite plates with relatively small thicknesses compared to length and width, diffusion is dominant in the thickness direction, and therefore, one-dimensional. If one also assumes steady state diffusion, then the simple model of Fick's law can be applied to describe moisture diffusion through a composite laminate (Colin and Verdu, 2014), Eq. (4.1).

$$\frac{\partial C}{\partial t} = D \frac{\partial^2 C}{\partial z^2} \qquad (4.1)$$

where C is the local water concentration, z is the depth in the thickness direction, and D is diffusivity.

The analytical solution to this equation is given below, Eq. (4.2) (Chilali et al., 2017).

$$\frac{M_t}{M_\infty} = 1 - \left(\frac{8}{\pi^2}\right)^3 \sum_{i=0}^{\infty}\sum_{j=0}^{\infty}\sum_{k=0}^{\infty} \frac{\exp\left(-\pi^2 t \left(D_1 \left(\frac{2i+1}{L}\right)^2 + D_2 \left(\frac{2j+1}{w}\right)^2 + D_3 \left(\frac{2k+1}{h}\right)^2\right)\right)}{((2i+1)(2j+1)(2k+1))^2} \qquad (4.2)$$

where L, w, and h are length, width, and thickness, respectively, M_∞ = moisture uptake at saturation, and M_t, the periodic weight gain, is expressed in Eq. (4.3).

$$M_t = \frac{W_t - W_0}{W_0} \qquad (4.3)$$

where W_t = specimen weight at time, t and W_0 = dry specimen weight.

Three separate diffusivity values (D) are present in Eq. (4.2), one for each primary direction in the orthotropic composite material. These parameters must be found via separate absorption tests in each of the primary directions. As diffusion through the thickness is dominant for a flat composite plate with relatively low thickness, it is sufficient for industrial purposes to assume a single diffusivity. The equation then simplifies to the form in Eq. (4.4) (Naceri, 2009):

$$\frac{M_t}{M_\infty} = 1 - \frac{8}{\pi^2} \sum_{k=0}^{\infty} \frac{\exp\left(-(2k+1)^2 \pi^2 \frac{Dt}{h^2}\right)}{(2k+1)^2} \qquad (4.4)$$

The first 6 terms ($k = 1$–6) of Eq. (4.4) will give sufficient accuracy for most industrial purposes.

Assuming a semi-infinite plate with low water concentration on the back surface, the above equation can be simplified further and rearranged to give the diffusivity, D, in terms of the initial linear rate of moisture uptake, moisture saturation level, and specimen thickness (Shen and Springer, 1975), Eq. (4.5).

$$D = \pi \left(\frac{h}{4M_\infty}\right)^2 \left(\frac{M_2 - M_1}{\sqrt{t_2} - \sqrt{t_1}}\right)^2 \tag{4.5}$$

The calculated value for D can then be inserted into Eq. (4.4) to calculate the predicted periodic weight gain according to Fick's law. Therefore, the periodic weight gain of a composite laminate can be predicted with only three variable inputs: Thickness, initial linear rate of moisture uptake, and moisture uptake at saturation. This greatly simplifies the analysis and allows for a fast theoretical prediction of the weight gain plot.

4.7.2 Effect of temperature on diffusivity

The previous analysis is independent of temperature. One can predict the effect of temperature on moisture absorption by assuming that diffusivity varies with temperature following the Arhhenius relationship in Eq. (4.6).

$$D = D_0 \exp\left(-\frac{E_D}{RT}\right) \tag{4.6}$$

where E_D is the activation energy of water diffusion, R is the universal gas constant, T is temperature in Kelvin, and D_0 is a constant. Using this equation, one assumes that diffusivity is independent of water activity.

Both E_D and D_0 can be found by reducing Eq. (4.6) to a linear form; Eq. (4.7).

$$\ln(D) = \frac{-E_D}{RT} + \ln(D_0) \tag{4.7}$$

Both E_D and D_0 can then be found by plotting $\ln(D)$ versus $1/T$. To do this, a minimum two values of D for the same laminate at two different temperatures are required. The accuracy of this method improves with the number of temperatures investigated (Starkova et al., 2013; Deroiné et al., 2014).

These values can then be inserted into Eq. (4.6) to estimate the diffusivity at different temperatures, which can then be inserted into Eq. (4.4) to estimate the periodic weight gain at different temperatures. Alternatively, one can use E_D values from other literature sources (Colin and Verdu, 2014), but it is important to note that this method is only valid for temperatures below the Tg of the laminate in question.

4.7.3 Effect of temperature on saturation level

Research has shown that the maximum moisture content diffused into a composite is dependent on temperature (Davies and Choqueuse, 2008), and data presented in Fig. 4.6 supports this.

Predictive tools should, therefore, account for this effect. A simple approach is to assume that water concentration follows Henry's Law, Eq. (4.8) (Merdas et al., 2001). The use of this equation is based upon the assumptions that water concentration (weight gain) remains low (less than 7%) and that the polymer matrix does not change state.

$$C = Sp \tag{4.8}$$

where C = equilibrium water concentration (equivalent to M_∞), S = coefficient of solubility, and p = water vapor pressure.

One can assume that S and p vary with temperature according to an Arrhenius law. The latter only follows this law between 20°C and 100°C (Merdas et al., 2001). These two relationships can be combined with Eq. (4.8) to form the following Arrhenius equation, Eq. (4.9) (Merdas et al., 2001):

$$C = C_0 \exp\left(\frac{-H_C}{RT}\right) \tag{4.9}$$

Eq. (4.9) can be made into linear form, and the values H_C and C_0 are found using the same method as described previously. The Arrhenius equation can then be used to calculate the predicted equilibrium water concentration (or saturated moisture level).

4.7.4 Application of the model

Fig. 4.9 compares the theoretical Fickian plots and experimental data at both 45°C and 55°C. The data shows that the process of moisture absorption into the glass/epoxy specimens can be described well by the simplified analysis based upon Fick's law. Indeed, in its simplest form, this method can be used in industry to check whether experimental data follows a Fickian absorption trend, and therefore, gain a basic understanding of the mechanisms by which moisture is absorbed into the specimen.

Fig. 4.9 also shows the predicted weight gain due to moisture absorption at 50 °C. Although the accuracy of the method still needs to be validated at lower temperatures with these materials, this illustrates how the Fickian model can be used to estimate the effect of conditioning at other temperatures once calibrated using experimental data. In future, it may be possible to combine this method with the relationship between knockdown in strength and moisture absorption (Fig. 4.5) to predict the knockdown in strength at various, untested temperatures. This would allow designers to conduct accelerated conditioning procedures at elevated temperatures, and then correct the observed knockdowns in laminate strength to produce design data representative of in-service temperatures.

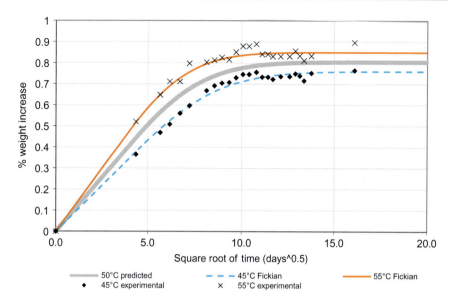

Fig. 4.9 Fickian prediction of weight gain of a glass/epoxy composite laminate due to moisture absorption, Effect of temperature.

This analysis technique can also predict the time taken for a specimen to fully saturate, but is limited by the need for some existing experimental data to which the Fickian curve can be fitted. Therefore, a prediction of saturation time can only be made once the specimen conditioning is underway. However, even this limited information can help companies predict project duration and budgets to greater accuracy. The accuracy of this prediction can be improved by also considering the behavior of other similar laminates that have previously been tested. Purnell et al. (2008) have presented a model that may be used to predict the duration of accelerated aging required to represent equivalent in-service lifetimes.

Since thickness is a variable in the Fickian analysis, it is possible to predict the rate of moisture absorption at different thicknesses. Fig. 4.10 displays the predicted weight gain curve for a specimen of 1 mm thickness at 45°C, compared to the actual specimen thickness of 2.7 mm. Predicting the effect of thickness in this way assumes that the diffusivity is unchanged. This assumption would only be valid for thin plates where diffusion through the thickness is dominant.

Limiting moisture absorption to only the through-thickness direction during conditioning allows measurement of the diffusivity in this direction. A combination of this test data and the above model can then be used to predict the long-term moisture penetration in composite structures operating in a typical marine environment. This can indicate to the designer what proportion of the laminate will experience moisture degradation over the design life, and therefore, help to avoid overconservative design safety factors being applied.

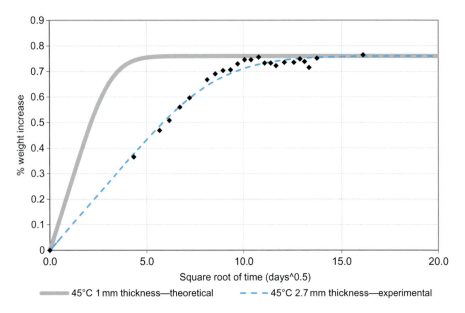

Fig. 4.10 Fickian prediction of weight gain of a glass/epoxy composite laminate due to moisture absorption, Effect of specimen thickness.

4.7.5 Limitations of the model and potential improvements

The simplicity of the Fickian diffusion model presented allows prediction of periodic moisture absorption for a wide range of composite laminates. Although a number of more advanced models are available to predict the behavior of specific laminates, as one focuses on more detailed diffusion behavior, the range of materials to which these models can be applied decreases. This limits their application, but may be of interest to companies who utilize a very small number of select materials.

Depending on the level of accuracy required, the Fickian modeling technique could be extended further to include all three diffusivities, making it applicable to specimens of other shapes and dimensions. Other absorption models may also be used when specimens exhibit non-Fickian behavior. Case II, sigmoidal, and two-stage absorption models can all be used to match experimental data more closely.

To improve the accuracy of the analysis presented in this chapter, it is proposed that the model is validated over a wider range of temperatures, such as 35°C, 25°C, 15°C, and 10°C. This will also generate a greater understanding of the level of conservatism built into current 45°C sea water conditioning procedures.

It should be noted that the above models are limited in their application as they map basic analytical models to experimental data; one must have the experimental data in order to apply the models. However, they give potential to significantly reduce the amount of physical test data required, and therefore, the time and cost involved in its acquisition. While the data presented both here and in other sources indicate apparent general trends for moisture absorption and mechanical degradation that may be used for top-level material selection, further research is needed to refine the accuracy of the predictive models.

4.8 Constituent-level predictive methods

The majority of research that has been conducted to this date considers the moisture degradation effect at a laminate level. Therefore, as the accuracy of theoretical predictions increases, the range of applicable materials reduces. For such tools to be applicable to industry, they must be valid for a wide range of composite materials.

Modeling of aging at a constituent level has been proposed as an alternative to laminate-level predictions (Echtermeyer et al., 2018). Predicting laminate mechanical properties from constituent data is commonly used by both academia and industry in the form of classical laminate analysis (CLA). It is thought that combining CLA with constituent-level degradation predictions will provide a much greater level of flexibility during product design, as well as greatly reducing the level of physical testing required.

Such models will also require accurate knowledge of the chemistry and formulation of each constituent part, and it can be difficult to persuade manufacturers to reveal this information. A laminate, in its simplest form, is a combination of three unique constituents; fiber, resin, and sizing. It is the combination of all three constituents that determines the laminate properties. The quality of the bond between all three components is determined by the chemical and physical properties of each as well as manufacturing parameters.

Of course, combining the aging effects of multiple constituents is a complex task, especially when one considers the combined effects of multiple environments such as moisture, temperature, and chemical attack in addition to applied stresses. Indeed, the aging behavior of each constituent may also be dependent on the behavior of the other constituents. For example, it has previously been discussed how resin acts as a protective barrier to shield glass fibers from moisture exposure. Complex analytical models will be required to predict the combined aging of constituents at a laminate level.

4.9 Summary and future work

The research presented in this chapter indicates that dry and wet material properties can vary significantly between materials marketed for the same purpose. Furthermore, manufacturing variations can exaggerate this effect. It is, therefore, important that composite manufacturers aiming to use such materials for high-performance applications perform extensive durability testing to better understand their properties and limitations.

Current accelerated conditioning practices are costly and time-consuming, resulting in longer project durations and increased project risk. Modeling techniques of various complexities may be used alongside testing programs to give more accurate predictions of conditioning duration. Currently, the most reliable method for obtaining long-term durability data is still through physical testing.

There is still further work remaining to develop accurate accelerated composite aging prediction methods across a range of different material types. Each unique laminate combination of fiber, resin, and sizing must be extensively tested, the results of which are only applicable to that certain combination. While improvements can be made to current processes in the form of accelerated conditioning and predictive methods using general trends, extensive testing cannot yet be avoided. While constituent-level models may allow further reduction in physical testing, this approach is at a very early stage. The development of such a tool will be a complex process and requires input from material suppliers, product designers, and manufacturers. Both laminate-level and constituent-level models could play a role in reducing the engineering cost of composite structures, the former for material selection and the latter for in-depth structural design.

Physical testing of materials is a reliable process, and so any analytical method that aims to replace this must be proven to be reliable over a wide range of materials. In the future, it may be possible to predict the effects of long-term submersion in sea water on a wide range of composite materials with much greater accuracy. This knowledge can then be used to improve the efficiency of structural designs and reduce project duration and cost.

References

Ashland (2016). Epoxy Vinyl Ester Resins Chemical Resistance Guide.
Beringhier, M., et al., 2016. Identification of the orthotropic diffusion properties of RTM textile composites for aircraft applications. Compos. Struct. 137, 33–43.
Boisseau, A., et al., 2011. Sea water ageing of composites for ocean energy conversion systems: influence of glass fibre type on static behaviour. Appl. Compos. Mater. 19 (3–4), 459–473.
Brødrene. http://www.braa.no/. Retrieved 27/11/2017, 2017, 2017.
Carlsson, L.A., Du, E., 2016. Water uptake in polymer composites with voids. In: Davies, P., Rajapakse, Y.D.S. (Eds.), Durability of Composites in a Marine Environment 2. Springer, Cham.
Cheour, K., et al., 2016. Effect of water ageing on the mechanical and damping properties of flax-fibre reinforced composite materials. Compos. Struct. 152, 259–266.
Chilali, A., et al., 2017. Effect of geometric dimensions and fibre orientation on 3D moisture diffusion in flax fibre reinforced thermoplastic and thermosetting composites. Compos. A: Appl. Sci. Manuf. 95, 75–86.
Choqueuse, D., Davies, P., 2008. Ageing of composites in underwater applications. In: Martin, R. (Ed.), Ageing of Composites. Woodhead publishing, Cambridge.
Clegg, N., 2006. How to Paint your Boat: Painting, Varnishing, Antifouling. Sheridan House, New York.
Colin, X., Verdu, J., 2014. Humid ageing of organic matrix composites. In: Davies, P., Rajapakse, Y.D.S. (Eds.), Durability of Composites in a Marine Environment. Springer, Dordrecht.
CSC (2009). Developments in fire protection of FRP composite vessels. Innovation in High Speed Marine Vessels. Freemantle, Australia.

Davies, P., Choqueuse, D., 2008. Ageing of composites in marine vessels. In: Martin, R. (Ed.), Ageing of Composites. Woodhead publishing, Cambridge.

Davies, P., et al. (1997). Composites underwater. Progress in Durability Analysis of Composite Systems, Balkema.

Dawson, M., et al. (2016). Effects of conditioning parameters and test environment on composite materials for marine applications. 2nd Ifremer-ONR Workshop—Durability of Composites in a Marine Environment. Ifremer Centre, Brest, France.

De Mony-Pajol, J., Cail, G., 2014. The first 100% jute reinforced biocomposite sailboat, a prototype of importance for Bangladesh. JEC Composites Magazine 51, 76–77.

Deroiné, M., et al., 2014. Accelerated ageing of polylactide in aqueous environments: Comparative study between distilled water and seawater. Polym. Degrad. Stab. 108, 319–329.

DNV (2008). Certification of tidal and wave energy converters. Offshore Service Specification. DNV-OSS-312.

Echtermeyer, A.T., et al., 2018. Multiscale modelling of environmental degradation—first steps. In: Davies, P., Rajapakse, Y.D.S. (Eds.), Durability of Composites in a Marine Environment 2. Springer, Cham.

Flaxcomposites, 2014. http://flaxcomposites.com/?p=231. Retrieved 27/11, 2017.

Guermazi, N., et al., 2016. On the durability of FRP composites for aircraft structures in hygrothermal conditioning. Compos. Part B 85, 294–304.

Humeau, C., et al., 2016. Moisture diffusion under hydrostatic pressure in composites. Mater. Des. 96, 90–98.

Imielińska, K., Guillaumat, L., 2004. The effect of water immersion ageing on low-velocity impact behaviour of woven aramid–glass fibre/epoxy composites. Compos. Sci. Technol. 64 (13–14), 2271–2278.

IMO (2012). International shipping facts and figures—information resources on trade, safety, security, environment London: IMO.

IMO (2014). Reduction of GHG emissions from ships—Third IMO GHG Study 2014, Final Report, MEPC 67/INF.3, IMO.

Jaksic, V., et al., 2018. Influence of composite fatigue properties on marine tidal turbine blade design. In: Davies, P., Rajapakse, Y.D.S. (Eds.), Durability of Composites in a Marine Environment 2. Springer, Cham.

Kennedy, C.R., et al., 2016. Immersed fatigue performance of glass fibre-reinforced composites for tidal turbine blade applications. J. Bio Tribocorrosion 2, 2.

Liu, M., et al., 2017. Targeted pre-treatment of hemp bast fibres for optimal performance in biocomposite materials: A review. Ind. Crops Prod. 108, 660–683.

Maxwell, A. S., et al. (2005). Review of accelerated ageing methods and lifetime prediction techniques for polymeric materials, National Physics Laboratory, Teddington.

Merdas, I., et al., 2001. Factors governing water absorption by composite matrices. Compos. Sci. Technol. 62.

Naceri, A., 2009. An analysis of moisture diffusion according to Fick's law and the tensile mechanical behaviour of a glass-fabric-reinforced composite. Mech. Compos. Mater. 45.

OpenHydro. "http://www.openhydro.com/." Retrieved 27/11, 2017, 2017.

Peters, L. (2016). Influence of glass fibre sizing and storage conditions on composite properties. 2nd Ifremer-ONR Workshop—Durability of Composites in a Marine Environment. Ifremer, Brest.

Purnell, P., et al., 2008. Service life modelling of fibre composites: A unified approach. Compos. Sci. Technol. 68 (15–16), 3330–3336.

QinetiQ (2013). Global marine trends 2030, QinetiQ Lloyd's Register, University of Strathclyde.

Renaud, C. M. and M. E. Greenwood (2005). Effect of glass fibres and environments on long-term durability of GFRP composites.

Sala, G., 2000. Composite degradation due to fluid absorption. Compos. Part B 31, 357–373.

Sanjay, M.R., et al., 2018. Characterization and properties of natural fiber polymer composites: A comprehensive review. J. Clean. Prod. 172, 566–581.

Schutte, C.L., 1994. Environmental durability of glass-fiber composites. Mater. Sci. Eng. 13, 265–323.

Searle, T.J., Summerscales, J., 1999. Review of the durability of marine laminates. In: Pritchard, G. (Ed.), Reinforced Plastics Durability. Woodhead Publishing Limited, Cambridge.

Shen, C., Springer, G., 1975. Moisture absorption and desorption of composite materials. J Compos. Mater. 10, 2–20.

Shen, C.-H., Springer, G.S., 1976. Effects of moisture and temperature on the tensile strength of composite materials. J. Compos. Mater. 11, 2–16.

Shen, C.-H., Springer, G.S., 1977. Environmental effects on the elastic moduli of composite materials. J. Compos. Mater. 11.

Smith, L.V., Weitsman, Y.J., 1996. The immersed fatigue response of polymer composites. Int. J. Fract. 82 (1), 31–42.

Stamenović, M., et al., 2011. Effect of alkaline and acidic solutions on the tensile properties of glass–polyester pipes. Mater. Des. 32 (4), 2456–2461.

Starkova, O., et al., 2013. Water transport in epoxy/MWCNT composites. Eur. Polym. J. 49 (8), 2138–2148.

Sun, L. and M. Dawson (2013). AEL Internal Report, MTMS013A.

Surendra Kumar M, et al. (2007). Acid degradation of FRP composites. National Conference on Developments in Composites. National Institute of Technology Rourkela, India.

Sutherland, L.S., Guedes Soares, C., 2005. Impact behaviour of typical marine composite laminates. Compos. Part B 37 (2–3), 89–100.

Thomason, J.L., 1995. The interface region in glass fibre-reinforced epoxy resin composites: 2, water absorption, voids and the interface. Composites 26, 477–485.

Tual, N., et al., 2015. Characterization of sea water ageing effects on mechanical properties of carbon/epoxy composites for tidal turbine blades. Compos. A: Appl. Sci. Manuf. 78, 380–389.

Wisnom, M.R., et al., 2009. Scaling effects in notched composites. J. Compos. Mater. 44 (2), 195–210.

Wright, W.W., 1981. The effect of diffusion of water into epoxy resins and their carbon-fibre reinforced composites. Composites 12 (3), 201–205.

Yan, L., Chouw, N., 2015. Effect of water, seawater and alkaline solution ageing on mechanical properties of flax fabric/epoxy composites used for civil engineering applications. Constr. Build. Mater. 99, 118–127.

Fire performance of maritime composites

Quynh Thuy Nguyen*, Phuong Tran[†], Xin Ren[‡], Guomin Zhang[†], Priyan Mendis*
*The University of Melbourne, Parkville, VIC, Australia, [†]Department of Civil and Infrastructure Engineering, RMIT University, Melbourne, VIC, Australia, [‡]Nanjing Tech University, Jiangsu, PR China

Chapter Outline

5.1 Introduction   115
5.2 Advanced polymer composites and design for maritime fire   117
    5.2.1 Use of advanced polymer composites in maritime structures   117
    5.2.2 Fire hazards for naval composite structures   118
5.3 Test methods and requirements for fire safety of maritime composites   126
    5.3.1 Fire testing of maritime composite structures   126
    5.3.2 Fire safety and protection requirements for ships and submarines   131
5.4 Fire reaction of maritime composites   132
    5.4.1 Pyrolysis reaction of composites   133
    5.4.2 Fire characteristics of composite materials for marine use   136
5.5 Structural performance of maritime composite during fire and postfire mechanical performance   141
    5.5.1 Structural responses of maritime polymer composites during fire   141
    5.5.2 Postfire mechanical properties of polymer composites   143
5.6 Numerical analysis of naval composite structure performance in fire   146
    5.6.1 Modeling of composite fire reaction: from bench-scale to full-scale testing   147
5.7 Enhancement of maritime composite structures subjected to fire   152
5.8 Conclusions   156
References   156
Further reading   159

5.1 Introduction

Most common books on polymer composite materials (PCMs) are essentially limited to the fabrication and mechanical performance of PCMs in relation to general structural applications. Since 1960s, PCMs have found a diverse range of applications from lightweight vehicles to large civil structures. PCMs, especially fiber-reinforced polymer composites (FRPCs), have replaced traditional

Marine Composites. https://doi.org/10.1016/B978-0-08-102264-1.00005-4
Copyright © 2019 Elsevier Ltd. All rights reserved.

materials such as aluminum alloys and steel at an impressive rate in the market of marine structures owing to their low density, low life cycle cost, outstanding thermal performance, high mechanical strength, and excellent corrosion resistance. The main drawback of FRPCs is their poor fire performance triggered by the combustible nature of the polymer matrix. Unlike civil applications, where comprehensive and well-structured safety standards have been developed, the current fire-related standards for PCMs in marine structures are very limited and complex. This chapter aims to provide insights into the fire performance of naval FRPC structures.

Artificially combined from different components such as the reinforcement and the polymer matrix, PCMs attain outstanding properties that cannot be achieved with an individual material. A representative example is high performance carbon fiber-reinforced polymer composites (Chung, 2010). Well-designed carbon FRPCs possess density lower than that of aluminum, stronger than high performance steel, and stiffer than titanium. What is more, carbon FRPCs offer high damage tolerance, corrosion/chemical resistance, electrical resistivity, and vibration damping ability. Recent developments in advanced PCM research have advanced PCMs to functionalized high performance application.

However, the combustibility of the composites still limits its use in marine structures. Some research projects on the fire behavior of marine composite structures have been conducted worldwide aimed at providing basic data on the safe use of this material. Overall, the research and basic understanding of composite structures in fire are limited compared to traditional materials, due to the difficulty in manufacturing and conducting fire tests. The developed design concepts and models are mostly limited to standard time-temperature exposure (e.g., ISO 834-1 and ASTM E119). Further need for studies of the fire behavior of marine composite structures is required, particularly with regard to the global fire behavior of composite structures, especially exposed to natural fires applicable to marine structures.

The impact of fire on offshore infrastructure is probably the most severe owing to its enormous harm to both life and economy. Offshore fire is destructive owing to the potential high temperature and the isolative manner of offshore structures. Fire and blast in the Piper Alpha platform in UK North Sea made 167 workers lost their life. The flame still continued over 17 hours after the incident (Dewhurst, 1997) and total lost were estimated to be $1.4 billion (Offshore Technology, 2014). Understanding the fire hazards in offshore environment and the behavior of composite in fire is ultimate to design passive/active fire protection for human and the structures.

This chapter provides a review on the behavior of maritime composite at elevated temperature, characterizations of fire occurring in marine environment, and safety codes of designing composite structures for marine applications. Work on simulation of composite materials in marine application is also reviewed to provide the insight of the materials' behavior subjected to fire. Finally, methods of improving the fire performance of maritime composite are presented.

5.2 Advanced polymer composites and design for maritime fire

5.2.1 Use of advanced polymer composites in maritime structures

5.2.1.1 Conventional materials for naval vehicles and offshore structures

Aluminum and alloys (such as Al–Cu and Al–Zn) have been widely applied as structural materials in maritime environment. 5xxx, 6xxx, and 7xx series aluminum alloys are most popularly used owing to their good strength and low density and fatigue resistance (Rosliza and Nik, 2010; Li et al., 2007). However, electrochemical reaction between aluminum alloys and seawater environment may result in inevitable corrosion to marine structures. Continuous or intermittent exposure of naval vehicles and structures to seawater or even mere moisture in the air can form marine corrosion (Nik et al., 2014). Several researches on inorganic and natural inhibitors to protect aluminum alloys from corrosive deterioration have been performed (Abdel-Gaber et al., 2008; Ezuber et al., 2008; Satapathy et al., 2009); however, further work should be done to address the toxicity, the environmental effect, availability, and cost-effectiveness.

In another application such as offshore platforms for oil and gas exploration, the increasing need of oil and gas supply has extended the industry from onshore and shallow waters to deep waters (Bhattacharyya et al., 2003). Oceania conditions set challenges of severe environmental loads to offshore platforms. Apart from dynamic loads of wind and wave and static structural-mass load, corrosion is one of the major causes of degradation during the service life of steel structural members in the platform. The deep draft of offshore platform element such as Buoyant leg structures even triggers the possibility of unusual and more critical corrosion (Chandrasekaran, 2016).

Mine countermeasure vessels (MCMVs) are used to locate and destroy sea mines. These were originally constructed from timber to work in areas protected by magnetic sea mines. The low availability of high-quality wood and high life cycle costs were the main limitations of timber-built mine countermeasure vessels. Naval superstructures made of aluminum alloy also experienced poor performance at elevated temperature and severe fatigue cracking which requires expensive repairs (Mouritz et al., 2001). FRP, on the other hand, has been a proven and widely used material for the construction of MCMVs for more than 30 years now, owing to its lightweight, nonmagnetic characteristics, and excellent shock resistance.

5.2.1.2 Advantages of advanced polymer composites for naval applications

Since early development of PCMs in the 1940s, FRPCs were considered limitedly to applications of small patrol boats (less than 10-m-long) operating in inland waterways or coastal waters; noncritical ship structures such as bow sonar domes and surveillance antennas owing to better acoustic transparency in comparison with traditional steel. Recent interest has changed to building larger offshore boat (over 50 m in length) from PCMs. The motivation of PCMs comes from their light weight (10% and 36% lighter than similar size aluminum or steel boat, respectively), lower maintenance cost, and fuel consumption with 7% life cycle cost lower than that of a comparable steel boat (Mouritz et al., 2001).

Other advantages of FRPCs in naval vessels are high strength-to-weight ratio, good damage resistance against underwater shock loading, excellent corrosion resistance, low infrared, magnetic signatures, and good noise dampening properties. Use of Kevlar fiber-reinforced composites in recent patrol corvette enhanced the resistance against high velocity impact. Sandwich structures with glass/carbon fiber-reinforced composite and thick poly (vinyl chloride) core are also applied in hull structure to utilize the high stiffness and strength of the skins as well as the high shear resistance of the foam core.

PCMs also start finding their applications as an alternative material to high corrosive and low radar signature steel truss masts. Early work on FRPC masts emphasized the advantages of 20%–50% reduction in weight, improved fatigue resistance, and better performance of mast sensor with no corrosion observed; however, FRPC mast are still 50% more expensive than those built of aluminum alloys. In a recent project of developing advanced enclosed mast/sensor by the US Navy, FRPC masts designed with advanced technology are proved to minimize corrosion, enhance the sensor performance, and lower radar cross section while providing weather protection for major antennas/sensors.

PCM applications are increasing rapidly in marine superstructures, submarines, propellers systems, or secondary structures of naval ships. A summary of these applications and advantages is given in Table 5.1.

5.2.2 Fire hazards for naval composite structures

Composite structures have experienced an increase in popularity during the recent few decades due to the superior properties and other credentials. However, the combustible nature of the polymeric resin makes composite flammable. Fire in maritime environment often occurs more severe than other inland conditions and potential higher damage. A fire broke out in the US Navy submarine docked in Miami in 2012, causing a damage of $400 million Fig. 5.1. Two out of five worst submarine disasters were triggered with maritime fire and others all involved secondary fire. The fire happened with the K-278 Komsomolets in 1989 sank the submarine which killed 42 of 69 crew members. The K-8 nuclear-powered attack submarine caught fire in 1970 and sank thereof with 52 crew members in heavy seas. In 2013, a fire broke out in a submarine of the Indian Navy at Mumbai harbor with 18 crew members trapped inside (Fig. 5.2). Secondary fire of the Deepwater Horizon platform in Gulf of Mexico, Louisiana, United States, sank the platform with 11 workers dead and 17 injured. This 2010 oil spill incident was considered as the largest environmental disaster in the US history (Fig. 5.3).

Table 5.1 Advantages of PCMs in maritime structures

Maritime structures	Advantages	References
Composite superstructure–La Fayette frigate (French Navy) and hulls of mine counter measures vessels	• Lower fatigue cracking as FRPC yield strain is 10 times higher than that of steel • 15%–70% lighter than similar size steel structure • Low maintenance/repair cost and high ship availability	(Smith, 2001), (Critchfield et al., 1991, 1994; Dow, 1995; Hoyning and Taby, 2000)
Composite propellers (compared to propellers made from nickel–aluminum–bronze alloys)	• Weight savings, reduced corrosion, improved fatigue performance • 25% reduction in magnitude of resonance vibrations in engines • 50% higher maximum in-plane bending and shear stresses • Lower life maintenance cost	(Mouritz et al., 2001)
Composite propulsors (against metal propulsors)	• Lower life cycle cost • Reduced mass and magnetic/electric signature • Improved corrosion resistance • Better fatigue performance • Reduced radiated noise	(Naval Technology, 1991)
Composite propulsion shafts (compared to steel shaft of similar size)	• 25%–80% weight saving • Lower noise transmissions from machinery • Reduced acoustic signatures • Less issues with corrosion and fatigue load-bearing • 25% lower life cycle cost • 40%–70% lighter	(Zimmerman, 1997) (Caplan, 1993)

Continued

Table 5.1 Continued

Maritime structures	Advantages	References
Naval ship secondary structures of machinery and engine components (compared to same size metal ones)	• 10%–40% reduction in engine acquisition cost • 5–20 dB lower structural/air-borne noise • Lower electromagnetic signature • Improved resistance to fatigue and corrosion	
Composite bow planes, fins, and rudders in submarine control (compared to aluminum parts)	• 50% lighter • 23% cheaper • Superior hydrodynamic and acoustic stealth performance • Feasible to be molded with complex shape with machining • Excellent corrosion resistance • Good shock resistance	(Smith, 1999)
Composite submarine masts (compared to steel masts)	• Lighter weight • No corrosion • Complex shape manufacturing without machining • Possibility to incorporate radar absorbing materials	(Mouritz et al., 2001)

5.2.2.1 Fire design: Background

Fire safety engineering can be defined as the application of scientific and engineering principles to the effects of fire in order to reduce the loss of life and damage to property by quantifying the risks and hazards involved and to provide an optimal solution to the application of preventive or protective measures. Fire safety is assured by using a performance-based fire protection approach.

The SFPE Engineering Guide to Performance-Based Fire Protection (2015) defines performance-based design as "an engineering approach to fire protection design based on (1) agreed upon fire safety goals and objectives, (2) deterministic and/or probabilistic analysis of fire scenarios, and (3) quantitative assessment of design alternatives against the fire safety goals and objectives using accepted engineering tools, methodologies, and performance criteria."

Fire performance of maritime composites

Fig.5.1 Fire broke out in US Navy submarine in May 2012 (Miami).
Source: AP.

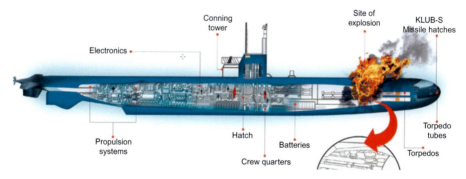

Fig. 5.2 A huge fire occurred in an Indian submarine in 2013 (Mumbai).
Source: theguardian.com

Performance-based design of a marine structure would involve the following steps:

- Identify goals,
- Identify the possible fire scenarios,
- Evaluate the likelihood and consequences of such scenarios,
- Establish appropriate performance criteria,
- Ensure effective evacuation, escape, and rescue,
- Predict the performance of the system based on available engineering data,
- Ensure robustness of the designs and reliability and durability of the protection system.

All fires in a marine environment could start with an event, such as an electrical failure, equipment failure such as rupture of hydrocarbon containing equipment, or

Fig. 5.3 Deepwater Horizon offshore platform caught on fire in Louisiana (2010).

smoking materials discarded inappropriately. Once the fire starts, it will grow in size, spread to other items, activate fire detection or suppression systems, or self-extinguish. Under certain conditions, structural failure or collision impact may also lead to fires.

Each combination of possible sequences of actions is a fire scenario. A hazards analysis would investigate the consequences of possible scenarios. A risk analysis would analyze the probability of the scenarios occurring and their consequences. For example, in the offshore risk assessment, usually two types of fire risk are considered: the topside fire and the fire on sea. In both cases, consequences could be a measure of threat to life, property loss, business continuity, or damage to the environment. Their final consequences are largely dependent on the escalating sequences.

5.2.2.2 Fire safety strategies

There is no doubt that fire safety has a significant influence on the overall design and the performance of the marine structure. Fire safety is a multidisciplinary task for the engineering profession. To conduct a fire safety analysis of a structure, a combination of different fields of knowledge needs to be used, such as physics, chemistry, fluid dynamics, material sciences, structural behavior, psychology, and toxicology, among other (Buchanan, 2001a).

Fire safety can be categorized in two ways: active fire protection (AFP) and passive fire protection (PFP). The AFP system controls a fire at any stage of its development, whether caused by a human or an automatic device. However, PFP systems do not involve any specific action at any stage of the fire development process. PFP is part of the fabric of the marine structure and protects its structural stability. A fire safety strategy is classified into several levels of management, as presented in Fig. 5.4.

5.2.2.3 Periods of fire development

Fire is defined as exothermic oxidation of a combustible substance, releasing heat and light in the form of a flame. The temperature vs. time relationship, or the process of fire development, can be represented by a curve as shown in Fig. 5.5. For most fires, the design fire curve is broken down into six main sections: Incipient period and

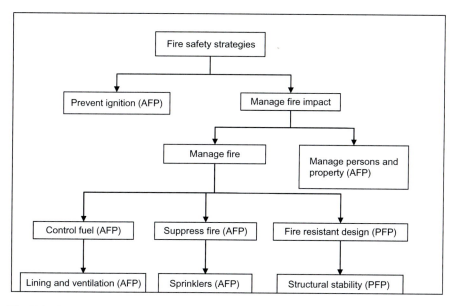

Fig. 5.4 Fire safety strategy plan for marine structures (Buchanan, 2001a).

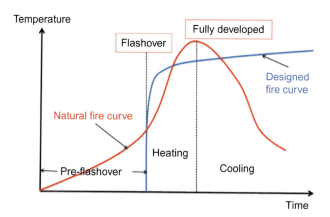

Fig. 5.5 Typical periods of fire development.

Ignition, growth, flashover, burning, fully developed, and decay. Each of these sections represents one particular aspect of the fire.

5.2.2.3.1 Incipient period and ignition

The term "incipient period" refers to the period before ignition of the fire. Fires in a marine structure can occur for various reasons, as described earlier. During the incipient period, fires can be prevented by an active fire control system, such as smoke detectors and alarms. The incipient period can be small or large, depending upon the temperature required for ignition of the fuel/combustible material (Buchanan, 2001a, b). Ignition is the point at which fuel, oxygen, and heat combine to begin combustion. During this stage, the heat release rate is negligible. For most deterministic approaches, ignition is just assumed to occur.

5.2.2.3.2 Growth period

The "growth period" is defined as the period before the flashover (defined below). During this stage, the fire ignites and starts to burn the combustible materials in its vicinity, reaching a larger scale. The amount of fuel/combustible material in the vicinity of the ignition determines the growth period. The growth period can be identified by the heat and smoke detectors and the fire can be controlled with human intervention in the form of fire extinguishers. Automatic sprinkler systems, which are considered to be active control measures, can also play an important role in preventing flashover. If sprinklers do not control the fire, further fire-fighting actions are required. As a PFP measure, it is recommended that noncombustible materials are used to build the marine structure in order to reduce the chance of a fire reaching the flashover point.

5.2.2.3.3 Flashover period

The "flashover period" is the shortest period in the whole fire development process. Flashover occurs when the active control measures (such as sprinkler systems) fail to operate during the growth period. During the growth period, noncombustible materials in the structural elements absorb heat energy, causing the overall temperature of the compartment to reach auto-ignition levels. At this stage, as a consequence of the flashover, most of the combustible materials in the compartment will ignite. Flashover can only be controlled by fire-fighting systems that are operated from outside the compartment (Buchanan, 2001b).

5.2.2.3.4 Burning period

Following the flashover period is a "burning period". This period reaches the highest temperatures of the fire development process, such that almost all of the combustible materials in the compartment are burnt. The length of the burning period is determined by the amount of fuel available and the ventilation. Like the flashover, fires during the burning period can only be controlled by fire-fighting measures operated from outside the compartment. During this period, only ventilation control or PFP measures can save the structure (Buchanan, 2001b).

5.2.2.3.5 Fully developed
Once the fire stops growing, it transitions to the third stage of a fire curve: the fully developed region. At this point, the heat release rate plateaus and will remain at this level until the fire grows again or starts to decay. How long it remains at this level depends on the amount of fuel remaining, ignition of other fuels, ventilation conditions, and initiation of manual or automatic suppression activities.

5.2.2.3.6 Decay period
The "decay period" begins after the burning period. This period relates to the availability of fuel around the fire. Usually, the decay period is a reverse of the process leading up to the burning period, that is, the flashover, growth, and incipient periods (Buchanan, 2001b). The fire may change from ventilation-controlled to fuel-controlled at the onset of this stage. This stage always contains the decline of the fire, but is not limited to just the end of a fire. Many fires are dynamic and have the ability to grow and decay multiple times due to the amount of fuel remaining, ignition of other fuels, changing ventilation conditions, and initiation of manual or automatic suppression activities.

5.2.2.4 Design fire curves
Standard fire curves are designed to determine the fire resistance of structural elements. The major fire curves are plotted graphically and then described mathematically as follows (see Fig. 5.6): Compared to a standard fire, hydrocarbon fire creates a significant increase in temperature in the initial period of the fire (i.e., it reaches 1000°C in only 8 minutes). As a result, hydrocarbon fire can lead to severe damage to the structure. Thus, the behavior of marine structures subjected to hydrocarbon fire is a very important issue which requires urgent research effort.

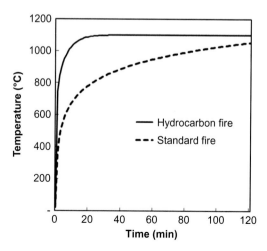

Fig. 5.6 Standard and hydrocarbon fire curves used for fire safety design.

5.2.2.4.1 Standard fire (ISO 834, 1975)

The ISO 834 (ISO 834, 1975) fire curve is the most popular fire curve for predicting the fire resistance of structural elements. Most standards around the world follow this time–temperature relationship, including the Australian Standard given in AS-1530.4 (1997). Mathematically, the ISO 834 fire curve is defined via the following Eq. 5.1:

$$T = T_0 + 345 \log 10 \left(8t + 1\right) \qquad 5.1$$

where t is time in minutes, $T0$ is ambient temperature (degrees centigrade), and T is temperature (degrees centigrade).

5.2.2.4.2 Hydrocarbon fire curve

The hydrocarbon fire curve is defined as follows:

$$T = T_0 + 1080\left(1 - 0.325 e^{-0.167t} - 0.675 e^{-2.5t}\right) \qquad 5.2$$

where t is time in minutes, $T0$ is ambient temperature (degrees centigrade), and T is temperature (degrees centigrade).

Among these fire curves, the hydrocarbon fire curve is the most frequent for marine structures. It is utilized in the design of members that are exposed to the burning of hydrocarbon products. In this case, the temperature rise is significantly more rapid when compared to a standard fire, such as those represented by the ISO 834.

5.3 Test methods and requirements for fire safety of maritime composites

5.3.1 Fire testing of maritime composite structures

5.3.1.1 "Room corner" test ISO 9705

ISO 9705 (ISO 9705-1:2016, 2016) describes a full-scale fire test of a room closure to evaluate both fire spread and smoke growth. The internal dimensions of the testing enclosure are 2400 × 2400 × 3600 mm. The enclosure is designed with a doorway of 800 × 2000 mm. A 170 × 170 mm propane gas burner located in a corner opposite to the doorway is utilized as the ignition source. The burner provides a heat of 100 kW in the first 10 minutes and 300 kW in the next 10 minutes. Parameters such as fire growth index (FGI) and smoke growth (SMOGRA) are calculated to evaluate the fire spread and smoke growth of the tested product as follows:

$$\mathrm{FGI}_{\mathrm{av}}(t) = \begin{cases} 0 & t < t_{\mathrm{ig}} \\ \overline{\mathrm{FGI}}\left[t_{\mathrm{ig}}..t\right] & t_{\mathrm{ig}} \leq t \leq t_{\mathrm{ig}} + 2 \\ \mathrm{FGI}_{6\mathrm{s}}(t) & t > t_{\mathrm{ig}} + 3 \end{cases} \qquad 5.3$$

The 6 s average of fire growth index $\mathrm{FGI}_{6\mathrm{s}}(t)$ is then calculated as:

$$\mathrm{FGI}_{6s}(t) = \frac{0.5\mathrm{FGI}(t-3) + \mathrm{FGI}(t-2) + \ldots + \mathrm{FGI}(t+2) + 0.5\mathrm{FGI}(t+3)}{6} \qquad 5.4$$

where fire growth at time t is given as:

$$\mathrm{FGI}(t) = \frac{d\mathrm{THR}(t)/dt}{t} \qquad 5.5$$

$$\mathrm{SPR}_{av}(t) = \begin{cases} 0 & t < t_{ig} \\ \overline{\mathrm{SPR}}\,[t_{ig}..t] & t_{ig} \leq t \leq t_{ig}+2 \\ \mathrm{SPR}_{6s}(t) & t > t_{ig}+3 \end{cases} \qquad 5.6$$

The 6 s average of fire growth index $\mathrm{SPR}_{6s}(t)$ is then calculated as:

$$\mathrm{SPR}_{6s}(t) = \frac{0.5\mathrm{SPR}(t-3) + \mathrm{SPR}(t-2) + \ldots + \mathrm{SPR}(t+2) + 0.5\mathrm{SPR}(t+3)}{6} \qquad 5.7$$

where fire growth at time t is given as:

$$\mathrm{SPR}(t) = kV_s(t) \qquad 5.8$$

$$\mathrm{SMOGRA} = 10,000 \frac{\mathrm{PeakSPR}_{av}}{t}, \mathrm{m^2/s^2} \qquad 5.9$$

where t is the time from the ignition of the burner, k is the optical density, and $V_s(t)$ is the volume flow rate measured at duct gas temperature.

Safety criteria recommended by the US Navy (Sorathia et al., 2001) for fire spread include peak of heat release rate (PHRR) less than 500 kW over any 30 s period of the test, the average heat release rate (HRR) less than 100 kW with no dripping, and observed flame spread not reaching 0.5 m above the floor (exception to the area of 1.2 m from the burning source). The requirements for smoke growth are peak of smoke production rate (PSPR) less than 8.3 m²/s over any 60 s period of test with average SPR less than 1.4 m²/s.

The full-scale room corner test is often applied in accessing the fire safety of materials owing to its ability to replicate the real fire conditions, especially flashover behavior of composite structures (Mouritz and Gibson, 2006). However, the high cost of the test is the major reason for its limited use.

Numerical works are being used to predict the behavior of composite structure in ISO 9705 and will be discussed further in Section 5.4.

5.3.1.2 Cone calorimeter test ISO 5660

The Cone calorimeter test according to ISO 5660 (ISO 5660:2015, 2015) is the most versatile bench-scale fire to access the burning behavior of composite materials. Similar standards for Cone calorimeter have been developed such as ASTM

E1354 (Chandrasekaran, 2016)—North America, BS 476.15—United Kingdom, AS/NZ 3837:2016—Australia and New Zealand. Cone calorimeter test can be used to accurately determine a wide range of material reactions to fire from heat release to smoke production rate/obscuration (refer to Eq. 5.10). The sample size is 100 × 100 mm which is exposed to the heat flux provided by a Cone-shaped heater. The heat flux can vary from 0 to 100 kW/m² and samples can be tested horizontally or vertically.

The heat release rate is calculated as shown in Eq. 5.10.

$$\dot{q}_A(t) = \frac{\dot{q}(t)}{A_S} = \frac{1}{A_S} \cdot \frac{\Delta h_C}{r_0} \cdot 1.10 C \sqrt{\frac{\Delta p}{T_e}} \cdot \frac{X_{O_2}^0 - X_{O_2}}{1.105 - 1.5 X_{O_2}} \qquad 5.10$$

where A_S initial area of the sample = 0.0088 m, $\frac{\Delta h_c}{r_o}$ corresponding value of the specimen (often taken as 13.1 × 103 kJ/kg or as stated for each material), 1.10 is ratio of the molecular masses of oxygen and air, C is calibration constant (determined from previous calibration tests) T_e is absolute temperature of gas at the orifice meter [K], Δp is orifice meter pressure differential [Pa], $X_{O_2}{}^0$ is initial value of oxygen analyser reading in the baseline measurement, and X_{O_2} is oxygen analyser reading (mole fraction of oxygen).

The cone calorimeter collects data from the smoke measurement systems and calculates the smoke production rate as follows:

$$P_{s,A} = \frac{P_s}{A}$$

where $P_{s,A}$ is the smoke production rate per unit area of exposed specimen, A is the exposed surface area of the test specimen, and P_s is the smoke production rate, which is calculated from the extinction coefficient as follows:

$$P_s = k \cdot \dot{V}_s$$

where $\dot{V}_s$ is volumetric flowrate at the smoke meter and k is extinction coefficient, defined as:

$$k = \ln\left(I_0/I\right) L^{-1}$$

where I_0 is beam intensity in the absence of smoke, I is attenuated beam intensity, and L is optical path length across the exhaust duct.

The small size sample and low heat flux do not allow to investigate realistic behavior of composites in large-scale fire. For instance, flashover point is often achieved at heat flux of 125–150 kW/m² (Wickström and Göransson, 1992). Results from Cone Calorimeter test otherwise are useful to effectively compare different composite materials and to generate inputs for numerical models.

5.3.1.3 ISO 1182 for noncombustible materials

ISO 1182:2010 (ISO 1182:2010, 2010) defines the test method to access the combustibility of composite materials used in maritime applications. The purpose of the test is to determine whether the material produces limited heat and volatile vapor when exposed to 750°C. This test method is limited to homogeneous material which means each member of the layered/laminated composite must be tested individually. The test specimen is tailored to the cylindrical shape with diameter of 45 mm and height of 50 mm. The specimen is tested in a vertical cylindrical furnace maintained at 750° C. Temperature at the surface and the centre of the specimen is recorded. The composite material will be accepted if it meets both of the following requirements:

- No sustainable flame lasting 5 s or longer observed
- Temperature rise of the specimen not exceeding 50°C at the end of the test determined by the point where temperature equilibrium between two thermocouples is established

5.3.1.4 Other applicable tests

The flame resistance test ASTM E119 is applied to large components of ships and offshore building. Other tests including surface flammability ASTM E84 and Smoke toxicity ASTM E662 are also intermittently applicable for maritime composites.

ASTM E119, also known as AS1530.4-2005 (Mouritz et al., 2004), specifies a testing method for an element when exposed to fire. In most cases, a standard fire curve is applied, whereas the temperature curve relevant to hydrocarbon fire is applied in special cases when there is higher risk of severe fire (such as in tunnels or with the presence of fuel and petrol). Three criteria of the material in terms of fire retardancy are identified as: insulation, integrity, and structural stability.

Insulation is the ability of the material to maintain an average temperature rise of less than 140°C and a temperature rise of less than 180°C at any recorded point on the unexposed surface of the element.

Integrity is the ability of the element to prevent flame and hot gases from passing through its depth to the unexposed surface. The integrity is measured using cotton pads or gap gauges. However, 'requirements for integrity are achieved as long as the tested element meets the requirements of insulation conditions.

Structural stability or load-bearing capacity reflects the ability of the element to maintain its load-carrying function when subjected to fire. The single burning item (SBI) test was used to evaluate the fire performance of the polymer composite panel in offshore buildings. The fire behavior of the specimen is assessed through the heat and smoke released from a specimen exposed to a propane gas burner of constant heat flux. The performance is evaluated over the first 20 minutes of exposure. The SBI tests are arranged in accordance with the testing standard EN 13823:2010 (EN 13823:2010, 2010). The standard specimen has two wings of nominal dimensions (length × height × thickness) of 500 × 1500 × 115 mm (short wing) and 1000 × 1500 × 115 mm (long wing), respectively. The wings are tested in a freestanding position with a standard mounting on the trolley. Pictures of the SBI

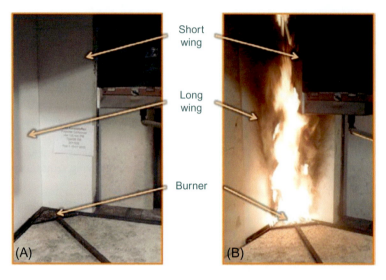

Fig.5.7 SBI: (A) setup, (B) testing of GFRP composite panels. *Arrowed lines* indicate the location of the burner and long and short wings of the specimen with dimensions specified by the fire testing standard EN 13823:2010.

experimental setup and the test are shown in Fig. 5.7, with arrowed lines indicating the wing's and the burner's locations.

Heat release from the specimen is evaluated through the heat release rate average (HRRav), total heat release (THRta), and fire growth rate index (FIGRAta). HRRav is calculated as the average value of the recorded HRR in 30 consecutive seconds. This parameter reflects the speed of the burning process of the material under tested conditions, while THRta shows the overall heat released into the surroundings. Parameter THRta is important when considering the influence of combustion on the occupants' safety and the damage of other nearby infrastructure. THRta is calculated for the first 600 s after exposed to the propane burner. FIGRAta is another proposed parameter in this study, which is proportional to the peak of HRRav and inversely proportional to the time to reach this peak. FIGRAta is meaningful in designing construction materials as it indicates how fast the combustion of the material develops, and therefore, the maximum time for safe evacuation of the occupants.

Other results obtained from the SBI experiment are the average smoke production rate (SPRav), total smoke production (TSPta), and smoke growth rate index (SMOGRAta). The smoke production rate is calculated from the total smoke production rate SPRtotal, which is computed from recorded values of volume flow in the exhaust duct (m3/s), and the average smoke production rate of the burner SPRav-burner computed from average SPRtotal(t) during the base line period. The equation to calculate SPR is given below:

$$\text{SPR}(t) = \begin{cases} \text{SPR}_{\text{total}} - \text{SPR}_{\text{av\_burner}} & t > 312\,\text{s} \\ 0\left(\text{m}^2/\text{s}\right) & t = 300\,\text{s} \\ \max\left[0, \text{SPR}_{\text{total}}(t) - \text{SPR}_{\text{av\_burner}}\right] & 300\,\text{s} < t \leq 312\,\text{s} \end{cases} \quad 5.11$$

where the average smoke production rate of the burner SPRav_burner is given as:

$$\text{SPR}_{\text{av\_burner}} = \overline{\text{SPR}}_{\text{total}}(210\,\text{s}\ldots270\,\text{s}) \qquad 5.12$$

The average smoke production rate (SPRav) is then calculated as:

$$\text{SPR}_{\text{av}}(t) = \begin{cases} 0\,(\text{m}^2/\text{s}) & t = 300\,\text{s} \\ \overline{\text{SPR}}[300\,\text{s}\ldots(2t-300\,\text{s})] & 300\,\text{s} < t < 330\,\text{s} \\ \text{SPR}_{60\text{s}}(t) & t \geq 330\,\text{s} \end{cases} \qquad 5.13$$

Here, SPR60s is the 60 s average of SPR and is given as:

$$\text{SPR}_{60\text{s}}(t) = \frac{0.5\text{SPR}(t-30\,\text{s}) + \text{SPR}(t-27\,\text{s}) + \ldots + \text{SPR}(t+27\,\text{s}) + 0.5\text{SPR}(t+30\,\text{s})}{20}$$

$$5.14$$

The total smoke production, TSPta, is calculated directly from the smoke production rate as:

$$\text{TSP}(t_a) = 3\sum_{300\,\text{s}}^{t_a} \max[\text{SPR}(t), 0] \qquad 5.15$$

Finally, the smoke growth rate index is calculated from the average smoke production rate as follows:

$$\text{SMOGRA}(t) = 10^4 \cdot \max\left(\frac{\text{SPR}_{\text{av}}(t)}{t - 300\,\text{s}}\right) \qquad 5.16$$

5.3.2 Fire safety and protection requirements for ships and submarines

5.3.2.1 Codes by international maritime organization

Fire safety of maritime composites is well-covered in international maritime organization (IMO) codes (Mouritz et al., 2006); however, IMO codes are referencing regulations for IMO and associate members. The High-Speed Code published by IMO determines the use of combustible composites through the "room corner" test ISO 9705, which is described in the previous section. The composite should also be tested in their end-use condition under the ASTM E119. The High-Speed Code requirements for passenger ships and other surface high-speed craft include:

1. 1 h of the insulation rating according to ASTM E119 (referred to Section 5.3.1 for the definition of insulation criteria)

Table 5.2 **Assessment of fire safety criteria used in US Naval structures (Mouritz and Gibson, 2006)**

Properties	Safety criteria
Fire growth (ISO 9705)	Heat release rate including $HRR_{av} \leq 100$ kW/m^2 and $HRR_{max} \leq 500$ kW/m^2
	Downwards flame spread rate not exceeding 0.5 m
Smoke production (ISO 9705)	Smoke production rate including $SPR_{av} \leq 1.4$ m^2/s and $SPR_{max} \leq 8.3$ m^2/s
Fire resistance (ASTM E119)	1 h rating of insulation and integrity criteria, further requirements on structural integrity under fire applied
Smoke toxicity (ASTM E662)	CO not exceeding 350 ppm, HCl not exceeding 30 ppm, HCN not exceeding 30 ppm, and fire gas IDLH* index not >1
Surface flammability (ASTM E84)	Flame spread index: 25 for both interior and exterior applications
	Smoke developed index: 15 for interior application

*Immediately Dangerous to Life or Health.

2. Heat release rate including $HRR_{av} \leq 100$ kW/m^2 and $HRR_{max} \leq 500$ kW/m^2 according to ISO 9705 (referred to Section 5.3.1 for the determination of HRR)
3. Smoke production rate including $SPR_{av} \leq 1.4$ m^2/s and $SPR_{max} \leq 8.3$ m^2/s according to ISO 9705 (referred to Section 5.3.1 for the determination of SPR)
4. Downwards flame spread rate not exceeding 0.5 m and no flaming debris further than 1.2 m from the burner

5.3.2.2 Other codes and regulations for naval ships and offshore structures

The Code of Federal Regulation covers the safety requirements of passenger ships, cargo vessels, shipbuilding materials, and offshore platforms (Mouritz and Gibson, 2006). The main requirement is the noncombustibility or equivalent which is described in Section 5.3.1.3. Another regulation issued by the US Coast Guard approves the exemption of combustible composite with low flame spread rate for certain vessels (Mouritz and Gibson, 2006). The US Navy also has several criteria for warships and submarines (Sorathia et al., 2001). This performance-based regulation accesses several fire properties of materials and structures including fire growth (ISO 9705), smoke production (ISO 9705), fire resistance (ASTM E119), smoke toxicity (ASTM E662), and surface flammability (ASTM E84) as shown in Table 5.2.

5.4 Fire reaction of maritime composites

Maritime composites often involve heavy sandwich panels for structural applications, thus a good insight on the fire reaction of each component helps to complete the understanding of the composite sandwich panels. In this section, the decomposition of composites at elevated temperature is studied through important characteristics of marine composites exposed to fire such as heat release rate and smoke production.

5.4.1 Pyrolysis reaction of composites

5.4.1.1 Behavior of matrix at elevated temperature

Organic-based polymer matrix is the key component that makes composites susceptible to high temperature exposure. In this section, thermo-setting resins are reviewed more extensively owing to their abundant application in maritime application. Polymer matrix is generally flammable; however, the response of the matrix subjected to fire is influenced by several processes from phase changes to material property variations (Goodrich and Lattimer, 2012). Goodrich and Lattimer have investigated the thermo-gravimetric response of E-glass vinyl ester composites when exposed to elevated temperature up to 600°C with heating rate of 5–40°C/min. The decomposition of the thermo-setting-based composite was found to be mainly governed by behavior of the vinyl ester matrix. As can be seen from Fig. 5.8, the resin started to decompose at 180°C; however, the main pyrolysis reaction occurs at the temperature range of 380–480°C. A real-time study on the morphology evolution of the resin was also conducted. Pores of 5–10 μm were observed as a result of the pyrolysis reaction and the diameter was increased to 20–30 μm with the development of the decomposition.

The pyrolysis reaction of unsaturated polyester resin is also studied without the presence of reinforced fiber (Nguyen et al., 2014). The pyrolysis reaction is quantified by mass fraction of the unsaturated polyester during the burning process and the reaction rate which is calculated by the derivation of mass residue against temperature (Eq. 5.17).

$$\left(\frac{dM}{dT}\right)_i = \frac{M_{i-1} - M_{i-1}}{T_i - T_{i-1}} \qquad 5.17$$

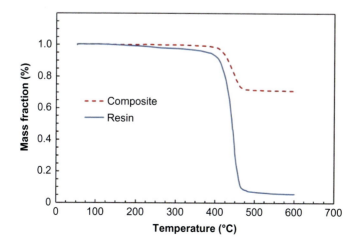

Fig. 5.8 Thermo-gravimetric response of E-glass vinyl ester composites at the heating rate of 20°C/min (Goodrich and Lattimer, 2012).

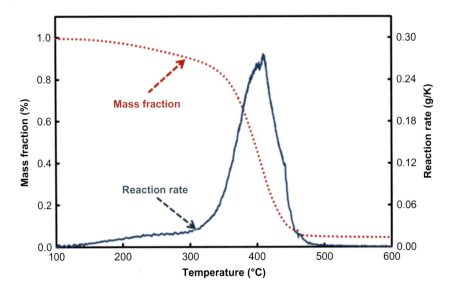

Fig. 5.9 Time history of relative mass reduction and the associated reaction rate of unsaturated polyester resin (Nguyen et al., 2014).

where *i* is the calculated point, *M* is the mass of the sample, and *T* is the absolute temperature of the sample. As can be seen from Fig. 5.9, the thermo-gravimetric analysis of unsaturated polyester resin shows one reaction. The pyrolysis reaction starts at around 300°C and reaches its maximum state at 415°C. The degradation reaction finishes at over 450°C.

The pyrolysis reaction of thermo-setting resins was comprehensive compared when testing up to 800°C at a heating rate of 25°C/min (Mouritz et al., 2006). This endothermic decomposition behavior of thermo-setting reflects the stability of the resin when exposed to the high temperature of the fire, but also has significant impacts on the heat release rate which will be discussed in Section 5.4.2. Polyester, vinyl ester, and epoxy have similar pyrolysis reactions from 350°C to 480°C, while phenolic resin decomposes mainly at 400 – 500°C. The high amount of char residual (around 60% wt.) after decomposition in phenol makes it less volatile than other thermo-settings. An early decomposition was observed with phenolic resin as a result of water evaporation. This early decomposition has a close relationship with the early deterioration and delamination of phenolic strength at this 100–200°C range (Michael Davies et al., 2006) (please refer to Section 5.5 for more details).

Michael Davies et al. (2006) summarized the behaviors of polymeric resin as the main agent triggering the changes of fiber-reinforced composite when exposed to elevated temperature. As can be seen from Fig. 5.10, the pyrolysis reaction of resin takes

Fig. 5.10 Behaviors of polymeric resin at elevated temperature.

place after initial stages where the heat is transferred by conduction through the thickness of the composite. The pyrolysis is a chemical reaction of the resin producing combustible gases and char. The formation of char is significantly dependent on the heat gradient and resin characteristics (Mouritz et al., 2006), and thus, the insulative effect and the thickness of the newly formed char layer vary accordingly. Michael Davies et al. (2006) also mentioned an important behavior of resin at high temperature of about 1000°C, which occurs in severe fire for maritime application. It is called carbon-silica reaction where the high carbon char chemically reacts with silica filler following the oxidation of char residues and burnt-out process of the reinforced glass fiber.

5.4.1.2 Behaviors of reinforcing fiber at high temperatures

The most common reinforcing fiber for maritime composites is E-glass which remains stable during standard fire test condition, while the polymeric resin burns out owing to the pyrolysis reaction (Fig. 5.11). Goodrich and Lattimer (Goodrich and Lattimer, 2012) used an environmental scanning electron microscope (ESEM) to capture real-time behavior of the composite at high temperature. Fig. 5.12 shows the image of the E-glass/vinyl ester composite laminate at 548°C with no melting/deformation of the fiber reinforcement observed.

As mentioned in the previous section (Section 5.4.1.1), the decomposition of the composite depends mostly on that of the polymeric resin. On the other hand, the thermal properties of the composite at high temperatures in turn are influenced significantly by the properties of the reinforcement fiber as the thermal characteristics of thermo-setting resins have minor dependence on temperature (Dodds et al., 2000). Thermal properties of the composite such as thermal conductivity and specific heat capacity are then calculated as in Eq. 5.18 and Eq. 5.19:

$$\frac{1}{k_{com}} = \frac{V_f}{k_f} + \frac{V_r}{k_r} \qquad 18$$

$$C_{com} = \frac{C_f \rho_f V_f + C_r \rho_r V_r}{\rho_f V_f + \rho_r V_r} \qquad 19$$

where k_{com}, k_f, k_r are the thermal conductivity of the composite, glass fiber, and resin, respectively, V_f and V_r are the volume fraction of the fiber and resin in the composite, respectively, C_{com}, C_f, C_r are the heat capacities of composite, fiber, and resin, respectively, ρ_f and ρ_r are the density of fiber and resin, respectively.

The behavior of other organic fiber such as polyethylene and aramid is also investigated (Mouritz et al., 2006). While the temperature range of polyethylene fiber decomposition is similar to that of thermo-setting resins, the aramid fiber's pyrolysis reaction undergoes multiple stages from 100°C to 575°C. The first mass loss appears at the low temperature range and is explained by the evaporation of moisture absorbed

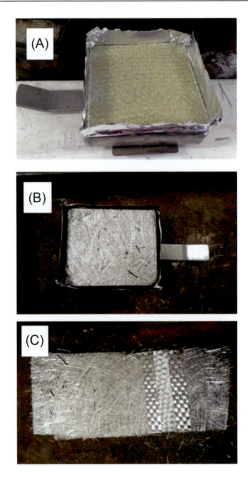

Fig. 5.11 Photos of glass fiber-reinforced composite (A) before and (B) after exposed to a heat flux of 50 kW/m^2, (C) layers of glass fiber being stable after exposure.

by aramid fiber. The second reaction accounts for the main pyrolysis decomposition and produces about 40%wt. of char.

5.4.2 Fire characteristics of composite materials for marine use

The pyrolysis reaction of composite is often studied by thermo-gravimetric analysis where the applied heat is relatively low and ideal in nitrogen atmosphere. The temperature range of decomposition provides essential information on the stability of the material at elevated temperature. However, real fire, especially in maritime applications, often involves high heat flux from combustion reaction of the composites with the presence of oxygen. The combustion of polymeric resin needs to be quantified by other parameters such as heat release rate, smoke production rate, and ignition point. Section 5.4.2 will focus on these characteristics of FRP composites.

Fire performance of maritime composites

Fig. 5.12 ESEM image of composite laminate at 548°C.

5.4.2.1 Ignitability

Ignitability or, quantitatively, the ignition time of composites reflects the ability of materials to resist from burning with sustainable flame when exposed to high temperature. Ignition time is calculated as the time from the start of high heat exposure to the point when composite burns. Egglestone and Turley conducted tests on the ignition of glass fiber-reinforced composites used in ship superstructures (Egglestone and Turley, 1994). Four types of composites were tested at five different heat fluxes from 25 kW/m^2, which is equivalent to a single burning item to a fully evolve fire 80 kW/m^2. It can be clearly seen from Fig. 5.13 that phenolic resin has excellent performance with the double time to ignition in comparison with other resins (polyester and vinyl ester). The magnitude of the applied heat flux also shows a prevalence effect on the time to ignition of all the tested resins. When the heat flux doubled from 25 to 50 kW/m^2, the time to ignition of all resins including phenol reduced to one third or one quarter. This shows the importance of fire conditions on the composite performance, especially when fire is often severe in maritime applications. A comprehensive data on the ignition time of composites, polymeric resins, and combustible fibers was also generated by Mouritz et al. (Mouritz et al., 2006), which can be used as a reference for selecting/designing of maritime composites.

5.4.2.2 Heat release rate

Heat release rate of materials subjected to a certain heat flux is recognized as the most important parameter to access the material combustion properties. Heat release rate is measured in most common fire tests from bench-scale (the cone calorimeter) to

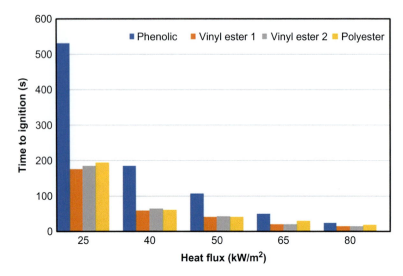

Fig. 5.13 Dependence of time to ignition on heat flux (Egglestone and Turley, 1994).

full-scale (single burning item test) ones. If composite absorbs energy in its pyrolysis reaction to produce volatile gases, the combustion of these gases releases a large amount of energy into the fire and in turn accelerates the burning process.

Heat release rate (HRR) is often measured as the energy produced per unit area of composite materials when exposed to fire/certain heat fluxes. There are two derivative parameters from heat release rate which are peak of heat release rate (PHRR) and total heat release rate (THR). PHRR and sometimes the time to reach PHRR are used to evaluate the highest heat flux that the structure will have to endure when composite combustion occurs. THR, on the other hand, shows the final heat contribution of the materials to the fire load.

A complete HRR curve of composite laminates includes four stages of initial induction, surface combustion, char formation, and resin depletion (Mouritz and Gibson, 2006). The PHRR is often achieved between surface combustion and char formation stages. A typical HRR and THR curve of polyester/glass fiber-reinforced composites exposed horizontally to a 50 kW/m^2 heat flux in cone calorimeter test is illustrated in Fig. 5.14 (referred to Section 5.3 for the test description).

The result is calculated from the data achieved from at least three runs with the fluctuation of no more than ±5%. The 100 × 100 mm composite samples are burnt mainly in the first 350 s of the experiment. HRR curve reaches the peak at 270 kW/m^2 after 131 s from the beginning of the burning test. The average THR of composite samples was 47.3 MJ/m^2 with effective heat of combustion of 17.9 MJ/kg.

Fig. 5.15 shows typical curves of HRR and THR of polyester glass fiber-reinforced sandwich panel exposed to a single burning item test (referred to Section 5.3 for details of the test) (Nguyen et al., 2014). This full-scale test was performed in accordance to EN 13823 and EN 13501 where another derivative parameter, FIGRA, is calculated to evaluate the intensity of the composite combustion. The calculation of this parameter

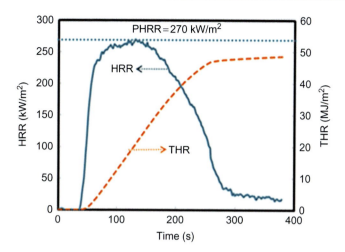

Fig. 5.14 Time evolution of Heat release rate (HRR), total heat release (THR) of the composite in cone calorimeter under 50 kW/m² heat flux (Ngo et al., 2016).

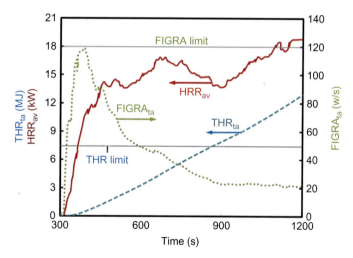

Fig. 5.15 Time evolution of average heat release rate (HRRav), total heat release (THRta), and fire growth rate index (FIGRAta) obtained from the single burning item test of a composite panel (Nguyen et al., 2014).

is described in Section 5.3. Two grey horizontal straight lines in Fig. 5.15 are plotted to illustrate the critical safety limits for materials subjected to fire. Both FIGRA and THR curves stay below the limit lines, especially for the first 600 s (from the 300th second to the 900th second), which evidently proves that the tested composite structure has successfully met the requirements for construction materials in compliance to the fire classification code EN 13501. In the first 100 s after material exposure to the propane burner, the development of combustion (HRR curve) to its half-peak at 12 kW was also achieved. Two other PHRR are observed after around 450 and 700 s, which correspond to the 750th second and 1000th seconds, respectively. While the main peak of 18.38 kW (HRR curve) is reached at the end of the test, the first half-peak is associated with the highest peak of the FIGRA curve approaching the FIGRA limit. After reaching the peak, FIGRA decreases and no longer poses an issue in terms of fire performance. From this study, it is important to note that one of the most critical periods of time is the first 200 s involving the burning of the coating and possibly the surface layer of the resin.

5.4.2.3 Smoke production and toxicity

Smoke production of composites is often quantified by two parameters: the specific extinction area (SEA) and the smoke production rate (SPR). Similar to HRR and THR, a derivative parameter of SPR is total smoke production (TSP). SEA is measured as the total obscuration area of smoke released by the total mass loss during combustion, while SPR shows the area of smoke obscuration released per second (Egglestone and Turley, 1994).

The dense smoke emitted from burning composites contains fine soot particles and microscopic fragments of noncombustible fiber (Mouritz et al., 2006). The value of SEA and SPR shows important characteristics of the released smoke, which affects significantly the possibility of human survival in a fire, especially in maritime environment. Egglestone and Turley compared SEA and SPR of common maritime composites exposed to different heat fluxes as illustrated in Fig. 5.16 (Egglestone and Turley, 1994). The resole phenol composite possessed lower SEA among all tested composites. Unlike the heat release and ignition properties, SEA value of phenol was observed to decrease with higher heat fluxes, which could be explained by the increase in char formation. Mouritz et al. (2006) revealed an important correlation

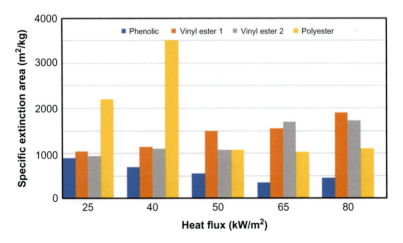

Fig. 5.16 Specific extinction area of common maritime composites (Egglestone and Turley, 1994).

between SEA and average HRR of several fiber-reinforced composites, which shows the close relationship between smoke production of burning composites and the heat release rate.

5.5 Structural performance of maritime composite during fire and postfire mechanical performance

Composites are often designed and used in marine construction as a structural element. Understanding the behavior of these elements exposed to fire load is of importance as the structural integrity of composites at elevated temperature is crucial to the safety of the structures and human on board. Structural performance or structural response is another characteristic of composites and is often misunderstood with the fire reaction. In this section, the structural response of composite in fire and their postfire mechanical performance will be investigated. Methods of defining structural response of composites under high temperature are defined in details by Dimitrienko (Dimitrienko, 2016).

5.5.1 Structural responses of maritime polymer composites during fire

5.5.1.1 Behavior of single skin laminates

Fiber-reinforced composites in naval and commercial surface ships have been well-designed to outperform conventional materials under normal operating conditions. However, the structural performance of composites is also known to change significantly with the high temperature gradient produced in standard fires and especially severe in hydrocarbon fires. In Section 5.4, we have identified the combustion/ignition temperature of most common maritime composites at around 300–500°C. The structural performance of composite decreases much earlier owing to the loss of strength after the glass transition point (Kootsookos, 2005). For example, polyester or vinyl ester used in ship structures has the glass transition temperature of as low as 120°C (Asaro et al., 2009; Lua, 2005).

Asaro et al. tested the structural response of a 12 mm single skin E-glass composite when exposed to IMO A.754 and more severe UL1709 fire conditions under different compression loading levels (Asaro et al., 2009). The applied load has been varied from 4.8 to 21.4 kN under the standard IMO A.754 curve and the vinyl ester laminate failed the insulation criteria after 45.8–98.9 minutes. When the severe UL1709 fire curve is applied with a constant load of 22.3 kN, the composite failed after only 11.3 minutes.

Both in-plane and out-plane deflections were measured at different specimen height. It was observed that these deflections accelerated significantly after the temperature through the laminate thickness passed the glass transition point. The reduction of vinyl ester/E-glass laminate stiffness versus temperature has also been studied (Bausano et al., 2006). As can be clearly seen from Fig. 5.17, the composite stiffness reduced drastically around the glass transition temperature, which is 105°C for

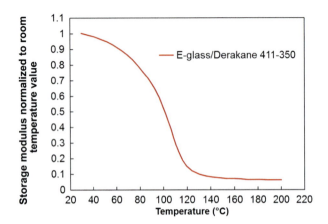

Fig. 5.17 Reduction of normalized stiffness of vinyl ester (Derakane 411-350) with the increase of temperature (Bausano et al., 2006).

Derakane resin. The glass transition temperature was found to play an important role on the failure mechanism of the composite laminate at higher heat fluxes when the temperature through the laminate thickness tended to pass the glass transition point. Failure caused by exposure to lower heat fluxes has been controlled by creep rather than the glass transition point and the temperature diffusion process within the thickness. Typical failure mode of glass fiber composite when subjected to fire and compression load is presented in Fig. 5.18 (Bausano et al., 2006).

The tensile performance of both glass- and carbon-fiber composites was investigated (Feih et al., 2007; Feih and Mouritz, 2012; Gibson et al., 2011). The tensile strength of glass fiber-reinforced composite was strongly influenced by the thermal softening of the fiber. The softening of the polymer matrix had a less impact on the composite tensile strength at elevated temperature.

The term "Weibull bundle strength" was used to better describe the effect of fiber strength at elevated temperature when the heated polymeric matrix lost the ability to maintain the so-called "composite action". Carbon fiber composites were found to significantly lose the strength above 550°C when carbon fiber underwent the oxidation and thinning process with thermally activated damage observed (Feih and Mouritz, 2012). Apart from the static mechanical strength, the dynamic structural response of fiber-reinforced polymer composite skin against low velocity impact also decreased significantly after a very short time (less than 100 s) exposed to a heat flux of 80 kW/m^2 or 800°C equivalent (Ulven and Vaidya, 2008).

5.5.1.2 Behavior of polymer sandwich composites

Understanding the risk of high heat flux involving an oil fire (up to 100 kW/m^2 or more) (Sorathia et al., 2001), the mechanical behavior of sandwich composites with focus on the failure mode was studied by Mouritz and Gardiner (Mouritz and Gardiner, 2002). Two main mechanisms of compression failure for sandwich panel were core shear

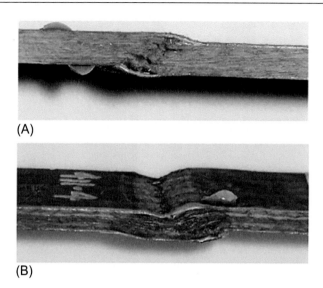

Fig. 18 Failure mode of glass fiber composite exposed to fire under compression load: (A) single kinking and (B) double kinking phenomena (Bausano et al., 2006)

failure and buckling of face skin. The former was found to be more applicable for pristine sandwich composite without fire damage, while global buckling reflected the condition of sandwich panels under damage from one-side exposure to fire (Figs. 5.19 and 5.20). Fig. 5.19 also showed the deflection recorded on balsa core sandwich composite during the exposure to IMO A.754 fire condition under a 10.0 kN compressive load. The sandwich composite possessed less in-plane and out-of-plane deflection than the single skin laminate which is studied in the previous section.

5.5.2 Postfire mechanical properties of polymer composites

Early applications of fiber-reinforced composites in maritime structures were mostly nonstructural and less critical elements. With the advancement in composite science and technology, composite members in modern naval submarines and maritime superstructures have now moved to structural applications. Their postfire integrity and structural performance are, thus, important for the design and use of composite elements.

Postfire flexural stiffness of 1350 × 550 × 13.5 mm woven glass-reinforced polyester ship panels was investigated after horizontal exposure to kerosene fire, which was equivalent to the fire load in large marine structures (Gardiner et al., 2001). This panel was used in the construction of the Huon class mine hunter ships of the Royal Australian Navy. The generated fire achieves temperature around 300°C located 280 mm underneath the composite panel. Data of the unburnt composite panel was taken as the baseline comparison. Fig. 5.21 showed the applied load vs. cross-head displacement curves of composite panel after getting exposed to the above-mentioned

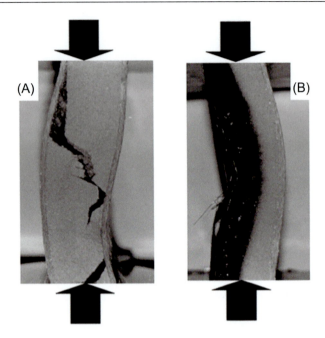

Fig. 5.19 Failure mode of sandwich composite (A) core shear failure and (B) buckling of surface skin (Mouritz and Gardiner, 2002).

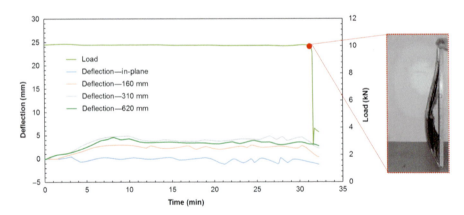

Fig. 5.20 In-plane and out-of-plane deflection of balsa core sandwich composite exposed to fire. Photo on the right hand-side showing buckling effect on the sandwich composite (Asaro et al., 2009).

fire load for durations varying from 60 to 600 s. The postfire flexural stiffness calculated from this data reduced by 50% and over 85% after 120 and 600 s exposure to kerosene fire. Further analysis conducted on the postfire flexural strength of the composite showed the degradation in flexural stiffness was a function of the charring depth. The formed char debonded from the unburnt part of the composite panel

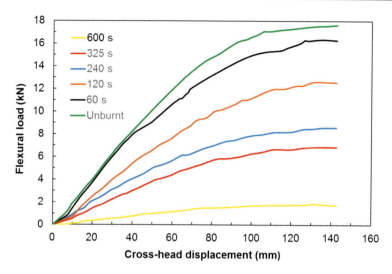

Fig. 5.21 Applied flexural load versus cross-head displacement of ship panels with different time exposed to fire load equivalent to that of large marine structures (Gardiner et al., 2001).

and resulted in a significant reduction in strength/stiffness of the unprotected layers (Gardiner et al., 2001).

Mouritz and Mathys (Mouritz and Mathys, 2001) investigated the postfire mechanical behavior after exposure to a 50 kW/m² heat flux in a cone calorimeter test. Critical reductions were observed with all the mechanical properties including tensile, compression, bending-induced compression, and bending-induced tension strength. As clearly seen in Fig. 5.22 (data generated from Mouritz and Mathys, 2001), marine composites remained with no strength after 1800 s of heat exposure. The development of char thickness was also studied as a function of heat exposure time. The relationship between the composite's mechanical strength and the char thickness was also observed from Fig. 5.22.

The mechanical characteristics of marine composite structures after exposure to a 50 kW/m² cone calorimeter heat flux for the duration of 30 minutes were studied by Mouritz and Mathys (Mouritz and Mathys, 1999). Polyester composite degraded after the exposure to a low heat flux of 20 kW/m², while phenolic resin was not affected if the exposed heat flux is less than 40 kW/m2. Mouritz and Mathys (Mouritz and Mathys, 1999) give important details on structural response of marine composites after subjected to fire, which can be used for prediction and design of naval and offshore composite structures.

The postfire low velocity impact response of marine sandwich composite was determined by the instrumented drop tower (Ulven and Vaidya, 2006). The marine sandwich panel was made of E-glass/vinyl ester faces and balsa wood core and tested against 175 kW/m² flame for 0, 50, 100, and 200 seconds. Fig. 5.23 plotted the low velocity impact force and energy versus heat exposure time. The peak load of the exposed surface, energy to peak load as well as the propagation of impact energy through the balsa core were revealed to decrease when longer exposure to fire was performed.

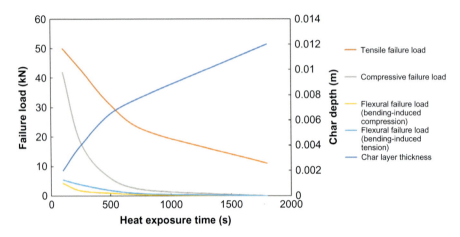

Fig. 5.22 Comparison of mechanical failure loads and char thickness development over exposure time.

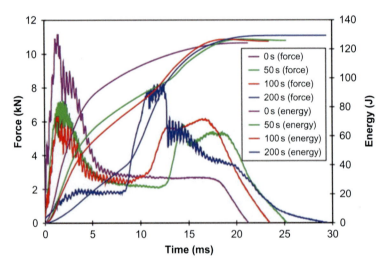

Fig. 5.23 Low velocity impact force and energy vs. time curves of marine grade sandwich composites after exposed to 175 kW/m² flame for 0–200 s (Ulven and Vaidya, 2006).

5.6 Numerical analysis of naval composite structure performance in fire

Fire testing of marine composite structures is technically difficult, expensive, and time-consuming (Anjang et al., 2014). Numerical analysis on fire performance of composite structures has been developed to predict the composite behavior in fire and postfire exposure. Both the fire reaction characteristics and structural response are modeled and will be presented in this section.

5.6.1 Modeling of composite fire reaction: From bench-scale to full-scale testing

5.6.1.1 Quintiere's fire growth model

Quintiere (Quintiere, 1993) was first to develop a numerical model that integrated the thermal feedback without empirical computation of flame spread. The outputs of his model gave comprehensive sets of fire reaction characteristics of composite materials from ignition, flame spread, and burn-out rate. Quintiere's model was designed based on the room corner test ISO 9705 described in Section 5.3. This standard has been developed based on several requirements in maritime composite design, but also applicable to other applications such as civil buildings and constructions. The model, therefore, can be utilized in the prediction of composite elements in a wide range of applications, not limited to marine structures. This method can also provide a quick prediction on fire growth properties as mentioned earlier. Prediction of composite ignition was conducted by comparing the surface temperature of composite with the ignition temperature T_{ig}. The surface temperature of composite laminate was governed as shown in Eq. 20.

$$T_{s,0} - T_\infty = \frac{1}{\sqrt{\pi k \rho c}} \int_0^t \frac{q(\tau)}{\sqrt{t-\tau}} d\tau \qquad (20)$$

where $T_{s,0}$ is the surface temperature responding to the ignition, T_∞ is the ambient temperature, k is the thermal conductivity of composite, ρ is the density of the materials, c is the composite-specific heat capacity, t is the simulation time, τ is a variable for time, $q(\tau)$ is calculated from the ignitor flame heat flux $\dot{q}''_{ig}$ and radiative heat as in the following Eq. 21:

$$q(\tau) = \dot{q}''_{ig} + \sigma\left(T^4 - T^4_{s,0}\right) \qquad (21)$$

The upward spread of the flame and burn-out position were calculated with the pyrolysis front position y_p with the assumption of a universal flame spread theory. Eqs. 22–25 described the methods of modeling the upward flame spread taken into account the flame length y_f, ignition time t_{ig}, and the associated flame heat flux beyond the burning region $\dot{q}''_f$.

$$\frac{dy_p}{dt} = \frac{y_f - y_p}{t_{ig}} \qquad (22)$$

$$t_{ig} = \frac{\pi}{4} k \rho c \left[\frac{T_{ig} - T_s}{\dot{q}''_f}\right]^2 \qquad (23)$$

$$y_f = y_b + \begin{cases} k_f \left[\dot{Q}'_{ig} + \dot{Q}''(y_p - y_b)\right]^n, & y_b \leq k_f \dot{Q}'''_{ig} \\ k_f \left[\dot{Q}'(y_p - y_b)\right]^n, & y_b \geq k_f \dot{Q}'''_{ig} \end{cases} \qquad (24)$$

$$\frac{dy_b}{dt} = \frac{y_b(t+t_b) - y_b(t)}{t_b} \qquad 25$$

where T_s is the surface temperature, y_b is the burn-out position, $\dot{Q}'_{ig}$ is the energy release rate for the burner, $\dot{Q}''$ is the total energy release rate per unit area determined in the cone calorimeter test.

5.6.1.2 Janssens' method to predicting ignition properties

The prediction of ignition property as mentioned in the previous section is important in modeling fire reaction of composite as it is the starting point of the burning process. Quintiere's method of ignition modeling has the solution shown in Eq. 26, and Janssen's work improved theory developed in Quintiere's model as an alternative solution (Alston, 2004, 2001) for marine composites. The governing equation of ignition prediction in Janssen's method is shown in Eq. 27.

$$\dot{q}''_{cr} = \dot{q}''_e \left[1 + A \left(\frac{k\rho c}{h^2 t} \right)^{\alpha} \right]^{-1} \qquad 26$$

$$(\dot{q}''_e - \dot{q}''_{cr}) t^n_{ig} = C \qquad 27$$

where A, α are constants and the first step of Janssen's method is to determine the material characterization from Eq. 26. The exponential factor n varies in the range from 0.5 to 1.0, depending on the thermal characteristics of the composite. Owing to the formation of char which acts as the insulation layer as discussed in Section 5.4, it is essential to define two sets of thermal constants to best fit the composite fire reaction. These two sets are corresponding to the initial state where no char involves, and the "blackened" state. For example, n value of untreated eight-layer glass fiber-reinforced composite and blackened eight-layer composite is 0.724 and 0.500, respectively. The untreated composite has a higher n value as it is thermally thinner. With the formation of char, the blackened composite becomes thermally thicker interpreting the lower n value.

5.6.1.3 Computational fluid dynamic models to predict fire behaviors

Several computational fluid dynamic (CFD) models have been developed with the aim of predicting fire behavior of fiber-reinforced polymer composite (Nguyen et al., 2014, 2016a, b; Ngo et al., 2016; Nguyen, 2013). The fire reaction model is built as the combination of large eddy simulations (LES) combustion (McGrattan et al., 2010) and the pyrolysis of composite under a heat flux of 50 kW/m^2. The LES model calculates the burning rate of the combustion and each volumetric element in the simulated zone contains a mixture of air, fuel materials, and combustion products. The combustion of material starts at the point of ignition when the concentration of fuel material and the oxidizing agent are mixed at the appropriate temperature. The pyrolysis model simulates the chemical reaction of the solid composite when exposed to the required heat flux. The pyrolysis reaction is calculated as the Arrhenius function (Coats and Redfern, 1964, 1965) and power function as follows:

$$\alpha_i(T) = \frac{M_{ib} - M_i(T)}{M_{ib} - M_{ia}} \qquad 28$$

where α is the degree of decomposition of reaction ith at temperature T, M_{ib} is sample mass before reaction ith, M_{ia} is sample mass after reaction ith, $M_i(T)$ is sample mass at temperature (T). The rate of the pyrolysis reaction, then, can be determined as follows:

$$\frac{d\alpha}{dt} = f(T) \cdot g(a) \qquad 29$$

where $\frac{d\alpha}{dt}$ is the rate of decomposition at time t, $f(T)$ and $g(a)$ are influence factors of temperature and reactant quantity, respectively, to the reaction rate. $f(T)$ and $g(\alpha)$ can be expressed in the following Eq. 30 and Eq. 31:

$$f(T) = Ae^{-\frac{E_A}{RT}} \qquad 30$$

$$g(\alpha) = (1-\alpha)^n \qquad 31$$

where β is the heating rate, A is the pre-exponential factor, E_A is the activation energy, n is the reaction order, R is the Boltzmann gas constant (8.314 kJ/kmol K).

From Eqs. 29–31, the rate of pyrolysis reaction can be written as follows:

$$\frac{d\alpha}{dt} = (1-\alpha)^n \cdot A \cdot e^{-\frac{E_A}{RT}} \qquad 32$$

It can be easily seen from Eq. 32 that if the kinetic parameters including A, EA, and n are known, the rate of reaction can be determined.

The kinetic parameters A, EA, and n are evaluated through thermo-gravimetric analysis (TGA). The materials are ground into powder form and filled in the ceramic holder with a sample mass of 2 ± 0.2 g for each run. TGA is performed by raising the temperature from 25°C to 700°C with a heating rate $\beta = 25$°C/min, using the PerkinElmer Diamond Thermo-gravimetric/Differential Thermal Analyser (TG/DTA). As in TGA, $dT/dt = \beta$ and Eq. 32 can now be expressed as follows:

$$\frac{d\alpha}{dT} = (1-\alpha)^n \cdot \frac{A}{\beta} \cdot e^{-\frac{E_A}{RT}} \qquad 33$$

An example of a composite filled with organoclay is taken to illustrate the calculation in this CFD simulation. The resin is mixed with organoclay before being infused in the reinforced glass fiber. The cured composite is called S3. A curve of degree of decomposition α vs. temperature is recorded for each material as plotted in Fig. 5.24. Three samples are tested in TGA including the resin, organoclay, and sample S3.

The modified Coats-Redfern method (Coats and Redfern, 1964, 1965) is introduced here to determine E_A and A (Eq. 34) and the reaction order n is determined from Kissinger method (Kissinger, 1957). In Coats-Redfern method, Eq. 17 can be written as follows:

$$\frac{d\alpha}{(1-\alpha)^n} = \frac{A}{\beta} . e^{-\frac{E_A}{RT}} . dT \tag{34}$$

Integrating Eq. 34 gives

$$\int_0^\alpha \frac{d\alpha}{(1-\alpha)^n} = \alpha + \frac{n\alpha^2}{2} + \frac{n(n+1)\alpha^3}{6} + \ldots = \frac{ART^2}{\beta E_A}\left(1 - \frac{2RT}{E_A}\right).e^{-\frac{E_A}{RT}} \tag{35}$$

In the case of small value of α, α^2 and higher order terms can be neglected to give.

$$\alpha \approx \frac{ART^2}{\beta E_A}\left(1 - \frac{2RT}{E_A}\right).e^{-\frac{E_A}{RT}} \tag{36}$$

Eq. 36 can be further transformed into Eq. 37 and Eq. 38 as follows:

$$\frac{\alpha}{T^2} = \frac{AR}{\beta E_A}\left(1 - \frac{2RT}{E_A}\right).e^{-\frac{E_A}{RT}} \tag{37}$$

$$-\ln\left(\frac{\alpha}{T^2}\right) = \frac{E_A}{RT} - \ln\frac{AR}{\beta E_A}\left(1 - \frac{2RT}{E_A}\right) \tag{38}$$

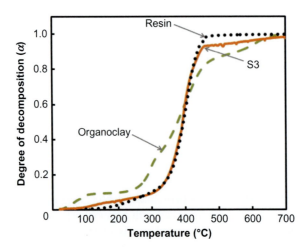

Fig. 5.24 Change in degree of decomposition $\alpha\alpha$ with respect to temperature.

Fire performance of maritime composites

For each reaction appearing in TGA curve, the slope of the curve correlating ($-\ln(\alpha/T^2)$) and ($1/T$) should give the value of E_A/R (Fig. 5.25). Once E_A is calculated, A can be determined from the Eq. 36.

To evaluate the reaction order n according to Kissinger method, the maximum rate of reaction occurs at critical temperature T_m (as shown in Fig. 5.26) when.

$$\frac{d^2\alpha}{dT^2} = 0 \qquad 39$$

The derivative of Eq. 40 at temperature T_m gives

$$\frac{E_A \beta}{RT_m^2} = An(1-\alpha_m)^{n-1} e^{-\frac{E_A}{RT_m}} \qquad 40$$

The reaction order n can be determined by substituting the calculated E_A and A into the Eq. 40. The results of these calculations are summarized in Table 5.3.

Fire growth index takes into account not only the magnitude of the PHRR, but also the point when PHRR occurs. Lower FGI value reflects less severe flammability of the material. FGI is calculated as the maximum value of FGIav(t) computed from the time of ignition $t$ig until flame out $t$out. The equations to calculate FGIav(t) are given in Section 5.3.

In Fig. 5.27, the fire growth index and final total heat release of each case are calculated. The excellent agreement between numerical results and experimental data of samples S0-5 further confirms the ability of this model to simulate the burning process of composite laminates integrated with organoclay particles.

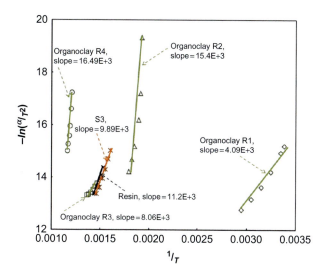

Fig. 5.25 Evaluation of *EA* for resin (*black line*), organoclay reaction R1–R4 (*green lines*), and sample S3 (*orange line*). Experimental data from the TGA tests are plotted and marked with corresponding symbols.

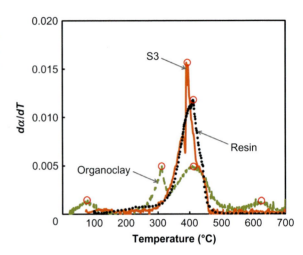

Fig. 5.26 Determination of critical temperature T_m (marked with *red open circles*) for each reaction based on the respective ($d\alpha/dT$) versus temperature curves.

Table 5.3 **Calculated kinetic parameters of resin, organoclay, and composite**

	Pre-exponential factor A (1/s)	Activation energy EA (kJ/kmol)	Reaction order n
Resin			
Reaction	439,304,188	93,632	1.06
Organoclay			
Reaction 1	2,969,127	34,039	1.32
Reaction 2	1.326×10^{13}	128,052	1.01
Reaction 3	3,839,190	66,973	2.32
Reaction 4	5935	40,577	3.96
Composite			
Reaction	49,925,183	82,250	2.03

5.7 Enhancement of maritime composite structures subjected to fire

Most polymers have higher flammability than the reinforcing glass fibers. In addition, the "candlewick effect" is a major drawback of fiber-reinforced composites exposed to fire (Liu et al., 2011; Zhao et al., 2008; Patrick Lim et al., 2012). The highly thermal-conductive glass fibers facilitate rapid heat transfer into deeper layers of the poor thermal-conductive polymer. Simultaneously, glass fibers act as capillary

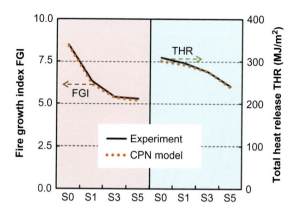

Fig. 5.27 Fire growth index (FGI) and total heat release (THR) of composites: comparison of calculated and experimental results.

objects that feedback the pyrolysis of the polymer to the flame. Therefore, flame retardants need to be added to increase the composite's flame retardancy. Only flame retardants for GFRP are considered in this section.

For glass fiber-reinforced polyester composites, carbon-based nanopaper was used as a coating (Tang et al., 2011; Zhuge et al., 2011, 2012; Zhao and Gou, 2009). Tang et al. (2011) used the carbon nanofiber paper consisting of carbon nanofiber and ammonium phosphate APP/clay or polyhedral oligomeric silsesquioxane POSS/APP to coat the laminate composite. Fire retardancy of the system was evaluated based on a bench-scale fire test of the cone calorimeter. Different doses of carbon nanopaper were investigated and it was concluded that carbon nanopaper with a carbon nanofiber/clay/APP ratio of 5/1/9 was the optimum among the tested specimens. However, no tests on the mechanical behavior of the coated GFRP were carried out in this research.

Zhuge et al. (2012) developed research based on Tang et al.'s (2011) work with the optimum dose of carbon nanopaper. A cone calorimeter test with a heat flux of 35 kW/m^2 and a postfire three-point bending test were launched. No detachment from the sample coated with carbon nanofiber/exfoliated graphite nano-platelets (GnP)/APP was observed, whereas the carbon nanofiber/clay/APP paper was eliminated from the base composite. However, when the heat release rate in both cases is taken into consideration, the peak of heat release rate is reduced to two thirds; nonetheless, the total heat release rate remained approximately the same (Fig. 5.28). It was not feasible to increase the thickness of the nanopaper to achieve a V0 rating in the UL-94 vertical fire test.

In another study by Zhao and Gou (Zhao and Gou, 2009), a comparison between using carbon nanofiber as a coating sheet and as an additive was carried out. The first group of samples consists of one ply of carbon nanofiber sheet, while the second group was moulded by fusing the mixture of carbon nanofiber and the unsaturated polyester resin. The same amount of carbon nanofiber was investigated for the two groups. Samples of 100×100 m^2 were cut out for the cone calorimetry test with a heat flux of 50 kW/m^2. The results showed that carbon nanofiber had no influence on the flame

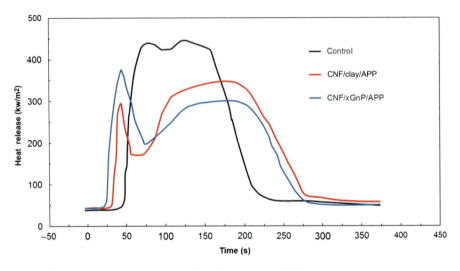

Fig. 5.28 Heat release rate of the samples (Zhuge et al., 2012).

retardancy of the tested unsaturated polyester. On the other hand, with 0.38% carbon nanofiber, the sheet acted as both a barrier on the exposed surface and an "activator" that promoted char formation, thereby helping to increase the flame retardancy of the laminate composite. Nevertheless, no mechanical tests were conducted with these studies.

A third group consisted of GFRP with epoxy matrix. Epoxy resin itself is better than polyester in terms of mechanical properties and fire retardancy, thereby making it an important polymer in civil applications. However, the glass fiber reinforcement in GFRP causes a "candlewick" effect (Zhao et al., 2008; Patrick Lim et al., 2012), which makes epoxy more sensitive to the flame. Patrick Lim et al. (2012) evaluated the application of ammonium polyphosphate and melamine cyanurate as intumescent flame retardants on the flame retardancy of epoxy/glass fiber composites. A boosting effect between these two additives was found in comparison with separated compounds (Table 5.4). Melamine cyanurate contributed to the formation of char, while ammonium polyphosphate increased the flexural strength of the GFRP composite. A synergist effect in flame retardancy was achieved with 2.5% ammonium polyphosphate and 2.5% melamine cyanurate. A fire rating of V1 according to the Ul-94 test was obtained, whereas samples with 20% ammonium polyphosphate or melamine cyanurate separately were able to pass the V0 rating only. This result needs to be developed for further understanding of the mechanism of this synergist effect and other mechanical characteristics.

Table 5.4 Flammability, thermal, and mechanical characteristics of epoxy/glass fiber composite with intumescent flame retardants (Patrick Lim et al., 2012)

Epoxy/glass fiber	Sample composition (%)					
	Melamine cyanurate	Ammonium polyphosphate	UL-94	LOI (%)	Temperature of 5% weight loss (°C)	Flexural strength (MPa)
100	–	–	NR	21.0	357.6	87.1
95	5	–	NR	24.4	354.6	97.9
80	20	–	V0	28.2	354.0	95.3
95	–	5	V0	26.0	326.7	98.0
80	–	20	V0	48.0	321.4	121.1
95	1	4	V0	27.8	327.1	119.4
95	2.5	2.5	V1	25.8	333.38	97.3

5.8 Conclusions

The book chapter presented a brief overview of the key issues on the behavior of naval polymer composites in fire. Key composite fire performance indicators were described including the thermal degradation and combustion mechanisms, which are related to thermal decomposition reactions, reaction rates, toxic gases and flammable volatiles, and polymers matrix and fibers. The fire response properties, which define the fire hazard and flammability of polymer composites, are also described including ignition time, heat release rate, and smoke and gaseous combustion products. General fire resistance design principles and fire safety codes such as International Maritime Organization (IMO) were presented focusing on the regulations for key parameters such as heat and smoke release rate, smoke toxicity, and surface flammability. Quantitative and numerical models were described for predicting the development and spread of fire together with models for calculating the decomposition and thermal response of composites to fire. An overview of methods for reducing the fire risk of composites is given in Section 5.7. The efficacy of various types of flame retardant polymer systems is described, including polymers with flame retardant fillers, halogenated polymers, char forming polymers, and polymers that are chemically and structurally modified to increase flammability resistance.

References

Abdel-Gaber, A.M., et al., 2008. Inhibition of aluminium corrosion in alkaline solutions using natural compound. Mater. Chem. Phys. 109 (2–3), 297–305.

Alston, J., Compartment fire dynamics model calibration data: marine composite screening specimens Interflam Proceedings. Vol. 1. 2001. 695.

Alston, J., Room/corner fire calibration data marine composite screening specimens. Master of Science Thesis, Worcester Polytechnic Institute. 2004.

Anjang, A., et al., 2014. Tension modelling and testing of sandwich composites in fire. Compos. Struct. 113, 437–445.

AS-1530.4, 1997. Methods for fire tests on building materials, components and structures. In: Part 4: Fire Resistance Tests of Elements of Building Construction. Standards Australia International Ltd, Sydney.

Asaro, R.J., Lattimer, B., Ramroth, W., 2009. Structural response of FRP composites during fire. Compos. Struct. 87 (4), 382–393.

Bausano, J.V., Lesko, J.J., Case, S.W., 2006. Composite life under sustained compression and one sided simulated fire exposure: Characterization and prediction. Compos. A: Appl. Sci. Manuf. 37 (7), 1092–1100.

Bhattacharyya, S.K., Sreekumar, S., Idichandy, V.G., 2003. Coupled dynamics of sea star mini tension leg platform. Ocean Eng. 20, 709–737.

Buchanan, A.H., 2001a. Structural Design for Fire Safety. John Wiley, Chichester.

Buchanan, A.H., 2001b. Structural Design for Fire Safety. John Wiley, Chichester.

Caplan, I.L. Marine composites the US Navy experience: Lessons learned along the way. In First International Workshop on Composite Materials for Offshore Operations. 1993. Houston, Texas.

Chandrasekaran, S., 2016. Advanced Marine Structures. CRC Press, Boca Raton.

Chung, D.D.L., 2010. Composite Materials Science and Applications, 2nd edn. Springer, New York.
Coats, A.W., Redfern, J.P., 1964. Kinetic parameters from thermogravimetric data. Nature 201 (4914), 68.
Coats, A.W., Redfern, J.P., 1965. Kinetic parameters from thermogravimetric data II. J. Polym. Sci. B Polym. Lett. 3 (11), 917–920.
Critchfield, M.O., Morgan, S.L., Potter, P.C., 1991. GRP deckhouse development for naval ships. Adv. Marine Struc. 2, 372–391.
Critchfield, M.O., Judy, T.D., Kurzweil, A.D., 1994. Low-cost design and fabrication of composite ship structures. Mar. Struct. 7, 475–494.
Dewhurst, D., The influence of fire on the design of polymer composite pipes and panels for offshore structures. 1997, Phd Thesis, University of Salford.
Dimitrienko, Y.I., 2016. Thermomechanics of composite structures under high temperatures. In: Solid Mechanics and Its Applications. Volume 224. Springer, Dordrecht.
Dodds, N., et al., 2000. Fire behaviour of composite laminates. Compos. A: Appl. Sci. Manuf. 31 (7), 689–702.
Dow, R. A technology demonstrator for a composite superstructure. International Maritime Defence Conference 1995. London.
Egglestone, G.T., Turley, D.M., 1994. Flammability of GRP for use in ship superstructures. Fire Mater. 18 (4), 255–260.
EN 13823:2010, Reaction to fire tests for building products—building products excluding floorings exposed to the thermal attack by a single burning item. 2010. BSI.
Ezuber, H., El-Houd, A., El-Shawesh, F., 2008. A study on the corrosion behavior of aluminium alloys in seawater. Mater. Des. 29 (4), 801–805.
Feih, S., Mouritz, A.P., 2012. Tensile properties of carbon fibres and carbon fibre–polymer composites in fire. Compos. A: Appl. Sci. Manuf. 43 (5), 765–772.
Feih, S., et al., 2007. Tensile strength modelling of glass fiber-polymer composites in fire. J. Compos. Mater. 41, 2387–2410.
Gardiner, C.P., et al., Post-fire flexural response of GRP composite ship panels. Proceedings of the Eleventh International Offshore and Polar Engineering Conference, 2001.
Gibson, A.G., Feih, S., Mouritz, A.P., 2011. Developments in Characterising the Structural Behaviour of Composites in Fire. In: Nicolais, L., Meo, M., Milella, E. (Eds.), Composite Materials: A Vision for the Future. Springer, London, pp. 187–218.
Goodrich, T.W., Lattimer, B.Y., 2012. Fire decomposition effects on sandwich composite materials. Compos. A: Appl. Sci. Manuf. 43 (5), 803–813.
Hoyning, B. and J. Taby. Warship design: the potential for composites in frigate superstructures. in International Conference on Lightweight Construction Latest Developments. 2000.
ISO 1182:2010, Reaction to fire tests for products—non-combustibility test. 2010, ISO: Switzerland.
ISO 5660:2015, Reaction-to-fire tests—heat release, smoke production and mass loss rate. 2015, ISO: Switzerland.
ISO 834, Fire resistance tests—elements of building construction. 1975, International Standard ISO 834: Geneva.
ISO 9705-1:2016, 2016. Reaction to fire tests—room corner test for wall and ceiling lining products. In: Part 1: Test Method for a Small Room Configuration. ISO, Switzerland.
Kissinger, H.E., 1957. Reaction kinetics in differential thermal analysis. Anal. Chem. 29 (11), 1702–1706.
Kootsookos A., Compression properties of marine sandwich composites in fire, ICCM Proceedings (13 edn), CD-ROM edition. 2005. 423.

Li, J.F., et al., 2007. Localized corrosion mechanism associated with precipitates containing mg in Al alloys. Trans. Nonferrous Met. Soc. Chin. 17 (4), 727–732.

Lua, J., Damage progression of a loaded marine composite structure subjected to a fire. Proceedings of the International Conference on Composites in Fire. 2005. 135.

McGrattan, K., et al., 2010. Fire Dynamics Simulator, User's Guide. (Version 5)NIST.

Michael Davies, J., Y.C. Wang, and P.M.H. Wong, Polymer composites in fire. Compos. A: Appl. Sci. Manuf., 2006. 37(8): p. 1131–1141.

Mouritz, A.P., Gardiner, C.P., 2002. Compression properties of fire-damaged polymer sandwich composites. Compos. A: Appl. Sci. Manuf. 33 (5), 609–620.

Mouritz, A.P., Gibson, A.G., 2006. Fire Properties of Polymer Composite Materials. Solid Mechanics and Its Applications. Vol. 143. Springer, Dordrecht.

Mouritz, A.P., Mathys, Z., 1999. Post-fire mechanical properties of marine polymer composites. Compos. Struct. 47 (1–4), 643–653.

Mouritz, A.P., Mathys, Z., 2001. Post-fire mechanical properties of glass-reinforced polyester composites. Compos. Sci. Technol. 61 (4), 475–490.

Mouritz, A.P., et al., 2001. Review of advanced composite structures for naval ships and submarines. Compos. Struct. 53 (1), 21–42.

Mouritz, A.P., Mathys, Z., Gardiner, C.P., 2004. Thermomechanical modelling the fire properties of fibre–polymer composites. Compos. Part B 35 (6–8), 467–474.

Mouritz, A.P., Mathys, Z., Gibson, A.G., 2006. Heat release of polymer composites in fire. Compos. A: Appl. Sci. Manuf. 37 (7), 1040–1054.

Naval Technology French pursue composites for SSBNNavy News Undersea Technologies 1991. 8, 3–4.

Ngo, T.D., Nguyen, Q.T., Tran, P., 2016. Heat release and flame propagation in prefabricated modular unit with GFRP composite facades. Build. Simul. (5), 607.

Nguyen, Q., et al., 2013. Composite materials for next generation building facade systems. Civil Eng. Architect. 1 (3), 88–95.

Nguyen, Q.T., et al., 2014. Experimental and computational investigations on fire resistance of GFRP composite for building façade. Compos. Part B 62, 218–229.

Nguyen, Q.T., et al., 2016a. Experimental and numerical investigations on the thermal response of multilayer glass fibre/unsaturated polyester/organoclay composite. Fire Mater. (8), 1047.

Nguyen, Q.T., et al., 2016b. Fire performance of prefabricated modular units using organoclay/glass fibre reinforced polymer composite. Construct. Build. Mater. 129, 204–215.

Nik, W.B.W., et al., 2014. Corrosion of aluminum alloy in seawater and development of green corrosion inhibitor for marine applications. In: Olanrewaju, O.S. et al., (Ed.), Marine Technology and Sustainable Development: Green Innovations. pp. 146–156.

Offshore Technology The world's worst offshore oil rig disasters. www.offshore-technology.com. 2014.

Patrick Lim, W.K., et al., 2012. Effect of intumescent ammonium polyphosphate (APP) and melamine cyanurate (MC) on the properties of epoxy/glass fiber composites. Composites Part B 43, 124–128.

Quintiere, J.G., 1993. A simulation model for fire growth on materials subject to a room-corner test. Fire Saf. J. 20 (4), 313–319.

Rosliza, R., Nik, W.B.W., 2010. Improvement of corrosion resistance of AA6061 alloy by tapioca starch in seawater. Curr. Appl. Phys. 10 (1), 221–229.

Satapathy, A.K., et al., 2009. Corrosion inhibition by *Justicia gendarussa* plant extract in hydrochloric acid solution. Corros. Sci. 51 (12), 2848–2856.

Smith, C.S., 2001. Design of Marine Structures in Composite Materials. Elsevier Applied Science, New York.
Smith, J., 1999. Novel submarine hulls. Naval Composite News 3, 13–14.
Sorathia, U., et al., 2001. Screening tests for fire safety of composites for marine applications. Fire Mater. 25 (6), 215–222.
Tang, Y., et al., 2011. Flame retardancy of carbon nanofibre/intumescent hybrid paper based fibre reinforced polymer composites. Polym. Degrad. Stab. 96, 760–770.
Ulven, C.A., Vaidya, U.K., 2006. Post-fire low velocity impact response of marine grade sandwich composites. Compos. A: Appl. Sci. Manuf. 37 (7), 997–1004.
Ulven, C.A., Vaidya, U.K., 2008. Impact response of fire damaged polymer-based composite materials. Compos. Part B 39 (1), 92–107.
Wickström, U., Göransson, U., 1992. Full-scale/bench-scale correlations of wall and ceiling linings. Fire Mater. 16 (1), 15–22.
Zhao, C.-S., et al., 2008. A novel halogen-free flame retardant for glass-fiber-reinforced poly(ethylene terephthalate). Polym. Degrad. Stab. 93, 1188–1193.
Zhao, Z.F., Gou, J., 2009. Improved fire retardancy of thermoset composites modified with carbon nanofibers. Sci. Technol. Adv. Mater. 10.
Zhuge, J., et al., Fire performance of composite laminates coated with hybrid carbon nanofiber paper, in SAMPE Spring Technical Conference and Exhibition—State of the Industry: Advanced Materials, Applications, and Processing Technology. 2011: Long Beach, CA.
Zhuge, J.F., et al., 2012. Fire performance and post-fire mechanical properties of polymer composites coated with hybrid carbon nanofiber paper. J. Appl. Polym. Sci. 124, 37–48.
Zimmerman, S., 1997. Submarine Technology for the 21st Century. Pasha Publications Inc., Arlington, VA.

Further reading

Dimitrienko, Y., 1997a. Modelling of the mechanical properties of composite materials at high temperatures: Part 1. Matrix and fibers. Appl. Compos. Mater. 4 (4), 219.
Dimitrienko, Y., 1997b. Modelling of the mechanical properties of composite materials at high temperatures: Part 2. Properties of unidirectional composites. Appl. Compos. Mater. 4 (4), 239.
Feih, S., et al., 2007. Tensile strength modeling of glass fiber—polymer composites in fire. J. Compos. Mater. 41 (19), 2387–2410.
Goodrich, T.W. and B.Y. Lattimer. Microscopic behavior of composite materials during heating and cooling. In 17th International Conference on Composite Materials. 2009. Edinburgh, UK.
Key, C.T., Lua, J., 2006. Constituent based analysis of composite materials subjected to fire conditions. Compos. A: Appl. Sci. Manuf. 37 (7), 1005–1014.
Liu, L., et al., 2006. Thermal buckling of a heat-exposed, axially restrained composite column. Compos. A: Appl. Sci. Manuf. 37 (7), 972–980.
Liu, Y., et al., 2011. An efficiently halogen-free flame-retardant long-glass-fiber-reinforced polypropylene system. Polym. Degrad. Stab. 96, 363–370.
Lua, J., 2007. Thermal–mechanical cell model for unbalanced plain weave woven fabric composites. Compos. A: Appl. Sci. Manuf. 38 (3), 1019–1037.
Lua, J., et al., 2006. A temperature and mass dependent thermal model for fire response prediction of marine composites. Compos. A: Appl. Sci. Manuf. 37 (7), 1024–1039.

Luo, C., DesJardin, P.E., 2007. Thermo-mechanical damage modeling of a glass–phenolic composite material. Compos. Sci. Technol. 67 (7–8), 1475–1488.

Luo, C., Lua, J., DesJardin, P.E., 2012. Thermo-mechanical damage modeling of polymer matrix sandwich composites in fire. Compos. A: Appl. Sci. Manuf. 43 (5), 814–821.

Mouritz, A.P., 2003. Simple models for determining the mechanical properties of burnt FRP composites. Mater. Sci. Eng. A 359 (1–2), 237–246.

Mouritz, A.P., et al., 2009. Review of fire structural modelling of polymer composites. Compos. A: Appl. Sci. Manuf. 40 (12), 1800–1814.

Zhang, Z. and S.W. Case. Finite element modelling for composites exposed to fire. in 54th International Symposium and Exhibition. 2009. Baltimore, MD.

Effective use of composite marine structures: Reducing weight and acquisition cost

Luis F. Sánchez-Heres*, Jonas W. Ringsberg*, Erland Johnson[†]
*Chalmers University of Technology, Department of Mechanics and Maritime Sciences, Division of Marine Technology, Gothenburg, Sweden, [†]RISE Research Institutes of Sweden, Department of Safety—Mechanics Research, Borås, Sweden

Chapter Outline

6.1 Introduction   161
6.2 General objective and methodology   162
6.3 Material safety factors   163
6.4 Material characterization   170
6.5 Structural design exploration   174
6.6 Conclusions   181
References   182

6.1 Introduction

Composite structures are a way to reduce the operational costs of a vessel or to increase its potential revenue, as lighter vehicles consume less fuel or are able to transport more cargo with the same amount of fuel. Considering the plethora of upcoming environmental regulations and economical challenges, both traits are highly desirable in a vessel. However, depending on the design of the vessel, its operational profile, and the business model of the owner, the benefits brought by a composite structure may not justify its acquisition cost.

In general, the acquisition cost of a composite structure is higher than the one of a similar structure made out of aluminium or steel (Hertzberg, 2009). To economically motivate the use of a composite structure in a vessel, one must weigh the premium cost of the structure against its added benefits through life cycle analyses (LCA), life cycle cost analyses (LCCA), or both.

International projects such as Breakthrough in European Ship and Shipbuilding Technologies (BESST) (BESST, 2013) and Lightweight Constructions Applications at Sea (LÄSS) (Hertzberg, 2009) had LCAs and LCCAs in their agendas because they are a fundamental part of the sales argument of composite marine structures. Regardless of the type of structure, fuel prices, material cost, or any other detail on the LCAs or LCCAs, one aspect is true for all composite marine structures: the smaller its premium cost is, the easier it is to sell it. There is always a need to reduce the acquisition cost.

The acquisition cost of a composite structure can be reduced through a wide range of approaches. In this work, we focus on one approach that is arguably neglected: efficient material utilization. Efficient material utilization is not about material selection, but about fully exploiting the properties of the selected material to reduce its use without compromising safety. Since one of the largest parts of the acquisition cost of a composite structure is the material itself (Hertzberg, 2009), it is of great importance to reduce its use as much as possible.

6.2 General objective and methodology

Marine structures are commonly designed with fast and simple tools and methods, as well as large safety factors, due to time and cost constraints. Simple and fast methods with large safety factors render robust structures: over-dimensioned structures with a degree of safety higher than the one considered necessary. This design methodology identified three areas of opportunity for increasing the material efficiency of composite materials:

(1) Material safety factors
(2) Material characterization
(3) Numerical optimization of large composite structures

Albeit the three of them are significantly different, they are intertwined in the design of a structure. To illustrate their relation, consider the following equation stated in DNV's Rules for Classification of High Speed Light Craft and Naval Surface Craft (DNV's HSLC) (DNV. Rules for Classification of High Speed, 2013a) for the "laminate calculation method",

$$\varepsilon_i \leq \frac{\varepsilon_{uf}}{R} \tag{6.1}$$

where ε_i is the strain in a composite laminate at direction i, ε_{uf} is the ultimate fiber strain of said laminate (a material design value), and R is a safety factor against fiber failure. The quotient of ε_{uf}/R is the maximum allowable strain (an operational limit) which must not be exceeded by ε_i.

This inequality determines, from a laminate strength perspective, whether or not a structure design is acceptable. If the inequality is not met, a structure designer can remediate the situation by: modifying the structure to reduce ε_i, selecting another laminate with a higher ε_{uf}, motivating the use of a smaller R value compared to the one stated in the rules, or combinations thereof. The first two options may or may not result in an increase of structural weight and material cost. The third option does not increase the weight of the structure, but does increase its design time, and therefore, its design cost. These three possibilities are directly linked to the three aforementioned opportunities.

The value of R handles all the uncertainties: the variability of the load, the material properties, and the geometry, as well as the differences between their true and predicted values. Strength reliability analyses are used, in some cases, to calibrate safety factors such as R and are instrumental for motivating lower values. Section 6.3 (of this chapter) summarizes the contents of two publications (Sánchez-Heres et al., 2013; Sánchez-Heres et al., 2014) aimed at exploring the possibility of reducing material safety factors of composite materials through improved reliability analyses and mechanical models. The publications focused on the probabilistic response of the material, as compared to the loads, a material, and its response, can be a common element in a wide variety of structures.

The value of ε_{uf} is a material property design value determined through experimental testing and measuring. Ideally, the value is representative of the material, and to improve it, the only option is to change the material itself. Unfortunately, in reality, laboratory test cases are not perfectly representative of the conditions a material is subjected to in a structure, and the experimental measurements from which material property design values are determined are imperfect. Better test designs and measuring technologies can improve the material property design values without changing the material itself. Section 6.4 (of this chapter) summarizes experimental work (Sánchez-Heres et al., 2016) aimed at obtaining more accurate measurements of the mechanical properties of composite materials through improved practices and nonstandard measuring technologies.

The value of ε_i is the result of a structural analysis. Potentially, it can be reduced by improving the accuracy of the structural analysis. For example, performing detailed finite element calculations instead of simplified analytical calculations known to render conservative results. Furthermore, it can also be reduced by modifying the design of the structure itself. In lightweight structure design, the challenge with respect to strength is to find, among the myriad of possibilities, a design that exhibits the lowest possible stresses and strains under the required loads and has the lowest weight and cost. Essentially, lightweight design is a constrained optimization problem. Section 6.5 (of this chapter) presents an investigation that compares the weight saving potential of searching the design space for the best possible design, against the weight saving potential of the two past approaches: reducing material safety factors and improving material characterization. The objective is to answer a simple question: for reducing the weight of a large composite structure, is searching for alternative designs (optimization) better than modifying the design constraints?

The work summarized in Sections 6.3 and 6.4 has been previously published in peer-reviewed publications, while the work summarized in Section 6.5 has not. Both published and unpublished works are jointly presented to provide a full picture of the PhD work of the first author.

6.3 Material safety factors

The aim of the work presented in this section was to explore the possibility of motivating lower material safety factors for marine composite structures. The combination of safety factors and material design values (e.g., ε_{uf}/R in Eq. 6.1) expresses the operational limit of the material: the threshold condition after which the desired safety level is not met.

Our methodology for exploring the possibility of motivating lower material safety factors was the following:

- Calculate the operational limit for a composite material under a deterministic load with material safety factors stated in design rules.
- Calculate the operational limit for the same composite material under the same deterministic load with structural reliability analyses.
- Compare the calculated operational limits, and discuss whether or not the results of the structural reliability analyses can motivate a reduction of the material safety factors.

Fig. 6.1 presents an overview of all the elements considered in our structural reliability analyses, as well as the calculation methodology. A structure may respond to a loading condition by exhibiting deformation, vibration, damage, and fracture. Calculating the probability of any of these structural responses requires two fundamental elements: a mechanical model and a set of models describing the variability of the input for the mechanical model. The upper part of Fig. 6.1 presents a simple diagram of how these two fundamental elements can be used to estimate the probability of a structural response through a Monte Carlo simulation (a suitable method because it does not require the analytical manipulation of the mechanical model). Instances of each of the random models (dimensions, loads, and material properties) are the input to the mechanical model. Each set of instances generates an output (black dots in Fig. 6.1). Overall, since the input to the structural model is random, so is its output. From the random output (consisting of large numbers of output instances), the empirical probability distributions related to the structural responses can be calculated.

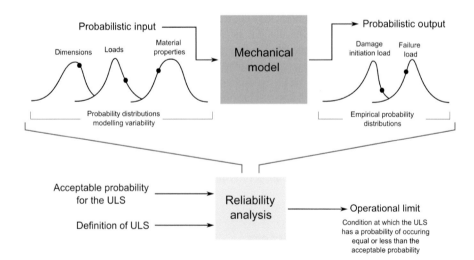

Fig. 6.1 Necessary elements for the estimation of operational limits. Dots represent instances of the input and output variables.

Exactly what the output is depends on the model. For the structural reliability analyses dealt with in this work, the main desired output is the load at which the structure presents the conditions that are considered as the Ultimate Limit State (ULS) of the structure.

From Fig. 6.1, it should be clear that the mechanical model is central to the calculation of operational limits. Clearly, the simplifications and assumptions in a mechanical model will affect the calculation of the operational limits. A key novelty of our work is that we performed structural reliability analyses with a mechanical model capable of accurately predicting matrix-cracking initiation, development, and its effects on laminate stiffness. Matrix cracks and their effects are generally dealt with crude simplifications in reliability analyses of composites (Chiachio et al., 2012), so an improved mechanical model held promise of gains in effective material utilization.

Fig. 6.2 presents a sketch of how four different mechanical models predict the degradation of the longitudinal stiffness (E_x^d/E_x) and first-ply fiber fracture stress ("X" marks) of a hypothetical laminate subjected to a uniaxial tensile stress. The simplest models are the linear elastic model and the linear elastic model with full stiffness degradation. In both models, the longitudinal laminate stiffness does not change as the stress increases. The linear elastic model is nonconservative because it disregards stiffness degradation due to matrix cracking. In contrast, the linear elastic model with full degradation is perhaps too conservative because it exaggerates the stiffness degradation due to matrix cracking. These extremes bound the two models incorporating

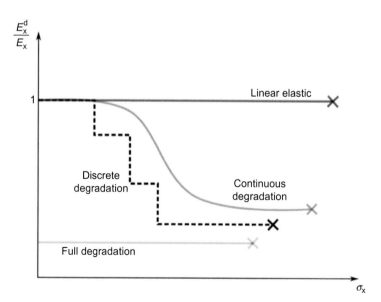

Fig. 6.2 Loss of laminate stiffness according to the laminate response models (E_x^d: degraded longitudinal laminate stiffness, E_x: undegraded longitudinal laminate stiffness; "linear elastic" is omitted in most names for clarity). Failure is predicted with the maximum strain criterion; hence, the different failure stresses.

progressive degradation. The discrete degradation model predicts a step-like degradation with the occurrence of matrix cracking in each of the plies in the laminate. The abrupt loss of laminate stiffness is the result of considering matrix cracking as a sudden and fully developed event. In contrast, the continuous degradation model predicts a smooth and steep loss of stiffness followed by a more gradual reduction. All the four models predict different first-ply fiber fracture stresses because of the different ways laminate stiffness is treated.

In Fig. 6.1, one of the inputs for a reliability analysis is the definition of ULS. Matrix cracking, fiber fracture, and delamination affect the integrity of the laminate, so one could define the ULS of a laminate as the state on which any of these damage mechanisms initiate; however, this definition would be very conservative. These damage mechanisms are progressive, and in fact, they may initiate at very low loads compared to the ones where a laminate completely fractures (Wang, 1984). First-Ply Failure (FPF) due to fiber fracture is typically considered as an ULS. FPF due to matrix cracking, however, is sometimes considered to be admissible. This possibility and the availability of mechanical models accounting for initiation and development of matrix cracks brought another possibility for increasing the utilization of composite materials: the definition of an ULS due to matrix cracking.

Limitations are a part of every investigation. The variety of composite materials is enormous, and the laminates that can be built with them can be subjected to a broad range of loading conditions. Studying the operational limits and material safety factors of all the different types of FRPs would be a monumental task; therefore, we limited our investigations to cross-ply laminates made with unidirectional plies of pre-impregnated FRPs (aka. "prepregs") subjected to a uniaxial tensile load. These laminates and loading condition were chosen because of the availability of two important things: extensive experimental data on progressive matrix cracking and loss of laminate stiffness, and a mechanical model capable of accurately modelling both events and still suitable for reliability analyses. Prepregs are not commonly used in marine structures, and uniaxial tensile loading is just one of the many loading conditions that a laminate may be subjected to in a marine structure; however, as it will be argued further on, the findings of our investigations can still be used to draw useful conclusions for the design and evaluation of composite structures.

A detailed description of the investigated laminates, mechanical models, and the definitions of laminate failure is presented in (Sánchez-Heres et al., 2013; Sánchez-Heres et al., 2014). Fig. 6.3 presents the *probabilistic* laminate stress vs. crack density curve of a carbon/epoxy cross-ply laminate. Deterministic versions of this type of curve are commonly used in research investigations concerning mechanical models capable of predicting matrix-cracking initiation and development (Varna et al., 1999; Nairn, 2000). In Fig. 6.3, the probabilistic aspect of the curve is presented through two sets of boxplots at four points on the curve. Each of the two sets of boxplots (dashed vs. solid whiskers) summarizes the scatter of five key values predicted by two different reliability analyses. The five key values are measured at four points along the curve (hence the labelling). In addition to the predicted probabilistic degradation and failure of the laminate, Fig. 6.3 includes the operational limit stated by design rules (Offshore Standard DNV-OS-C501, 2013b), maximum

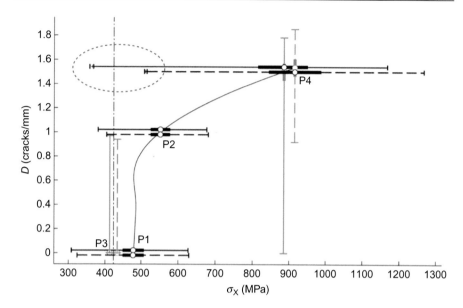

Fig. 6.3 Average crack density vs. laminate stress curve of a [0/90]s carbon/epoxy cross-ply laminate (solid line), and the scatter (probabilistic response) of five measurements of interests (boxplots). Dashed-dot line indicates the maximum allowable stress determined by design rules (Offshore Standard DNV-OS-C501, 2013b). Ellipse used for emphasis.

allowable stress, for that particular laminate (vertical dotted-line). The operational limit is meant to guarantee a probability of laminate failure equal or less than 10^{-5}. The idea was that if our improved reliability analyses showed that the operational limits stated by design rules were too conservative, we could then lower material safety factors.

The leftmost or topmost value indicated by the boxplots is the quantile where the key values have a probability of being observed equal to or less than 10^{-5}. The operational limit is then the leftmost whisker for the horizontal boxplots at P4: the stress where the probabilty of ULS is lower or equal to 10^{-5}. Then necessary comparison is between these values and the operational limit stated by the design rules. In the figure, these three values are emphasized with an ellipse. This comparison is of great importance as it shows that, depending on the choices made in the formulation of the reliability analyses, currently used material safety factors may be deemed conservative or unconservative. The two reliability analyses differ only with respect to the probability distribution used to model the matrix-cracking toughness and the ultimate longitudinal tensile strain. Since none of the probability distribution types can be considered strictly right or wrong, the operational limits estimated with the design rules methodology are in an uncertainty region. This observation repeated itself in all the investigated laminates. A simple choice dominated the conclusions.

Operational limits estimated through the reliability analyses are an integral part of structural design based on goal-based standards and methods for motivating lower safety factors. These estimates are considered so trustworthy that Reliability-Based Design Optimization has emerged as a discipline (Valdebinito and Schuëller, 2010) and is sometimes used as a method for structural weight reduction. Considering our observation about the sensitivity of operational limit estimates to the choice of probability distributions, we investigated how other choices may affect the operational limits of a broader range of laminates.

Figs. 6.4 and 6.5 show how the modelling of laminate stiffness degradation and choice of probability distributions affect the estimation of maximum allowable stresses. It is remarkable that the estimates in the figures vary mostly not due to the different types of laminate or desired safety level, but due to the mechanical and probabilistic models used in their calculation. Fig. 6.4 corresponds to glass/epoxy cross-ply laminates with "first ply failure due to fiber fracture" as the definition of laminate failure. This definition allows for matrix cracking in the laminate, and therefore, for the degradation of the laminate stiffness. Fig. 6.5 corresponds to carbon/epoxy cross-ply laminates with "first ply failure due to matrix cracking or fiber fracture". This definition does not allow for the development of matrix cracking. In this figure, it is clear that the most critical aspect of the estimation of the maximum allowable stresses is the mechanical model. The main difference between the two mechanical models is the definition of matrix cracking as a failure. In the linear elastic model with the Tsai-Wu criterion model (TW in Fig. 6.5), matrix cracking is a sudden fully developed event which can be predicted with this strength-based criterion. In the linear elastic model with progressive continuous degradation model (progD-MS in

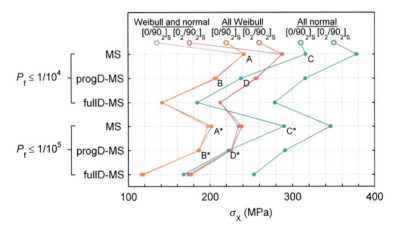

Fig. 6.4 Comparison of the maximum allowable stress estimates of two glass/epoxy cross-ply laminates with "first ply failure due to fiber fracture" as the definition of laminate failure. The maximum allowable stresses were calculated for two safety levels ($P_f \leq 1/10^4$ and $P_f \leq 1/10^5$) with three different sets of probability distributions ("All Weibull", "Weibull and normal", and "all normal"), and three different mechanical models (linear elastic: MS, linear elastic with progressive continuous degradation: progD-MS, and linear elastic with full degradation: fullD-MS).

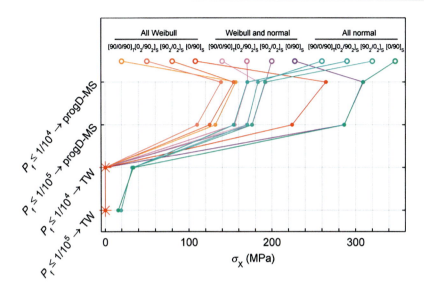

Fig. 6.5 Comparison of maximum allowable stress estimations of four carbon/epoxy cross-ply laminates with the "first ply failure due to matrix cracking or fiber fracture" as the definition of laminate failure. The maximum allowable stresses were calculated for two safety levels ($P_f \leq 1/10^4$ and $P_f \leq 1/10^5$) with three different sets of probability distributions ("All Weibull", "Weibull and normal", and "all normal"), and two different mechanical models (linear elastic Tsai-Wu: TW and linear elastic with progressive continuous degradation: progD-MS).

Fig. 6.5), matrix cracking as a failure is a certain crack density considered as unacceptable. This distinction results in widely different reliability estimates.

Our final study on reliability of composites presented three major conclusions. First, for laminates with a definition of failure that allows for matrix cracking, mechanical models that account for the progressive loss of stiffness may be useful for motivating higher operational limits than the ones obtained with a full degradation model. Second, for laminates with a definition of failure that does not allow for matrix cracking, the exact definition of "failure due to matrix cracking" is a key on motivating higher operational limits. Third, the choice of probability distributions for the mechanical properties is not trivial.

Arguably, the most significant contribution of this work is that it demonstrates how frail and uncertain are strength reliability analyses of FRP laminates in tension. Changes in material and layup, factors which one can be certain of, have a relatively small effect on the estimated operational limits compared with changes in the mechanical and probabilistic models, and definition of laminate failure. Of these three high-influence factors, probabilistic modelling is the most prominent one. A very large number of investigations have dealt with the question of assigning probability distributions to the mechanical properties of FRPs (Hwang et al., 2003; Lekou and Philippidis, 2008; Shaw et al., 2010; Zureick et al., 2006). Nevertheless, despite decades of research, there is no consensus on which probability distributions should be used. Considering that in a structure very low probabilities of failure are sought, and how sensitive are strength reliability analyses of FRP structures to probability distributions, engineers should be conservative and use the Weibull distribution for strength properties. Furthermore, a sensitivity analysis should be a requirement for the acceptance of the results of any kind of reliability analysis; the uncertainty and effect of every choice should be evaluated. Can lower material safety factors for composite marine structures be motivated through reliability analyses? Maybe. The investigations presented in this chapter are not exhaustive. Despite that for the analyzed cases higher operational limits could not be motivated, there is the possibility that for other cases the conclusion might be the opposite. The second most significant contribution of this work is the identification of the uncertainties that highly influence strength reliability analyses of FRPs. If these uncertainties are resolved, significant increases in material utilization could be achieved.

6.4 Material characterization

The work presented in this section aimed at exploring the possibility of improving material utilization by using nonstandard measuring techniques in material characterization. Material properties are fundamental information in all structural analyses, whether they are deterministic or probabilistic. Essentially, material characterization consists of experimentally measuring the properties of a representative sample of material specimens, and subsequently, determining their probabilistic characteristics through statistical inference. For a certain material property, an increase in the mean of experimentally measured values or a reduction on their scatter can strongly influence its design value in a deterministic analysis or its probabilistic modelling in a reliability analysis, potentially allowing for lighter and cheaper structure designs.

A series of mechanical tests related to the characterization of an ideal NCF ply and a quasi-isotropic NCF laminate were carried out with a nonstandard measuring technique: Digital Image Correlation (DIC) (Sánchez-Heres et al., 2016). This measuring technique held promise of more accurate measurement of stiffness and strength properties.

DIC is an optical method for measuring the 2D or 3D position and displacement of points on a surface. Each surface point is identified through a facet—a rectangular area of a predefined size (typically 15×15 pixels) containing a particular distribution of grey-scale pixels. As the surface deforms, the location of the facets changes. Compared to bonded strain gauges and physical extensometers, a DIC system is an advanced measuring technique because it allows for the study of the deformation of a whole surface at a local level. At a given instance, a physical extensometer or a bonded strain gauge renders only one measurement: strain in the gauge direction. In contrast, a DIC system renders for all the facets it recognizes: their current location in a 2D or 3D global coordinate system, their displacement with respect to the previous or initial measurement, the strains in the global coordinate system, the magnitude and direction of the major and minor strains, and more. Clearly, the amount of information obtained from a DIC system is extensive. The challenge is to make it useful.

Effective use of composite marine structures: Reducing weight and acquisition cost 171

Fig. 6.6 presents an example of the experimental setup: a DIC system measuring the deformation of a laminate sample in a tensile rig. The figure also shows a video extensometer; this device was only used in some of the tests. Figs. 6.7 and 6.8 present how the DIC system measurements were used to measure laminate strain in the specimens. For details regarding the experimental setup, the reader is referred to (Sánchez-Heres et al., 2016).

Two basic measurement techniques are shown in Fig. 6.7: facet strains and virtual extensometer. Facet strains are calculated by the DIC system through a deformation tensor determined from the relative displacement of adjacent facets. These measurements are not used for determining the laminate strain (more on that later). A virtual extensometer consists of the identification of two viable facets aligned in the longitudinal or transverse direction, and on the recording of the Euclidian distance between them as the load increases. Virtual extensometers are the building blocks for the next strain measuring technique.

Fig. 6.8 presents the measurement and location of two virtual strain *gauges*. A virtual strain gauge is an arrangement of horizontal and vertical virtual extensometers set to mimic the measurement capacity of a bonded strain gauge or a clip-on extensometer; the height and width of the virtual strain gauge can be adjusted to match the ones of either device. A crucial detail regarding the characterization of textile composites is shown in this figure: the strain measured by identical devices on the *same* specimen at different locations is different. The difference is because the surface strain field is not homogenous (due to the material's heterogeneity) and because the virtual

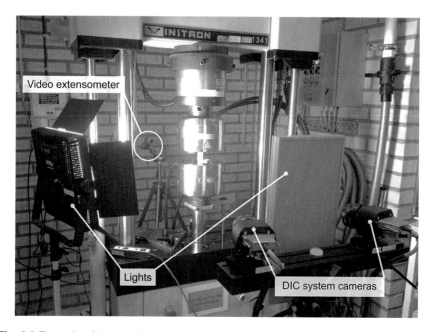

Fig. 6.6 Example of the experimental setup.

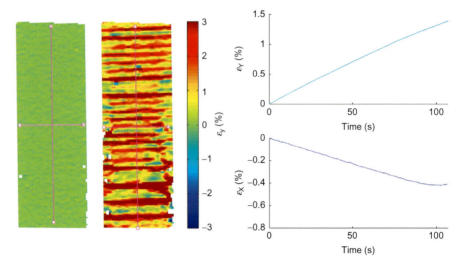

Fig. 6.7 The surface plots show the facet strains measured with the DIC system at the beginning (left) and end (right) of the test, as well as the location of a virtual extensometer (magenta lines with white-dotted ends). The line plots show measurements of the virtual extensometer. (For interpretation of the references to colour in this figure legend, the reader is referred to the web version of this article.)

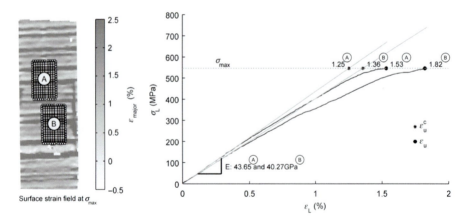

Fig. 6.8 Example of two sets of mechanical properties measured on a quasi-isotropic laminate test specimen at different location with virtual strain gauges (Subscript L: longitudinal).

strain gauge averages the strains within its gauge. As a consequence, the difference between two local strain measurements will depend on the dimensions and location of the strain gauge, as well as on the characteristics of the laminate surface (Ifju et al., 1995; Masters and Ifju, 1997).

In material characterization, local strain measurements (e.g., bonded strain gauges and clip-on extensometers) are used as proxies of global laminate strain. When treated as such, the previously mentioned difference between local strain measurements is essentially a random strain measurement error. Then, the question is whether or not this error is significant. In Fig. 6.8, one can see that because of this error, considerably different material property instances can be derived from the same specimen. Having this question in mind, we performed a study to assess current practices in the characterization of Non-Crimp Fabric laminates.

Fig. 6.9 presents boxplots summarizing the scatter of fifty design values for a laminate property. Each one of the fifty design value instances is obtained from a set of mechanical properties determined through single local strain measurements on a fixed set of specimens (i.e., for all the design value instances, the specimens are the same). The only thing that changes is the location of the strain measurement in each specimen, and therefore, the strain measurement error in each specimen. The strain measurement error in each specimen causes the scatter of the design values in Fig. 6.8. It pollutes the measurement of the mechanical property on each specimen, rendering different value combinations each time the design value is determined (for a more thorough explanation see (Sánchez-Heres et al., 2016)). An important characteristic of the strain measurement error is that its size is inversely proportional to the gauge size of the strain measuring device, so the larger the gauge size is, the smaller the error, and consequently, the smaller the scatter of design values. The dashed horizontal line in Fig. 6.9 represents the "best estimate" of the design value obtained when the strain measurement error is averaged out with multiple strain measurement replications in each specimen. For the design value shown in Fig. 6.9, the larger the gauge size is, the lower are the chances that the design value determined without strain measurement replications will be smaller than the "best estimate" one.

These results demonstrate that the random strain measurement error can significantly reduce the accuracy of characterization methodologies for NCF laminates.

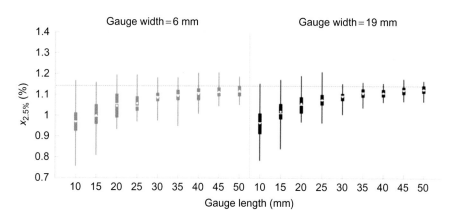

Fig. 6.9 Relation between the scatter of a design value and the gauge size of a strain measurement. The design value is the 2.5-percentile of the corrected ultimate laminate strain (ε_u^c in Fig. 6.8) of a carbon/vinylester non-crimp fabric laminate. The boxplots summarize the scatter as whiskers: 1% to 99%, box: 25% to 75%, circle: average, white line: median.

The strain measurement error pollutes the mechanical property measurements on laminate test specimens, leading to inaccurate statistical inferences. For the design value shown in Fig. 6.9, with single local strain measurements, one is more likely to get a value that is lower than it should be. How much smaller (or larger) is related to the gauge size of the strain measuring device; therefore, strain measurements should be performed with a device with the largest possible gauge size. Furthermore, if a DIC system is available for the characterization of a laminate, strain measurement replications on the specimens could be performed to average out the strain measurement error and obtain the "best estimate" design values.

Can material characterization with a DIC system lead to higher material efficiency? Yes. The repeatability and noncontact of the DIC system are valuable advantages compared to standard strain measuring techniques. However, for the sole purpose of averaging out the strain measurement error, video extensometers and other types of strain measuring devices with large gauge lengths are likely to perform nearly as good as a DIC system without the additional postprocessing of data and additional costs.

6.5 Structural design exploration

Lightweight structural design is essentially a constrained optimization problem. As such, it can be stated as,

$$\text{find } \mathbf{X} = \begin{bmatrix} x_1 \\ x_2 \\ \vdots \\ x_n \end{bmatrix} \text{ that minimizes } f(\mathbf{X}) \tag{6.2}$$

constrained by,

$$\begin{array}{l} g_j(\mathbf{X}) \leq 0 \ \ j=1,2,\ldots,m \\ l_j(\mathbf{X}) = 0 \ \ j=1,2,\ldots,p \end{array} \tag{6.3}$$

where $\mathbf{X}$ is the design vector representing design variables (e.g., laminate thickness, spacing between longitudinals, web height), $f(\mathbf{X})$ is the objective function (e.g., weight and cost of the structure), and $g_j(\mathbf{X})$ and $l_j(\mathbf{X})$ are inequality and equality constraints, respectively (e.g., operational limits stated by design rules and manufacturing constraints). The design space is a n-dimensional Cartesian space where each coordinate represents the value of one of the design variables, and consequently, each point in the design space is a possible structure design. If the structure design point meets the constraints, the structure design is an acceptable one.

Considering this mathematical description of a constrained optimization problem, an engineer has essentially two approaches for reducing the weight of a structure design:

(1) Search the design space for a lighter acceptable version of the design.
(2) Modify the design constraints to directly obtain a lighter acceptable version of the design.

Typically, in the design of a structure, both approaches are used, albeit to different extents depending on the characteristics of the problem, the available resources, and the benefits associated with the objective value (e.g., manual design exploration vs. optimization algorithms). A simple example of modifying a design constraint to reduce weight is using carbon-fiber instead of glass-fiber reinforced polymers.

The relative effectiveness of both approaches was investigated on a study case structure with the following design exploration methodologies:

(1) Sample the design space within a set of fixed constraints.
(2) Evaluate the objective function.
(3) Implement hypothetical sets of modifiable design constraints.
(4) Plot histograms of the objective values of the structure designs deemed as acceptable by the different sets of modifiable design constraint.
(5) Analyze the histograms to determine the potential benefits of the two approaches.

Fig. 6.10 presents the procedure for performing the first four steps of the design exploration methodology for our study case: a catamaran ferry hull girder built with sandwich construction. The fixed constraints are the ones related to the construction of the catamaran hull girder (i.e., side constraints), while the modifiable constraints are the ones related to the acceptance of the structure design based on its performance (i.e., behavioral constraints). The histogram obtained in Step 4 gives an estimate of the distribution of weights of the acceptable structure designs available in the constrained design space. If the histogram is representative (a quality depending on the shape of the objective function and the number of samples), optimization algorithms will only find structure designs with weights within the range of values shown by the histogram. Therefore, the range of values gives us an idea of the weight saving capacity of optimization algorithms. Similarly, comparing histograms corresponding to different modifiable constraint values gives us an idea of the weight saving capacity of modifying design constraints.

A catamaran ferry hull girder built with sandwich construction was chosen as the study case because it has been built for some time now (Phillips, 2008), and the interest for it is increasing (Mosgaard et al., 2014). Considering that the findings of this investigation are to a large extent coupled to the analyzed structure, it is reasonable to choose one which is of interest today and of relevance for the future. The catamaran ferry hull girder has the additional benefit of being a complex structure subjected to multiple and challenging load cases; therefore, the different behavioral constraints of the material are likely to limit the design. Fig. 6.11 shows the geometry of the structure, Fig. 6.12 shows the local loads, and Fig. 6.13 shows the Load Cases (LC).

For this study case, 1200 different structure designs were generated with latin-hypercube sampling. Out of these 1200, only 758 structure designs are used in the design exploration analysis. The other structure designs were scrapped for one of three following reasons:

(1) A configuration of the local loads resulting in zero-trim could not be found for the structure design.
(2) The nonlinear structural analysis did not converge.
(3) Ansys could not mesh appropriately the structure design.

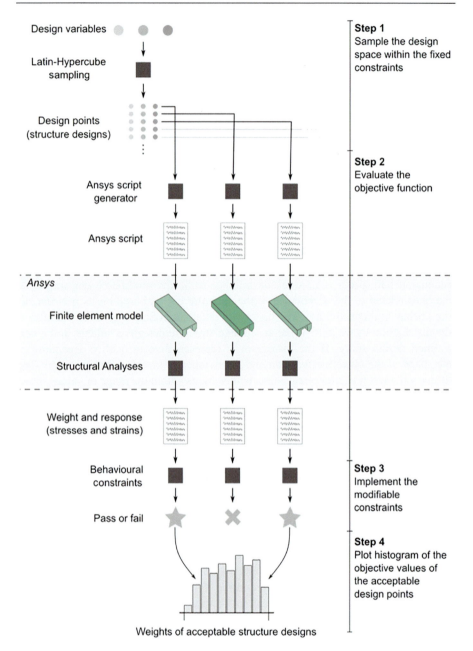

Fig. 6.10 Procedure for performing the first four steps of the design exploration methodology for the study case.

Effective use of composite marine structures: Reducing weight and acquisition cost

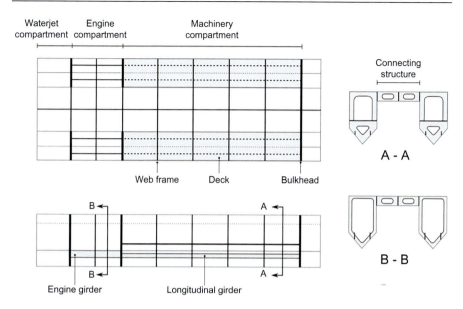

Fig. 6.11 Example of the catamaran hull girder scantlings.

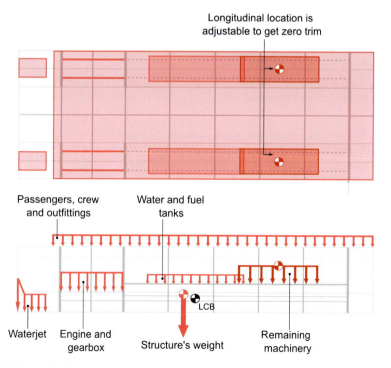

Fig. 6.12 Local loads on the catamaran hull girder.

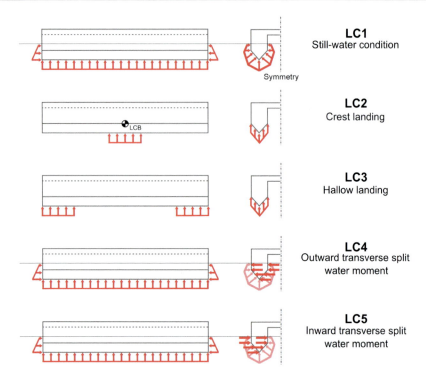

Fig. 6.13 Load Cases (LC) for the structural analyses. Only the sea loads are pictured. For the local loads, see Fig. 6.12.

The Behavioral Constraints (BC) assessed in the study case are the following:

- BC1. Tensile strength of the quasi-isotropic laminate
- BC2. Compressive strength of the quasi-isotropic laminate
- BC3. Tensile strength of the unidirectional laminate
- BC4. Compressive strength of the unidirectional laminate
- BC5. Core shear strength
- BC6. Local skin buckling

All these behavioral constraints can be expressed as inequalities of the form,

$$x \leq x_{max} = \frac{mp}{sf} \tag{6.4}$$

where x_{max} is the maximum allowable stress or strain, sf is the safety factor, and mp is a material property or a number derived from material properties (as in the local skin buckling constraint). Hypothetical beneficial changes on these behavioral constraints can be expressed as,

$$x \leq F \cdot x_{max} = \frac{F_{mp} \cdot mp}{F_{sf} \cdot sf} \tag{6.5}$$

Effective use of composite marine structures: Reducing weight and acquisition cost   179

where F_{mp} and F_{sf} are factors representing a potential increase on the values of the material property and a reduction of the safety factor, respectively, while F represents the quotient of the division of these two factors. For the study case, we considered possible a maximum beneficial change of 10% for both F_{mp} and F_{sf}. These beneficial changes translate to a range of values for F between 1.00 and 1.20.

Fig. 6.14 shows the histogram of the weights of the acceptable structure designs within the 758 samples for the case when $F = 1.00$ for all behavioral constraints. The x-axis range of the histogram plot (approximately 16–80 tons) is the range of weights of all the sampled structure designs, both acceptable and unacceptable. Fig. 6.14 also shows pie charts indicating the fraction of the evaluated samples deemed as acceptable and unacceptable, as well as an analysis of the unacceptable

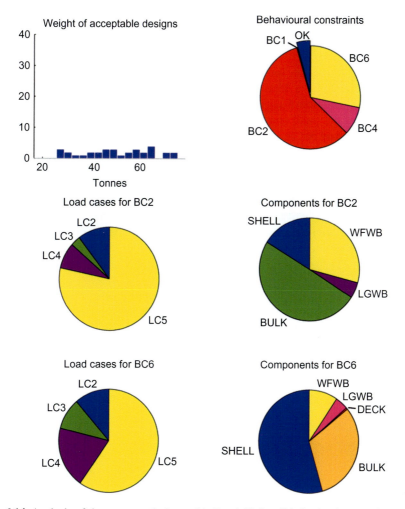

Fig. 6.14 Analysis of the structure designs with $F = 1.00$ for all behavioral constraints.

ones. The analyses of the unacceptable structure designs informs which behavioral constraint is the foremost limiting one (by calculating the fraction between x/x_{max}), as well as in which loading condition and structural element. For example, in the figure, one can observe that less than one eigth of the evaluated samples were deemed as acceptable and that most of the unacceptable ones failed due to BC2 and BC6. BC2 was violated mostly in LC5 and on the bulkheads (BULK), while BC6 was violated mostly also LC5 and in the hull shell (SHELL).

For LC4 and LC5, some zones (see Fig. 6.15) had to be removed from the evaluation of the behavioral constraints. The zones were removed because the stresses and strains at these areas under these loading conditions are unrealistically large, and therefore, the structure designs were too far from meeting the behavioral constraints. The stresses and strains are unrealistically large because the finite element model of the joints in these zones does not include the extra core material, adhesive, and over-lamination that, in real structures, spread the loads. Removing these zones, instead of increasing the complexity of the model, was deemed as reasonable simplification for this first study case.

In Fig. 6.14, one can see that most of the structure designs are rejected when $F = 1.00$ for all the behavioral constraints. Fig. 6.16 shows how changing BC2 affects the weights. Going from $F_{BC2} = 1.00$ to 1.20 does not decrease the lower bound of the histogram, but the small number of acceptable designs close to the bound indicates that the bound may not be so trustworthy (i.e., the bound is not thoroughly explored, so there might be in fact some change in the bound). However, increasing the F-factor for the BC2 does increase the number of structure designs that are acceptable. Further analyses showed that the same is true for all the other evaluated behavioral constraints: positively modifying them increased the number of acceptable designs, but not the range of weights (at least not in a manner detectable through our methodology).

Our design exploration methodology indicates that searching the design space has the largest potential, as modifying the design constraints did not enable previously

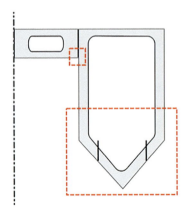

Fig. 6.15 Zones excluded from the evaluation of the behavioral conditions for loading cases 4 and 5.

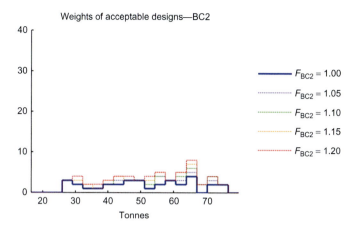

Fig. 6.16 Changes in the weights histogram caused by different F-factors for BC2 (the F-factors of all other behavioral constraint are set to 1.00).

unattainable lighter structure designs. This result does not mean that modifying the design constraints is not beneficial. Weight reduction can definitely be achieved by modifying the design constraints; however, the results indicate that for a large hull built with sandwich-construction, any weight reduction achieved by modifying the constraints is also achievable by searching the design space.

The applicability of this conclusion regarding the relative potential of both approaches to other composite marine structures is uncertain. To the author's knowledge, this study is original and no other study comparing searching the design space and modifying the design constraints of marine structures has been published. Without any other similar investigation, the only evidence supporting this conclusion is the one presented here, and it consists of a study case with some notable limitations. Considering the complexity of the structure, and the authors' first estimate of how the results would turn out, it seems futile to try to predict how well a less simplified hull or more load cases will affect the conclusion. An improved study case and additional study cases are necessary to strengthen this conclusion and broaden its applicability.

6.6 Conclusions

The objective of this work was to reduce the weight of composite marine structures, so as to make them more economically attractive. Three approaches for reaching this objective were presented, investigated, and discussed.

Reliability of composites has been investigated for decades; DIC systems are commonly used in modern experimental investigations; and searching for better structure designs through optimization algorithms is an approach currently researched and available to industry. The novelty of this work lies in the details of the presented approaches.

The use of a mechanical model accounting for the progressive degradation of the composite material is an important step forward in the composite reliability field according to a recent survey (Chiachio et al., 2012). The results show that strength reliability analyses of FRP laminates in tension are frail and uncertain, because uncertain details of the probabilistic and mechanical models, as well as definitions of laminate failure, highly influence the reliability estimates. Considering that in a structure very low probabilities of failure are sought, with the current knowledge, engineers should be conservative. Our studies on the estimation of operational limits do not provide a mature and straightforward approach for reducing the weight of composite structures, but they do elucidate the aspects which hold the largest opportunities for weight reduction (aspects such as choice of probability distributions and definition of laminate failure). These aspects indicate research areas of interest.

The experimental part of our work provides a mature and straightforward approach for reducing the weight of composite structures: for the characterization of NCF laminates (or other textile laminates), use a strain measuring technique with a large strain gauge, or even better, a DIC system with a script for taking multiple virtual extensometer strain measurements per coupon. Doing so will reduce or nearly eliminate the strain measurement error, and therefore, improve the accuracy of the characterization method.

Weight reduction can definitely be achieved by modifying design constraints; however, the results of our design exploration methodology indicate that for a large hull built with sandwich- construction, any weight reduction achieved through lower safety factors or more accurate mechanical properties could alternatively be achieved by searching the design space. This conclusion may or may not be bound to the study case or the simplifications made in the structural analyses; further investigations are necessary to strengthen it. The work presented in Section 6.5 does not provide a new approach for reducing the weight of marine composite structures; it indicates where the largest opportunities lie.

Each of these opportunities has two characteristics of great importance: cost and worth. Cost should be interpreted as the effort, time, and money necessary to exploit the opportunity, while worth should be interpreted as the possible weight reduction achievable through it. Obtaining more accurate mechanical properties through better laminate strain measuring techniques is a low hanging fruit. This opportunity is arguably the one worth the least, but the cost of taking it is minimal. On the other hand, the cost of searching for better structure designs is high, but this opportunity is arguably the one worth the most. The high cost comes from building robust parametric structure designs. Finally, motivating higher material operational limits through reliability analysis is the opportunity with the lowest worth to cost ratio. Its cost is high because the mechanical and probabilistic models required by this opportunity are not mature enough, and therefore, need to be further researched.

References

BESST, 2013. Breakthrough in European Ship and Shipbuilding Technologies. http://www.besst.it (web archive link, 16 May 2010). (Accessed 16 May 2010).

Chiachio, M., Chiachio, J., Rus, G., 2012. Reliability in composites—a selective review and survey of current development. Compos. B Eng. 43 (3), 902–913.

DNV, 2013a. Rules for Classification of High Speed, Light Craft and Naval Surface Craft. Det Norske Veritas, Høvik.

DNV, 2013b. Offshore Standard DNV-OS-C501: Composite Components. Det Norske Veritas, Høvik.

Hertzberg, T., 2009. LASS, Lightweight Construction Applications at Sea. SP Technical Research Institute of Sweden, Borås.

Hwang, T.K., Hong, C.S., Kim, C.G., 2003. Probabilistic deformation and strength prediction for a filament wound pressure vessel. Compos. B Eng. 34 (5), 481–497.

Ifju, P.G., Masters, J.E., Jackson, W.C., 1995. The use of moiré interferometry as an aid to standard test-method development for textile composite materials. Compos. Sci. Technol. 53, 155–163.

Lekou, D., Philippidis, T.P., 2008. Mechanical property variability in FRP laminates and its effect on failure prediction. Compos. B Eng. 39 (7–8), 1247–1256.

Masters, J.E., Ifju, P.G., 1997. Strain gage selection criteria for textile composite materials. J. Compos. Technol. Res. 19 (3), 152–167.

Mosgaard, M.A., Riisgaard, H., Kerndrup, S., 2014. Making carbon-fibre composite ferries a competitive alternative: the institutional challenges. Int. J. Innov. Sustain. Dev. 8 (3), 290–310.

Nairn, J.A., 2000. Matrix microcracking in composites. In: Kelly, A., Zweben, C. (Eds.), Comprehensive Composite Materials. In: vol. 2. Elsevier, Amsterdam, pp. 403–432.

Phillips, S.J., 2008. Jane's High-Speed Marine Transportation 2008–2009. Jane's Information Group, Coulsdon.

Sánchez-Heres, L.F., Ringsberg, J.W., Johnson, E., 2013. Study on the possibility of increasing the maximum allowable stresses in fibre-reinforced plastics. J. Compos. Mater. 47 (16), 1931–1941.

Sánchez-Heres, L.F., Ringsberg, J.W., Johnson, E., 2014. Influence of mechanical and probabilistic models on the reliability estimates of fibre-reinforced cross-ply laminates. Struct. Saf. 51 (1), 35–46.

Sánchez-Heres, L.F., Ringsberg, J.W., Johnson, E., 2016. Characterization of non-crimp fabric laminates—loss of accuracy due to strain measuring techniques. J. Test. Eval. 44 (6), 2321–2337.

Shaw, A., Sriramula, S., Gosling, P.D., Chryssanthopoulos, M.K., 2010. A critical reliability evaluation of fibre reinforced composite materials based on probabilistic micro and macro-mechanical analysis. Compos. B Eng. 41 (6), 446–453.

Valdebinito, M.A., Schuëller, G.I., 2010. A survey on approaches for reliability-based optimization. Struct. Multidiscip. Optim. 42 (5), 645–663.

Varna, J., Joffe, R., Akshantala, N.V., Talreja, R., 1999. Damage in composite laminates with off-axis plies. Compos. Sci. Technol. 59 (14), 2139–2147.

Wang, A., 1984. Fracture mechanics of sublaminate cracks in composite materials. Compos. Technol. Rev. 6 (2), 45–62.

Zureick, A.H., Bennet, R.M., Ellingwood, B.R., 2006. Statistical characterization of fibre-reinforced polymer composite material properties for structural design. J. Struct. Eng. 132 (8), 1320–1327.

Section B

Sandwich structures

Core materials for marine sandwich structures

Nikhil Gupta*, Steven Eric Zeltmann*, Dung D. Luong*, Mrityunjay Doddamani[†]
*Composite Materials and Mechanics Laboratory, Department of Mechanical and Aerospace Engineering, New York University Tandon School of Engineering, Brooklyn, NY, United States, [†]Advanced Manufacturing Laboratory, Department of Mechanical Engineering, National Institute of Technology Karanataka, Surathkal, India

Chapter Outline

7.1 Introduction 187
    7.1.1 Fabrication of foam core sandwich structures 189
    7.1.2 Marine loading conditions 190
7.2 PVC foams 190
    7.2.1 Microstructure 191
    7.2.2 Compressive properties of PVC foams and sandwiches 192
    7.2.3 Impact properties of PVC foams and sandwiches 198
    7.2.4 Moisture effects 201
7.3 Syntactic foams 203
    7.3.1 Hollow particles and their properties 203
    7.3.2 Compressive properties 206
    7.3.3 Impact properties of syntactic foams 209
    7.3.4 Moisture effects 214
    7.3.5 Tailoring the properties of syntactic foams 216
7.4 Summary 219
Acknowledgments 219
References 219

7.1 Introduction

Development of the USS *Zumwalt*, also known as DDG 1000, has launched a new era of materials for the US Navy. This ship has used a variety of composites for construction of large scale parts such as the deckhouse. Such advancements have become possible due to intense research efforts over the past three decades in developing composite materials suitable for marine conditions. Marine structures encounter some unique conditions during their service life, which include the combined effects of moisture, temperature, and mechanical loads. Composite materials for marine structures are optimized to provide high performance under such rigorous loading conditions.

Marine Composites. https://doi.org/10.1016/B978-0-08-102264-1.00007-8
Copyright © 2019 Elsevier Ltd. All rights reserved.

Low density and the capability to keep damage localized are two important characteristics of marine composite materials. Initially, natural materials such as balsa and cork were used in weight-sensitive applications (Gil, 2009; Castro et al., 2010). These materials provide natural and sustainable options in construction of core materials in sandwich structures (Le Duigou et al., 2012; Reis and Silva, 2009). *E*-glass/balsa wood sandwiches are currently used in the Advanced Enclosed Mast/Sensor on the US Navy DDG-968 surface ships (Ulven and Vaidya, 2006), among many other smaller naval applications (Mitra, 2010). Such applications necessitate close study of their impact properties in addition to the standard tests (Atas and Sevim, 2010; Shir Mohammadi and Nairn, 2017). Moisture uptake and degradation mechanisms of such sandwich structures have been studied due to their relevance to marine applications (Legrand et al., 2015). Large dimensional changes due to moisture uptake can be problematic for structural applications in moist environment. Since Navy ships and boats widely used balsa wood, studies are also available on ballistic and blast response of balsa core sandwich structures (Chen et al., 2011; Jover et al., 2014). However, over the past several decades, desire to develop materials with greater control over the density and mechanical properties has resulted in development of polymer foams for the same applications.

Foams are among the materials that are known to have low density and impact damage tolerance. Therefore, foams are widely used as core materials in marine sandwich structures. Using glass or carbon fabric skins (also called face-sheets) helps in increasing the bending stiffness and impact strength of foam core sandwich structures. There are several examples for use of polymer foams in marine sandwich structures. Hulls of small- and medium-sized boats and yachts have been constructed using PVC foam core sandwich. The hull of the Visby class corvette developed by the Swedish Navy is made of foam core sandwich. Large scale sandwich panels have been used in the construction of the deckhouse and deck of USS Zumwalt. Foams are used in fenders of small boats for impact energy absorption and for buoyancy.

Foams are cellular materials that are classified into two categories of open- and closed-cell foams. In open-cell foams, the cells are interconnected and a fluid can penetrate through the thickness of the foam. The cells of such foams are made of struts that provide a very high level of porosity. Open-cell foams are not very suitable for marine sandwich structure applications for many reasons: (a) their strength and stiffness are low and so their load bearing capacity is small, (b) damage to skins can lead to water uptake in the entire foam, making it very heavy, (c) attaching face-sheets (skins) is challenging due to very small cross section area in the foam to bond with the face-sheets. For these reasons, closed-cell foams are preferred in marine sandwich structures as core materials.

This chapter is focused on presenting an overview of the polymeric core materials used in sandwich structures. Two types of polymeric foams are discussed, which include polyvinyl chloride (PVC) foams and syntactic foams. Metal matrix sandwich structures containing open-cell foam and truss cores have also been studied in the published literature for marine structures (Jing et al., 2016; Liu et al., 2017; Magnucka-Blandzi, 2018; Magnucka-Blandzi et al., 2017; Wang et al., 2017; Radford et al., 2006). However, metallic materials are not discussed in this chapter because the fabrication methods, deformation mechanisms, and mechanical behavior of metallic materials are very different from those of polymeric materials. Specialized literature available on metal foams, including syntactic foams, can be consulted for more information on those materials (Gupta and Rohatgi, 2014; Dukhan, 2013).

7.1.1 Fabrication of foam core sandwich structures

Several methods are used for manufacturing foam core sandwich composites. These methods are briefly described below.

7.1.1.1 Adhesive bonding

In this method, the face-sheets are adhesively bonded to the foam slabs. The face-sheets may be fiber or fabric reinforced laminates, which are fabricated separate from the foam slab. Metal or polymer sheets can also be taken as the face-sheets depending on the application of the sandwich. Factors that affect the properties of sandwich include:

- The chemical compatibility of the adhesive with the materials used in both the foam and face-sheets, to provide a strong bond at the interface.
- The adhesive layer thickness plays an important role in defining the behavior of the sandwich composites. Shear failure within thick adhesive layers can be a problem.
- The surface roughness is an important property in ensuring a high-quality joint. In foams, the surface roughness is defined by the cell size and cell wall thickness and cannot be directly controlled.

Adhesive bonding provides the flexibility of fabricating the face-sheets separately from the core and is a popular method for sandwich fabrication.

7.1.1.2 Hand layup

The use of laminated face-sheets provides the possibility of fabricating them directly on the foam core material. In this case, the laminate fabrication and bonding with the core can be achieved in one step. Vacuum bagging may or may not be applied during the hand layup depending on the type of materials and objective of the process. The vacuum bagging technique has been used commonly as a method to fabricate sandwich composites.

Very large cell size and highly porous foams are difficult to process by this method because the resin from the skins seeps into the core and makes the whole sandwich heavier. Such methods are suitable with closed-cell foams because resin penetration is limited to the top layer of cells and the inner core remains intact. Marine sandwich structures using closed-cell foams can use this process.

7.1.1.3 Infiltration techniques

These techniques are applicable to syntactic foam core sandwich structures. In this method, the skin fabric layers and particles that are used in syntactic foam are poured in a mold in the desired configuration. The matrix resin is infiltrated into the prepared mold using a pressure difference (Huber and Klaus, 2009). This kind of process eliminates the presence of interface between the core and the face-sheets. Similar processes have been used for metal matrix syntactic foam core sandwich structures (Omar et al., 2015; Yaseer Omar et al., 2015; Lamanna et al., 2017). Elimination of interface reduces the possibility of shear failure at the interface between core and face-sheets.

7.1.2 Marine loading conditions

Marine structures are subjected to dynamic, and often complex, loading conditions. Materials used in boat and ship hulls are exposed to water immersion conditions as well as hydrodynamic loading due to repeated water exit and entry at different rates. Temperatures may vary from arctic conditions to over 45°C. In addition, the submerged components are subjected to hydrostatic compression. All these effects may take place individually or in tandem. Numerous possible combinations of loading and environmental conditions have made the testing of marine composites very challenging.

Apart from the routine operating conditions, Navy relevant structures are also designed for dynamic loading conditions such as high strain rate deformation and blast loading. There are many challenges for testing materials at strain rates of 10^3–10^6 s^{-1} that are generated by such loading conditions. The heat generated in the material during high strain rate deformation changes the measured mechanical properties. In addition, test methods such as the split-Hopkinson pressure bar technique have several assumptions such as neglecting the radial dispersion effects of the elastic waves, stress equilibrium in the specimen during deformation, and neglecting the temperature-dependent change of the bar properties, which can compromise the results and make the calculation of mechanical properties from these techniques a rigorous exercise.

Given the variety of possible test conditions for marine relevant structures, the discussion in the present chapter is mainly focused on the compressive and impact properties of core materials and their sandwich composites. In addition, effects of moisture on these properties are also discussed. Strain rate and temperature-dependent properties are available in the literature for some material systems and should be considered while designing marine structures where such conditions are relevant to the actual application.

7.2 PVC foams

The use of PVC foam as a core material in sandwich composites dates back to the 1930s, when manufacturers began tailoring the composition of PVC foams to meet the requirements of marine applications by blending polyvinyl carbonate with isocyanate. DIAB and Airex/Herex are among the main suppliers of PVC foams for the marine industry. PVC foams with densities in the range of 60–350 kg/m^3 are most commonly found in the literature and in applications (Colloca et al., 2012; Kidd et al., 2012). Linear PVC, polystyrene (PS), styreneacrylonitrile (SAN), polyurethane (El-Hadek and Tippur, 2003), polyisocyanurate (PIR), polymethylmethacrylate (PMMA), and polyetherimide (PEI) have also been used to produce lightweight foams for sandwich cores in the recent past. One major approach of current research for improving the properties of foam cores is exploration of modification of the polymer network by the addition of various additives or nano-fillers. A polyurea network modified by epoxy resin cross-linked PVC has been reported (Jiang et al., 2014), showing significant improvement in energy absorption, particularly in shear. Special blowing agents and additives can be utilized to gain exact control over the foam microstructure while maintaining the same density (Saha et al., 2005). The degree of cross-linking during polymerization is controlled by the additives, which affects cell size without affecting the total blown volume. This control allows for manipulation of the aspect ratio of the cell walls, which can be used to control the failure mechanisms that will dominate under a given loading condition. However, such modification often involves the use of potentially harmful additives and increased processing costs, which must be weighed against the benefits due to the modification of the foam (Demir et al., 2008).

7.2.1 Microstructure

Closed-cell PVC foams are widely studied for use as core in marine sandwich structures. Fig. 7.1 shows PVC foams of two different densities, 60 and 250 kg/m^3. The closed-cell structure of these foams helps in obtaining high compressive strength and modulus compared to open-cell foams. The gas entrapped in the foam cells applies back pressure, which helps in improving the compressive and impact properties of these materials. The cells are polygonal in shape. Although the cell walls are thin, the cell junctions can be thick and provide strength. Usually, there is a large deviation in the cell size and wall thickness in these foams as shown in Fig. 7.2 for foams of four different densities (Luong et al., 2013). In general, the average values of cell size and wall thickness are representative of their properties relative to each other.

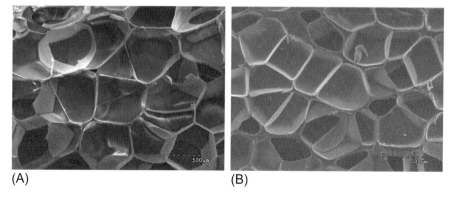

Fig. 7.1 Closed-cell PVC foams: (A) HP60 and (B) HP250. The numbers in the designation denote their nominal density in kg/m^3. Both images are acquired at the same magnification on specimens sectioned using sharp razor blades. HP250 has thicker walls.

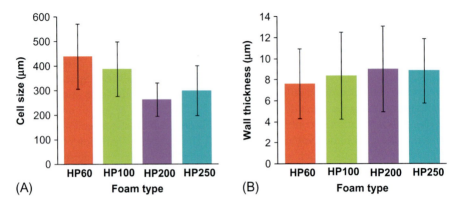

Fig. 7.2 Measurements of PVC foam (A) cell size and (B) wall thickness (Luong et al., 2013). Standard deviations are shown as error bars. Usually standard deviations are large in foam microstructure measurements.

7.2.2 Compressive properties of PVC foams and sandwiches

PVC foam exhibits elastic-plastic behavior under compression. A representative compressive stress-strain graph for a PVC foam is shown in Fig. 7.3 (Russo and Zuccarello, 2007). There is an initial linear elastic region (segment 0A) that is associated with the elastic bending and stretching of the cell walls. This region ends at the onset of cell collapse or cell wall micro-buckling. The collapse process continues throughout the segment BC. As the strain increases beyond point B, cell walls begin to consolidate inside the foam cells and densification begins. In the densification

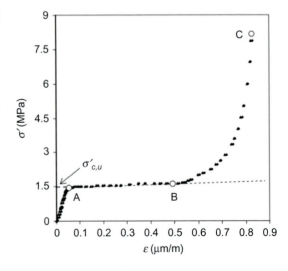

Fig. 7.3 Compression test result of a representative PVC foam (Russo and Zuccarello, 2007). Note that the strain axis appears to have incorrect units in the figure taken from the original source. The strain values on the axis are likely to have units of m/m.

region, the response is dominated by the mechanical properties of the cell wall material. Energy absorption for the foam is defined as the area underneath this curve, usually up to the densification strain. The densification strain has been defined by various means, such as the intersection between tangents drawn in the plateau and densification regions or as the strain at which stress rises 20% above the plateau stress. Behavior in the plateau region depends strongly on the foam density. Collapse in low-density foams is dominated by micro-buckling of the cell walls due to their high aspect ratio. The buckling process is sudden and leads to a drop in stress following the peak. Higher density foams tend to show a smaller decrease in stress following the peak due to plastic collapse of the walls (Kidd et al., 2012). This difference is highlighted in Fig. 7.4, which shows stress-strain curves for two different densities of PVC foams. Discussion of the critical density at which the primary mechanism of collapse switches between these two modes can be found in (Gibson and Ashby, 1999).

PVC foams also show anisotropy in their mechanical properties. Uniaxial compression and tension experiments carried out in different loading directions on Divinycell PVC foam with 250 kg/m^3 density (DIAB Inc.) are shown in Fig. 7.5 (Daniel and Cho, 2011). Both modulus and strength were found to be better in the through-thickness direction than in-plane. While the tensile modulus is close to the compressive modulus for each loading direction, the tensile strengths are higher. In marine applications, sandwich core materials can be expected to bear compressive loads in both directions and in-plane tensile loading due to bending stresses. Thus, properties must be determined in the appropriate directions for use in design calculations. Fig. 7.6 shows how slightly off-axis loading can further decrease the properties of the foam core. The directionality in the mechanical properties of the foam is an important consideration in using them in marine structures. Among the reasons for directionality in mechanical properties is the gravity-induced drainage during stabilization of foam in liquid condition and time taken in curing.

Fig. 7.7A shows representative compressive stress-strain curves for Divinycell PVC foams having densities from 60 to 250 kg/m^3 (HP60, HP100, HP200, and HP250) (Colloca et al., 2012). The cell size and wall thickness for these foams are

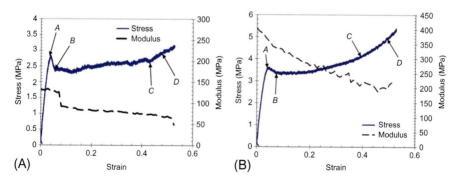

Fig. 7.4 Stress and modulus for (A) lower and (B) higher density PVC foams (Kidd et al., 2012). The y-axes of the two figures have different scales.

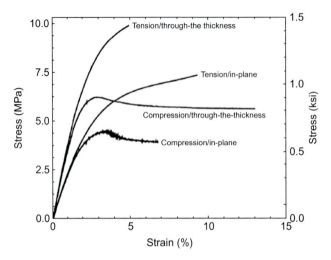

Fig. 7.5 Response of PVC foams under tensile and compressive loading in two different orientations (Daniel and Cho, 2011). The properties of gas blown PVC foams are dependent on the test orientation, which represents anisotropy in their microstructure.

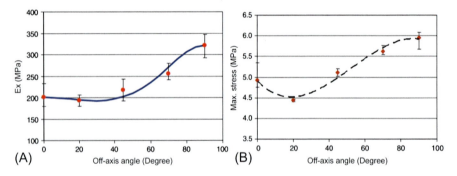

Fig. 7.6 Variation of (A) axial modulus and (B) critical stress with load orientation from the in-plane direction (1–2 plane) (Daniel and Cho, 2011).

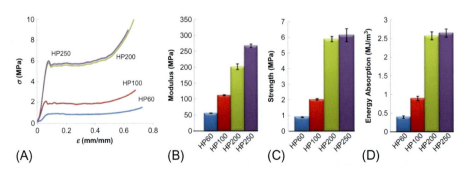

Fig. 7.7 (A) Compressive stress-strain curves for PVC foams of five different densities. (B) Modulus, (C) strength, and (D) energy absorption of PVC foams (Colloca et al., 2012).

shown in Fig. 7.2. These foams fail by cell wall micro-buckling rather than plastic collapse. Due to the high densification strains, energy absorption of such foams can still be high. As shown in Fig. 7.7B–D, the modulus, strength, and energy absorption increase with increasing foam density.

Various mathematical models have been proposed to describe the mechanical behavior of foams in terms of the properties of the bulk polymer and the geometry of the cells. Properties of open-cell foams in compression are usually described by a power law model:

$$\frac{E}{E_b} = C \left(\frac{\rho}{\rho_b}\right)^m \tag{7.1}$$

where E and ρ are the foam modulus and density, and E_b and ρ_b are the modulus and density of the bulk polymer, respectively. C and m are fitting coefficients, which are usually taken as 1 and 2, respectively. This model is also applicable to some low-density closed-cell foams, where the cell walls are so thin that the back pressure causes them to rupture at low strain. For closed-cell foams that contain sufficient material in their walls to contain the back pressure, the model given by Gibson and Ashby (1999) is often applied

$$\frac{E}{E_b} = \phi^2 \left(\frac{\rho}{\rho_b}\right)^2 + (1-\phi)\frac{\rho}{\rho_b} + \frac{\rho_0(1-2\nu)}{E_b(1-\rho/\rho_b)} \tag{7.2}$$

where ϕ is the fraction of material in the cell walls, and usually taken as 0.8 for rigid foams, ν is the Poisson's ratio of the foam, which is usually assumed to be 1/3, and ρ_0 is the atmospheric pressure. The first term represents the cell bending, similarly to the open-cell model. The second is related to the membrane stress in the cell wall, and the final term is the contribution of the back pressure of the enclosed gas. For PVC foams, this last term has been shown to be negligible, and good fitting to the experimental data is achieved when ignoring this term (Colloca et al., 2012). A similar model exists for predicting the strength of closed-cell foams, σ_y, given as

$$\frac{\sigma_y}{\sigma_{y,b}} = 0.3 \left(\phi \frac{\rho}{\rho_b}\right)^{3/2} + (1-\phi)\frac{\rho}{\rho_b} \tag{7.3}$$

when ignoring the contribution of the gas pressure inside the cells. Here, $\sigma_{y,b}$ is the yield strength of the bulk polymer. Other relations, such as for the shear modulus and tensile properties, can be found in Gibson and Ashby (1999).

Since PVC is a viscoelastic material, there is also significant strain rate sensitivity in the mechanical response. Experiments on various densities of PVC foams have shown that the strain rate sensitivity is well-described by a simple power law model, with an exponent of about 0.03 depending on the density (Luong et al., 2013). A more accurate description of the viscoelastic strain rate sensitivity can be obtained using dynamic mechanical measurements combined with time-temperature superposition to estimate the relaxation function of the polymer (Zeltmann et al., 2016, 2017a). Efforts have been made to combine the strain rate sensitivity and anisotropy measurements to create a single constitutive model for syntactic foams based on a potential function which combines deviatoric and dilational deformation components (Daniel et al., 2013). The aim of such studies is to develop a unified nonlinear elastic-plastic model for a general state of strain (or stress) (Gielen, 2008). Underwater structures are also generally subjected to multiaxial states of stress due to structural and pressure loading, and appropriate failure criteria have been experimentally determined for PVC foams of various densities (Gdoutos et al., 2002).

Measurement of foam properties at high strain rates is complicated due to the limited test methods available for such experiments. One of the major techniques that can measure stress-strain behavior at high rates is the split-Hopkinson pressure bar. This apparatus consists of two long elastic bars fitted with strain gages. The test specimen is sandwiched between the bars as the free end of one of bars is struck by a striker bar. An elastic pulse travels along the bars and interacts with the specimen. The reflection and transmission of the wave by the specimen is treated using one-dimensional wave theory to determine its state of stress and strain during the event. This information is used to develop the stress-strain curve. A detailed description of this test method, and the considerations needed for characterization of soft materials (such as PVC foam), can be found in the book by Chen and Song (2010). Experiments on high strain rate compression of PVC foams using this technique have found that the strength follows a power law relationship with respect to strain rate (Luong et al., 2013). The combined effect of temperature and strain rate on the compressive properties of PVC cores has also been studied. Denser foams show more sensitivity to temperature at high and low strain rates, and all foams show higher temperature sensitivity at low strain rates than at high strain rate (Thomas et al., 2002). Part of the temperature effect was attributed to the increased gas pressure at elevated temperatures exerting a restoring force on the cell walls.

Variation in the failure mechanism is also observed in high strain rate loading. Quasi-static (10^{-3} s^{-1}) failure behavior of a PVC foam of 250 kg/m^3 density is shown in Fig. 7.8. The compression occurs uniformly throughout the specimen and failure occurs due to buckling and subsequent folding of the cell walls. A PVC foam of the same density compressed at high strain rate (≈ 2000 s^{-1}) is shown in Fig. 7.9. In this specimen, nonuniform densification is observed, with many cells experiencing relatively little deformation while others have collapsed. In both cases, failure initiates with the weakest cell walls, which causes a stress concentration around it as neighboring cells are made to bear the load, propagating the failure across the specimen. In the high strain rate case, there is insufficient time for this damage to spread before the target strain is reached, resulting in multiple sites where failure has initiated and locations where cells are not yet deformed significantly.

In sandwich constructions, the mechanical properties of the sandwich are determined by the properties of the face-sheet, the stiffness and strength of the core, and the strength of core-to-face-sheet bonding. Stitching and z-pinning are among the approaches used to improve core-to-face-sheet bonding characteristics. Recently Yalkin et al. (2015), a foam core with drilled holes (3.5 mm diameter holes spaced

Core materials for marine sandwich structures

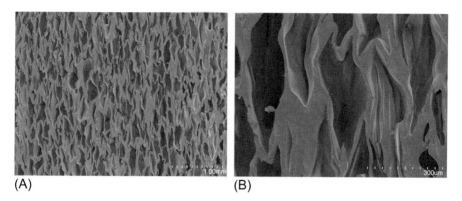

Fig. 7.8 Quasi-static compression failure of HP250 PVC foams showing cell wall buckling and folding at densification strain in the material. The specimen is compressed to 65% strain. The specimen compression is uniform through the thickness.

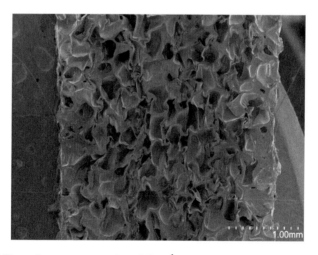

Fig. 7.9 HP250 specimen compressed at 1950 s^{-1} strain rate. It appears that the weaker cells have collapsed first in this specimen that has lower strain than the densification strain.

with 1 hole per 2 cm^2) was studied in sandwich composites using glass-epoxy face-sheets for flatwise compressive properties as shown in Fig. 7.10. Glass fibers were also stitched through the perforations in the composite and infiltrated with epoxy to clamp the face-sheets to the core as well as to create struts that resist compression. Fig. 7.11 presents compressive behavior for all the sandwich samples. The foam core (curve R) shows typical behavior for PVC foam, while the foam with perforations filled with epoxy (curve P) shows similar behavior but at higher strength and modulus. Addition of glass fibers to the perforations (curves S1-S4) substantially improves the compressive properties. Sandwiches with 2400 tex stitch show a 16% weight gain, but almost

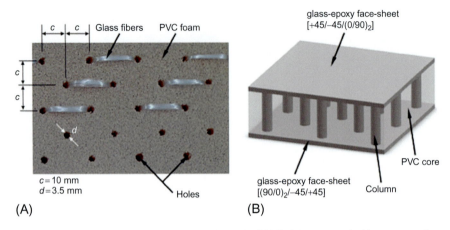

Fig. 7.10 Stitched sandwich structure (Yalkin et al., 2015). Numerous stitching patterns have been studied in published literature.

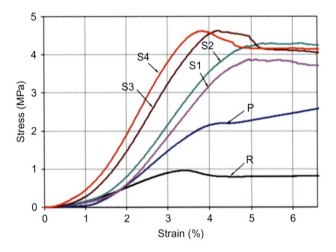

Fig. 7.11 Stress-strain curves of stitched sandwich structures (Yalkin et al., 2015). R, Non-perforated; P, Perforated; S1, Perforated 600 tex stitched; S2, Perforated 1200 tex stitched; S3, Perforated 1800 tex stitched; S4, Perforated 2400 tex stitched.

five times higher strength and modulus as compared to non-perforated sandwiches. Such innovative approaches can greatly improve the compressive properties of PVC sandwiches and allow their use in applications where stiffer cores are usually required due to the need to withstand high compressive load.

7.2.3 Impact properties of PVC foams and sandwiches

Loading conditions in marine environments are rarely static, but rather dynamic and often locally concentrated. Impact testing is a common laboratory technique for determining the response of materials and structures to dynamic loading. The most common technique for composites is the drop weight impactor, which uses a falling load attached to a tup. The tup (usually hemispherical or flat) impacts a clamped plate specimen and its acceleration throughout the impact event is measured. The acceleration is used to determine the impact force, displacement, and absorbed energy. Various results on the composite's response can be generated from such experiments, particularly the peak impact load and the energy to rupture.

Various typical load-deflection curve profiles that are observed in impact testing are shown in Fig. 7.12 (Atas and Sevim, 2010). In curve (a), the deflection decreases at the end of the experiment, indicating that the tup rebounded off the specimen rather than penetrating. Curve (b) shows a curve typical of the case where the impactor penetrates the material and may pass through and break out of the back face. Curve (c) shows the case where the material perforates after impact. Up to the peak, this curve appears the same as (b), but differs in the post-peak behavior. Perforation causes nonuniform deceleration of the tup after penetration.

Representative load-deflection curves of PVC foams and their energy absorption in drop weight impact tests are presented in Fig. 7.13 (Colloca et al., 2012). The peak

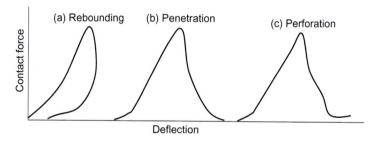

Fig. 7.12 Typical load-deflection curves for sandwich structures under impact loading (Atas and Sevim, 2010).

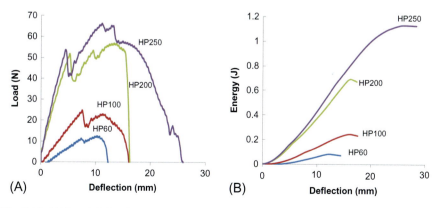

Fig. 7.13 (A) Load-deflection curves and (B) energy-deflection curves recorded during impact tests on PVC foams (Colloca et al., 2012).

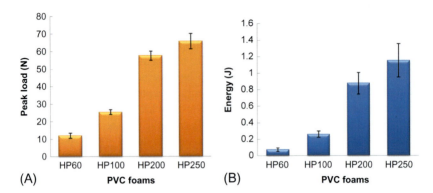

Fig. 7.14 (A) Peak load and (B) energy of PVC foams during impact testing (Colloca et al., 2012). The values show increasing trend as the foam density increases.

load and energy are plotted in Fig. 7.14. As one would expect, the denser foams show higher peak load and energy absorption. The presence of multiple peaks may indicate a multi-stage damage process due to nonuniform deformation and growth of the damage zone. Postimpact analysis by imaging and ultrasonic inspection revealed penetration of the impactor in lower density foams, while shearing of the damaged zone due to partial impactor penetration was observed in higher density foams (Colloca et al., 2012).

Improved understanding of the impact behavior of sandwiches can be obtained with knowledge of the strain rate sensitivity of the properties and failure mechanisms of the core and skin. The energy balance model can be used to predict the maximum impact force, P_{max}, for a given impact energy of a sandwich structure by (Akil Hazizan and Cantwell, 2002).

$$\frac{1}{2}mv^2 = \frac{P_{max}^2}{2}\left(\frac{L^3}{48D} + \frac{L}{4AG}\right) + \frac{C\left(\frac{P_{max}}{C}\right)^{\frac{n+1}{n}}}{n+1} \quad (7.4)$$

where m is the mass of the tup, v is its velocity at impact, L is the span, D is the flexural rigidity of the skins, G is shear modulus of the foam core, and C and n are the indentation constants used in the Meyer contact law which are determined experimentally from load-indentation curves. The Meyer contact law is given as

$$P = C\alpha^n \quad (7.5)$$

where P is the applied load and α is indentation depth. Comparisons of the predictions of the energy balance model with experimental data for a woven glass fiber phenolic resin skin PVC foam core sandwich are shown in Fig. 7.15 (Akil Hazizan and Cantwell, 2002) for a variety of impactor weights and energies. Good agreement is

Core materials for marine sandwich structures

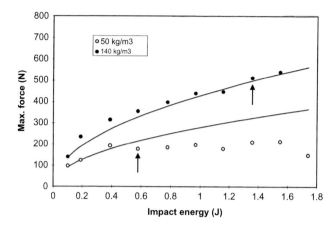

Fig. 7.15 Variation of maximum impact force and energy (Akil Hazizan and Cantwell, 2002). Damage threshold energy is shown by arrows and the solid lines show predictions of the energy-balance model.

observed between the model and the experimental results. Postimpact examination revealed all the samples fail due to buckling failure of top surface skin.

Failure in sandwiches is governed by fracture properties of the core and its degree of support to the skins. In the case of brittle cores, shear cracking through the thickness is prominent. Shear cracking can be avoided by having tougher cores. However, tougher cores result in buckling failure of the skins on the front face, as commonly seen in low modulus foams. In other cases, tensile tearing of the skins may occur on the back face. Delamination dominates for higher modulus foams. Careful observation using high-speed photography reveals that all of these failure mechanisms may be present in a single experiment. Experiments on the impact response PVC foam core sandwich beams observed that while ultimate failure was due to tensile rupture of the back face-sheet, cracking of the core and delamination of face-sheets were also observed simultaneously, as shown in Fig. 7.16 (Tagarielli et al., 2007). Similar observations of the deformation mechanism were reproduced using finite element modeling (Tagarielli et al., 2010). Together, these studies indicate that the shock resistance of a sandwich composite is maximized by selecting face-sheets with high strain to failure to delay total rupture of the sandwich. Underwater shock loading experiments have been conducted in both air-backed and water-backed conditions for PVC core sandwich composites, and it was found that the damage mechanism is strongly dependent on the backing condition (Huang et al., 2016a).

7.2.4 Moisture effects

An obvious concern for materials used in marine applications is moisture absorption. Mechanical, electrical, and thermal insulation properties are severely affected by moisture uptake, and the large free volume of foams exacerbates the problem. In

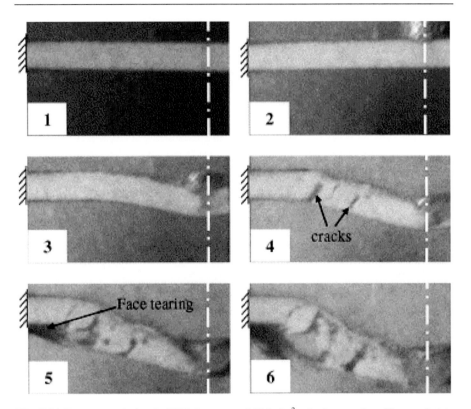

Fig. 7.16 Damage evolution in PVC foam core (100 kg/m^3) vinyl ester glass fiber sandwich composites observed by high speed photography. Impact occurs in frame 2, and the flexural wave can be seen in frame 3. Cracks form in the core in frame 4, which coalesce as damage progresses. In frame 5, face-sheet tearing causes final failure of the sandwich and the projectile continues unimpeded (Tagarielli et al., 2007).

sandwiches, the face-sheet provides some protection; however, face-sheet damage can subject the core to water or humidity causing high moisture absorption. Seawater resistance needs to be evaluated for naval ships (Siriruk et al., 2012). Experiments have been conducted to examine the effect of seawater exposure on PVC foam samples. The specimens were immersed in seawater and tap water, periodically removed, weighed, measured, and subjected to compression tests. Dry PVC foam, with density of 116 kg/m^3, was also tested in compression for comparison (Sadler et al., 2009). During saltwater experiments, evaporation losses need to be avoided to keep the salt concentration constant throughout the test.

The dry samples had strength and modulus of 2.1 and 60 MPa, respectively, in compression. The effect of 100 days of saltwater immersion is studied in (Sadler et al., 2009). PVC is a hydrophobic material, which results in slow moisture uptake. However, after 100 days of immersion, the samples were observed to absorb 57% of their original weight in water and 70% of their original weight in 500 days. The effect

Core materials for marine sandwich structures

on strength is relatively small, with specimens losing 15% of the dry strength after freshwater immersion and 12% of the dry strength in saltwater. However, the modulus suffers substantially and so must be controlled by appropriate sealing of the sandwich from moisture exposure (Karthikeyan and Sankaran, 2001). Elevated temperature also causes substantial acceleration of the moisture uptake by promoting diffusion through the polymer (Avilés and Aguilar-Montero, 2010).

7.3 Syntactic foams

Syntactic foams are also called composite foams. These materials are fabricated by embedding hollow particles in a matrix material. The fabrication methods for these foams are similar to those used for particulate composites. The hollow particles result in closed-cell porosity in these foams, which provides several advantages such as low moisture absorption, high compressive modulus, ability to keep damage localized, and high surface area to bond with the skins in sandwich structures. Significant research has been conducted to explore the potential of reinforcing syntactic foams to provide improved tensile properties. However, typically poor bonding of hollow particles to the matrix and the lack of space within the syntactic foam microstructure for reinforcing fibers due to very high packing of hollow particles have also been described in the literature (Gupta et al., 2013; Huang et al., 2016b). In such cases, nanoscale reinforcement is considered promising for syntactic foams (Gupta et al., 2013; John et al., 2010; Poveda and Gupta, 2014; Guzman et al., 2012).

Representative microstructures are presented in Fig. 7.17 for two types of syntactic foams. The particles are uniformly distributed in these foams providing them with isotropic mechanical properties. A three-dimensional solid model representation of these microstructures is presented in Fig. 7.18, where the matrix is assigned transparent properties to observe the particle distribution inside the material. The hollow particles located at the surface of the model are cut open. However, due to the ceramic nature of particle shell, the moisture absorption in the particle material is usually neglected and only the limited exposed area of the matrix contributes to water uptake. This results in low moisture uptake in syntactic foams and also provides improved dimensional stability in humid conditions for even long-term exposure.

7.3.1 Hollow particles and their properties

Hollow particles of a large number of different materials have been used in fabricating syntactic foams, including glass, carbon, phenol (Bunn and Mottram, 1993), silicon carbide (Labella et al., 2014a), and alumina. These particles are engineered as per the requirement of an application to have a given range of diameter and wall thickness. Hollow glass particles, commonly called glass microballoons (GMBs) or hollow glass microspheres (HGMs), are most widely used as fillers in syntactic foams. The mechanical properties of hollow particles depend on their wall thickness to diameter ratio. In the published studies, the properties of hollow particles are modeled to

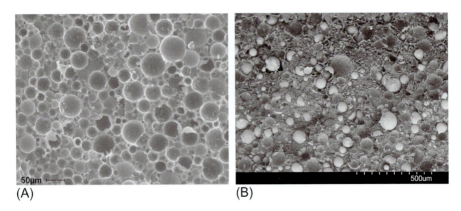

Fig. 7.17 Syntactic foam microstructure: (A) vinyl ester matrix filled with glass hollow particles and (B) high density polyethylene (HDPE) filled with fly ash cenospheres.

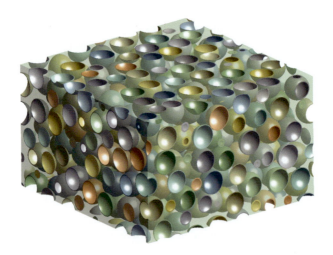

Fig. 7.18 A three-dimensional solid model representation of a syntactic foam containing hollow particles dispersed in a matrix material. Matrix is represented as translucent in this model to clearly show the internal distribution of particles in the syntactic foam.

incorporate wall thickness effects using a parameter named radius ratio defined as (Gupta et al., 2004).

$$\eta = \frac{R_i}{R_o} \qquad (7.6)$$

where R_i and R_o are the internal and outer radii of hollow particles, respectively. Radius ratio is related to the particle wall thickness (t) by

$$t = R_o(1 - \eta) \qquad (7.7)$$

The ratio t/R_0 defines the mechanical properties of a particle. Average values of η and t for a batch of particles can be determined through measurement of true particle density of hollow particles (ρ_{mb}) and the density of the particle material (ρ_g) by

$$\eta = \left(1 - \frac{t}{R_0}\right) = \left(1 - \frac{\rho_{mb}}{\rho_g}\right)^{1/3} \qquad (7.8)$$

where ρ_{mb} and ρ_g can be experimentally measured using a pycnometer. In addition, particle size can be measured using a particle size analyzer. Eq. (7.7) shows that thin-walled particles have η close to 1, while thick-walled particles have η close to 0. Thin-walled particles are commonly used as fillers to benefit from the possibility of developing low-density syntactic foams.

The properties of hollow particles depend on their wall thickness. The effective modulus of the particle (E_{mb}) is different from that of the particle material (E_g) and is related to η by

$$E_{mb} = \frac{E_g(1 - 2\nu_g)(1 - \eta^3)}{(1 - 2\nu_g) + \left(\frac{1 + \nu_g}{2}\right)\eta^3} \qquad (7.9)$$

where ν_g is the Poisson's ratio of the particle material. The dependence of particle modulus on the wall thickness can be used to design syntactic foams with desired properties.

An example of commonly used GMBs is shown in Fig. 7.19. GMBs of sodalime-borosilicate or borosilicate glass are most commonly used in large scale bulk applications of syntactic foams due to their low cost and wide availability compared to the hollow particles of other materials. These particles are mostly in the size range of 10–250 µm. The most commonly used GMBs have density in the range of 0.15–0.6 g/cm^3. Hollow particles of carbon and phenolic polymers are also available. In addition, demand for high-performance syntactic foams is leading to the development of hollow particles of higher grade ceramics such as SiC, Al$_2$O$_3$, and SiO$_2$. However, these particles are several times more expensive than the glass particles and their applications are limited to certain high value specialty applications. Their use in marine composites is not very common due to their high cost.

Although most of the engineered particles have uniform spherical structure, some of the particles have defects as shown in Fig. 7.20. Presence of such defects significantly reduces the mechanical properties of the particles (Carlisle et al., 2007; Koopman et al., 2004). In addition, matrix material may fill the inside of the defective particles, leading to higher than expected density of the syntactic foam. Separation of some of the defective particles is possible using flotation methods, but some of the particles having internal defects may survive such separation methods. These defective particles are the ones that fracture first when the syntactic foam is loaded under compression. Mechanical tests on individual particles have been conducted to

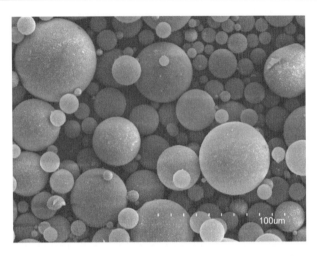

Fig. 7.19 Glass hollow particles that are widely used in fabricating syntactic foams.

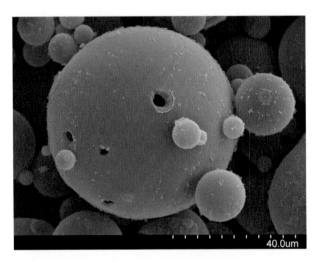

Fig. 7.20 An engineered hollow SiC particle having some defects in the shell. The strength of such particles is very low compared to the intact particles.

understand their mechanical properties with respect to their size distribution and failure mechanisms under compression (Koopman et al., 2004; Shunmugasamy et al., 2014).

7.3.2 *Compressive properties*

Compared to the gas-blown foams, syntactic foams provide greater control over their properties because of the use of hollow particles to incorporate porosity in their structure. Two parameters are available in syntactic foams that can be varied independent of each other in order to tailor the properties of syntactic foams:

- Hollow particle volume fraction
- Hollow particle radius ratio

Particles of any radius ratio can be selected and then used in any desired volume fraction to obtain the required set of properties in syntactic foams. In fact, the combination of these two parameters allows tailoring multiple properties of syntactic foams simultaneously (Gladysz et al., 2006).

Compressive properties of syntactic foams have been studied in detail for a number of marine-relevant syntactic foam compositions (Gupta et al., 2010). A representative set of compressive stress-strain curves for syntactic foams shows the effect of GMB volume fraction in Fig. 7.21A and the effect of GMB wall thickness variation in Fig. 7.21B. In both cases, vinyl ester resin is used as the matrix material. One of the important features of the stress-strain curve is the long stress plateau that results in high energy absorption in syntactic foams. The plateau stress is higher for syntactic foams containing low volume fraction of GMBs and also GMBs of thicker shell.

Particle crushing is the main failure mechanism in polymer syntactic foams. The initial failure of the specimen happens through shearing and the failure is observed along ~45° plane (Song et al., 2004; Gupta et al., 2001), as can be seen in Fig. 7.22 in the image of the specimen near the end of the linear region. The peak is observed due to the onset of particle fracture. The stress-plateau region corresponds to crushing of GMBs and consolidation of debris in the cavity and the specimen compresses uniformly as observed in Fig. 7.22.

The compressive properties of marine syntactic foams are summarized in Fig. 7.23 from the published literature (Gupta et al., 2010; Labella et al., 2014b; Luong et al., 2014; Pellegrino et al., 2015). It is observed that the foams of a wide range of density values from about 500 to 1100 kg/m^3 are available. In the data set shown in Fig. 7.23, syntactic foams with modulus and strength as high as about 3.2 GPa and 95 MPa, respectively, can be found. Vinyl ester resins are widely used in marine composites

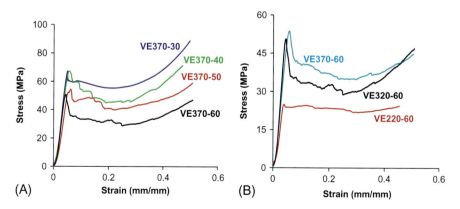

Fig. 7.21 Representative quasi-static compressive stress-strain curves for GMB reinforced vinyl ester matrix syntactic foams showing the effect of variation in (A) GMB volume fraction and (B) GMB wall thickness. VE refers to vinyl ester, followed by true particle density in kg/m^3 and then the volume% of GMBs in syntactic foam.

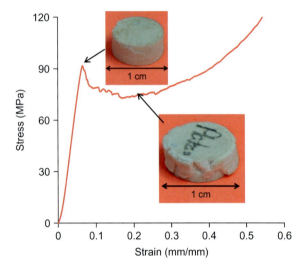

Fig. 7.22 Failure initiation in a specimen through shear crack formation at the end of the linear region. The plateau region corresponds to crushing of particles.

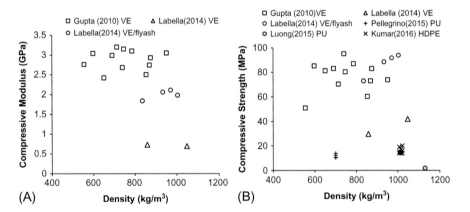

Fig. 7.23 Comparison of compressive (A) modulus and (B) strength for marine syntactic foams. Vinyl ester (VE), high density polyethylene (HDPE), and polyurethane (PU) matrix syntactic foams are included in the figure. All syntactic foams contain glass hollow particles unless fly ash is marked in the nomenclature.

Data points are obtained from Gupta, N., Ye, R., Porfiri, M., 2010. Comparison of tensile and compressive characteristics of vinyl ester/glass microballoon syntactic foams. Compos. Part B, 41(3), 236–245; Labella, M., Zeltmann, S.E., Shunmugasamy, V.C., Gupta, N., Rohatgi, P.K., 2014b. Mechanical and thermal properties of fly ash/vinyl ester syntactic foams. Fuel 121, 240–249; Luong, D.D., Shunmugasamy, V.C., Strbik III, O.M., Gupta, N., 2014. High strain rate compressive behavior of polyurethane resin and polyurethane/Al2O3 hollow sphere syntactic foams. J. Compos. 2014, 795984; Pellegrino, A., Tagarielli, V.L., Gerlach, R., and Petrinic, N., 2015. The mechanical response of a syntactic polyurethane foam at low and high rates of strain. Int. J. Impact Eng. 75, 214–221.

because of low moisture absorption and retention of mechanical properties in moist environment. The low cost of these resins is also a benefit. However, high cure shrinkage of vinyl ester resin is a challenge for ensuring that the particles do not fracture during syntactic foam fabrication. Epoxy is also commonly reported as a matrix for syntactic foams, and its properties can be effectively tailored using toughening additives (Garg and Mai, 1988).

There is significant interest in understanding the properties of syntactic foam core sandwich structures. Glass and carbon fabric skins have been used to create sandwich structures with syntactic foam core, as an example shown in Fig. 7.24. The fabric configuration in the sandwich skin can be optimized using standard theories for sandwich mechanics.

An existing study has focused on a variety of syntactic foam core sandwiches with glass fabric skins (Woldesenbet et al., 2005). Edgewise compression was conducted on the specimens, where the skins were oriented parallel to the loading direction. The initial linear region of the stress-strain curves for syntactic foam core sandwich is divided into three parts as shown in Fig. 7.25 (Woldesenbet et al., 2005). The slopes are calculated for each of these parts and are reported in Table 7.1 (Woldesenbet et al., 2005). Due to the fracture of both skins during the first two stages of the linear portion, the third stage corresponds to the compression of the foam core material only. The slope of this stage is found to be close to the modulus of the syntactic foam material tested separately. The modulus of sandwich composites is found to vary from 1140 to 1590 MPa for a variation in η from 0.922 to 0.866. The ultimate compressive strength of sandwich composites is found to vary from 33.9 to 66.4 MPa for the compositions tested in this study. Increasing the skin thickness can change the load partitioning between the skin and core. For sufficiently thick skins, the core failure may occur before the skin failure, leading to overlap between linear stages in the stress-strain diagram.

7.3.3 Impact properties of syntactic foams

Impact response of vinyl ester, polyester, polypropylene, and polyvinylchloride (PVC) matrix syntactic foams has been studied under notched Izod and Charpy impact conditions. Some of the general characteristics observed in these studies are summarized here. These observations can be augmented with the high strain rate compression studies available on syntactic foams (Gupta and Shunmugasamy, 2011).

Fig. 7.24 A syntactic foam core sandwich structure having glass fabric skins.

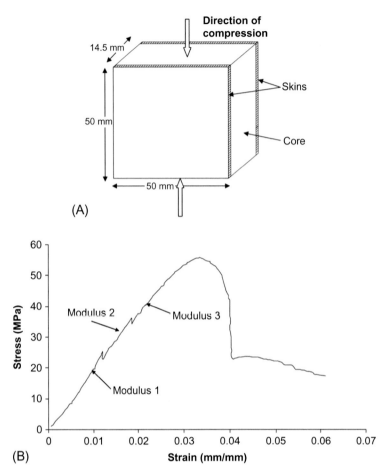

Fig. 7.25 (A) The orientation of the sandwich structure test specimen during compression testing. (B) A typical stress-strain curve for the syntactic foam core sandwich structure. The initial linear region is a combination of three distinct regions. The slopes of each of these regions are calculated and represented as Modulus I, II, and III and are reported in Table 7.1 (Woldesenbet et al., 2005).

Table 7.1 **Compressive strength and modulus of syntactic foam core sandwich composites (Woldesenbet et al., 2005)**

| Foam core type | Sandwich composite (MPa) | | | Ultimate compressive strength (MPa) |
	Modulus I	Modulus II	Modulus III	
SF22	1595	–	1140	33.9
SF32	2064	1720	1290	41.1
SF37	2120	1880	1331	48.2
SF38	2198	2073	1568	57.0
SF46	2585	2375	1590	66.4

Size and volume fraction of the filler are the two main study parameters in the existing literature. The addition of 5 vol% of GMB decreased the impact strength by 20% and 70% for epoxy (Wouterson et al., 2004) and PVC (Liang, 2002) matrix syntactic foams, respectively, in comparison to the neat resin. The addition of phenolic hollow microspheres resulted in increase of the impact strength by up to 300% at 10 wt% of the filler (Yusriah and Mariatti, 2013). The addition of 5 wt% of hollow epoxy microspheres in polyester matrix resulted in 32% increase in the impact strength (Low and AbuBakar, 2013). A clear trend between filler volume fraction and the impact strength of syntactic foam is not observed in the existing literature.

The effect of particle size on the impact properties of the syntactic foams has also been analyzed. An increase in HGM diameter from 10 to 70 μm resulted in a linear decrease in the notched impact strength by a maximum of 12% as observed in Fig. 7.26 (Liang, 2007). However, large standard deviations are observed in the impact strength data and the results are not conclusive within the range of parameters tested in this study. The hollow particle size effects may be contributed by the fact that the impact test specimen dimensions remain constant and do not scale up with the particle size. In such case, the total interfacial area between particle and matrix increases as the particle diameter decreases. Higher interfacial area will be useful if the particle-matrix interfacial bonding is strong, but specimens with a poor particle-matrix adhesion will benefit from using larger size particles.

Another study conducted impact testing on syntactic foams with GMBs of three different sizes (10, 35, and 70 μm) in polypropylene matrix. The results presented in Table 7.2 show that the 35 μm particles showed the lowest U-notched, V-notched, and unnotched impact strength (Liang and Wu, 2009). Table 7.2 shows the effect of particle size on the impact strength is very small when the testing is conducted on U or V-notched specimens. Such difference may be within the standard

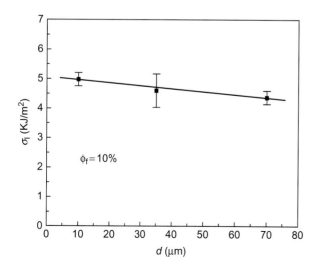

Fig. 7.26 Effect of glass hollow particle diameter on the notched Izod impact strength of polypropylene matrix syntactic foam (Liang, 2007). Reproduced with permission.

Table 7.2 Impact strength of glass hollow particle filled polypropylene matrix syntactic foams (Liang and Wu, 2009)

Impact strength (kJ/m^2)	HGM diameter		
	10 μm	35 μm	70 μm
V-notched	5.3	5.1	5.3
U-notched	6.5	5.6	6.1
Unnotched	56.1	30.3	48.5

deviation for the multiple specimens tested for the same material type. Usually, the impact properties are very sensitive to the presence of defects along the crack path and the standard deviation values are large and small differences may not provide conclusive results.

In the case of plain syntactic foams, drop weight impact has been conducted using a flat end impactor (Kim and Oh, 2000; Kim and Khamis, 2001). The effect of GMB volume fraction and variation of the specimen diameter on the impact strength of syntactic foam is evaluated. The 5 mm diameter specimens are found to have 75% lower peak load in comparison to the neat resin (Kim and Oh, 2000) and the increase in GMB volume fraction resulted in decrease of the peak load (Kim and Khamis, 2001).

The variation of the impact strength for the various vinyl ester/GMB syntactic foam compositions is shown in Fig. 7.27A (Shunmugasamy et al., 2015). The results show that the impact strength decreases with increasing GMB volume fraction. The HGM volume fraction is found to have a more profound effect on the impact strength than the wall thickness. All the syntactic foam compositions show lower impact strength in

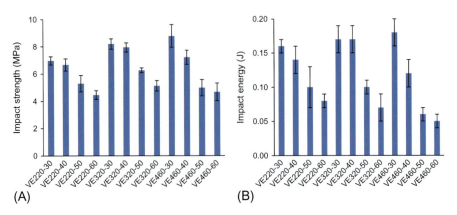

Fig. 7.27 Experimentally measured Izod (A) impact strength and (B) energy for unnotched glass/vinyl ester syntactic foams. The specimen nomenclature includes VE for vinyl ester, the following three digits representing the GMB true particle density in kg/m^3 and the last two digits represent volume % of GMBs in the syntactic foam (Shunmugasamy et al., 2015).

Core materials for marine sandwich structures 213

comparison to the neat vinyl ester impact strength of 15 MPa. For the impact strength, a maximum decrease of 70% is observed for VE220–60 type syntactic foam in comparison to the neat vinyl ester resin. The impact strength for syntactic foams containing the same type of particles increases almost linearly with GMB volume fraction. The energy absorbed until the specimen failure is presented in Fig. 7.27B. The impact energy shows an increasing trend with increasing HGM wall thickness at low GMB volume fraction of 0.3. However, at GMB volume fraction of 0.6, a decrease in impact energy is observed as the HGM wall thickness increases.

The impact failure mechanism of syntactic foams showed a single crack that initiated from the tensile side of the specimen and propagated to the compression side as shown in Fig. 7.28A. In each specimen, consistently a kink is observed near the compression edge of the specimen, which provides evidence of change in the crack propagation direction. Finite element analysis (FEA) is used to understand the mechanism that results in this kind of crack pattern. Previous studies have shown that the compressive and tensile moduli for syntactic foams are different (Gupta et al., 2010). Therefore, the compressive and the tensile moduli values were taken as 2.8 GPa and 3.0 GPa, respectively, in the simulation. A difference in the modulus values causes the neutral axis to shift. Fig. 7.28B shows the stress profile obtained under the applied load and the boundary condition for elastic analysis. Based on the stress profile, it can be expected that the crack that originates from the tensile side encounters a compression zone of high stress near the compression edge of the specimen. The crack is deflected around this region of high compressive stress and results in the crack profile that is observed experimentally.

Fig. 7.29 shows the failure mechanisms in syntactic foams that lead to energy absorption. The total energy absorption depends on the relative extents of these

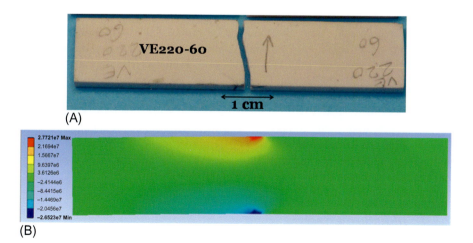

Fig. 7.28 (A) Macroscopic failure features of representative specimens of VE220-60. (B) The normal stress distribution found in the finite element analysis result. The stress values shown in (B) are in Pa.

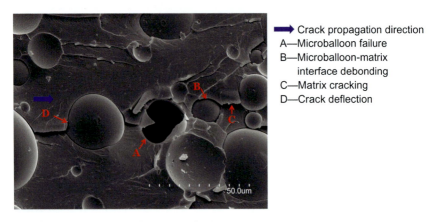

Fig. 7.29 Failure mechanisms observed in syntactic foams under impact loading conditions (Shunmugasamy et al., 2015).

mechanisms. For example, for syntactic foams containing low volume fraction of thick-walled GMBs, the energy absorbed in matrix cracking will dominate, whereas in syntactic foams containing high GMB volume fraction of thin-walled particles, particle-matrix debonding and particle crushing may absorb significant amount of energy. The failure features of a syntactic foam are shown in Fig. 7.30. This syntactic foam contains thin-walled particles. Tensile failure of the specimen leads to a matrix dominant fracture mechanism in this syntactic foam, including particle-matrix debonding. Fracture of particles without crushing and debris formation is observed.

In the case of glass fiber reinforced syntactic foams containing a constant 43 vol% GMBs, the absorbed impact energy increased with the fiber content (Ferreira et al., 2010). Crack bridging effect of fibers is useful in delaying the failure of the specimen and results in increased energy absorption before failure. Such studies can be extended to determine the effect of fiber material, volume fraction, and orientation.

Syntactic foam core sandwich composites with laminated glass fiber skin (Kim and Mitchell, 2003), carbon fiber skin (Scarponi et al., 1996), and with interleaved syntactic foam cores (Hiel and Ishai, 1992) have also been studied for impact characteristics. Both of these approaches minimize the damage caused by impact and improve the residual strength of the composite. Grid stiffened syntactic foam core sandwich structures have also been studied for impact properties (Li and Chakka, 2010; Li and Muthyala, 2008). The nodes of grids are the strong regions of these structures and their performance can be improved by optimization processes.

7.3.4 Moisture effects

Syntactic foams have long been used in marine applications because the hollow particle walls prevent moisture from filling in the pore space, even when the matrix is saturated with moisture. However, long-term moisture exposure research on syntactic foams has exposed that the commonly used sodalime-borosilicate glass hollow

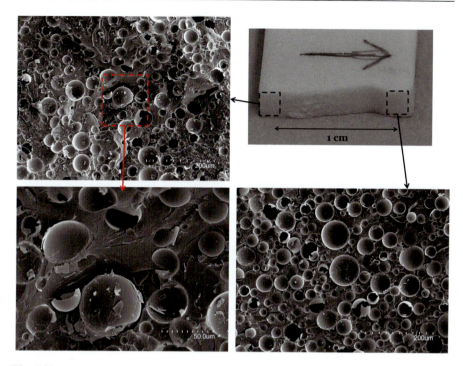

Fig. 7.30 Fractured surface on impact tested VE220–30 syntactic foam. The arrow drawn on the specimen represents the direction of the crack propagation (Shunmugasamy et al., 2015).

particles are subject to degradation within the composite, leading to their eventual collapse and filling with water. This phenomenon causes significant water uptake and degradation of the mechanical properties in long term or at high temperatures.

A schematic moisture uptake graph for a syntactic foam is shown in Fig. 7.31 (Poveda et al., 2013). The profile is divided into five regions: In region 1, cut microspheres on the surface are filled with water, leading to a small but nearly instantaneous

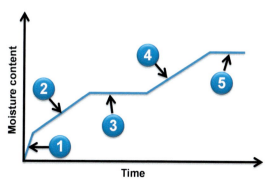

Fig. 7.31 Schematic representation of specimen weight gain over time. Five stages are identified in the graph (Poveda et al., 2013).

weight gain. The weight gain in this region is usually subtracted from the total quoted weight gain since this moisture can be easily removed by drying the surface and does not affect the mechanical properties. In region 2, diffusion of water takes place through the matrix and along the particle-matrix interface. This region is usually described well by a Fickian model (De Wilde and Frolkovic, 1994; Shen and Springer, 1976), and the presence of the hollow particles makes the diffusion path more tortuous and thus may slow the uptake in this region. Region 3 corresponds to saturation of the matrix, which for most polymers used in syntactic foams occurs around 5%–10% weight gain. If the hollow particles were inert to degradation by water, this would be the final portion of the response. However, in this region sodium ions from glasses containing sodalime diffuse from the particle wall into the water. This causes structural collapse of the hollow sphere, which opens the cavity. In region 4, water fills in the newly opened cavities, until these are finally saturated and the composite has reached its maximum water uptake in region 5. This process is also illustrated schematically in Fig. 7.32 (Poveda et al., 2013). All hollow spheres do not have the same wall thickness and size, so they fail at different times (Gupta et al., 2010). Therefore, in a real syntactic foam specimen, where particles of a wide range of sizes and wall thicknesses may exist, the stages indicated by Fig. 7.31 overlap with each other and may not be well-resolved. Accelerated testing also causes significant overlap of the regions, often making them indistinguishable (Zeltmann et al., 2015). A number of studies are available on the residual strength of weathered reinforced and unreinforced syntactic foams (Zeltmann et al., 2015; Tagliavia et al., 2012; Kumar and Ahmed, 2016). More recently, glass hollow particles consisting of borosilicate glass have been studied and have shown resistance to degradation in water while providing the same level of mechanical properties (Zeltmann et al., 2017b, 2018).

7.3.5 Tailoring the properties of syntactic foams

In solid particle-filled composites, only particle volume fraction can be selected as an independent parameter. Once the volume fraction is selected, all the mechanical and thermal properties of that composition are locked down. However, in syntactic foams,

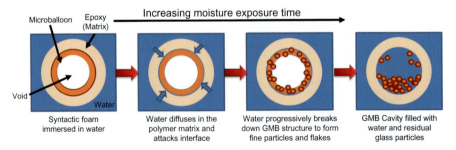

Fig. 7.32 Illustration of moisture damage sequence in syntactic foams in various stages of water uptake and material degradation (Poveda et al., 2013).

Core materials for marine sandwich structures   217

even if the volume fraction is maintained constant at a given level, a change in the radius ratio provides the ability to further tailor the properties. The possibility of independently tailoring the wall thickness and volume fraction has been used to design novel functionally graded syntactic foams with very high compressive energy absorption (Caeti et al., 2009; Gupta, 2007).

Figs. 7.33 and 7.34 show the possibility of tailoring syntactic foams for multiple properties. Contour plots of the modulus, coefficient of thermal expansion (CTE), and density of vinyl ester syntactic foams are shown in Fig. 7.33. Young's modulus is estimated using the experimentally validated Porfiri-Gupta model (Porfiri and Gupta, 2009) and CTE is estimated using the modified Turner model adapted for syntactic foams by (Shunmugasamy et al., 2012). The Young's modulus, Poisson ratio, density, and CTE for vinyl ester are taken as 2.8 GPa, 0.35, 1160 kg/m^3, and 76.5 μ/K (Gupta et al., 2014), respectively, and for borosilicate glass, are taken as 70 GPa, 0.21, 2540 kg/m^3, and 0.75 μ/K (Bharath Kumar et al., 2016a, b). Fig. 7.34 shows predictions for HDPE/borosilicate glass syntactic foams, where the Young's modulus,

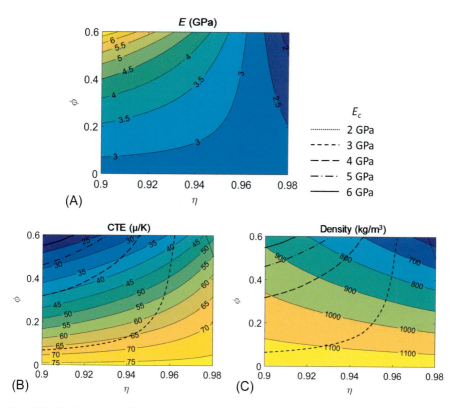

Fig. 7.33 Predictions of (A) elastic modulus by the Porfiri-Gupta model (Porfiri and Gupta, 2009), (B) coefficient of thermal expansion by the modified Turner model (Shunmugasamy et al., 2012), and (C) composite density by the rule of mixtures, for vinyl ester with borosilicate glass hollow particles.

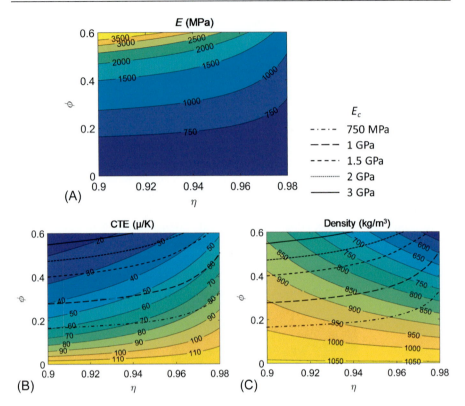

Fig. 7.34 Predictions of (A) elastic modulus by the Porfiri-Gupta model (Porfiri and Gupta, 2009), (B) coefficient of thermal expansion by the modified Turner model (Shunmugasamy et al., 2012), and (C) composite density by the rule of mixtures, for HDPE with borosilicate glass hollow particles.

Poisson ratio, density, and CTE of HDPE are taken to be 529 MPa, 0.425, 1056 kg/m^3, and 120 μ/K, respectively (Bharath Kumar et al., 2016a, b). In both figures, selected level curves of syntactic foam modulus are overlaid on the plots for CTE and density to demonstrate the ability to tailor multiple properties of the syntactic foam simultaneously. For example, we can consider syntactic foams with modulus of 3 GPa. Using vinyl ester as the matrix material, we can simultaneously achieve density as low as 650 kg/m^3 or CTE as low as 35 μ/K. Increasing the density of the syntactic foam at this modulus requires using thicker-walled hollow particles in a lower volume fraction, which causes a simultaneous increase in the CTE. If, instead, we choose HDPE as the matrix, the range of properties achievable at this modulus is more limited: the density can be varied from 700 to 850 kg/m^3 and the CTE will be 18–21 μ/K. However, for the HDPE matrix syntactic foams at this modulus, increasing the density decreases the CTE.

To take another example, if a submersible application required a Young's modulus of 4 GPa and a density of 1000 kg/m^3 for neutral buoyancy, the plots indicate that this can be achieved using hollow particles of 660 kg/m^3 true particle density in 33 vol% added to vinyl ester. The CTE that results is about 62 μ/K. Using the typical range of wall thicknesses and volume fractions, this set of properties is not possible using HDPE/borosilicate glass syntactic foams.

7.4 Summary

Lightweight materials provide benefits of buoyancy, speed, and maneuverability for marine vessels. Foam core sandwich structures are used as lightweight materials in the construction of parts of several types of marine vessels. This chapter discussed two types of closed-cell foams, including gas-blown polyvinyl chloride (PVC) foams and hollow particle-filled syntactic foams. Cell size and shape are important parameters for PVC foams. Due to the random nature of the gas bubbles, the cell size and shape vary widely in PVC foams. In addition, the properties of PVC foams are found to be different in the thickness direction compared to the longitudinal and transverse directions. Syntactic foams are fabricated by filling spherical hollow particles in a matrix material. There is greater control over the properties of syntactic foams because particles of required size and wall thickness can be selected for use as filler. In addition, the possibility of independently tailoring the particle wall thickness and volume fraction allows precise control over a number of properties simultaneously, which is not possible in gas-blown foams. Continued demand of lightweight core materials for marine structure is driving the research in various aspects of these materials. Development of low cost, high-quality particles will be of great interest for this field.

Acknowledgments

The Office of Naval Research grant N00014-10-1-0988 is acknowledged to support this work. The authors also thank the Dean of Engineering at NYUAD for partially funding MRD's visit to NYUAD. The views expressed in this article are those of the authors, not of the funding agencies. The authors thank the ME Department at NIT-K and MAE Department at NYU for providing facilities and support.

References

Hazizan, M. Akil, Cantwell, W.J., 2002. The low velocity impact response of foam-based sandwich structures. Compos. Part B 33 (3), 193–204.

Atas, C., Sevim, C., 2010. On the impact response of sandwich composites with cores of balsa wood and PVC foam. Compos. Struct. 93 (1), 40–48.

Avilés, F., Aguilar-Montero, M., 2010. Moisture absorption in foam-cored composite sandwich structures. Polym. Compos. 31 (4), 714–722.

Kumar, B.R. Bharath, Doddamani, M., Zeltmann, S.E., Gupta, N., Ramesh, M.R., Ramakrishna, S., 2016a. Data characterizing tensile behavior of censophere/HDPE syntactic foam. Data Brief 6, 933–941.

Kumar, B.R. Bharath, Doddamani, M., Zeltmann, S.E., Gupta, N., Ramesh, M.R., Ramakrishna, S., 2016b. Processing of cenosphere/HDPE syntactic foams using an industrial scale polymer injection molding machine. Mater. Des. 92, 414–423.

Bunn, P., Mottram, J.T., 1993. Manufacture and compression properties of syntactic foams. Composites 24 (7), 565–571.

Caeti, R., Gupta, N., Porfiri, M., 2009. Processing and compressive response of functionally graded composites. Mater. Lett. 63 (22), 1964–1967.

Carlisle, K.B., Lewis, M., Chawla, K.K., Koopman, M., Gladysz, G.M., 2007. Finite element modeling of the uniaxial compression behavior of carbon microballoons. Acta Mater. 55 (7), 2301–2318.

Castro, O., Silva, J.M., Devezas, T., Silva, A., Gil, L., 2010. Cork agglomerates as an ideal core material in lightweight structures. Mater. Des. 31 (1), 425–432.

Chen, W.W., Song, B., 2010. Split Hopkinson (Kolsky) Bar: Design, Testing and Applications. Springer Science & Business Media, New York.

Chen, A., Kim, H., Asaro, R.J., Bezares, J., 2011. Non-explosive simulated blast loading of balsa core sandwich composite beams. Compos. Struct. 93 (11), 2768–2784.

Colloca, M., Dorogokupets, G., Gupta, N., Porfiri, M., 2012. Mechanical properties and failure mechanisms of closed-cell PVC foams. Int. J. Crashworthiness 17 (3), 327–336.

Daniel, I.M., Cho, J.-M., 2011. Characterization of anisotropic polymeric foam under static and dynamic loading. Exp. Mech. 51 (8), 1395–1403.

Daniel, I.M., Cho, J.-M., Werner, B.T., 2013. Characterization and modeling of stain-rate-dependent behavior of polymeric foams. Compos. A: Appl. Sci. Manuf. 45, 70–78.

De Wilde, W.P., Frolkovic, P., 1994. The modelling of moisture absorption in epoxies: effects at the boundaries. Composites 25 (2), 119–127.

Demir, H., Sipahioğlu, M., Balköse, D., Ülkü, S., 2008. Effect of additives on flexible PVC foam formation. J. Mater. Process. Technol. 195 (1–3), 144–153.

Dukhan, N., 2013. Metal Foams: Fundamentals and Applications. DEStech Publications, Inc, Lancaster, PA.

El-Hadek, M.A., Tippur, H.V., 2003. Dynamic fracture behavior of syntactic epoxy foams: optical measurements using coherent grading sensing. Opt. Lasers Eng. 40, 353–369.

Ferreira, J.A.M., Capela, C., Costa, J.D., 2010. A study of the mechanical behaviour on fibre reinforced hollow microspheres hybrid composites. Compos. A: Appl. Sci. Manuf. 41 (3), 345–352.

Garg, A.C., Mai, Y.-W., 1988. Failure mechanisms in toughened epoxy resins—a review. Compos. Sci. Technol. 31 (3), 179–223.

Gdoutos, E.E., Daniel, I.M., Wang, K.A., 2002. Failure of cellular foams under multiaxial loading. Compos. A: Appl. Sci. Manuf. 33 (2), 163–176.

Gibson, L.J., Ashby, M.F., 1999. Cellular Solids: Structure and Properties. Cambridge University Press, Cambridge.

Gielen, A.W.J., 2008. A PVC-foam material model based on a thermodynamically elasto-plastic-damage framework exhibiting failure and crushing. Int. J. Solids Struct. 45 (7–8), 1896–1917.

Gil, L., 2009. Cork composites: a review. Materials 2 (3), 776.

Gladysz, G., Perry, B., Mceachen, G., Lula, J., 2006. Three-phase syntactic foams: structure-property relationships. J. Mater. Sci. 41 (13), 4085–4092.

Gupta, N., 2007. A functionally graded syntactic foam material for high energy absorption under compression. Mater. Lett. 61 (4–5), 979–982.

Gupta, N., Rohatgi, P.K. (Eds.), 2014. Metal Matrix Syntactic Foams: Processing, Microstructure, Properties and Applications. DEStech Publications, Lancaster, PA.

Gupta, N., Shunmugasamy, V.C., 2011. High strain rate compressive response of syntactic foams: trends in mechanical properties and failure mechanisms. Mater. Sci. Eng. A 528 (25–26), 7596–7605.
Gupta, N., Kishore, Woldesenbet, E., Sankaran, S., 2001. Studies on compressive failure features in syntactic foam material. J. Mater. Sci. 36 (18), 4485–4491.
Gupta, N., Woldesenbet, E., Mensah, P., 2004. Compression properties of syntactic foams: effect of cenosphere radius ratio and specimen aspect ratio. Compos. A: Appl. Sci. Manuf. 35 (1), 103–111.
Gupta, N., Ye, R., Porfiri, M., 2010. Comparison of tensile and compressive characteristics of vinyl ester/glass microballoon syntactic foams. Compos. Part B 41 (3), 236–245.
Gupta, N., Pinisetty, D., Shunmugasamy, V.C., 2013. Reinforced Polymer Matrix Syntactic Foams: Effect of Nano and Micro-Scale Reinforcement. Springer, New York.
Gupta, N., Zeltmann, S.E., Shunmugasamy, V.C., Pinisetty, D., 2014. Applications of polymer matrix syntactic foams. JOM 66 (2), 245–254.
Guzman, M.E., Rodriguez, A.J., Minaie, B., Violette, M., 2012. Processing and properties of syntactic foams reinforced with carbon nanotubes. J. Appl. Polym. Sci. 124 (3), 2383–2394.
Hiel, C., Ishai, O., 1992. Damage tolerance of a composite sandwich with interleaved foam core. J. Compos. Technol. Res. 14 (3), 155–168.
Huang, W., Zhang, W., Ye, N., Gao, Y., Ren, P., 2016a. Dynamic response and failure of PVC foam core metallic sandwich subjected to underwater impulsive loading. Compos. Part B 97, 226–238.
Huang, C., Huang, Z., Qin, Y., Ding, J., Lv, X., 2016b. Mechanical and dynamic mechanical properties of epoxy syntactic foams reinforced by short carbon fiber. Polym. Compos. 37 (7), 1960–1970.
Huber, O., Klaus, H., 2009. Cellular composites in lightweight sandwich applications. Mater. Lett. 63 (13–14), 1117–1120.
Jiang, Z., Yao, K., Du, Z., Xue, J., Tang, T., Liu, W., 2014. Rigid cross-linked PVC foams with high shear properties: the relationship between mechanical properties and chemical structure of the matrix. Compos. Sci. Technol. 97, 74–80.
Jing, L., Wang, Z., Zhao, L., 2016. The dynamic response of sandwich panels with cellular metal cores to localized impulsive loading. Compos. Part B 94, 52–63.
John, B., Nair, C.P.R., Ninan, K.N., 2010. Effect of nanoclay on the mechanical, dynamic mechanical and thermal properties of cyanate ester syntactic foams. Mater. Sci. Eng. A 527 (21–22), 5435–5443.
Jover, N., Shafiq, B., Vaidya, U., 2014. Ballistic impact analysis of balsa core sandwich composites. Compos. Part B 67, 160–169.
Karthikeyan, C.S., Sankaran, S., 2001. Effect of absorption in aqueous and hygrothermal media on the compressive properties of glass fiber reinforced syntactic foam. J. Reinf. Plast. Compos. 20 (11), 982–993.
Kidd, T.H., Zhuang, S., Ravichandran, G., 2012. In situ mechanical characterization during deformation of PVC polymeric foams using ultrasonics and digital image correlation. Mech. Mater. 55, 82–88.
Kim, H.S., Khamis, M.A., 2001. Fracture and impact behaviours of hollow micro-sphere/epoxy resin composites. Compos. A: Appl. Sci. Manuf. 32 (9), 1311–1317.
Kim, H.S., Mitchell, C., 2003. Impact performance of laminates made of syntactic foam and glass fiber reinforced epoxy as protective materials. J. Appl. Polym. Sci. 89, 2306–2310.
Kim, H.S., Oh, H.H., 2000. Manufacturing and impact behavior of syntactic foam. J. Appl. Polym. Sci. 76 (8), 1324–1328.

Koopman, M., Gouadec, G., Carlisle, K., Chawla, K.K., Gladysz, G., 2004. Compression testing of hollow microspheres (microballoons) to obtain mechanical properties. Scr. Mater. 50 (5), 593–596.

Kumar, S.A., Ahmed, K.S., 2016. Effects of ageing on mechanical properties of stiffened syntactic foam core sandwich composites for marine applications. J. Cell. Plast. 52 (5), 503–532.

Labella, M., Shunmugasamy, V.C., Strbik, O.M., Gupta, N., 2014a. Compressive and thermal characterization of syntactic foams containing hollow silicon carbide particles with porous shell. J. Appl. Polym. Sci. 131 (17), 8593–8597.

Labella, M., Zeltmann, S.E., Shunmugasamy, V.C., Gupta, N., Rohatgi, P.K., 2014b. Mechanical and thermal properties of fly ash/vinyl ester syntactic foams. Fuel 121, 240–249.

Lamanna, E., Gupta, N., Cappa, P., Strbik III, O.M., Cho, K., 2017. Evaluation of the dynamic properties of an aluminum syntactic foam core sandwich. J. Alloys Compd. 695, 2987–2994.

Le Duigou, A., Deux, J.-M., Davies, P., Baley, C., 2012. PLLA/flax mat/balsa bio-sandwich—environmental impact and simplified life cycle analysis. Appl. Compos. Mater. 19 (3), 363–378.

Legrand, V., TranVan, L., Jacquemin, F., Casari, P., 2015. Moisture-uptake induced internal stresses in balsa core sandwich composite plate: modeling and experimental. Compos. Struct. 119, 355–364.

Li, G., Chakka, V.S., 2010. Isogrid stiffened syntactic foam cored sandwich structure under low velocity impact. Compos. A: Appl. Sci. Manuf. 41 (1), 177–184.

Li, G., Muthyala, V.D., 2008. Impact characterization of sandwich structures with an integrated orthogrid stiffened syntactic foam core. Compos. Sci. Technol. 68 (9), 2078–2084.

Liang, J.-Z., 2002. Tensile and impact properties of hollow glass bead-filled PVC composites. Macromol. Mater. Eng. 287 (9), 588–591.

Liang, J.-Z., 2007. Impact fracture toughness of hollow glass bead-filled polypropylene composites. J. Mater. Sci. 42 (3), 841–846.

Liang, J.Z., Wu, C.B., 2009. Gray relational analysis between size distribution and impact strength of polypropylene/hollow glass bead composites. J. Reinf. Plast. Compos. 28 (16), 1945–1955.

Liu, C., Zhang, Y.X., Ye, L., 2017. High velocity impact responses of sandwich panels with metal fibre laminate skins and aluminium foam core. Int. J. Impact Eng. 100, 139–153.

Low, L.F., AbuBakar, A., 2013. Fracture toughness and impact strength of hollow epoxy particles toughned polyester composites. Sains Malays. 42 (4), 443–448.

Luong, D.D., Pinisetty, D., Gupta, N., 2013. Compressive properties of closed-cell polyvinyl chloride foams at low and high strain rates: experimental investigation and critical review of state of the art. Compos. Part B 44 (1), 403–416.

Luong, D.D., Shunmugasamy, V.C., Strbik III, O.M., Gupta, N., 2014. High strain rate compressive behavior of polyurethane resin and polyurethane/Al_2O_3 hollow sphere syntactic foams. J. Compos.. 2014795984.

Magnucka-Blandzi, E., 2018. Bending and buckling of a metal seven-layer beam with crosswise corrugated main core – Comparative analysis with sandwich beam. Compos. Struct. 183, 35–41.

Magnucka-Blandzi, E., Walczak, Z., Jasion, P., Wittenbeck, L., 2017. Buckling and vibrations of metal sandwich beams with trapezoidal corrugated cores—the lengthwise corrugated main core. Thin-Walled Struct. 112, 78–82.

Mitra, N., 2010. A methodology for improving shear performance of marine grade sandwich composites: sandwich composite panel with shear key. Compos. Struct. 92 (5), 1065–1072.

Omar, M.Y., Xiang, C., Gupta, N., Strbik III, O.M., Cho, K., 2015. Syntactic foam core metal matrix sandwich composite under bending conditions. Mater. Des. 86, 536–544.

Pellegrino, A., Tagarielli, V.L., Gerlach, R., Petrinic, N., 2015. The mechanical response of a syntactic polyurethane foam at low and high rates of strain. Int. J. Impact Eng. 75, 214–221.

Porfiri, M., Gupta, N., 2009. Effect of volume fraction and wall thickness on the elastic properties of hollow particle filled composites. Compos. Part B 40 (2), 166–173.

Poveda, R.L., Gupta, N., 2014. Carbon-nanofiber-reinforced syntactic foams: compressive properties and strain rate sensitivity. JOM 66 (1), 66–77.

Poveda, R.L., Dorogokupets, G., Gupta, N., 2013. Carbon nanofiber reinforced syntactic foams: degradation mechanism for long term moisture exposure and residual compressive properties. Polym. Degrad. Stab. 98 (10), 2041–2053.

Radford, D.D., McShane, G.J., Deshpande, V.S., Fleck, N.A., 2006. The response of clamped sandwich plates with metallic foam cores to simulated blast loading. Int. J. Solids Struct. 43 (7–8), 2243–2259.

Reis, L., Silva, A., 2009. Mechanical behavior of sandwich structures using natural cork agglomerates as core materials. J. Sandw. Struct. Mater. 11 (6), 487–500.

Russo, A., Zuccarello, B., 2007. Experimental and numerical evaluation of the mechanical behaviour of GFRP sandwich panels. Compos. Struct. 81 (4), 575–586.

Sadler, R.L., Sharpe, M., Panduranga, R., Shivakumar, K., 2009. Water immersion effect on swelling and compression properties of eco-core, PVC foam and balsa wood. Compos. Struct. 90 (3), 330–336.

Saha, M.C., Mahfuz, H., Chakravarty, U.K., Uddin, M., Kabir, M.E., Jeelani, S., 2005. Effect of density, microstructure, and strain rate on compression behavior of polymeric foams. Mater. Sci. Eng. A 406 (1–2), 328–336.

Scarponi, C., Briotti, G., Barboni, R., Marcone, A., Iannone, M., 1996. Impact testing on composites laminates and sandwich panels. J. Compos. Mater. 30 (17), 1873–1911.

Shen, C.-H., Springer, G.S., 1976. Moisture absorption and desorption of composite materials. J. Compos. Mater. 10 (1), 2–20.

Mohammadi, M. Shir, Nairn, J.A., 2017. Balsa sandwich composite fracture study: comparison of laminated to solid balsa core materials and debonding from thick balsa core materials. Compos. Part B 122, 165–172.

Shunmugasamy, V.C., Pinisetty, D., Gupta, N., 2012. Thermal expansion behavior of hollow glass particle/vinyl ester composites. J. Mater. Sci. 47 (14), 5596–5604.

Shunmugasamy, V.C., Zeltmann, S.E., Gupta, N., Strbik, O.M., 2014. Compressive characterization of single porous SiC hollow particles. JOM 66 (6), 892–897.

Shunmugasamy, V.C., Anantharaman, H., Pinisetty, D., Gupta, N., 2015. Unnotched Izod impact characterization of glass hollow particle/ vinyl ester syntactic foams. J. Compos. Mater. 49 (2), 185–197.

Siriruk, A., Penumadu, D., Sharma, A., 2012. Effects of seawater and low temperatures on polymeric foam core material. Exp. Mech. 52 (1), 25–36.

Song, B., Chen, W., Frew, D.J., 2004. Dynamic compressive response and failure behavior of an epoxy syntactic foam. J. Compos. Mater. 38 (11), 915–936.

Tagarielli, V.L., Deshpande, V.S., Fleck, N.A., 2007. The dynamic response of composite sandwich beams to transverse impact. Int. J. Solids Struct. 44 (7–8), 2442–2457.

Tagarielli, V.L., Deshpande, V.S., Fleck, N.A., 2010. Prediction of the dynamic response of composite sandwich beams under shock loading. Int. J. Impact Eng. 37 (7), 854–864.

Tagliavia, G., Porfiri, M., Gupta, N., 2012. Influence of moisture absorption on flexural properties of syntactic foams. Compos. Part B 43 (2), 115–123.

Thomas, T., Mahfuz, H., Carlsson, L.A., Kanny, K., Jeelani, S., 2002. Dynamic compression of cellular cores: temperature and strain rate effects. Compos. Struct. 58 (4), 505–512.

Ulven, C.A., Vaidya, U.K., 2006. Post-fire low velocity impact response of marine grade sandwich composites. Compos. A: Appl. Sci. Manuf. 37 (7), 997–1004.

Wang, T., Qin, Q., Wang, M., Yu, W., Wang, J., Zhang, J., Wang, T.J., 2017. Blast response of geometrically asymmetric metal honeycomb sandwich plate: experimental and theoretical investigations. Int. J. Impact Eng. 105, 24–38.

Woldesenbet, E., Gupta, N., Jerro, H.D., 2005. Effect of microballoon radius ratio on syntactic foam core sandwich composites. J. Sandw. Struct. Mater. 7 (2), 95–111.

Wouterson, E.M., Boey, F.Y.C., Hu, X., Wong, S.-C., 2004. Fracture and impact toughness of syntactic foam. J. Cell. Plast. 40 (2), 145–154.

Yalkin, H.E., Icten, B.M., Alpyildiz, T., 2015. Enhanced mechanical performance of foam core sandwich composites with through the thickness reinforced core. Compos. Part B 79, 383–391.

Yaseer Omar, M., Xiang, C., Gupta, N., Strbik III, O.M., Cho, K., 2015. Syntactic foam core metal matrix sandwich composite: compressive properties and strain rate effects. Mater. Sci. Eng. A 643, 156–168.

Yusriah, L., Mariatti, M., 2013. Effect of hybrid phenolic hollow microsphere and silica-filled vinyl ester composites. J. Compos. Mater. 47 (2), 169–182.

Zeltmann, S.E., Poveda, R.L., Gupta, N., 2015. Accelerated environmental degradation and residual flexural analysis of carbon nanofiber reinforced composites. Polym. Degrad. Stab. 121, 348–358.

Zeltmann, S.E., Kumar, B.B., Doddamani, M., Gupta, N., 2016. Prediction of strain rate sensitivity of high density polyethylene using integral transform of dynamic mechanical analysis data. Polymer 101, 1–6.

Zeltmann, S.E., Prakash, K.A., Doddamani, M., Gupta, N., 2017a. Prediction of modulus at various strain rates from dynamic mechanical analysis data for polymer matrix composites. Compos. Part B 120, 27–34.

Zeltmann, S.E., Chen, B., Gupta, N., 2017b. Mechanical properties of epoxy matrix-borosilicate glass hollow-particle syntactic foams. Mater. Perform. Charact. 6 (1), 1–16.

Zeltmann, S.E., Chen, B., Gupta, N., 2018. Thermal expansion and dynamic mechanical analysis of epoxy matrix–borosilicate glass hollow particle syntactic foams. J. Cell. Plast. 54 (3), 463–481.

Section C

Manufacture

Resin infusion for the manufacture of large composite structures

Ned Popham
Sunseeker International Limited, Poole, Dorset, United Kingdom

Chapter Outline

8.1 Introduction 228
    8.1.1 What is resin infusion? 228
    8.1.2 Why resin infusion? 229
    8.1.3 Why not resin infusion 230
    8.1.4 Challenges in a production environment 230
8.2 Physics of resin infusion 231
    8.2.1 Permeability 231
    8.2.2 Pressure 232
    8.2.3 Cross sectional area 235
    8.2.4 Viscosity 235
8.3 Materials selection and characterization 237
    8.3.1 Resin selection 237
    8.3.2 Reinforcement 240
    8.3.3 Core materials 241
    8.3.4 Consumables 242
8.4 Tooling 244
    8.4.1 Arrangement for infusion 245
    8.4.2 Heated tooling 246
8.5 Plant equipment, setup, and redundancy 246
    8.5.1 Equipment 246
    8.5.2 Vacuum ring main 247
    8.5.3 Vacuum receiver 247
    8.5.4 Back-up generators and redundancy 247
    8.5.5 Resin traps 248
    8.5.6 Connectors 248
    8.5.7 Vacuum gauges 248
    8.5.8 Leak detectors 248
    8.5.9 Hand tools 249
8.6 Infusion prediction, strategy, and setup 250
    8.6.1 Infusion setup 251
    8.6.2 A note on inclined surfaces 253
    8.6.3 3D resin flow 254
    8.6.4 Resin flow management 254
    8.6.5 Resolving dry patches 256
    8.6.6 Gravity 257

Marine Composites. https://doi.org/10.1016/B978-0-08-102264-1.00008-X
Copyright © 2019 Elsevier Ltd. All rights reserved.

8.7 **Resin delivery and management** 258
    8.7.1 Hand mixing 258
    8.7.2 Machine mixing 259
    8.7.3 Resin management 260
8.8 **Manufacturing process** 261
    8.8.1 Wet phase 261
    8.8.2 Fiber placement 261
    8.8.3 Core fit 262
8.9 **Process control and preinfusion checks** 263
    8.9.1 Measuring variables 263
    8.9.2 Understanding variables 263
    8.9.3 Dry layup phase 264
    8.9.4 Raw materials and ambient conditions 264
    8.9.5 Preinfusion checks 264
    8.9.6 Vacuum 265
    8.9.7 Resin 265
    8.9.8 Back-up systems 265
    8.9.9 In-process monitoring 265
    8.9.10 Leaks in the vacuum bag 265
8.10 **Postinfusion management** 266
8.11 **Conclusion/summary** 267
References 267

8.1 Introduction

8.1.1 What is resin infusion?

Resin Infusion is a manufacturing process whereby liquid resin is drawn into a consolidated dry reinforcement stack, using atmospheric pressure to drive the resin flow and consolidate the reinforcement.

Resin infusion has history dating back to the 1940s (Muskat, 1950) and has been in and out of fashion until the late 1990s.

Bill Seeman's SCRIMP (Seeman Composites Resin Infusion Molding Process) was an important development for the industry. His 1990 US patent (Seeman, 1990) publicized the process. However, efforts to get around the patent yielded new materials and process variants. Although the patent did not apply in Europe due to "prior art," the publicity also stimulated developments in Resin Infusion there.

Suppliers of reinforcements, resin, and consumable materials developed products more suited to resin infusion and the demand from producers of composite components grew. This allowed advanced composites manufacturing methods to grow beyond research institutions and displace traditionally high cost techniques (prepreg, autoclave processing).

There are a number of academic reviews (Williams et al., 1996; Cripps et al., 2000; Summerscales and Searle, 2005; Beckwith, 2007a) and brief technical articles describing the resin infusion process (Beckwith, 2007b,c; Summerscales, 2012). This chapter addresses the practical and commercial issues. Resin infusion has been described by a variety of different names and abbreviations. These have been reduced to four variants in the Summerscales and Searle taxonomy (Summerscales and Searle, 2005). The most common names are Resin Infusion under Flexible Tooling (RIFT), SCRIMP, and Vacuum Assisted Resin Transfer Molding (confined to North America and misleading as it infers two-hard tools).

8.1.2 Why resin infusion?

Performance: Due to the greater consolidation pressure, infused laminates have higher fiber content when compared to hand laminated components. This produces stiffer, stronger, lighter laminates when properly designed. Because of the high consolidation pressure, infusing a laminate specified for hand lamination (especially a monolithic section) would result in a more flexible laminate due to the decreased thickness.

Productivity: Hand lamination is limited by factors associated with resin chemistry. If too much material is put down at once, high exotherm temperatures can lead to shrinkage distortions and prerelease. There is also a waiting time between lamination stages when each layer is curing. Productivity cannot be increased by simply adding more labor. Resin Infusion allows the dry materials to be loaded without any regard for such delays. Resins tailored for resin infusion allow thick laminates to cure without excessive exotherm, aided by a lower resin content in the laminate. Due to lower viscosity resin, heavier fabrics can be used leading to fewer plies for a given weight or thickness of laminate. Application of the consumable materials and vacuum bag is the main additional activity, but can be offset by the removal of the wetting out and rolling that takes place during open molding.

Quality Control: Hand lamination relies on the skill of each operator to ensure correct resin:glass ratio, resin catalyst level, void content, etc. These variations can have an impact on the weight, mechanical performance, and cost of a component.

Resin infusion, by its nature, provides repeatable laminate properties. Process variables such as consolidation pressure and resin viscosity can be measured and adjusted to ensure they conform. Although operator skill is required to ensure correct laminate staggers (changes in number of plies) and fiber placement, these can be checked at stages in the build.

Environment: Eliminating the large areas of wet resin improves the working environment greatly. Workshops become less "sticky" as the majority of work is done with dry materials. Operator exposure to styrene is greatly reduced which not only improves the workplace, but also helps with conforming to ever tightening styrene (worker and environment) exposure legislation and reduces the need for much of the personal protective equipment (PPE) associated with hand lamination.

Waste: In a large-scale open molding, it is inevitable that significant amounts of waste will be generated. This waste can include excess resin left in buckets, wetting out rollers, and the buckets used to mix the resin themselves. The infusion process has the ability to eliminate the majority of this waste, although the consumable materials, especially saturated flow media, that are vital for the process are generally removed and as they are not reusable will have to be disposed of.

8.1.3 Why not resin infusion

Although resin infusion offers many advantages over open molding, there are factors that could make it less appealing. While these are often outweighed by the benefits infusion offers, they must be considered.

Training/skill set: Resin infusion requires a different set of skills when compared to open molding. Staff will need to be well-trained in tooling requirements, dry materials placement, infusion setup, and managing the infusion process once resin is introduced.

Process control: In an open molding workshop, it is a requirement to ensure that certain parameters are correct. These would generally include workshop temperature and humidity, material and mold tool temperature, and the recording of material weights and correct mixing ratios. Infusion will require further understanding and recording of parameters such as resin viscosity and vacuum levels to ensure process robustness.

Cost: There is significant cost involved adopting resin infusion. Besides the required investment into the plant equipment (vacuum pumps, catch pots, tooling, etc.), in some cases material and consumable materials costs will increase. In the example of a motor yacht hull bottom that would be made from a monolithic laminate if hand laminated, an infused structure would call for a high density core to add the required stiffness.

Risk: Open molding is generally low risk. When providing a good level of staff competence and robust quality control (QC), significant defects are rare. With resin infusion, once the resin is introduced into the part, the point of no return has passed. After this, a system failure, such as loss of vacuum due to a power cut or the introduction of unmixed resin, can be catastrophic for the part. Providing the risks are understood and managed, infusion can be a very robust process but there is certainly more to think about!

8.1.4 Challenges in a production environment

In a production environment, where component delivery times are crucial to the overall program, resin infusion can offer additional challenges. In an open molding process, provided the workshop is at a reasonable temperature/humidity and the materials are at workshop temperature, there is little to stop the process.

Resin infusion relies on much tighter process control and any one of the process parameters out of specification can have serious consequences. For example, tool damage can affect vacuum integrity; a small change in ambient temperature can affect resin viscosity; and incorrect material loading can change the resin flow path. Some of these parameters are easily checked and corrected. However, if unchecked or found at the wrong time (tool vacuum integrity is a good example), incorrect parameter selection can delay the manufacturing process or result in serious quality issues with the finished part.

The infusion process is defined by a number of stages:

- Prepare the mold tool by application of release agent,
- Load dry fabric reinforcement onto the tool,

- Position consumables, and seal the bag,
- Draw vacuum and compress the laminate stack,
- Check vacuum integrity,
- Flow resin to fill all pore space in the laminate,
- Cure the resin matrix, then demold.

8.2 Physics of resin infusion

Resin infusion is governed by the physical properties of the materials used, the available vacuum, the arrangement of resin and vacuum supply, and the geometry of the component.

The work of Henry Darcy in 1800s can be used to define the physical principles of the infusion process. He discovered that the flow of fluid through a porous medium is proportional to the driving pressure, the permeability of the medium, and the cross section and is inversely proportional to the fluid viscosity and the distance over which the pressure gradient acts.

Although Darcy's work was based around saturated or wetted flow, it is used in resin infusion to describe unsaturated or wetting flow. Wetting flow may be different to the wetted flow and is dependent on the permeating fluid (resin) used (Summerscales, 2004). Darcy's law as used for the manufacture of isotropic composites is given in Eq. (8.1):

Darcy's Law

$$Q = \frac{-k \cdot \Delta P \cdot A}{\mu \cdot l} \tag{8.1}$$

where

Q = flow rate
K = permeability
ΔP = pressure differential
A = cross sectional area
μ = dynamic viscosity
l = distance from inlet to flow front

8.2.1 Permeability

Permeability defines how easily a fluid flows through a porous material. Materials with a high permeability allow easy flow, while materials with a low permeability resist flow.

When selecting materials for resin infusion, it is important to balance permeability with the properties of the finished composite.

Materials with a high permeability, while readily allowing resin flow, are likely to have large pore spaces which become resin-rich volumes in the cured laminate and

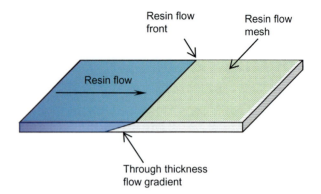

Fig. 8.1 Representation of mesh-assisted infusion showing flow fronts.

these may contain a high proportion of voids. Materials with a low permeability are difficult for the fluid to penetrate, are likely to be starved of resin, and may not wet out before the resin begins to cure.

When describing the permeability of a laminate stack, there are two principal directions of flow; in-plane and through-thickness. In-plane flow describes the resin path between the inlet and the vacuum source. Through-thickness flow describes the resin path from the top to the bottom of the laminate stack. In-plane permeability can be further defined as some materials (especially noncrimp fabrics) that will have different values depending on fiber orientation in the respective layers.

The measured permeability may be different depending on the chosen permeating fluid (Summerscales, 2004), so when determining the permeability of the reinforcement, it is vital that the resin type to be used in the part is also used in any experiments.

Typically, in-plane flow will be limited to ~1.5 m (while maintaining reasonable flow rates), while through-thickness flow will be laminate thickness.

To maximize the distance between resin inlet and vacuum point, while keeping the resin content in the laminate under control, a disposable flow mesh (also known as distribution mesh or transport layer) is used. This allows resin to travel along the surface of the laminate with little resistance. For cored laminates, the core can be scored to allow flow, but as the resin in the groove remains in the core there is a weight penalty.

Due to the relatively short "through-thickness" distance, the resin can flow through unaided and achieve good wet out and high fiber content. Careful material selection must still take place as certain styles of fabric will not allow easy through-thickness flow (Figs. 8.1 and 8.2).

8.2.2 Pressure

Pressure provides two functions in the resin infusion process when using a flexible bag. It is the force consolidating the dry reinforcement and the driving mechanism of the resin flow.

Resin infusion for the manufacture of large composite structures

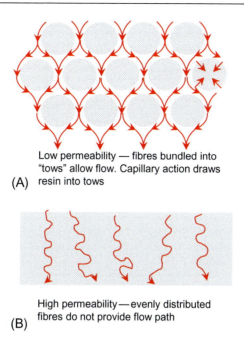

Fig. 8.2 Showing low (A) and high (B) permeability fabrics.

8.2.2.1 Driving force

Fig. 8.3 shows the pressure distribution through the infusion process. The pressure difference between the open resin pot (atmospheric pressure) and the vacuum pump drives resin through the laminate.

8.2.2.2 Consolidation and fabric compressibility

Unlike infusion processes that use a rigid back (fixed cavity) tool, the vacuum bag itself is unable to exert any consolidation pressure in its own right. For this reason, exact thickness control is not possible. The resin inlet pressure must also not exceed the applied vacuum—otherwise the vacuum bag will balloon as it overfills with resin. When infusing from an open container, it is crucial that this container is placed lower than the lowest point of the part to avoid a syphoning effect.

The consolidation of the laminate is achieved from the column of air acting directly above it. At sea level, there is approximately 10 tons/m^2 of air between the ground and the edge of space.

Fig. 8.4A shows the bagged laminate before the vacuum has been applied. The inside of the bag is open to the atmosphere and the pressure is equal on either side of the bag.

Fig. 8.4B shows the bagged laminate after the vacuum has been applied. The pressure inside the bag is lowered by removing the air inside it and the atmospheric pressure outside the bag provides the consolidating force.

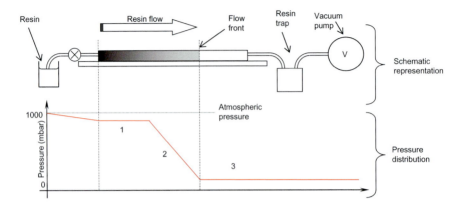

Fig. 8.3 Schematic illustration (not to scale) of resin infusion process showing pressure difference through the laminate. 1—fully wetted reinforcement, 2—through thickness flow gradient, 3—dry reinforcement.

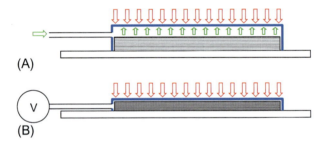

Fig. 8.4 Showing the effect of consolidation pressure.

8.2.2.3 Measuring pressure

The standard SI unit for pressure is Pascal (N/m^2), with atmospheric pressure formally expressed as integer values of hectoPascal (100 Pa) with three significant figures (SF) for low pressure and four SF for higher pressure. However, in industry many other units of pressure are used including psi, bar, in. Hg (inches of mercury), and m H$_2$O (meters of water).

There are two fundamentally different ways of displaying pressure:

- Absolute measurements use a fixed "zero" point (the reference is a perfect vacuum).
- Gauge measurement take the "zero" point to be atmospheric pressure.

Fig. 8.5 shows how gauge measurements vary depending on the atmospheric conditions on the day.

It does not matter which units are selected, but it is important that everyone uses the same ones!

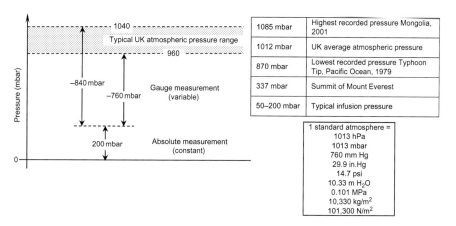

Fig. 8.5 Pressure range and measurement.

8.2.3 Cross sectional area

Resin is introduced into the laminate via a hose, typically between 12 and 25 mm in diameter for a large structure. To maximize the flow speed, the effective cross section of the flow front is increased using a resin gallery. The gallery allows the resin to be carried away from the inlet with little resistance before it makes its way into the laminate. Resin galleries can be made from spiral cut hose, similar to that used to organize electrical cables, or are available commercially as molded polymer sections. Other products that resemble a heavy flow medium are also available.

8.2.4 Viscosity

Viscosity describes the internal resistance to the flow of a fluid. Fluids with a low viscosity, such as water, flow easily. Fluids with a high viscosity, such as honey, flow slowly.

Infusion resins are specially formulated with low viscosity, ~200 centipoise (cP) = 200 mPa s, to ensure they flow through the fibers, but these resins are too thin to be used for hand lamination.

Resin viscosity and temperature are related to one another. As a rule-of-thumb, when a resin is heated by 10°C, its instantaneous viscosity will half, whereas if it cooled by 10°C, the viscosity will double.

Temperature will also affect the gel time of the resin, so it is vital that the viscosity and gel time are understood at given temperatures.

Viscosity can be measured using a number of methods. In the absence of high-specification rheometers, the most reliable technique is to use a spindle viscometer. This piece of equipment uses a rotating spindle that is submerged in the resin sample. The torque required to rotate it is converted into viscosity.

A viscosity cup can be used for shop-floor checking. They are less accurate than a spindle viscometer and are used to simply verify that the resin is within a viscosity range rather than finding the absolute viscosity.

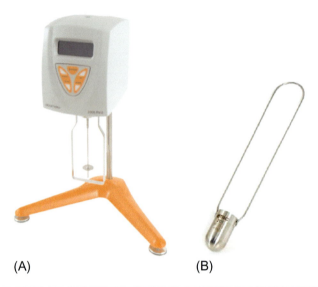

Fig. 8.6 (A) Spindle viscometer, (B) A range of typical fluid viscosities, and (C) Some typical liquid viscosities.

Viscosity cups work by measuring the time taken for the resin to drain from a calibrated cup with a specific size of hole. Different cups are available for different viscosity ranges (Fig. 8.6).

8.2.4.1 Distance from inlet to flow front

As the infusion progresses, the flow front moves away from the inlet. As this distance increases, the pressure gradient and hence flow rate into the part decreases, slowing the progress of the flow front. For the part to fill before the resin gels, the arrangement of resin feeds must be considered (Fig. 8.7).

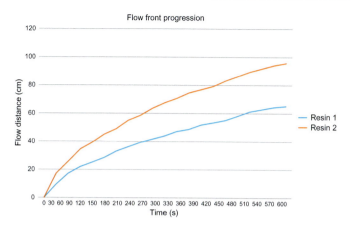

Fig. 8.7 Showing flow front progression vs time for two different resins. Note that as the flow distance progresses, it advances more slowly. Resin 1 is high viscosity, Resin 2 is low viscosity.

8.3 Materials selection and characterization

In the previous section, the physics of the resin infusion process were introduced. Two of the physical properties, viscosity and permeability, are determined by the selection of materials used. However, in addition to simply ensuring the effective flow of resin into the laminate, there are additional considerations that can influence the materials selected.

Mechanical properties of the cured composite will be defined at the design stage depending on mechanical and environmental demands to which the component will be subjected. It is likely that a number of resin systems and reinforcements can be identified that will be suitable, but also that each will have different characteristics that may present challenges during the build of the part. While a "paper" study can aid comparison, product datasheets provide guidelines, but rarely give all of the information needed to understand how the materials will behave and how the resin, reinforcement, and flow promoters perform when brought together. Thorough testing and evaluation is often required to ensure the material selection is suitable for the project.

The characterization of the raw materials does not require a high-tech laboratory; however, some investment in measurement equipment and good experimental practices are vital when embarking on large complex infusions to ensure the material behavior is understood.

8.3.1 Resin selection

The rheological (viscosity) properties of the resin can affect the infusion process. The resin curing behavior must also be considered as it can present additional challenges during the cure of the part and can also affect the finished component. Each characteristic is measurable and the effects can be determined with some simple trial work.

8.3.1.1 Viscosity

A resin's viscosity is affected by temperature. In a manufacturing environment, the ambient temperature should be controlled, but may still vary by 5–10°C depending on the effectiveness of the factory heating and the external temperature. A 10°C change in temperature can double/half the viscosity, so if the relationship between temperature and viscosity are not understood, then the effect on the infusion can be disastrous.

8.3.1.2 Gel time

A resin's gel time, or pot life, indicates the time before the resin begins to cure.

The working time of epoxy resins is determined by the hardener and can only be altered by using a different hardener or changed process temperature. Polyester and vinylester resins use a peroxide catalyst to initiate the curing reaction. Changing the amount and type of catalyst can change the gel time.

The figures quoted on product datasheets will be the time at which the resin turns to a jelly-like state and will be determined on a relatively small liquid sample, often held at a constant temperature in a beaker placed in a water bath. In practical terms, this information is a good starting point, but needs to be better understood to be useful.

Resin gel time when held in bulk is shorter than that for a small sample. This must be considered when mixing large quantities to place in the feed drum. Conversely, gel time in thin laminates may be longer than expected.

As well as ensuring the resin remains liquid for the duration of the infusion, it is also important that the resin does not remain liquid for an excessive time after the infusion has finished. The component is in a delicate state during this time. Any leaks in the bag are difficult to deal with as the flow front is no longer moving, so air will remain trapped in the laminate. There is also a requirement to "babysit" the infusion to ensure there are no problems such as loss of vacuum. Excessive gel times require unnecessary resource to be consumed.

8.3.1.3 Exotherm temperature

The resin curing reaction is exothermic. The temperature will begin to climb as soon as the reaction begins (while the resin is still liquid) and will reach its maximum shortly after the resin has solidified. As with gel time, an indication of peak exotherm temperature may be given on the resin datasheet, but the temperature seen when manufacturing the part will be dependent on a number of factors.

High exotherm temperatures are of concern and can damage tooling, consumables, and the component. They can also cause a greater degree of resin shrinkage and will, if very high, have a detrimental effect on the properties of the resin. High temperatures are often seen in areas of thick laminate and around resin inlet galleries where there is a greater bulk of resin. Careful selection of resin and hardener/catalyst can help to keep exotherm temperatures under control, but there are often trade-offs with the cured resin properties and extended gel times.

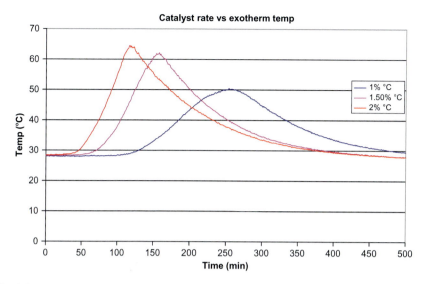

Fig. 8.8 Showing the effect of catalyst addition on peak exotherm temperature.

The cured properties of polyester/vinylester systems are less affected by catalyst selection, but choice of levels does control high exotherm temperatures in thick sections, or expedite the cure of thin laminate (Fig. 8.8).

Epoxy resins can be more challenging as they must be used as stoichiometric mixtures. Hardeners form a defined proportion of the structure of the cured resin. Changing proportions may adjust gel time or exotherm temperature, but will also affect cured mechanical properties and thermal performance as unreacted sites produce pendant groups which lower the glass transition temperature.

8.3.1.4 Shrinkage

Resin shrinkage is difficult/expensive to avoid. Different resin types shrink by different amounts. Linear shrinkage is typically <2% for epoxy resins and <3% for unsaturated polyester resins (Summerscales and Grove, 2014). Data sheets for epoxy resins tend to quote linear shrinkage, while datasheets for polyester tend to quote (higher) volumetric shrinkage! As a guideline, epoxy resins shrink the least, polyester resins shrink the most, and vinylester resins sit somewhere in between. Shrinkage can not only influence the dimensional accuracy of the component, but may also cause gross distortion. If a molded, cosmetic finish is required, then a resin with a lot of shrinkage will show "fiber print" (also known as print-through) on the surface of the part. As with gel time and exotherm, there are ways to control shrinkage that may have an effect on the properties of the cured resin.

Fillers can be used to prevent shrinkage, but at high loading levels they increase viscosity. They could include calcium carbonate or aluminum trihydrate (ATH) as two commercially used products, typically found in filled resin systems used for enhanced surface finish (Scott Bader, 2005).

Cure rate can also affect shrinkage. Rapid cure will often result in excessive shrinkage. Retarding the cure can minimize shrinkage.

Shrinkage can be assessed in a number of ways.

A shrinkage trough is a channel section mold in which a casting can be made. The amount by which different resins shrink can be measured using a Vernier calliper and feeler gauge once the casting is cured. The casting should not be so large that excessive exotherm is generated that could give an unrealistic result.

To establish the effect of resin shrinkage on the cosmetic performance, laminate samples are often made on a flat, high gloss mold tool using different resins. These can be assessed once demolded, often after they are exposed to some heat to simulate aging.

Resin properties not only relate to the liquid state (viscosity) or mechanical properties of the cured resin, the behavior during and after cure are also important to consider.

With many epoxy resin systems, manufacturers will provide a base resin and a range of hardeners. A single resin may have a range of hardeners that can provide a range of gel times, but will also affect cure characteristics and ultimate properties. Epoxy resins generally require postcuring to achieve ultimate properties. Even those that cure at ambient temperature can benefit from a post-cure before they develop enough cure to allow demold.

Resin testing can include "pot" tests to determine gel time, exotherm, and viscosity. Resin shrinkage is also an important factor that can have an effect on the finished part both in terms of its dimensional accuracy (resin shrinkage can mean that the part is a different size to the mold tool once released) and its cosmetic appearance if a molded finish is required.

8.3.2 Reinforcement

The principal characterization required for reinforcements used in infused laminates is permeability.

As previously mentioned, permeability needs to be considered both in-plane and through-thickness.

Due to the increase in closed mold processing, some reinforcement manufacturers are providing permeability data for their fabrics. As with resin properties on product datasheets, this can give a good starting point; however, each laminate stack differs from application to application, so in order predict the flow through a particular laminate stack, a permeability experiment can be done very simply for a flat laminate. It is less straightforward to establish the local permeability of sheared reinforcements or reinforcements on multicurvature tools.

In-plane permeability is relatively easy to measure.

Using a flat mold surface, a rectangular test panel is set up with the laminate as specified.

In-plane permeability is determined by measuring the progression of the flow front with time. If simulation software is not being used, then this is usually enough to allow the infusion strategy to be developed. If simulation software is being used, then the resin viscosity, pressure differential, and fiber volume fraction will also be required to allow the permeability "K" value to be calculated. The quality of the data input to the simulation may be critical to the quality of the output.

Through-thickness permeability can be assessed in a similar way, but as it is much more difficult to measure the progression of the flow front through the laminate accurately, it is often a comparative test. Trial infusions can be set up on a glass plate so that the flow front can be seen emerging on the mold face of the laminate. Often, it is necessary to set up a representation of the infused structure, which is set up and infused on the bench, to verify good flow.

If required, resin infusion can use fewer heavier plies than hand lamination to build up the required fiber weight. Stitched multiaxial fabrics have the benefit of being constructed from layers of fiber orientated at different angles and can replace multiple layers of woven fabric (Fig. 8.9).

There are no standardized permeability tests. The liquid composite molding community has conducted two rounds of round-robin tests to establish reproducibility of permeability data (Arbter et al., 2011; Vernet et al., 2014).

8.3.3 Core materials

Core materials that are suitable for infusion are available from a variety of suppliers. Two types of core can be found: Structural core to make sandwich panels and nonstructural "former" foam for stringers and girders.

Core used in sandwich panel manufacture has to be able to allow resin to flow from the top face into the laminate against the mold tool. This is done by selecting a perforated core. Typically, perforations are ~3 mm in diameter. They are usually spaced, in a grid pattern, between 20 and 50 mm apart.

Core materials can also be used to promote in-plane flow. As well as perforations, the core has a network of grooves machined into, usually, one face. Grooved core

Fig. 8.9 Example of through-thickness permeability. The two laminates are made from the same weight of fabric and fiber orientation. The picture on the left shows the flow fronts on the top surface to be equal. The picture on the right shows the difference in flow through the thickness of the laminate.

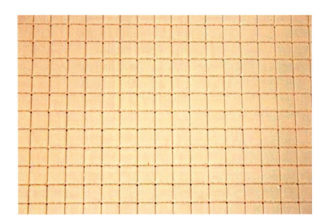

Fig. 8.10 Infusion core with grooves and perforations.

tends to facilitate flow more quickly and slightly further than mesh-driven flow and can save some setup time as mesh does not need to be placed on the part. It does, however, add weight (a 3 mm × 3 mm groove on 20 mm centers will add ∼700 g/m^2) due to additional resin uptake.

Core materials can also be used for formers to build stiffeners. In many hand laminating applications, this foam would be a low density PU (polyurethane) or PE (polyethylene) foam that only serves as a former for the laminate. The problem with some of these foams is that they will deform under vacuum. Foams for formers must be an appropriate density so that they do not deform under vacuum. Core materials with an open cell structure, including honeycombs, will fill with resin unless mitigation measures are in place (Fig. 8.10).

8.3.4 Consumables

Consumable materials are an integral part of the infusion process, but do not form part of the finished component. They are removed after the part has cured and are not reusable.

8.3.4.1 Vacuum bag

Vacuum bags are available in a range of widths and grades depending on the application. They can range from 2 to 16 m wide and are available to suit a range of temperatures. The size of bag is likely to be the first consideration, although if a large enough bag is not available "off the shelf", it is possible to join bags together using mastic "tacky tape". High temperature bags are available, but tend to be expensive and only available in narrower rolls. Low temperature bags are most common for ambient cure resin systems and are normally capable of withstanding temperatures up to 120°C.

8.3.4.2 Peel ply

Peel ply use is not limited to resin infusion and they are typically used to provide a textured surface ready for secondary bonding, especially where epoxy resin systems are used as they require abrasion before any secondary bonding occurs. As well as leaving a textured surface, peel ply can also be placed against the laminate to ease the removal of flow mesh and resin or vacuum galleries. Release coated peel plies are available to allow easier removal, although these should not be used when secondary bonding is to take place in case they transfer some of the release coating to the part. As with vacuum bags, a range of peel ply is available and should be selected depending on processing requirements.

8.3.4.3 Flow mesh

Flow mesh is placed on top of the laminate and provides the mechanism for in-plane resin flow. There are two types of mesh available. 2D mesh is an extruded net material. It is low cost, but only suited to flat panel manufacture. 3D mesh is a knitted product with a degree of drape (can conform to a 3D shape) but is typically more expensive than 2D mesh. Flow mesh from different suppliers will have different flow characteristics and should be assessed along with the reinforcements when considering permeability.

8.3.4.4 Tubing

In order to deliver resin to the part, and connect the part to the vacuum supply, plastic tubing is required.

The tubing must be stable under vacuum, not react with the resin system, and should be selected to be compatible with any fitting and connectors used. Crimping the pipes to temporarily stop resin flow of vacuum supply is common and the tube should be able to do this without splitting.

Nylon and LDPE are common tube materials which are stable up to $\sim$20 mm diameter. If larger pipe is required, for a vacuum ring main or resin feed, spiral reinforced pipe is used so that it does not collapse under vacuum.

8.3.4.5 Release film

Release film can be used to further aid removal of consumables, especially flow mesh and resin galleries, that can become very well-adhered to the laminate. Infusion grade release films are available that contain perforations to allow resin to flow through them. Even with the perforations, they can slow the flow of resin into the part and so should be assessed before use.

8.3.4.6 Resin galleries

In order to make the flow front as long as possible, a resin gallery is used. The gallery provides a means to take resin from a single inlet point and carry it, with little resistance, to lengthen the flow front.

Spiral cut tube is a common product; however, it can leave an indent on the finished laminate. If not careful, the vacuum bag can fall between the spirals and block the flow of resin. Other products are also available that can resolve these problems, but tend to carry the resin from the inlet point more slowly, sometimes requiring more resin inlets.

A note on mold release: Mold release coat is vital to ensure the component does not adhere to the mold surface. If infusing directly to the mold surface with epoxy or vinyl ester resin, a very high slip release agent will be required to guarantee release. High slip release agents can lead to problems when it comes to vacuum bagging as the mastic bagging tape will find it difficult to stick to the mold surface. This will make it very difficult to achieve an adequate vacuum. In this case, release agent should never be applied to the section of tool where the tacky tape is applied.

8.4 Tooling

Tooling for resin infusion shares many similarities with those used for hand laminating. There are a few key points that must be noted.

Materials selection, just as for mold tools used for open molding, must take into account durability, surface finish, thermal performance (relating to the expansion coefficient if postcuring is required and the ability to withstand cure temperatures), and dimensional accuracy.

The fundamental difference when considering tooling for infusion is the tools' ability to maintain a vacuum. Above all else, tooling used for resin infusion must not leak when a vacuum is applied to it. A leaking tool can be disastrous for the successful infusion of the part. If the leak is discovered during the infusion, it can be almost impossible to find the source of the leak from the rear surface of the tool, whereas a bag face leak can be sealed with local application of tacky tape.

For brand new single-piece tooling, vacuum integrity is not normally an issue provided the tool has been well-made.

Single use "direct" tools can be problematic. Often, these are made in a similar way to a tooling pattern (wooden frames, plywood/MDF facing) and are then sheathed with a thin layer of glass scrim and resin to achieve vacuum integrity.

Any repairs or modifications to the tool must also be done is such a way that the vacuum integrity is maintained.

It is common to use multipart tools in order to mold complex shapes in a single operation. Multipart mold tools require that every join is sealed. The most basic way to achieve this is with mastic "tacky tape" placed on the joining flanges. While this can work, it is not guaranteed and the tape can make disassembling the mold tool difficult.

Using a light RTM arrangement for sealing multipart tools is one option, but requires good alignment between sections and the channels for the seals must be considered when making the tooling. In addition, if not perfectly sealed, resin can be drawn from the part into the seal! As well as providing a seal, this type of flange arrangement can provide the clamping force to hold the mold tool sections together. If this is the case, the vacuum must be maintained throughout the dry layup process to

Resin infusion for the manufacture of large composite structures   245

ensure the tool sections remain aligned. More often, external mechanical clamping is employed to hold the sections together. Fig. 8.11 shows a schematic representation of a vacuum-sealed mold tool join.

Checking the vacuum integrity of the mold is achieved by placing a vacuum bag, with breather material to ensure even pressure distribution, onto the section of mold to be tested. A vacuum pump and vacuum gauge are connected so that the leak rate can be assessed. This process is best broken into sections so that, if a leak is found, it can be pinpointed to a specific area. The leak source may be obvious or it may be that smaller and smaller areas are tested until the leak is pinpointed.

8.4.1 *Arrangement for infusion*

To be suitable for infusion, the mold tool should have a flange that extends at least 150 mm beyond the edge of the part. This is to allow for the bagging tape and resin brake required between the part and the vacuum line (Fig. 8.12).

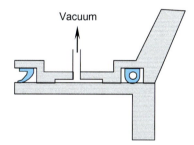

Fig. 8.11 Schematic representation of a vacuum-sealed mold tool join.

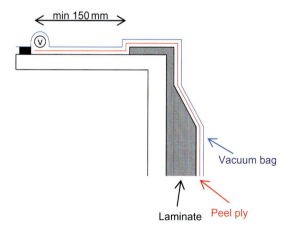

Fig. 8.12 Mold flange detail.

8.4.2 Heated tooling

Heated tooling can serve a number of functions.

It enables the moldings to be postcured in the mold tool while under vacuum. Postcuring in this way will minimize shrinkage at the gel face, ensuring high cosmetic quality providing no prerelease has occurred.

The postcuring will also complete the cure, meaning that any distortion after the part has been released will be minimized. By elevating the cure temperature, the T_g of the resin and gel can be taken to their respective maximum values (providing the postcure temperature is high enough).

It also reduces the time the part has to stay under vacuum. Since no secondary work can commence until the bag is removed, postcuring can improve productivity considerably.

There are a number of methods of achieving in-mold postcure including use of:

- Ovens
- Embedded electric heating
- Embedded water heating
- External electric heater blankets

Each method has its own pros and cons and the method selected will depend on the application and requirement.

Heating the tool, as opposed to using an oven, has the benefit of heating only the sections that require the postcure, reducing energy requirements. However, tooling must be built with embedded heating from the outset as it cannot be retro-fitted. Providing the tooling can withstand the postcure temperature, oven heating does not require special tooling.

8.5 Plant equipment, setup, and redundancy

Whether large or small, the fundamental equipment needed to infuse a component is the same. When scaling up to large infusions, some additional factors need to be considered to ensure a successful and robust infusion.

8.5.1 Equipment

8.5.1.1 Vacuum pumps

A vacuum pump is essential to provide the pressure differential needed to drive the flow of resin. Many different types of vacuum pump are available depending on the flow rate and pressure required ranging from dust extraction (high flow, relatively high absolute pressure) to specialist vacuum pumps (low flow, low absolute pressure).

Often, especially for large infusions, a high volume pump will be required to draw air from the bag quickly. A lower volume pump will then be used to achieve the required vacuum level.

Vacuum pumps will be specified by minimum operating pressure (mbar) and flow rate (m^3/h). Certain pumps will also have a recommended operating pressure; some types of low pressure pumps do not work well for extended periods above a certain pressure. For example, some oil lubricated vein pumps will be unable to scavenge the oil when air flow is too high, resulting in damage to the pump.

Where multiple vacuum systems are running, for example infusion (vacuum bag) and secondary vacuum (mold splits), each should be independent from one another, with the option to combine if one pump fails.

8.5.2 Vacuum ring main

Vacuum systems can be permanently installed or mobile depending on the application and the size of the part.

Permanent vacuum systems should be run in a "ring main" arrangement to ensure good pressure distribution throughout the system.

Ring mains should be made with a large bore (>40 mm) pipe to ensure no flow restriction. Quick connect couplings and shut-off valves will provide flexibility when using the ring main. Systems should be leak-tested prior to first use, and periodically after that, to ensure they are operating correctly. Where multiple infusions are connected to a ring main, users should ensure that other processes connected to the system are not at a critical stage before making a connection.

8.5.3 Vacuum receiver

Vacuum receivers, or reservoirs, can be added to the vacuum system. These can be useful if multiple jobs are connected to a single vacuum circuit. Although they will not change the ultimate vacuum available, they act as a buffer to prevent local pressure changes affecting the overall system pressure if, for example, one job is being set up while another is infusing.

8.5.4 Back-up generators and redundancy

Having invested large amounts of resource and material into the part, a loss of vacuum part way through the infusion would almost certainly result in the component being written off. It is wise to ensure that uninterrupted vacuum supply can be guaranteed, even in the event of a mechanical failure or loss of power.

Vacuum pumps should be well-serviced as most low pressure pumps rely on a vane compressor with wearing parts. Two pumps should be connected so that, in the event of a mechanical failure, a back-up pump is available.

Loss of power is something that is impossible to predict and can be more serious as not only will vacuum be lost, but lighting could also go. Back up generators, ideally ones that auto start when power is lost, are essential for large infusions and should not only supply the vacuum pumps, but also provide lighting and other power requirements so that the infusion can continue in the result of a power failure.

It is vital that back-up systems are tested before each infusion to ensure they work when needed. Trying to sort out back-up power when the lights have gone out is a recipe for disaster!

8.5.5 Resin traps

If resin drawn from the job enters the ring main or vacuum pump, costly repair will be required. To prevent this from happening, resin traps are used to collect resin before it enters the pipe work. Resin traps can be emptied during the infusion, provided they are isolated from the vacuum pump, to prevent air entering the part (Fig. 8.13).

8.5.6 Connectors

Connectors may be required to join vacuum and resin pipes. These should be able to provide an adequate seal under vacuum and be compatible with the resin system.

8.5.7 Vacuum gauges

Vacuum gauges are vital to understand processing parameters. As discussed in Section 8.2, there are 2 types of measurement; absolute and gauge measurement.

Typically, analogue gauges will measure gauge pressure. These types of gauges should be used for indication only and not relied on for process critical measurements (Fig. 8.14).

Calibrated digital (preferably absolute) gauges are to be used for final vacuum checking and leak rate measurement.

8.5.8 Leak detectors

Pinpointing small leaks and pinholes in the vacuum bag can be almost impossible without the aid of a leak detector.

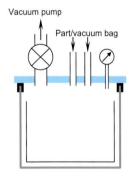

Resin trap requirements
- Removable lid with seal
- Transparent lid to see resin level
- Vacuum gauge
- Ability to drain resin collected while still liquid
- Valve to isolate from vacuum pump to allow emptying
- Stable under vacuum
- Fittings to suit plumbing pipe

Fig. 8.13 Resin trap schematic.

Fig. 8.14 Showing digital (left) and analogue (right) vacuum gauges.

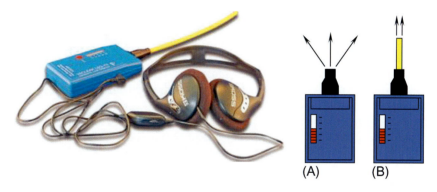

Fig. 8.15 Ultrasonic leak detectors can be used without the tube (A) for wide detection or with the tube (B) for more focused use.

Ultrasonic leak detectors are used to track down leaks. They are very sensitive listening devices that amplify the noise of the leak (Fig. 8.15).

These devices are also very good at picking up background noise from air tools and leaking airlines…use with care!

8.5.9 Hand tools

Hand tools used during infusion could include scissors, pipe cutters, dry brushes for smoothing fabric, and specialist rollers for pushing fabric into corners (e.g., the Dibber (Jones et al., 2017)) (Fig. 8.16).

Use sharp tools with care when working around the vacuum bag!

Fig. 8.16 A selection of hand tools.

8.6 Infusion prediction, strategy, and setup

So far, we have discussed the physical principles of resin infusion, selection of resins, reinforcements and core materials, and the plant equipment required. The next challenge is to take all of that, apply it to a physical component, and complete a successful infusion.

When deciding how to set out an infusion, there are a number of considerations and many possible scenarios.

The geometry of the component, speed of flow through selected laminate, gel time of the resin system as well as production requirements will all have a part to play in deciding the optimal infusion strategy.

As soon as the resin is mixed, the clock begins, so not only must the part completely fill, but it must do so before the resin begins to cure in the laminate.

As we know, the flow rate slows as the flow front progresses. In addition, the geometry and engineering of the component will play a part. Different laminate stacks flow at different rates and will need to be considered independently from one another. Corners and features that may act as race tracks need to be factored into the infusion plan so that they can work in favor of the infusion and not lead to a surprise once the infusion is under way.

It is also vital that the demand for resin is known. The total resin consumption can be calculated from engineering drawings and data collected when characterizing laminate stacks. The demand for resin can be harder to predict. For a large infusion, it may not be sensible to mix all of the resin required at once. Resin in bulk will gel more quickly and if the infusion time exceeds the gel time, then the resin in the feed container will be wasted. Enough resin must be available to cope with the initial flow as the feed pipes and early parts of the infusion will demand a lot of resin.

There are generally two ways to determine infusion setup strategy and resin demand.

Commercially available software has been developed to allow users to predict infusion fill times, flow patterns, and resin consumption, e.g., LIMS, myRTM, PAM-RTM, and RTMWorx.

These programs use Finite Element (FE) techniques, developed for structural analysis, and the work of Henry Darcy.

Users are required to characterize the materials they are using experimentally. The results of the experiments are fed into the software so that the behavior of the materials can be accurately modeled.

Although an experienced infusion team can predict flow, resin uptake, and have an idea of fill time, the layout is unlikely to be optimal. Flow modeling software allows the process to be optimized without the risk associated with full-scale tests.

The software comes at a cost. For relatively straightforward components, using the software will be of little benefit, but for large, complex components, flow modeling can be essential.

8.6.1 Infusion setup

Whether using software or not to set up and predict an infusion, some knowledge of the correct way to approach it is essential. It is up to the engineer to place resin galleries and vacuum points. Software simulation will then predict the outcome.

In order to be successful, the following must occur: The dry materials must be wet out before the resin gels; dry materials must be connected to a vacuum at all time.

The first point goes without saying. If the resin cures in the part before the infusion has finished, then flow will stop. Ensuring the dry materials are constantly connected to a vacuum can be more difficult and requires careful set-up to ensure the dry material does not become isolated. This usually happens as the result of an unexpected race track that cuts off a section of the laminate. When this happens, the dry area becomes disconnected from the vacuum supplied by the pump. Initially, the resin will continue to flow, but as the dry patch fills, the pressure differential will decrease until flow stops. This can be rectified by introducing vacuum into the center of the dry patch by puncturing the vacuum bag and applying local vacuum, but this can result in air entering the part.

Ensuring the part fills before the resin cures requires adequate resin to be fed into the part so that it can fill in the required time.

There are a number of ways this can be done. The picture below shows three typical infusion approaches. Center fill is uncommon as it is by far the slowest and offers no advantage. Linear and peripheral feeds are common and each has pros and cons. These arrangements have been confirmed by a simplified RTM process analysis based on the one-dimensional resin flow model by using Darcy's law for flow through porous media (Cai, 1992a,b) (Fig. 8.17).

As mentioned previously, the maximum distance the resin can practically flow from the resin gallery is limited. For large components, it is therefore required to fit multiple galleries and vacuum points to achieve complete wet out in the required time.

8.6.1.1 Sequential feeds

A sequential feed strategy splits a large panel up into a number of small infusions that follow one another.

In the case below, a large panel is set up with three resin galleries. The vacuum is positioned at the furthest end. The first feed is opened and the resin begins to flow.

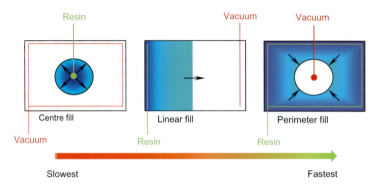

Fig. 8.17 Plumbing arrangements and approximate fill times.

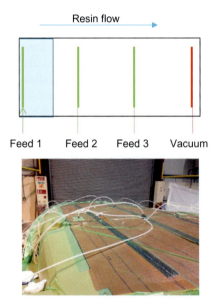

Fig. 8.18 A component that is set up with sequential feeds. Note that the flow front is approaching the outermost gallery.

When the flow front has passed the second gallery, feed 2 can be opened. This can continue until the part is filled (Fig. 8.18).

8.6.1.2 Parallel feeds

A parallel feed strategy sets up the infusion so that all inlets are opened at once.
 There are two approaches to setting up a parallel infusion.
 The "fishbone" setup shortens the distance the resin needs to feed and connects all resin galleries together. Resin inlets are placed into the gallery; the number and spacing of the feeds depend on the size of the part. The vacuum is placed around the

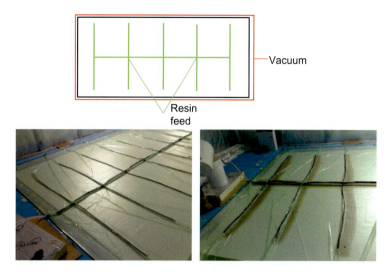

Fig. 8.19 A flat panel infusion using a fishbone setup. The picture on the right shows the flow front progressing.

perimeter of the part. The direction of flow tends, as in the sequential setup, to be toward the edge of the part (Fig. 8.19).

In a peripheral feed setup, the resin galleries are set up around the edge of the part. Large panels are then broken down further with additional resin galleries. Vacuum is then taken from the center of the panel.

Of the three approaches to setting up an infusion, there is no right or wrong method. There are times where a combination of each or all setups can be used on a single part to overcome geometry of a particular laminate arrangement. There are some considerations for each.

Sequential—this is the slowest method as the flow front is the smallest. It does, however, use the least amount of consumables and is generally quick to set up. Because the feeds are opened in sequence rather than all at once, it can be good for resins with short gel times; one section of the laminate can begin to cure, while resin is still being fed in upstream.

Both parallel setups will tend to result in a faster infusion, but often require more consumables and can take longer to set up. In addition, the resin gel time must be more than the time taken to complete the infusion. With a peripheral feed, careful set up of vacuum is required as the vacuum is placed directly on the component. This can lead to void formation and resin being drawn away from the part if not careful (Fig. 8.20).

8.6.2 A note on inclined surfaces

When setting up infusions, it is good practice to always flow "uphill". The reason for this is to ensure any gas or bubbles in the resin are drawn toward the vacuum rather than back toward the resin feed. If flowing a vertical panel downhill, bubbles will want

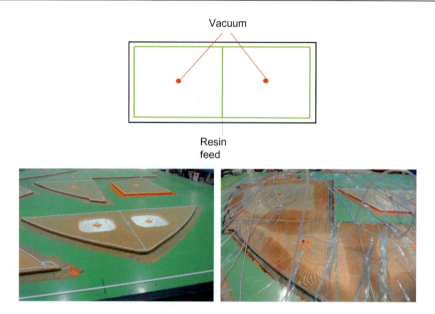

Fig. 8.20 Bulkhead panels set up with peripheral feeds. Note resin galleries and central vacuum feeds. The image on the right shows the flow front progressing.

to rise and will be stuck in the viscous matrix resin, rather than being pushed up into the dry fiber ahead of the flow front. Although this may not always be possible, it should be the aim when setting up the infusion.

8.6.3 3D resin flow

3D resin flow refers to a scenario where the in-plane flow of resin is required to flow out from the surface as well as along it. Coinfused stiffeners are an example of a 3D resin flow. Consider the cross section view below showing a laminate panel with a stiffener placed on top. The resin is flowing from point A toward point E. In addition, the resin also needs to wet out the stiffener laminate and needs to flow to point D. This type of infusion presents a number of challenges. Depending on the flow enhancer used, the stiffener could create interruption. Flow mesh can be butted up to the edge of the stiffener, but a means of transporting the resin from point B to point C is required. Resin will reach point B first and begin flowing up the side of the stiffener before it reaches point C, so a flow break will be required at the top of the stiffener to allow the top of the stringer to be wet out fully. Vacuum will also have to be placed on the top of the stiffener to maintain the pressure differential (Fig. 8.21).

8.6.4 Resin flow management

Controlling the flow of resin so that dry fiber is always connected to a vacuum can be a challenge. Problems can occur for a number of reasons. The setting out of resin galleries and vacuum points can result in uneven resin flow, race tracks can occur around

Resin infusion for the manufacture of large composite structures

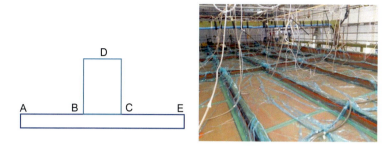

Fig. 8.21 A cross section of an example of 3D resin flow (left) and a deck panel incorporating coinfused stiffeners with 3D resin flow (right). Note additional vacuum required on the top of the stiffeners.

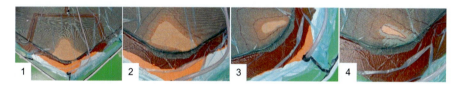

Fig. 8.22 Example of a race track at the edge of the component. By the fourth photo (left to right), the resin flow has stopped.

Fig. 8.23 Example of a race track caused by an overlap of flow mesh. The resin flow increased where the mesh was overlapped, resulting in the dry patch.

core edges, and at the end of the infusion, changes in thickness can leave thicker laminates isolated and the fitting of flow mesh incorrectly can cause unexpected problems (Figs. 8.22 and 8.23).

Resin breaks can be a useful way of controlling resin flow around complex areas where unbalanced in-plane flow or a through-thickness flow gradient could result in dry laminate. A resin break is simply a region where the in-plane flow is slowed to allow other areas to catch up. Fig. 8.24 below shows an example of this.

Resin flow also needs to be controlled around vacuum points. There are generally two arrangements for vacuum.

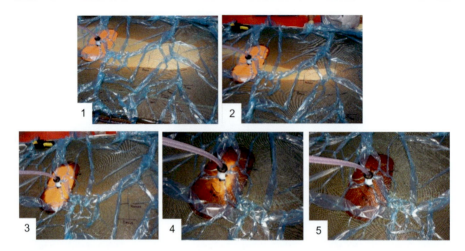

Fig. 8.24 The photos above show how resin flow is managed using a flow brake. The tube in the center of the panel is a vacuum point. Flow mesh is removed around this vacuum point. The first two photos show an uneven flow front approaching the vacuum point. When one flow front meets the break, it slows, allowing the other flow front to catch up. The resin can then flow toward the vacuum point in a more controlled manner.

Peripheral vacuum that runs around the perimeter of the part is usually placed "off the job", but connected to the laminate by way of a breather material. This breather should be able to allow the flow of air easily so that it can be removed from the laminate, but should not allow the flow of resin easily to prevent the vacuum system drawing resin away from the part.

The second type of vacuum point are those that are placed directly onto the laminate. These are common in peripheral fed infusions and also on the top of laminate in a 3D setup. In this case, the vacuum point has to be managed so that excess resin is not drawn away. As with the peripheral vacuum, a low permeability breather can be placed, in a pleat of vacuum bag, to connect the vacuum to the laminate while restricting the flow of resin. There are also semipermeable membrane materials available that have been developed for resin infusion. These materials are similar to the breathable membranes found in waterproof clothing. They will allow air to pass through them, but not liquid resin, and can be very useful as they remove the potential for drawing resin into the vacuum system. They do, however, allow air to flow more slowly than traditional breathers and so need to be tested and validated before use. They are also relatively expensive, so are usually used sparingly.

8.6.5 Resolving dry patches

If an area of laminate becomes cut off from vacuum and is large enough that the residual vacuum contained within the dry patch is not enough to close it, then it is necessary to introduce an additional vacuum into the dry material. This will reestablish the

Fig. 8.25 Using a needle connected to a vacuum supply to close out a dry patch of reinforcement.

pressure gradient and cause resin flow to begin. Introducing vacuum into these voids can remove potential rework, but must be done carefully to prevent a greater problem occurring. Introducing vacuum into areas that have locked off is achieved by inserting a local vacuum supply through the bag into the dry laminate. This is best done using a small diameter tube or needle that is connected to a vacuum source. Something similar to a needle valve used to pump up a football can be used, but dispensing needles that can be attached to syringe bodies are also available.

Using a small piece of mastic tape to ensure a seal, the needle is pushed through the vacuum bag into the dry material and can be removed once the dry patch is full. Great care must be taken to ensure no air enters the laminate (Fig. 8.25).

8.6.6 Gravity

When considering the setup of an infusion strategy, we have already mentioned that, when faced with an inclined surface, it is preferable to flow up hill.

Gravity also has the effect of lowering the pressure gradient as the flow front rises from the resin inlet as a function of the resin density and the force of gravity acting against it. Not only does the pressure gradient have to drive the resin through the fiber, it also has to overcome gravity attempting to pull the resin back down as shown in the equation below.

The slowing of the flow must be considered and may result in resin galleries being placed closer to one another.

The effect of gravity will also limit the maximum vertical gain that can be achieved as no positive resin pressure can be applied because the vacuum bag will simply balloon at the resin inlet.

$$\Delta p = \rho \cdot g \cdot h$$
or
$$h = \frac{\Delta p}{\rho \cdot g} \quad (8.2)$$

where

Δp = pressure difference (N/m^2)
ρ = fluid density (kg/m^3)
g = acceleration due to gravity (9.81 m/s^2)
h = vertical height (m).

Rearranging the Eq. (8.2) to make h the subject shows that the height a fluid can be raised depends on the driving pressure, the density of the fluid, and the force of gravity (on earth, at sea level, this is a constant value of 9.81 m/s^2).

A typical infusion resin has a density of ~1200 kg/m^3.

A "good" vacuum level to infuse with is 50 mbar absolute. This would provide a driving pressure of ~96,300 N/m^2.

Using these parameters, the resin would only be able to climb 8.2 m.

In practice, the maximum vertical height is limited to approximately 6 m to prevent resin pooling in the bottom of the bag and to maintain some consolidation pressure on the wetted fibers.

8.7 Resin delivery and management

Delivering the required amount of mixed resin to the infusion requires careful management. It is vital that enough resin is available to keep up with the demand, but also important that the bulk quantity is managed; large quantities of mixed liquid resin have a reduced pot life and can exotherm violently if allowed to gel.

It is also important that excess resin is not mixed as it is likely that this will end up being wasted.

8.7.1 Hand mixing

The simplest method of resin delivery is to mix by hand and place into a large bucket from which the resin feed pipes can draw it. Resin is decanted from its bulk container into smaller buckets to allow mixing of appropriate quantities. If using epoxy resin, the hardener should also be decanted (ensuring the resin and hardener are weighed so that the correct ratio has been decanted) as correct mix ratios are crucial. If using polyester/vinylester resins, it is also vital that the resin quantity is weighed so that catalyst can be added at the correct ratio. Depending on the size of the mix, the catalyst can be predecanted, but for mixes of under 10 kg, a dosing bottle can be used to quickly and accurately dose catalyst.

Thorough mixing is essential so that the resin and catalyst or hardener are completely combined. Although this can be done by hand using a "mixing stick," it is not the most efficient method for larger resin quantities. Rotary mixers connected to a drill are a good method of quickly and completely mixing resin. For even larger mixes, drum mixers can be used. Care should be taken not to entrain air in the mixed resin (Fig. 8.26).

Fig. 8.26 Hand mixing a large (200 kg) vat of resin.

8.7.2 Machine mixing

Meter mix machinery can be a worthwhile investment for large-scale resin infusions. Meter mix machines consist of a pair of pumps that will "meter" the correct quantity of resin and catalyst or hardener. These tend to be pneumatically driven displacement pumps, where the resin pump drives the catalyst or hardener pump with a "slave" arm. More complex gear pumping systems are also available, but these require complex electronic control to ensure the resin and catalyst/hardener ratios are correct.

After the correct ratio of resin and catalyst/hardener has been pumped, they are passed together through a mixer element—typically a static mixer. The mixer element also has the ability to be flushed with a suitable cleaning solution to clean the mixed resin that would cure and block the mixer if left.

Machine mixers can be used in two ways. The first is to use the machine to mix and dispense resin into a vessel from which it can be drawn into the part. Although the machine can be more efficient compared to hand mixing resin, the management of resin in the vessel must be carefully monitored. As previously mentioned, the demand for resin at the beginning of the infusion is great, so a buffer of resin must be placed in the vessel, especially if the machine dispense rate is less than the resin demand. The machine can then be used to maintain the resin level in the vessel.

Machine mixers can also be plumbed so that they feed resin directly into the vacuum bag, removing the requirement for the vessel. Direct injection offers a number of benefits. Firstly, losses due to pipe resistance are overcome due to the positive pumping pressure. Exposure to VOCs is also reduced further as there is no liquid resin outside of the vacuum bag. Resin wastage is further reduced; when drawing from a vessel, there will always be an amount of resin left in the bottom, but when feeding resin directly into the part this is eliminated.

There are, however, some additional controls that are required. Because the vacuum bag is flexible, the pressure at the point the resin enters the bag must be regulated to below 1 atm to stop the part overfilling with resin. A pressure transducer, common on injection equipment used for RTM (Resin Transfer Molding) or VRTM

(Vacuum-assisted Resin Transfer Molding), can be incorporated to provide feedback, but this becomes complex for parts with multiple injection points.

When using a machine, whether to dispense into a vessel or to inject directly into the vacuum bag, the comments made earlier regarding redundancy and back up become relevant, as a breakdown in resin delivery during an infusion would be disastrous.

8.7.3 Resin management

Large resin infusions will consist of multiple resin feed lines. Each will need to be opened and closed at different times in a particular order. It is necessary to organize the lines so that they are clearly identified and are able to be opened or closed independently from one another.

The infusion strategy will have a predetermined opening sequence. Whatever convention is used on the infusion plan should be used when resin lines are marked and organized. This plan should be clearly displayed at the point where the resin feeds are controlled.

Opening and closing resin feeds can be done in one of two ways. If relatively small diameter (<18 mm) nonreinforced hose is used, it is often possible to simply crimp the hoses to close them. They can be uncrimped when opened and crimped again to close. For larger diameter, reinforced hose, it will be necessary to include shut-off valves to allow the control of resin (Fig. 8.27).

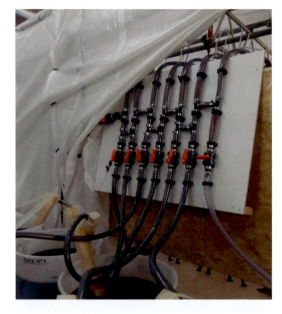

Fig. 8.27 Shut-off valves used to control resin from the mixing vessel into a large infusion.

8.8 Manufacturing process

8.8.1 Wet phase

While it is possible to place dry fiber directly on the mold tool, there are times where it may be necessary to carry out a "wet phase" before the component is infused.

The application of gelcoat, skin-coat, and print barriers is common for cosmetic/visual components. Gel-coating the part will deliver a cosmetically finished part straight out of the mold without the need to paint following demold. Resin infusion has a tendency to make parts that suffer from more print-through (the fabric pattern is visible on the gelcoat surface). To combat this, additional skin-coat of print barriers may be required to make cosmetically acceptable parts.

There may also be times where a resin-rich hand-laminated layer is applied to the mold surface where the component may be exposed to an aggressive environment (composite fuel tanks for example). This layer can be made using a different resin type depending on the requirements for the part.

Although epoxy bonding gelcoats and skin coats are available, it is more common to apply a wet phase coating with polyester or vinyl ester infusions as they are able to make a good secondary bond without abrasion or peel ply and are more tolerant to long delay times.

It is important to ensure adhesion between the infused laminate and the cured "wet phase". With large components, the delay between applying the wet phase and infusing the part may be long—many weeks for very large parts.

There have been recent attempts to develop in-mold gel-coating, but the systems are not yet commercially available (Rogers et al., 2014; Di Tomasso et al., 2014; Gombos and Summerscales, 2016).

8.8.2 Fiber placement

Loading dry fiber into a mold tool presents a number of challenges. While there is no doubt that it is a far cleaner process compared to "wet" laminating, or even prepreg, it can be difficult to get the fabric to stay where it is placed. This can make it difficult when working with vertical surfaces or attempting to apply reinforcement around complex detail.

Spray adhesives have been used for a long time to hold dry fiber. At first, off-the-shelf spray adhesives were used for this application. There are now a number of products available that are developed specifically for placing dry fiber during infusion layup. The adhesives used have been developed to be compatible with specific resin systems. Although compatible, care must be taken to ensure they are not overapplied. An excessive application of these adhesives can lead to incomplete wet out of the fibers. In addition, some spray adhesives contain VOCs that require PPE and can contaminate the surrounding area with a sticky overspray.

The health and safety problems associated with spray adhesives can be removed by using a self-adhesive fabric for resin infusion. These fabrics come with an adhesive preapplied to either one or both sides (two-sided application is used to help hold core

Fig. 8.28 Dry fiber placed on a large deck tool. Note blue backing film for self-adhesive fabric on vertical faces.

materials in place). These self-adhesive fabrics come with the benefit of knowing that a controlled amount of adhesive has been applied, so there is no danger of over-application. These self-adhesive fabrics come with a backing film so that they do not stick together on the roll. It is important to ensure the backing film is carefully removed. If any is left behind in the laminate, it will result in a bond failure between layers.

When placing the dry fiber into the mold tool, it is vital that care is taken to ensure the fibers are not wrinkled, but also are worked into corners and channels. Reinforcement that bridges an internal corner or channel will not only result in a void or resin-rich volumes, but could also create a race track for the resin (Fig. 8.28).

8.8.3 Core fit

The biggest challenges when fitting core in an infused part are keeping the core in position and controlling core gaps, especially on (multi)curved surfaces.

As with the reinforcement the core is fitted dry, so a light adhesive can be used to hold core to vertical surfaces. When fitting a large part, where the core kit is made up of multiple sheets, it is vital to dry fit the core so that the fit and alignment can be checked before any is stuck down. If adhesive is used to secure the core and the core sheet needs to be removed, it can pull the reinforcement away with it.

Large core gaps are a prime source of "race tracks"—easy paths for the resin to take. If these are not controlled, they have the ability to create flow control problems which can lead to areas of incomplete wet out. Accurate core cutting will help to keep the core joints as tight as possible. CNC core kits are readily available and are a very good option to ensure accuracy and also minimize core wastage as each part can be nested when cut (Fig. 8.29).

Securing the edges of core sheets together will ensure the core sheets do not move once positioned. Some adhesives can be used for this, but care must be taken so that

Resin infusion for the manufacture of large composite structures

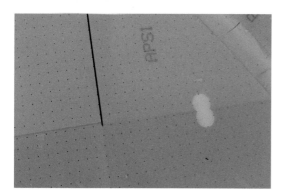

Fig. 8.29 An assembled section of core. Note the locking key (right) and a sheet gap that will require filling.

they do not contaminate the dry fibers. Hot glue can be used, but the working window is short. Contact adhesives can also be used, but they must be compatible with the resin and core and do not allow for repositioning if the alignment is not right first time. An effective method that can be incorporated into a CNC cut core kit are small locking keys. These mechanical fasteners are machined from pieces of core (of the same thickness as the core sheets) and locate into machined keyways in the core kit. These allow core sections to be secured without the need for adhesives.

Sometimes a core sheet will be too small and a small gap will require filling. In this case, thin pieces of core can be cut and used to fill these gaps.

8.9 Process control and preinfusion checks

One of the benefits of resin infusion is the repeatable nature of the process. However, for this repeatability to be realized, an understanding of the process variables, and their effects, is required and each variable must be measured to ensure it meets the required specification.

8.9.1 Measuring variables

In order to ensure that process variables are under control and within the required specification, it will be required to measure them.

Some, such as pressure, temperature, or resin viscosity, are easy to measure (Summerscales, 2003; Konstantopoulos et al., 2014). Others, such as dry fiber placement, are harder to measure and will require checking and rely on operator skill.

8.9.2 Understanding variables

The variables shown in Darcy's equation are fundamental to the infusion process. Pressure drives the resin flow. Reinforcement and flow media permeabilities allow the resin to pass through the system. Resin viscosity ensures the resin is able to flow and the flow length and cross sectional area are defined by the infusion plan. Resin gel time, which is not considered as a variable that normally dictates resin flow, is also important as it can change depending on the ambient conditions and, to a degree, from batch to batch of resin. Accurate catalyst (if using polyester resins) or hardener (for epoxy resins) addition is also vital to ensure good cure and the expected gel time.

Some of these variables can change depending on ambient conditions or machinery performance, some are dependent on the correct fitment and setup of the reinforcement, core material, resin inlets, galleries, and vacuum points, and some can be affected by the care and attention of the operator.

A quality system that prompts inspection at each relevant stage of manufacture will ensure that the process variables are kept within specification and the infusion is a success.

8.9.3 Dry layup phase

During the dry layup/materials placement phase, each part of the process should be inspected to ensure quality. Key quality points include fiber bridging (ensuring the dry fibers are pushed well into internal corners), overlaps, wrinkles/kinks and, of course, that the correct weight/orientation/fiber type has been used.

Core fit must also be checked to ensure core gaps are sufficiently close and that edges have the correct bevel. Checking correct core density and thickness used is also important as it will affect the properties of the finished component.

Resin galleries and vacuum points, as well as any flow media, should be checked off against the infusion plan to ensure they have been correctly placed and are connected to the reinforcement.

8.9.4 Raw materials and ambient conditions

Ambient conditions can affect the way some of the raw materials behave and must be controlled within the manufacturing area. Low temperatures will increase resin viscosity and, if too low, will prevent the resin from flowing. Low temperatures also inhibit or even stop the progress of cure. High temperatures will shorten gel times and, although the resin viscosity will also instantaneously drop, speeding up the resin flow rate, this may not be enough to offset the reduction in gel time. Resins will be selected to work within a particular temperature range and this range must be maintained during the infusion.

If resin is stored outside of the factory, it must also be allowed to come to the correct temperature before use. This is an important consideration during colder seasons when raw material stores may be outside of the temperature-controlled factory.

8.9.5 Preinfusion checks

As soon as the resin begins to enter the part, the point of no return has been passed. Before that happens, it is vital that all relevant checks take place.

8.9.6 Vacuum

Once the part is under vacuum, the quality of the vacuum must be assessed. The absolute vacuum within the bag must be measured. In addition, the leak rate should also be determined. The leak rate will indicate that the bag is well-sealed and no leaks are present. The leak rate is tested by connecting a vacuum gauge to the component, usually via a resin inlet near the center of the part, and isolating the bag from the vacuum pump. A drop in vacuum can be measured and is usually recorded as the pressure gain/minute. The test should be carried out for at least 5 min and the change should be as close to zero as possible. If leaks are present, they should be isolated and rectified and the test should be repeated. In reality, there is likely to be some leaks present in the part and a threshold is often set. There is no absolute rule regarding threshold and it will be up to the engineer to decide.

8.9.7 Resin

Resin viscosity and gel time should be tested before large infusions take place. Provided the ambient temperature is within specification, there should be no surprises regarding viscosity and gel time, but there can be batch-to-batch variation with resins which are worthwhile confirming before the resin is used.

8.9.8 Back-up systems

Back-up systems (generators, back-up pumps, etc.) should be checked to ensure they are functional before the infusion begins. If a power failure or machine breakdown does occur, the back-up systems should operate without delay.

8.9.9 In-process monitoring

Once the infusion has begun, it is still necessary to monitor the process.

Pressure should be constantly observed. Ideally, the vacuum ring-main and the pressure within the part should be measured. Drops in vacuum level can indicate pump failure or a breach due to a hole in the vacuum bag or a disconnected pipe.

Resin temperatures should be monitored, especially if resin has been mixed in bulk. If the bulk resin temperature begins to rise unexpectedly, then it may be necessary to dispose of it and replace it with a fresh mix before it gels.

8.9.10 Leaks in the vacuum bag

For large infusions, it is very unlikely that the part is 100% leak-free. The drop test performed before the infusion would verify that the leak rate is acceptable, but even then there may be some small holes in the vacuum bag. Fortunately, these leaks can be easy to find and fix, provided the part is closely monitored. The sequence of photos shown in Fig. 8.30 shows the appearance of a small leak in the vacuum bag as the flow front meets, then passes the bag defect.

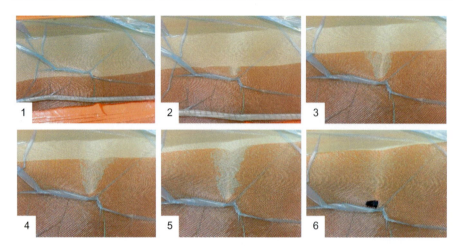

Fig. 8.30 Development of a pinhole leak in the vacuum bag as the flow front progresses.

Fig. 8.30(2) shows the first sign of the leak as the flow front passes it. As the flow front progresses, the stream of bubbles entering the part can be seen. A small piece of tacky tape placed on top of the leak is enough to reseal the bag and stop the leak. When caught early, the leak can be stopped and the part is unaffected, but it is vital that these types of issues are caught quickly and resolved so that they do not affect the quality of the part.

8.10 Postinfusion management

Once the part has filled with resin, the feed lines are clamped. This will stop the flow of resin into the part and prevent the vacuum lines from filling with resin.

Pressure management following the infusion can take a number of forms.

Provided the part is stable and the resin flow has stopped, the vacuum levels can be left as they were during the infusion.

In some cases, the vacuum can be "backed off". This involves raising the pressure in the vacuum system. This can help to stabilize the part and prevent excess resin from being drawn into the vacuum system. If the vacuum is backed off, this must be done carefully. Vacuum levels must be maintained so that resin does not drain from the laminate.

Visual checks should be made to the part during the cure stage to make sure that no leaks occur. The part is still vulnerable to leak-up until the point the resin has gelled.

Vacuum should be maintained until a point that the resin has developed sufficient cure. This will be well beyond the gel time and will vary from resin to resin. Temperature monitoring of the part during this phase will indicate the initiation of the resin gelation. Following gelation, the part can be safely left, but as mentioned, vacuum should be maintained until the resin has developed sufficient cure, which may include postcuring.

8.11 Conclusion/summary

The infusion process relies on many variables to ensure its success. Not only does the process rely on vacuum to achieve consolidation, maximize fiber content, and drive resin flow, the vacuum must be of good quality (no leaks in the mold tool or associated consumables) to ensure voids in the form of introduced air are not present in the finished part and be maintained throughout the infusion and during cure. It also relies on careful materials selection and consumables placement to ensure the resin flows completely through the dry laminate and fully wets out all reinforcement fully. The working time of the liquid resin must be such that the part can completely fill within the gel time, but cure in a reasonable time after the infusion is complete.

Resins must also be properly mixed and dosed with the appropriate catalyst or hardener at the correct level.

Failure of any of the factors above will result in rework or possibly scrap. The most common cause of rework or scrap is incomplete resin flow that could not be resolved while the resin was still liquid or the introduction of air through a leaking tool, pipe, or bag face that was not isolated and resolved during the infusion. Depending on the value of the part and the extent of the rework, it is often possible to recover the part, but this would often result in removal of the affected area and then a repair to the laminate, which may or may not be acceptable.

Total loss of vacuum, especially before the resin has cured, or improper mixing of resin that results in undercure would certainly result in the loss of the part as the performance of the finished composite would be severely compromised.

However, when the process has been well setup, resin infusion is a very repeatable process that is capable of delivering complex laminates to a very high quality.

References

Arbter, R., Beraud, J.M., Binetruy, C., Bizet, L., Bréard, J., Comas-Cardona, S., Demaria, C., Endruweit, A., Ermanni, P., Gommer, F., Hasanovic, S., Henrat, P., Klunker, F., Laine, B., Lavanchy, S., Lomov, S.V., Long, A., Michaud, V., Morren, G., Ruiz, E., Sol, H., Trochu, F., Verleye, B., Wietgrefe, M., Wu, W., Ziegmann, G., 2011. Experimental determination of the permeability of textiles: a benchmark exercise. Compos. A: Appl. Sci. Manuf. 42 (9), 1157–1168.

Becker D.W., Tooling for Resin Transfer Moulding, Wichita State University, Wichita, KS, n.d.

Beckwith, S.W., 2007a. Resin infusion technology part 1—industry highlights. SAMPE J. 43 (1), 61.

Beckwith, S.W., 2007b. Resin infusion technology part 2—process definitions and industry variations. SAMPE J. 43(3), 46 and 43(4).

Beckwith, S.W., 2007c. Resin Infusion Technology Part 3—a detailed overview of RTM and VIP infusion processing. SAMPE J. 43 (4) 6 and 66–70.

Cai, Z., 1992a. Analysis of mold filling in RTM process. J. Compos. Mater. 26 (9), 1310–1338.

Cai, Z., 1992b. Simplified mold filling simulation in resin transfer moulding. J. Compos. Mater. 26 (17), 2606–2630.

Cripps, D., Searle, T.J., Summerscales, J., 2000. Open mould techniques for thermoset composites. In: Talreja, R., Månson, J.-A. (Eds.), Comprehensive Composite Materials Encyclopædia. In: Polymer Matrix Composites, vol. 2. Elsevier Science, Oxford. ISBN 0-08-043725-7, pp. 737–761. (Chapter 21).

Scott Bader, 2005. Crystic Composites Handbook. Scott Bader, Wellingborough.

Di Tomasso, C., Gombos, Z.J., Summerscales, J., 2014. Styrene emissions during gel-coating of composites. J. Clean. Prod. 83, 317–328.

Gombos, Z.J., Summerscales, J., 2016. In-mould gel-coating for polymer composites. Compos. A: Appl. Sci. Manuf. 91 (1), 203–210.

Jones, H., Roudaut, A., Chatzimichali, A., Potter, K., Ward, C., 2017. The Dibber: designing a standardised handheld tool for lay-up tasks. Appl. Ergon. 65 (1), 240–254.

Konstantopoulos, S., Fauster, E., Schledjewski, R., 2014. Monitoring the production of FRP composites: a review of in-line sensing methods. Express Polym. Lett. 8 (11), 823–840.

Muskat I.E., Method of Molding, United States Patent 2495640, 1950.

Pearce, N.R.L., Guild, F.J., Summerscales, J., 1998. An investigation into the effects of fabric architecture on the processing and properties of fibre reinforced composites produced by resin transfer moulding. Compos. A: Appl. Sci. Manuf. A29 (1), 19–27.

Rogers, W., Hoppins, C., Gombos, Z., Summerscales, J., 2014. In-mould gel-coating of polymer composites: a review. J. Clean. Prod. 70, 282–291.

Seeman W.H., Plastic Transfer Molding Techniques for the Production of Fiber Reinforced Plastic Structures, United States Patent US 4902215 A, 1990.

Summerscales, J., 2003. In-process monitoring for control of closed-mold techniques for the manufacture of thermosetting matrix composites. In: Shonaike, G.O., Advani, S.G. (Eds.), Advanced Polymeric Materials–Structure and Property Relationships. CRC Press LLC, Boca Raton, FL. ISBN 1-58716-047-1, pp. 57–101. (Chapter 2).

Summerscales, J., 7–9 July 2004. The effect of permeant on the measured permeability of a reinforcement.The 7th International Conference on Flow Processes in Composite Materials (FPCM-7), Newark DE, USA Paper no. 74.

Summerscales, J., 2012. Resin infusion under flexible tooling (RIFT). In: Encyclopedia of Composites. second ed. John Wiley & Sons, pp. 2648–2658.

Summerscales, J., Grove, S., 2014. Manufacturing methods for natural fibre composites. In: Hodzic, A., Shanks, R. (Eds.), Natural Fibre Composites: Materials, Processes and Properties. Woodhead Publishing, Cambridge, pp. 176–215. (Chapter 7).

Summerscales, J., Searle, T.J., 2005. Review: low pressure (vacuum infusion) techniques for moulding large composite structures. Proc. IMechE P L: J. Mater. Des. Appl. L219 (1), 45–58.

Vernet, N., Ruiz, E., Advani, S., Alms, J.B., Aubert, M., Barburski, M., Barari, B., Beraud, J.M., Berg, D.C., Correia, N., Danzi, M., Delavière, T., Dickert, M., Di Fratta, C., Endruweit, A., Ermanni, P., Francucci, G., Garcia, J.A., George, A., Hahn, C., Klunker, F., Lomov, S.V., Long, A., Louis, B., Maldonado, J., Meier, R., Michaud, V., Perrin, H., Pillai, K., Rodriguez, E., Trochu, F., Verheyden, S., Wietgrefe, M., Xiong, W., Zaremba, S., Ziegmann, G., 2014. Experimental determination of the permeability of engineering textiles: benchmark II. Compos. A: Appl. Sci. Manuf. 61, 172–184.

Williams, C.D., Summerscales, J., Grove, S.M., July 1996. Resin infusion under flexible tooling (RIFT): a review. Compos. A: Appl. Sci. Manuf. 27A (7), 517–524.

Section D

Advanced concepts and special systems

Smart composite propeller for marine applications

H.N. Das*, S. Kapuria[†,‡]
*Naval Science & Technological Laboratory, DRDO, Visakhapatnam, India, [†]CSIR–Structural Engineering Research Centre, Taramani, Chennai, India, [‡]Department of Applied Mechanics, Indian Institute of Technology Delhi, New Delhi, India

Chapter Outline

9.1 Introduction   271
9.2 Flow solution   273
   9.2.1 RANS equations   273
   9.2.2 Domain and mesh for flow analysis   274
9.3 Deformation of composite propeller   274
9.4 Modeling of shape memory alloy   277
   9.4.1 Constitutive equation for SMA fiber   277
   9.4.2 Recovery stress in lamina with SMA   277
   9.4.3 Recovery stress at different temperatures   277
9.5 Fluid-structure interaction   279
9.6 Material failure   279
9.7 Analysis of different propellers   279
   9.7.1 Real marine propeller: prop1   279
   9.7.2 Simple composite propeller: prop2   279
   9.7.3 Pre-pitched propeller: prop3   283
   9.7.4 Smahc propeller: prop4   285
9.8 Conclusions   296
References   296
Further reading   297

9.1 Introduction

A scientific concept and design of a screw propeller was first introduced by Robert Hooke through his presentation on "Philosophical Collections" at the Royal Society in 1681. The ideas of "water screw" or "screw propulsion based on fan blades" were even older, as conceived by Archimedes in 250 BCE and Leonardo da Vinci in the fifteenth century, respectively (Carlton, 2007). Robert Hooke, better known for the Hooke's law of elasticity, went on to make a device in 1683 to measure the current

Marine Composites. https://doi.org/10.1016/B978-0-08-102264-1.00009-1
Copyright © 2019 Elsevier Ltd. All rights reserved.

of flow and explained how this can be used to propel a ship if powered adequately. A marine propeller was first developed only in the end of the eighteenth century. Joseph Bramah in 1785 and Edward Shorter in 1802 developed propellers for ships (Carlton, 2007). Initially, cast iron was widely used to make propellers. Gradually, materials like aluminum, nickel-aluminum bronze, etc. came into use. Only towards the end of the twentieth century, composite materials started being used for marine propellers. Except for a few developments in USSR, all such usage of composites began only in the last decade of the twentieth century. USSR, however, conducted sea trials to compare the performance of 0.25–3 m diameter composite and metal propellers with the same geometry on commercial ships with displacements of 2–5000 tons travelling at speeds of 5–35 knots (Ashkenazi et al., 1974). Molland and Turnock (1991) developed a model ship propeller with hybrid composite for testing in wind tunnel. In the middle of the last decade, composite propellers started being seen in practice for actual ships (Marsh, 2004). Until this time, composites were used for their light weight, lack of corrosion, less magnetic/electrical signature, dampening of vibrations/noise, etc. Other advantages of composite materials, like its flexibility, were not exploited initially. In very recent times, attempts are being made to exploit flexibility of composite materials to improve the properties of propellers (Das and Kapuria, 2016). The improvements of properties are in terms of reduction of dynamic stress (He et al., 2012), adapting toward spatially varying wake (Mulcahy et al., 2010), improvement of hydrodynamic efficiency (Motley et al. 2009), etc.

Smart materials like shape memory alloys (SMA) have been used to manufacture actuators, mechanical couplings for automobiles and aircrafts, and electrical connector, where the change of shape is used for connecting/disconnecting an electrical circuit (Duerig et al., 1990). SMA is also used in civil engineering structures against earthquakes or other environmental impacts (Song et al., 2006). SMA is extensively used in the field of medical science (Otsuka and Wayman, 1998). However, the use of SMA for marine propeller is not observed in literature. Park et al. (2011) have used SMA hybrid composite (SMAHC) material for the design of a proprotor of a "tiltrotor aircraft".

This chapter will describe the concept of improving hydrodynamic efficiency of a marine propeller using traditional and smart composite materials. A propeller is generally designed to operate for a particular advance velocity and revolutions per minute, where it delivers the maximum hydrodynamic efficiency. In many situations, propellers are required to operate at off-design conditions, where their efficiency is compromised. The angle of attack of the incoming flow to the propeller depends on the ratio of advance velocity to its rotational speed. The pitch of the propeller blade is designed such that it receives flow at the optimum angle of attack. At off-design conditions, the angle of attack becomes suboptimum and its hydrodynamic efficiency drops. It is well-known that the metallic propellers (fixed pitch) are rigid enough to hold its shape under hydrodynamic loading (Das et al., 2013) and can only give the maximum efficiency for a particular operating condition. The pitch can be altered in a controllable pitch propeller and it can remain efficient at different operating conditions (Das et al., 2012). In layered fiber reinforced composite materials, however,

Smart composite propeller for marine applications

the fiber orientations of the layers can be designed such that it can generate bending-twisting coupling. This coupling can be used to get a twist of the propeller blade from the bending deformation under the hydrodynamic loading. The twist alters the pitch of the propeller, and hence, the angle of attack. The bend-twist coupling of orthotropic material is thus used to introduce automatic adjustment in pitch angle of marine propellers, operating under different advance ratios.

Active control of the deformation of a propeller through the use of SMA may become an additional option and is also examined here. Nitinol fibers are used as SMA element. Resistive heating of these fibers can transform itself from martensite to austenite state, which causes distinct changes in mechanical properties and generates some internal stresses to create deformation and twist in the propeller blade. This twist is used advantageously to adjust the pitch of the propeller in different operating conditions. A full-scale 4.2 m diameter propeller is studied, as the loading and deformation of model propellers that are typically studied for this purpose cannot be extrapolated to a full-scale prototype propeller. The open water performance is estimated by solving the pressure-based Reynold's averaged Navier-Stokes (RANS) equation for steady, incompressible, turbulent flow using the finite volume method. The deformation analysis is done using the finite element method based on the first order shear deformation theory (FSDT) for composite laminates. The fluid-structure interaction is incorporated in an iterative manner. The numerical study reveals that it is possible to generate sufficient deformation and twist to cause tangible improvement in the hydrodynamic performance of the propeller in off-design conditions.

9.2 Flow solution

The propeller-flow is solved computationally using the finite volume method. The turbulent flow is analyzed using the RANS equations through a commercially available software, ANSYS Fluent 12.1. The solution of steady and viscous flow gives the pressure distribution over the propeller blade surface.

9.2.1 RANS equations

The Navier Stokes equations governing the fluid motion can be written in general form in the Cartesian coordinate system as:

Continuity:

$$\frac{\partial \rho}{\partial t} + \frac{\partial (\rho U_i)}{\partial x_i} = 0 \tag{9.1}$$

Navier Stokes equation

$$\frac{\partial (\rho U_i)}{\partial t} + \frac{\partial}{\partial x_j}\left[(\rho U_i U_j) - \mu\left(\frac{\partial U_i}{\partial x_j} + \frac{\partial U_j}{\partial x_i}\right) + p\right] = S_{U_i} \tag{9.2}$$

where U_i, p, and ρ denote the ith component of velocity, pressure, density, respectively, and S_{Ui} is any source term other than the pressure gradient. Now, instantaneous velocities can be decomposed to time-averaged and fluctuating components for turbulent flow, i.e., $U = \bar{U} + u$, where $\bar{U}$ is the time-averaged velocity vector, and u is the fluctuating component of velocity vector. Substitution of U, followed by time averaging, transforms the mass and momentum conservation equations into:

$$\frac{\partial \rho}{\partial t} + \frac{\partial \rho \bar{U}_k}{\partial x_k} = 0 \qquad (9.3)$$

$$\frac{\partial (\rho \bar{U}_i)}{\partial} + \frac{\partial}{\partial x_j}(\rho \bar{U}_i \bar{U}_j) = -\frac{\partial \bar{p}}{\partial x_i} + \mu \nabla^2 \bar{U}_i + \frac{\partial}{\partial x_j}(-\rho \overline{u_i u_j}) + \overline{S_{Ui}} \qquad (9.4)$$

In the present analysis, the additional unknown term, $-\rho \overline{u_i u_j}$, is evaluated through k-ε turbulence model. Eqs. (9.3) and (9.4) are solved computationally to get the flow around the propeller which gives the pressure distribution over the propeller blade.

9.2.2 Domain and mesh for flow analysis

Four different propellers of the same diameter are analyzed here. As the size of the propeller is same, similar mesh and domains are used for all the cases. A suitable domain size is considered around the propeller to simulate ambient condition. A circular cylindrical domain of diameter ~4D and length of ~7D is used for flow solution. A multi-block structured grid is generated for the full domain using ICEM CFD Hexa module. Around 1.2 million hexahedral cells are deployed to descritize the domain. Necessary boundary conditions are set. For the present solution, standard scheme is used for pressure and a SIMPLE (Semi-Implicit Pressure Link Equations) procedure is used for linking pressure field to the continuity equation.

9.3 Deformation of composite propeller

The deformation analysis of the propeller blade is performed using the commercial software ANSYS Mechanical 12.1. The software employs finite element method (FEM). The discretized system of equations of motion for the blade can be written as:

$$[M]\{\ddot{u}\} + [C]\{\dot{u}\} + [K + K_g]\{u\} = \{F_{co}\} + \{F_{ce}\} + \{F_h\} \qquad (9.5)$$

where $\{\ddot{u}\}, \{\dot{u}\}$, and $\{u\}$ denote the nodal acceleration, velocity, and displacement vectors, respectively. [M], [C], and [K] are the global consistent mass, damping, and stiffness matrices, respectively. [K_g] is the geometric stiffness matrix accounting for the effect of large displacements and is a function of $\{u\}$. The terms at the right hand side are the nodal force vectors due to the Coriolis force, centrifugal force, and the hydrodynamic force, respectively. For the present steady state analysis, only static loading is modeled and Eq. (9.5) becomes

$$[K + K_g]\{u\} = \{F_{co}\} + \{F_{ce}\} + \{F_h\} \tag{9.6}$$

The blade is assumed to be made of linear elastic, perfectly bonded, unidirectional laminas with different fiber orientations. For a thin lamina, the transverse normal stress σ_3 is neglected in comparison with the other stress components. The stress-strain relationship of linear elasticity for a specialy orthotropic material in the material coordinate system, x_1 and x_2, becomes (Jones, 1975).

$$\begin{Bmatrix}\sigma_1\\\sigma_2\\\tau_{12}\end{Bmatrix} = \begin{bmatrix}Q_{11}&Q_{12}&0\\Q_{12}&Q_{22}&0\\0&0&Q_{66}\end{bmatrix}\begin{Bmatrix}\varepsilon_1\\\varepsilon_2\\\gamma_{12}\end{Bmatrix}, \begin{Bmatrix}\tau_{23}\\\tau_{34}\end{Bmatrix} = \begin{bmatrix}Q_{44}&0\\0&Q_{55}\end{bmatrix}\begin{Bmatrix}\gamma_{23}\\\gamma_{34}\end{Bmatrix} \tag{9.7}$$

where Q_{ij} are reduced stiffness coefficients defined as

$$Q_{11} = \frac{E_1}{1-\nu_{12}\nu_{21}}, Q_{22} = \frac{E_2}{1-\nu_{12}\nu_{21}}, Q_{12} = \frac{\nu_{12}E_2}{1-\nu_{12}\nu_{21}}, Q_{44} = G_{23},$$

$$Q_{55} = G_{31}, Q_{66} = G_{12} \tag{9.8}$$

and ε_i, σ_i denote the normal strain and stress components, and γ_{ij} and τ_{ij} denote the shearing strain and stress components, respectively. E_i, ν_{ij}, and G_{ij} denote, respectively, the Young's modules, Poisson's ratios, and shear modules of the lamina.

After transforming the stresses and strain components in Eq. (9.7) from material (x_1, x_2) to laminate (x, y) coordinate system, the constitutive relations in the laminate coordinate system are obtained as.

$$\begin{Bmatrix}\sigma_x\\\sigma_y\\\tau_{xy}\end{Bmatrix} = \begin{bmatrix}\overline{Q}_{11}&\overline{Q}_{12}&\overline{Q}_{16}\\\overline{Q}_{12}&\overline{Q}_{22}&\overline{Q}_{26}\\\overline{Q}_{16}&\overline{Q}_{26}&\overline{Q}_{66}\end{bmatrix}\begin{Bmatrix}\varepsilon_x\\\varepsilon_y\\\gamma_{xy}\end{Bmatrix}, \begin{Bmatrix}\tau_{yz}\\\tau_{zx}\end{Bmatrix} = \begin{bmatrix}\overline{Q}_{44}&\overline{Q}_{45}\\\overline{Q}_{45}&\overline{Q}_{55}\end{bmatrix}\begin{Bmatrix}\gamma_{yz}\\\gamma_{zx}\end{Bmatrix} \tag{9.9}$$

where, for a lamina with its fiber axis x_1 oriented at angle θ with respect to the x-axis, $\overline{Q}_{ij}$ are given by

$$\overline{Q}_{22} = Q_{11}\sin^4\theta + 2(Q_{12} + 2Q_{66})\sin^2\theta\cos^2\theta + Q_{22}\cos^4\theta,$$
$$\overline{Q}_{12} = (Q_{11} + Q_{22} - 4Q_{66})\sin^2\theta\cos^2\theta + Q_{12}(\sin^4\theta + \cos^4\theta),$$
$$\overline{Q}_{66} = (Q_{11} + Q_{22} - 2Q_{12} - 2Q_{66})\sin^2\theta\cos^2\theta + Q_{66}(\sin^4\theta + \cos^4\theta),$$
$$\overline{Q}_{16} = (Q_{11} - Q_{12} - 2Q_{66})\sin\theta\cos^3\theta + (Q_{12} - Q_{22} - 2Q_{66})\sin^3\theta\cos\theta,$$
$$\overline{Q}_{26} = (Q_{11} - Q_{12} - 2Q_{66})\cos\theta\sin^3\theta + (Q_{12} - Q_{22} - 2Q_{66})\cos^3\theta\sin\theta,$$
$$\overline{Q}_{44} = Q_{44}\cos^2\theta + Q_{55}\sin^2\theta, \overline{Q}_{55} = Q_{44}\sin^2\theta + Q_{55}\cos^2\theta, \overline{Q}_{45} = (Q_{55} - Q_{44})\cos\theta\sin\theta.$$
$$\tag{9.10}$$

According to first order shear deformation theory (FSDT), the force and moment resultants of the laminates are related to the strain components ε_{ij}^0 and γ_{ij}^0, and curvature components κ_x, κ_y, and κ_x at the reference surface of the plate by

$$\begin{bmatrix} N_x \\ N_y \\ N_{xy} \end{bmatrix} = \begin{bmatrix} A_{11} & A_{12} & A_{16} \\ A_{12} & A_{22} & A_{26} \\ A_{16} & A_{26} & A_{66} \end{bmatrix} \begin{bmatrix} \varepsilon_x^0 \\ \varepsilon_y^0 \\ \gamma_{xy}^0 \end{bmatrix} + \begin{bmatrix} B_{11} & B_{12} & B_{16} \\ B_{12} & B_{22} & B_{26} \\ B_{16} & B_{26} & B_{66} \end{bmatrix} \begin{bmatrix} \kappa_x \\ \kappa_y \\ \kappa_{xy} \end{bmatrix}$$

$$\begin{bmatrix} M_x \\ M_y \\ M_{xy} \end{bmatrix} = \begin{bmatrix} B_{11} & B_{12} & B_{16} \\ B_{12} & B_{22} & B_{26} \\ B_{16} & B_{26} & B_{66} \end{bmatrix} \begin{bmatrix} \varepsilon_x^0 \\ \varepsilon_y^0 \\ \gamma_{xy}^0 \end{bmatrix} + \begin{bmatrix} D_{11} & D_{12} & D_{16} \\ D_{12} & D_{22} & D_{26} \\ D_{16} & D_{26} & D_{66} \end{bmatrix} \begin{bmatrix} \kappa_x \\ \kappa_y \\ \kappa_{xy} \end{bmatrix}$$

(9.11)

and

$$\begin{Bmatrix} Q_x \\ Q_y \end{Bmatrix} = \begin{bmatrix} A_{55} & A_{54} \\ A_{45} & A_{44} \end{bmatrix} \begin{Bmatrix} \gamma_{zx}^0 \\ \gamma_{yz}^0 \end{Bmatrix} \qquad (9.12)$$

where A_{ij}, D_{ij}, and B_{ij} ($i, j = 1, 2, 6$) denote the laminate extensional stiffness, bending stiffness, and extension-bending coupling stiffness coefficients, respectively, which are given for a laminate of L layers by

$$A_{ij} = \sum_{k=1}^{L} \overline{Q}_{ij}^k (z_k - z_{k-1}), \; B_{ij} = \frac{1}{2}\sum_{k=1}^{L} \overline{Q}_{ij}^k (z_k^2 - z_{k-1}^2), \; D_{ij} = \frac{1}{3}\sum_{k=1}^{L} \overline{Q}_{ij}^k (z_k^3 - z_{k-1}^3).$$

(9.13)

A_{ij} ($i, j = 4, 5$) are the transverse shear stiffness coefficients, defined by

$$A_{ij} = k_i k_j \sum_{k=1}^{L} \overline{Q}_{ij}^k (z_k - z_{k-1}) \qquad (9.14)$$

where k_i denotes the shear correction factors. The coefficients D_{16} and D_{26} are responsible for coupling between the bending moments (M_x and M_y) with twisting curvature (κ_{xy}). The presence of these two terms in the laminate stiffness matrix introduces the twist in the plate due to bending loads.

The middle surface of the blade is modeled with shell elements in ANSYS for the structural analysis. The pressure distributions on the two surfaces of the blade obtained from the flow solution are applied together on the middle surface. The rotational speed of the propeller is specified to account for the centrifugal and Coriolis forces. A four-node shell element, SHELL 181, based on the FSDT, is chosen for the analysis. The calculation of inter-laminar shear stresses is based on simplifying assumptions of uncoupled bending in each direction. The propeller blade is modeled as a cantilever by restraining all degrees of freedom at the root.

9.4 Modeling of shape memory alloy

9.4.1 Constitutive equation for SMA fiber

In this study, the Brinson model (Brinson and Lammering, 1993) is used to simulate the thermomechanical behavior of the SMA wire. In this model, the martensite volume fraction, ξ, during the phase transformation is divided into two parts: the stress-induced martensite volume fraction, ξ_S, and the temperature-induced martensite volume fraction, ξ_T, such that $\xi = \xi_S + \xi_T$. The constitutive equation is then expressed as

$$\sigma - \sigma_0 = E(\xi)\varepsilon - E(\xi_0)\epsilon_0 + \Omega(\xi)\xi_S - \Omega(\xi_0)\xi_{S0} + \Theta(T - T_0) \qquad (9.15)$$

where σ, ε, and T denote the normal stress, normal strain, and temperature, respectively, and the subscript 0 against a variable represents its value in the initial state.

9.4.2 Recovery stress in lamina with SMA

Initially, SMA fiber is considered at low temperature and fully at martensite state. If it is deformed plastically (beyond first yield point) and then unloaded, it will recover the elastic component of the strain and retain the plastic strain ε_P. The loading is done below second yield point (Chopra and Sirohi, 2014). If the material is now heated to a temperature higher than austenite finish temperature, it will recover its original "memorized" shape, i.e., zero strain. For NiTinol up to 8% of plastic strain, ε_P is fully recoverable. Now if a SMA wire, which was initially loaded to ε_t and unloaded to a pre-strain ε_P, is restrained against further deformation by embedding inside matrix, on heating to a temperature above austenite finish temperature, it will try to recover the pre-strain. However, because the ends are restrained, stress will be induced in the wire. This stress is called the "recovery stress" of the SMA, which is normally much higher than the stress that generated the pre-strain. This constrained or restrained recovery stress can be used to adaptively change the structural response.

9.4.3 Recovery stress at different temperatures

Consider that an SMA fiber, which is in a 100% martensite state (twinned), is loaded plastically at the ambient temperature $T_0 < A_s$ up to a stress below the second yield point and unloaded to a pre-strain, ε_0. Thus, at $T = T_0$, $\sigma_0 = 0$, $\xi_0 = 1$, and $\xi_{S0} = \varepsilon_0/\varepsilon_L$. As the SMA fiber is heated now, it will undergo martensite to austenite (M → A) transformation. But, since the fiber is constrained, there will be stress generated due to heating and the austenite start and finish temperatures will be increased to A_s^m and A_f^m, respectively, which can be obtained from stress influence coefficient C_A:

$$A_s^m = A_s + \frac{\sigma}{C_A}, A_f^m = A_f + \frac{\sigma}{C_A} \qquad (9.16)$$

There is no phase transformation until the temperature reaches to A_s^m. Hence, the recovery stress-temperature relation in this interval can be obtained from Eq. (9.16) as

$$\sigma^r = \Theta(T - T_0) \tag{9.17}$$

For $A_f^m > T > A_s^m$, there will be M → A transformation, and during this, the martensite volume fraction (ξ) is given by

$$\xi = \frac{\xi_0}{2}\left\{\cos\left[a_A\left(T - As - \frac{\sigma}{C_A}\right)\right] + 1\right\} \tag{9.18}$$

The recovery stress during this phase can be expressed as

$$\sigma^r = E(\xi)(1-\xi)\varepsilon_o + \Theta(T - T_0) \tag{9.19}$$

where

$$\xi = \frac{1}{2}\left\{\cos\left[a_A\left(T - As - \frac{\sigma^r}{C_A}\right)\right] + 1\right\} \text{ as } \xi_0 \text{ is } 1.0 \tag{9.20}$$

The recovery stress-temperature relation given by Eqs. (9.19) and (9.20) is not explicit. It can be solved iteratively. Beyond a certain temperature A_f^m, there will be no phase transformation when $\xi = 0$ and recovery stress can be calculated from

$$\sigma_{A_f}^r = E_A \varepsilon_o + + \Theta\left(A_f^m - T_0\right) \tag{9.21}$$

This recovery stress can be added to Eqs. (9.11) and (9.12) to get the laminate behavior with pre-strained SMA. So the force and moment resultants of the laminate corresponding to this theory can be related to the strain components ε_{ij}^0, γ_{ij}^0 and curvature κ_{ij} at the reference surface of the blade by

$$\begin{bmatrix} N_x \\ N_y \\ N_{xy} \end{bmatrix} = \begin{bmatrix} A_{11} & A_{12} & A_{16} \\ A_{12} & A_{22} & A_{26} \\ A_{16} & A_{26} & A_{66} \end{bmatrix} \begin{bmatrix} \varepsilon_x^0 \\ \varepsilon_y^0 \\ \gamma_{xy}^0 \end{bmatrix} + \begin{bmatrix} B_{11} & B_{12} & B_{16} \\ B_{12} & B_{22} & B_{26} \\ B_{16} & B_{26} & B_{66} \end{bmatrix} \begin{bmatrix} \kappa_x \\ \kappa_y \\ \kappa_{xy} \end{bmatrix} + \begin{bmatrix} N_x^r \\ N_y^r \\ N_{xy}^r \end{bmatrix}$$

$$\begin{bmatrix} M_x \\ M_y \\ M_{xy} \end{bmatrix} = \begin{bmatrix} B_{11} & B_{12} & B_{16} \\ B_{12} & B_{22} & B_{26} \\ B_{16} & B_{26} & B_{66} \end{bmatrix} \begin{bmatrix} \varepsilon_x^0 \\ \varepsilon_y^0 \\ \gamma_{xy}^0 \end{bmatrix} + \begin{bmatrix} D_{11} & D_{12} & D_{16} \\ D_{12} & D_{22} & D_{26} \\ D_{16} & D_{26} & D_{66} \end{bmatrix} \begin{bmatrix} \kappa_x \\ \kappa_y \\ \kappa_{xy} \end{bmatrix} + \begin{bmatrix} M_x^r \\ M_y^r \\ M_{xy}^r \end{bmatrix} \tag{9.22}$$

$$\begin{Bmatrix} Q_x \\ Q_y \end{Bmatrix} = \begin{bmatrix} A_{55} & A_{54} \\ A_{45} & A_{44} \end{bmatrix} \begin{Bmatrix} \gamma_{zx}^0 \\ \gamma_{yz}^0 \end{Bmatrix}$$

A_{ij}, D_{ij}, and B_{ij} are already defined at Eqs. (9.13) and (9.14).

9.5 Fluid-structure interaction

The deformed shape of the propeller blade under each operating condition is transferred to ANSYS ICEM-CFD software. After developing the actual blade around this deformed surface, the mesh is regenerated and exported to ANSYS Fluent 12.1 for the flow solution. A new pressure distribution now develops over the blade, which is used in ANSYS Mechanical 12.1 for the deformation analysis. The process is repeated till the pressure distributions obtained in two successive iterations do not differ significantly. In the present study, the convergence is found to occur rapidly within five iterations.

9.6 Material failure

The failure of composite laminas is checked using the Tsai-Hill criteria (Jones, 1975), given below.

$$\frac{\sigma_1^2}{X_T^2} - \frac{\sigma_1 \sigma_2}{X_T^2} + \frac{\sigma_2^2}{Y_T^2} + \frac{\tau_{12}^2}{S^2} < 1.0 \tag{9.23}$$

where X_T, Y_T, S denote, respectively, the tensile strengths in x_1 and x_2 directions and the in-plane shear strength. Wherever the normal stresses are compressive in nature, the corresponding compressive strengths, X_C and Y_C, are used in Eq. (9.23).

9.7 Analysis of different propellers

Four different propellers are analyzed and described below.

9.7.1 Real marine propeller: Prop1

The first one, Prop1, is a real marine propeller of an existing ship. It consists of five blades and its diameter (D) is 4.2 m, hub diameter is 1.3146 m, and pitch to diameter ratio at 0.7D is 1.547. Fig. 9.1 shows the propeller. This is made with aluminum nickel bronze which is stiff enough to hold its shape during operation, and thus, does not show tangible difference in its performance after deformation (Das et al. 2013) as seen at Fig. 9.2. The open water characteristics of this propeller before and after deformation are shown in Fig. 9.2. The results are also validated against experimental results (NSTL, 2010) to yield good agreement.

9.7.2 Simple composite propeller: Prop2

The geometry of a marine propeller is very complex. To ascertain the effect of fiber orientations of the composite layers on the bend-twist coupling, a propeller of simpler geometry is considered next. A tapered wing with aerofoil cross section is chosen to be

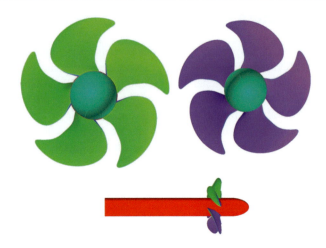

Fig. 9.1 Surface model of propeller, Prop1.

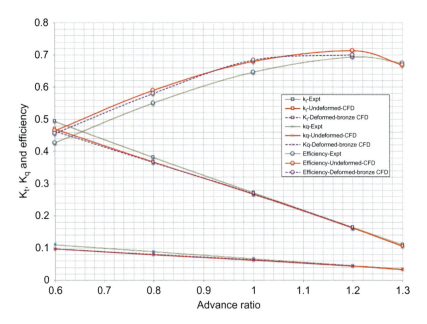

Fig. 9.2 Open water characteristics for deformed and undeformed propeller, Prop1.

the propeller blade. This wing is placed over a hub of 1.315 m diameter. The span of the wing is taken as 1.443 m, which makes the diameter of the propeller as 4.2 m (Fig. 9.3). A constant pitch is maintained throughout the blade to make the pitch ratio (P/D) 1.547. Graphite-epoxy is taken as composite material and its properties are given in Table 9.1.

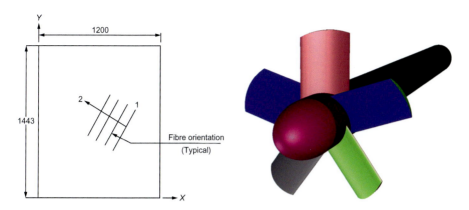

Fig. 9.3 Geometry of simple propeller, Prop2.

9.7.2.1 Design of thickness

To simulate the maximum loading, the propeller is revolved at 200 rpm with 15.4 m/s (30 knots) speed of advance. This typically corresponds to the fast attack mode of a naval ship. As the deformations are small, a geometrically linear analysis is performed in this section. The maximum values of the deflection and stresses in different plies obtained for graphite-epoxy blades with different thicknesses and laminate configurations are calculated. The failure of the plies is checked against the Tsai-Hill criteria. It is observed that a minimum thickness of 80 mm is required to avoid failure. An 80 mm thick laminate with 90°/0°/0°/90°/90° layup satisfies the failure criteria. The corresponding maximum twist angle is 0.45°. Many other stacking sequences and fiber orientation angles are also tried out. Two of these layups are as taken by Mulcahy et al. (2010) and Motley et al. (2009). It is observed that none of these layups satisfies the failure criteria and the presence of off-axis plies in these laminates causes only a marginal improvement in the twist angle from 0.45° to 1.6°. Therefore, 80 mm thick blade with 90°/0°/0°/90°/90° layup is considered as the safe design. The open water characteristics of this propeller before and after deformation are plotted in Fig. 9.4.

However, it is observed that at higher advance ratio, the efficiency is reducing and, at lower advance ratio, there is no significant change in efficiency. This may be explained with the angle of attack of incidental flow of the propeller (Fig. 9.5). At higher advance ratio, angle of attack is less than the optimum and vice versa. So, the positive twist in blade, which tries to reduce the pitch angle and hence the angle of attack of incidental flow, will only be beneficial at lower advance ratio and expected to reduce the hydrodynamic efficiency at higher advance ratio. At lower advance ratio, the propeller-rpm is large, which generates higher loads and (bending) deformation. The twist in the blade may be beneficial, but large bending deflection alters the shape of the propeller and thus hampers its performance.

Table 9.1 Material properties

Material/Properties	X_T (MPa)	Y_T (MPa)	X_C (MPa)	Y_C (MPa)	S (MPa)	E_1 (GPa)	E_2 (GPa)	ν_{12}	ν_{23}	G_{12} (GPa)	G_{23} (GPa)
Glass epoxy: S2–449 17 k/SP 381 Uni Dir Tape (Composite Materials Handbook, 2002)	1760	60	1190	250	136	47.8	12.7	0.3	0.02	4.70	2.9
Graphite-epoxy composite lamina (vf = 0.3) (Jones, 1975)	1035	41	689	117	69	207	5	0.25	0.354	2.6	1.97

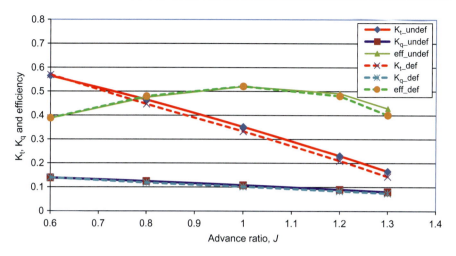

Fig. 9.4 Open water characteristics for Prop2.

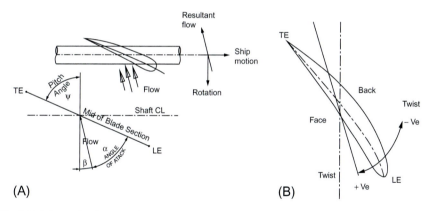

Fig. 9.5 Schematic diagrams for flow over propeller blade section. (A) Effect of twist on pitch. (B) Sign convention for twist in blade.

9.7.3 *Pre-pitched propeller: Prop3*

To get the benefits of the twist at low advance ratio, the propeller is now designed with a larger pitch than the requirement, so that after deformation the pitch can get reduced to provide optimum performance. The increased pitch ratio (P/D) is 2.2 and the propeller becomes a pre-pitched one, the Prop3. The twist in propeller blade due to hydrodynamic loading is expected to reduce the pitch (Das and Kapuria, 2016) and improve its hydrodynamic performance at low advance ratio. Notwithstanding the material failure, the phenomenon is studied here to examine the effect of shape adaption on the improvement of hydrodynamic efficiency of the propeller. To study the effect of ply-angle on the twist of the blade, a three-layer angle-ply graphite-epoxy laminate

(θ/ − θ/θ) is considered. A detailed analysis of this propeller with four different materials is reported in authors' earlier work (Das and Kapuria, 2016). In brief, it is stated that propellers with graphite-epoxy and glass-epoxy laminates exhibit larger deformation/twist, which in turn helped improving hydrodynamic efficiency. Fig. 9.6 shows deformed shape of glass-epoxy propeller-blade when operating at different rpms. Hydrodynamic performance of the deformed propeller made of graphite-epoxy (75°/−75°/75°) and glass-epoxy (45°/−45°/45°) laminates is plotted in Fig. 9.7. It is observed that the graphite-epoxy blade shows improvement in the efficiency at off-design conditions for a range of advance ratio between 1.0 and 1.5. The corresponding twist in the blade ranges between 6.8° and 1°. A maximum gain of 5.2% in the efficiency (in comparison to efficiency of undeformed propeller) is observed at $J = 1$, when the maximum blade deflection is 207 mm. As J is lowered further, the twist angle increases, but the bending deformation also increases, which causes reduction in the hydrodynamic efficiency. The glass-epoxy blade shows only a marginal improvement in the efficiency at the same range of the advance ratio. The values of the twist angle, observed here, are comparable to those reported in the literature. A twist of 6°–9° has been reported for a full-scale propeller (Motley et al. 2009) and 0.598° for a model propeller (Mulcahy et al., 2010). The corresponding improvement in the hydrodynamic efficiency is 9% (Motley et al. 2009) and 15% (Mulcahy et al., 2010). The large gain in hydrodynamic efficiency, reported by Mulcahy et al. (2010), may be due to the small size of the scaled-down model propeller. The present study reveals that a gain in the hydrodynamic efficiency by tailoring

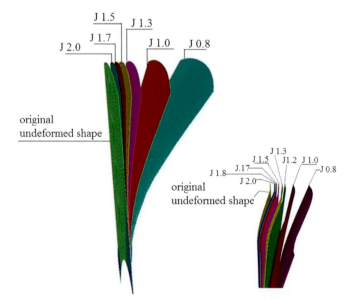

Fig. 9.6 Two different views of undeformed and deformed shape of blade at different operating conditions: Prop3, Glass epoxy.

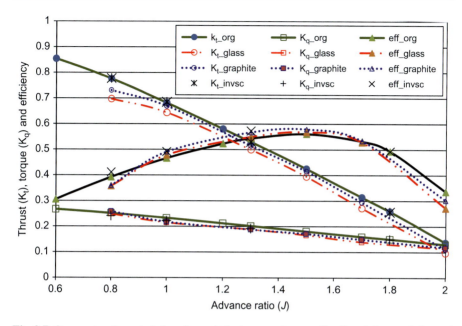

Fig. 9.7 Open water characteristics of pre-pitched composite propeller, Prop3 Org: undeformed propeller. Glass: deformed propeller with glass-epoxy composite. Graphite: deformed propeller with graphite-epoxy composite.

the layup of a laminated composite blade is possible with the available composite materials only if the material failure is ignored. This gain is also hampered due to the excessive bending deformation that accompanies a large twist.

9.7.4 SMAHC propeller: Prop4

9.7.4.1 Design of thickness

Before checking out the different positions of SMAHC, the thickness of composite propeller blade is designed. The propeller blade is primarily made of graphite-epoxy composite. The thickness of the blade is different at different locations. To form the (hydrodynamic) shape of the blade, relative distribution of thickness is maintained as far as possible. However, the maximum thickness and distribution of thickness are varied and internal stresses are calculated to check its failure against Tsai-Hill criterion to ascertain the required thickness. A graphite-epoxy ply with 75°/15°/75°fiber orientation is used in the entire blade except a 240 mm wide strip at trailing edge, where additional NiTinol fiber with volume fraction 0.3 is used at 90° orientation at the top layer. Fig. 9.8 shows the arrangement and Fig. 9.9 the final thicknesses.

For designing the thickness, a heavily loaded condition, corresponding to $J = 1.2$, when advance velocity is 12.3 m/s and propeller revolution is 147 rpm, is considered. The recovery stress due to change of state of embedded NiTinol fibers from martensite

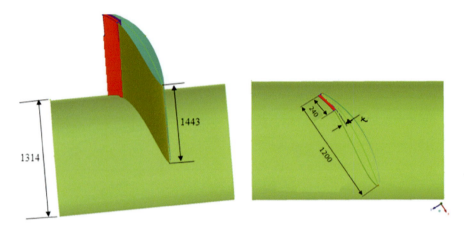

Fig. 9.8 Views of propeller blade (Prop4) over hub showing one typical way (Configuration number 4 of Table 8) of placing SMA layer.

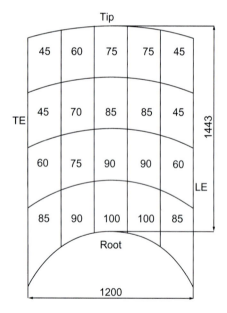

Fig. 9.9 Thickness distribution (mm) of propeller blade.

to austenite stage is considered as 270 MPa. A pre-strain of 2% and temperature of 172 °C give this condition. Appropriate centrifugal and Coriolis forces are applied.

Using different thicknesses of the blade, ANSYS Mechanical is used to solve stresses at different layers and locations. Failure is checked against Tsai-Hill criterion. Tsai-Hill index, which indicates if there is any failure, is calculated from these stress

Smart composite propeller for marine applications 287

Table 9.2 **Maximum Tsai-Hill index within the blade for different thickness**

Sl no	Maximum thickness (mm)	Minimum thickness (mm)	Max Tsai-Hill index	Thickness variation in both in-plane directions/only radial direction
1	90	45	1.92	Both
2	100	40	2.01	Radial
3	100	45	1.63	Radial
4	100	45	1.02	Both
5	100	45	0.99	Both with larger thickness near nose
6	120	45	0.91	Both
7	120	40	0.95	Both

solutions. Maximum Tsai-Hill index within the blade for its different thickness is shown in Table 9.2. To avoid failure, the maximum thickness of the blade is thus designed to 100 mm.

9.7.4.2 Positioning SMAHC inside propeller blade

The graphite-epoxy ply with 75°/15°/75° fiber orientations is consistently used within the blade. However, it is required to ascertain the best position of SMAHC element inside the blade to give optimum effect. Park et al. (2011) have used SMAHC layers spanning from 40% to 93% of the radius of the blade. They placed SMAHC layer near the leading edge and kept fiber orientation angle as 45°. At first, same configuration is maintained in the present study.

However, many different location and orientation of SMA fibers are also tried out here. Table 9.3 shows different arrangements. When NiTinol fibers are heated above austenite finish temperature, it gets transformed to austenite state and the layer gets subjected to a recovery stress of 270 MPa. The same pre-stress is applied in the fiber direction in the respective SMAHC layer. In this section, the relative merit of different arrangements for positioning of SMAHC layer is studied. No hydrodynamic load is applied here; only the pre-stress is applied. The twist and deformations for different conditions are also listed in Table 9.3. This table shows that when SMA fibers are laid in a similar way as described by Park et al. (2011), the blade undergoes a twist of 0.88°. This twist is quite small, though twist per unit deformation (42.1°/m) is reasonably good. The shape and size of present propeller blade is, of course, quite different from the proprotor blade analyzed by Park et al. (2011). Configurations 2–5 of the same table show arrangements of SMAHC layer with NiTinol fiber running at 90° orientation (along Y axis). Recovery stresses for these conditions are in Y direction and, due to oblique introduction of SMAHC layer (either at the face, i.e., first layer, or at the back, i.e., third layer of the propeller), this creates an additional bending moment. When SMAHC layer is placed away from support (configuration number 2), the blade

Table 9.3 Twist and deformation in blade with different arrangement of SMA layer

Config. number	Position of SMA layer — Along radius (span)	Position of SMA layer — Along chord (LE to TE)	Position of SMA layer — In thickness direction extent and location	Nitinol fiber angle (°)	Pictorial representation	Max deformation (mm)	Twist (°)	Twist per deformation (°/m)
1*	40%–93% of R	70%–90% from trailing edge	33% of t at back of propeller, 3rd layer	45		21.0	0.88	42.10
2	75%–100% of (R-r)	0%–20% from trailing edge	33% of t at back of propeller, 3rd layer	90		9.5	−0.45	−47.54
3	0%–25% of (R-r)	0%–20% from trailing edge	33% of t at back of propeller, 3rd layer	90		15.2	−0.84	−54.99
4	100%–0% of (R-r)	0%–20% from trailing edge	33% of t at back of propeller, 3rd layer	90		56.6	−2.96	−52.29

5	0%–100% of (R-r)	80%–100% from trailing edge	33% of t at back of propeller	3rd layer	90		6.7	0.27	40.50
6	100%–0% of (R-r)	0%–20% and 80%–100% from trailing edge	33% of t at face and back of propeller	Near TE, 1st & near LE 3rd layer	90		60.8	3.47	56.99
7	100%–75% of (R-r)	0%–20% from trailing edge	33% of t at back of propeller	3rd layer	0		3.4	−0.09	−24.04
8	75%–100% of (R-r)	0%–100%	33% of t at back of propeller	3rd layer	0		11.4	0.18	15.96

Continued

Table 9.3 Continued

| Config. number | Position of SMA layer ||||| Pictorial representation | Max deformation (mm) | Twist (°) | Twist per deformation (°/m) |
	Along radius (span)	Along chord (LE to TE)	In thickness direction extent and location		Nitinol fiber angle (°)			
9	0%–25% of (R-r)	60%–100% from trailing edge	33% of t at back of propeller	3rd layer	0		4.8	0.24	49.34
10	0%–100% of (R-r)	80%–100% from trailing edge	33% of t at back of propeller	3rd layer	0		17.7	1.04	58.62
11	0%–100% of (R-r)	0%–20% and 80% to100% from trailing edge	33% of t at face & back of propeller	1st & 3rd layer	0		15.3	−0.70	−46.03

12	100%–0% of (R-r)	0%–100%	33% of t at back of propeller	3rd layer	55		29.5	0.89	30.13
13	100%–0% of (R-r)	0%–100%	33% of t at back of propeller	3rd layer	−55		35.2	−0.36	−10.29
14	0%–75% of (R-r)	0%–100%	33% of t at back of propeller	3rd layer	42		30.5	0.55	17.84
15	0%–50% of (R-r)	0%–100%	33% of t at back of propeller	3rd layer	30		32.1	0.99	30.95

*Placing SMA layer inside blade; after Park et al. (2011).

gets twisted by very small angle (0.45°), whereas when it is applied near the support (configuration number 3), twist increases to 0.84°. A continuous strip of SMAHC layer near trailing edge (configuration number 4) gives good twist of 2.96°. When the SMAHC strip is applied near leading edge, twist decreases (configuration number 5). The direct twist due to oblique application of pre-stress and twist obtained due to bend-twist coupling of orthotropic material might work in opposite direction to cancel out each other. For sixth case, two SMAHC layers are applied, one at first layer near trailing edge and the other one at third layer near leading edge. Twist in the blade is increased further to 3.47°. This is one of the best configurations in terms of the twist. It is observed that if SMAHC layer is applied at first layer near leading edge as well as at third layer near trailing edge, the direction of twist gets reversed without much change in magnitude. This condition is similar to the sixth one, and hence, it is not reported separately.

Cases with orientation of NiTinol fibers at 0° (along X axis) are reported in configuration numbers 7–11 of Table 9.3. Because the fibers are running in X direction, recovery stress is in X direction and oblique application of SMAHC layer creates direct twist to the blade. Unfortunately, there is no support for the blade along the edges running parallel to Y axis. Cases reported in configurations 7, 8, and 9 show very small twist angles. When a continuous strip of SMAHC layer is used near leading edge 1.04°, twist is obtained. This case gives maximum twist per deformation, and hence, will be interesting for further hydrodynamic studies. When two strips of SMAHC layer are tried near leading and trailing edge at first and third layer, the twist obtained in blade, 0.7°, is not encouraging. The lack of support in Y parallel edge might hamper attaining larger twist.

Few arrangements of inclined band of SMAHC layer are, therefore, studied as reported in configurations 12–15. The orientation of NiTinol fiber is kept same as the inclination of the SMAHC layer, so that recovery stress works in this direction. Each of the inclined layers is started from the support at root section so that blade gets desirable twist/rotation. However, maximum twist observed for these cases is below 1° only.

It is observed from Table 9.3 that maximum twist per deformation is obtained for configuration 10 and maximum twist with large twist per deformation is obtained for configuration 6. Further studies on these two arrangements are carried out next.

9.7.4.3 Deformation of propeller blade under different operating condition

After finalizing the position of SMAHC layer, analysis is carried out for propeller blade with hydrodynamic loads of different operating conditions, Coriolis force for respective rotation, and the pre-stress when SMA is in austenite state. At high J, SMA fibers will be kept in low temperature to maintain it in martensite state and, at low J, it will be heated above austenite finish temperature to transform the fibers into austenite state. In this condition, entire 270 MPa recovery stress will work in the SMAHC layer.

Table 9.4 Twist and deformation of blade at different operating conditions

Arrangement of SMA layer	Operating conditions					Austenite				Martensite		
	Advance Ratio (J)	Advance Velocity ($U\infty$)	Propeller rpm	Deformation (mm)	Twist °	Twist per deformation (°/m)	Deformation (mm)	Twist °	Twist per deformation (°/m)
Configuration number 6 of Table 9.3	1.8	12.3	98.0	88.8	4.9	54.7	36.89	−0.92	52.0
	1.7	12.3	103.7	91.6	4.9	53.9	40.40	2.04	50.4
	1.5	12.3	117.6	100.4	5.2	51.9	49.85	2.35	47.1
	1.3	12.3	135.7	114.1	5.6	49.3	65.98	2.90	43.9
	1.2	12.3	147.0	123.8	5.9	47.8	77.30	3.28	42.5
Configuration number 10 of Table 9.3	1.8	12.3	98.0	34.4	2.4	71.0	30.66	−0.61	52.6
	1.7	12.3	103.7	37.2	2.5	68.0	33.51	1.70	50.8
	1.5	12.3	117.6	46.0	2.8	61.2	42.33	1.99	47.0
	1.3	12.3	135.7	60.0	3.3	54.6	56.46	2.45	43.5
	1.2	12.3	147.0	69.9	3.6	51.5	66.49	2.78	41.9

The deformation and twist of blade at different J is reported at Table 9.4. Analysis is carried out for two configurations of SMAHC layer (configuration numbers 6 and 10) and considering two conditions of NiTinol fibers either at martensite or austenite state.

Table 9.4 shows that when NiTinol is kept at martensite state (i.e., pre-stress is absent), twist per deformation for the blade is varying between 41 and 52°/m. When SMAHC layer is arranged like configuration number 6, at austenite state, blade shows twist per deformation is around 48 to 55°/m. In martensite state, twist is originated from the bend-twist coupling of orthotropic material, and in austenite state, unsymmetrical placement of SMAHC layer and pre-stress also contributed to it. Amount of twist per deformation, obtained here, is reasonably good. A sudden increase in twist per deformation (up to 71°/m) is observed when SMAHC layer is arranged like configuration number 10 and is kept in austenite state. In this arrangement, because of 0° orientation of SMA fibers, the pre-stress directly generates a twist in the blade. As the hydrodynamic load increases (i.e., **J reduces), effect of bend-twist coupling also starts contributing to twist. However, this twist comes in association with some bending deformation and hence twist per deformation gradually decreases. The advantage of using SMA is to get high twist per deformation, which alters the pitch of the blade with minimum change in overall shape of the blade. The study of para 7.3 on Prop3 showed that excessive bending deflection hampers the gain in hydrodynamic efficiency, otherwise arising from the alteration of twist of the blade.

CFD study of flow around the deformed propeller is carried to get its hydrodynamic performance. Fig. 9.10 shows the open water characteristics of the propeller both before and after deformation. It is observed that after deformation, the SMAH composite propeller shows noticeable change in its hydrodynamic performance. Keeping SMAHC layer in austenite state and in an arrangement of configuration number 6 (Table 9.3), which gave maximum twist in blade, maximum improvement in the hydrodynamic efficiency is obtained at low advance ratios. Fig. 9.10 shows an improvement of around 7% for advance ratio, $J = 1.2$ or 1.3. However, the positive twist in blade which reduces the pitch of the propeller is reducing its efficiency at higher advance ratios. When SMA layer is arranged as configuration number 10 (Table 9.3), the gain in efficiency is observed to be around 3 to 5% at $J = 1.3$ and 1.2.

Table 9.4 shows that when SMA fibers are kept in martensite state, pre-stress is absent and twist in blade is much smaller. For both the blades with SMAHC layer arranged as configuration numbers 6 and 10, at $J = 1.8$ twist even becomes negative, which increases the pitch. Fig. 9.10 shows a minor increase in hydrodynamic efficiency of the propeller at this condition.

When advance ratio is near 1.5, there is not much change in efficiency. Fig. 9.10 clearly indicates that if the SMA fibers are kept at martensite state for operating conditions of $J \geq 1.5$ and at austenite state for operating conditions of $J < 1.5$, hydrodynamic efficiency of this propeller increases, and over a wide range of advance ratio, the propeller exhibits improved efficiency.

Smart composite propeller for marine applications

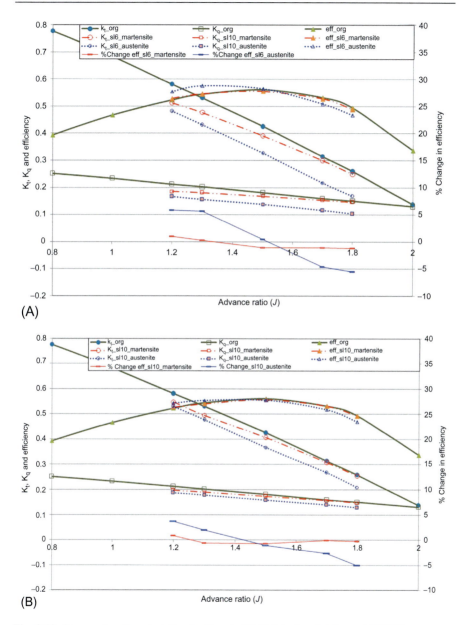

Fig. 9.10 Open water characteristics for Prop4. (A) Original and deformed SMAH propeller: SMA arranged as configuration number 6. (B) Original and deformed SMAH propeller: SMA arranged as configuration number 10.

9.8 Conclusions

The analysis is carried out on a full-scale propeller of diameter 4.2 m, considering viscous flow, as the loading and deformation of a model propeller cannot be extrapolated to a full-scale prototype propeller.

The study shows that a bronze propeller is stiff enough to hold its shape under normal operating conditions and its hydrodynamic performance is not affected due to structural deformations.

When operating within the limits of material failure using the commonly used composite materials, the twist generated in the deformed propeller, Prop2, is found inadequate to create any noticeable change in the hydrodynamic efficiency.

The numerical study reveals that the twist angle of blades made of angle-ply layers under the hydrodynamic loading changes its sign as the advance ratio is increased. The negative twist occurring at a higher advance ratio is due to the asymmetric nature of hydrodynamic pressure loading, and the positive twist occurring at a lower advance ratio is due to the bend-twist coupling. However, both can be made to keep the angle of attack of flow toward the optimum value by applying a suitable pre-twist in the blade. The induced twist will then reduce the effective pitch angle and tend to keep the angle of attack at the optimum. This is achieved for the propeller Prop3, and when the material failure is ignored, it is possible to generate sufficient deformation and twist that can cause some noticeable change in the open water characteristics.

The use of SMA shows certain improvement in the hydrodynamic performance of the propeller, Prop4. Its efficiency is improved over a wide range of operating conditions. When NiTinol fibers are kept at austenite state, its efficiency increases at low advance ratio. When these fibers are kept in martensite state, change in propeller efficiency is observed to be small. Only a hint of improvement is observed at $J = 1.8$. Even this slight improvement is ignored; it may be safely concluded that, for higher advance ratios, propeller efficiency does not change much when SMA fibers are kept at martensite state.

The SMA layer gets subjected to a recovery stress when NiTinol fibers are transformed to austenite state. It is observed that this recovery stress in SMA layer is more effective in twisting the blade than the bend-twist coupling of orthotropic material under hydrodynamic bending.

Some configuration of laying SMAHC has been shown to achieve improvement in hydrodynamic efficiency of a marine propeller at off-design condition, which is also safe from material failure criteria.

References

Ashkenazi, Y., Golfman, I., Rezhkov, L., Sidorov, N., 1974. Glass-Fiber-Reinforced Plastic Parts in Ship Machinery. Sudostroyenniye Publishing House, Leningard.

Brinson, L.C., Lammering, R., 1993. Finite element analysis of the behavior of shape memory alloys and their applications. Int. J. Solids Struct. (23), 3261–3280.

Carlton, J.S., 2007. Marine Propellers and Propulsion, second ed. Elsevier Ltd., Burlington, USA.

Chopra, I., Sirohi, J., 2014. Smart Structures Theory. Cambridge University Press.
Composite Materials Handbook, 2002. MIL-HDBK-17-2F. In: Polymer Matrix Composites Material Properties, Vol. 2. Department of Defense, United States of America.
Das, H.N., Kapuria, S., 2016. On the use of bend-twist coupling in full-scale composite marine propellers for improving hydrodynamic performance. J. Fluids Struct. 61, 132–153.
Das, H.N., Nagalakshmi, T., Kapuria, S., 2012. CFD Study for a controllable pitch propeller.3rd Asian Conference on Mechanics of Functional Materials and Structures, IIT, Delhi.
Das, H.N., Rao, P.V., Suryanarayana, C., Kapuria, S., 2013. Effect of structural deformation on performance of marine propeller. J. Mar. Res. 10 (3), 47–50.
Duerig, T.W., Melton, K.N., Stockel, D., Wayman, C.M., 1990. Engineering Aspects of Shape Memory Alloys. Butterworth-Heinemann, Essex, UK.
He, X.D., Hong, Y., Wang, R.G., 2012. Hydroelastic optimisation of a composite marine propeller in a non-uniform wake. Ocean Eng. 39, 14–23.
Jones, R.M., 1975. Mechanics of Composite Materials, second ed Taylor & Francis, Virginia, USA.
Marsh, G., 2004. A new start for marine propellers? Reinf. Plast. 48 (11), 34–38.
Molland, A.F., Turnock, S.R., 1991. The design and construction of model ship propeller blades in hybrid composite materials. J. Compos. Manuf. 2 (1), 39–47.
Motley, M.R., Liu, Z., Young, Y.L., 2009. Utilizing fluid-structure interactions to improve energy efficiency of composite marine propellers in spatially varying wake. Compos. Struct. 90, 304–313.
Mulcahy, N.L., Prusty, B.G., Gardiner, C.P., 2010. Hydroelastic tailoring of flexible composite propellers. Ships and Offshore Structures 5 (4), 359–370.
NSTL Internal Report on "Hydrodynamic Model Tests For New Design Frigate (Open Water, Self Propulsion & 3d Wake Survey Tests)". Report Number NSTL/HR/HSTT/221/2 November 2010.
Otsuka, K., Wayman, C.M., 1998. Shape Memory Materials. Cambridge University Press, Cambridge, UK.
Park, J.S., Kim, S.H., Jung, S.N., 2011. Optimal design of a variable-twist proprotor incorporating shape memory alloy hybrid composites. J. Compos. Struct. 93, 2288–2298.
Song, G., Ma, N., Li, H.N., 2006. Applications of shape memory alloys in civil structures. Eng. Struct. 28, 1266–1274.

Further reading

Blasques, J.P., Christian, B., Andersen, P., 2010. Hydro-elastic analysis and optimization of a composite marine propeller. Mar. Struct. 23, 22–38.
Hu, N. (Ed.), 2012. Composites and Their Properties. InTech, Janeza Trdine, Croatia. https://doi.org/10.5772/2816.

Part Two

Naval architecture and design considerations

A structural composite for marine boat constructions

Alexandre Wahrhaftig[*], Henrique Ribeiro[†], Ademar Nogueira[‡]
[*]Department of Construction and Structures, Polytechnic School, Federal University of Bahia, Salvador, Brazil, [†]Bahia Federal Institute of Education, Salvador, Brazil, [‡]Department of Mechanical Engineering, Polytechnic School, Federal University of Bahia, Salvador, Brazil

Chapter Outline

10.1 Introduction 301
10.2 Basic core materials 304
10.3 Composite structure concepts 305
10.4 Economic viability 307
    10.4.1 Cost 307
    10.4.2 Weight comparison with Divinycell structure 308
10.5 Case study: a vessel structural computational design 309
    10.5.1 Structural arrangement and geometry 309
    10.5.2 Computational modeling 310
    10.5.3 Computational results and discussion 312
10.6 Conclusions 313
References 313

10.1 Introduction

Steel is traditionally used in the hull construction of ships and submarines, although aluminum is often used as an alternative material in applications in which the minimization of structural weight can have a great impact on energy efficiency or providing other advantages. The use of lightweight structures in shipbuilding can, for example, allow an increase in payload, a higher speed, and a reduction in both fuel consumption and environmental emissions (Crupi et al., 2013). In addition, such replacement is often necessary when steel would make a surface ship top-heavy, and if aluminum is not a viable option. In these cases, for instance, composite materials offer a reasonable alternative (Chalmers, 1994). The use of composite and sandwich-structured materials in modern engineering applications, such as civil and military aircraft, launch vehicles, wind turbine blades, and assorted marine structures (Cerracchioa et al., 2015), has grown significantly over the past few decades due to their high strength-to-weight and stiffness-to-weight ratios (Yang et al., 2013).

Density of a composite can be easily controlled through arrangement, which can be also particularly valuable when increasing structural buoyancy (Craugh and Kwon, 2013). However, many other features have also been lauded, including greater material strength, flexibility, environmental resistance, and damage tolerance, as well as reductions in weight, size, and cost (Kimpara, 1991).

Traditionally, the preferred geometrical arrangement of fiber reinforcement for marine applications has been woven fabric, which is often combined with layers of chopped strand mat. The selection of such currently available materials for fast vessels such as the surface effect ship and hydrofoils, both surface piercing and fully submerged, is quite advanced, as in these applications weight savings are critical and structural optimization is essential (Davies et al., 1994).

Since the middle of the last century, fiberglass has also proven to be particularly popular in boat construction as it has the advantages of being chemically inert (for both general applications and in marine environments), lightweight, strong, easily moldable, and competitively priced. On the other hand, it also has a low modulus of elasticity and low fatigue strength when compared to steel and aluminum. Of course, these properties are somewhat dependent on the macrostructure of the material and the two most commonly used for marine applications are single-skin sheets and sandwich panels.

Sandwich panels address some of the aforementioned problems as they improve the section modulus and thickness of the hull without increasing their weight, but the fact that the material is still relatively flexible means that framing is necessary in most applications. Such modification increases the weight and reduces space inside a boat, the former being a problem in high-speed boats, and the latter causing issues in cargo transport ships. In addition, sandwich panels made of Divinycell and polyurethane are not only expensive, but are also only available in a limited range of thicknesses.

Given the characteristics of fiberglass with respect to its structural behavior, it is not surprising that significant research has been devoted to developing fiberglass composites which retain the advantageous features of fiberglass, while eliminating the problematic ones.

Both aramid and carbon fiber have a high tensile strength, modulus of elasticity, and specific resistance in addition to their low density. They are well-suited to being coupled with fiberglass to create new, more highly functional materials. The use of fiberglass, meanwhile, helps to offset the relatively high cost of these advanced materials. Cost concerns can be further offset by introducing an additional core material, which typically consists of a cellular of balsa wood, Divinycell, polyurethane, or polystyrene. A review of the advances made in composite structures for naval ships and submarines, up to 2001, can be found in a paper by Mouritz et al. (2001).

In this chapter, we investigate a new composite material in which fiberglass is impregnated with epoxy resin by the lamination of external and internal surfaces that are separated by a core of expanded polystyrene (EPS). Both outer and inner surfaces are linked through a web also made by epoxy resin and roving wire. While the mechanical properties EPS are low, the idea behind this composite is to create a netted structure made by "I" beams in two orthogonal directions (see Fig. 10.1). Truthfully, EPS is here just to give the shape necessary to make the "I" beam happen. The use of

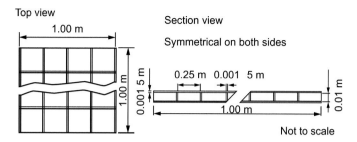

Fig. 10.1 Structural arrangement of a base plate made from the new composite.

EPS is a cheap way to make this structure happen, as the cost of a square meter sheet of EPS in the thickness of 5 mm is just US$ 0.10. Other thicknesses have a proportionally cheap price.

Another interesting characteristic of this material is the use of shear webs hidden inside the plate in order to save space. These webs are structural reinforcements. Once installed in the boat's hull, they act like the frames and stringers, but as they hide inside the plate, all of the space inside the boat is saved to make general arrangement easier.

This material has all the characteristics of an FRP plate, where the fibers are necessarily impregnated by epoxy resin. This system is inappropriate for unsaturated polyester resin matrix as the styrene dissolves the EPS core. If it would be desired to use polyester resin with a core, it would be necessary to use a much more expensive core, reducing the disadvantage of epoxy resin regarding cost. Also, epoxy resins have superior abrasion resistance, greater bonding, higher laminate strength, and less water absorption.

Regarding the use of this composite, though epoxy resin can be presented in a variety of chemical formulations, resulting also in a wide range of mechanical properties, this composite is more adequate to be applied to marine boat construction. There is still significant vulnerability to fire propagation, and the low modulus of elasticity is a large downside, impacting vibration issue. These problems make fiberglass-derived materials better suited to small- to medium-sized boat construction. These reasons indicate a better performance when applying this composite to the marine field, instead of naval ships.

On the other hand, when applied to marine boats, fiberglass reinforced by epoxy resin has a lot of advantages, and the disadvantages are not prohibitive. Some of these advantages are high specific strength related to weight, ability to orient fiber strength, competitive cost, ease of repair, and durability. If some small changes are made in the material arrangement (see Fig. 10.2), it still has the ability to mold complex shapes. Finally, some of the epoxy resins in market can make very flexible materials.

This chapter begins with an overview of core materials, followed by some composite structure concepts and their economic viability. Finally, a computational simulation is made in order to evaluate the mechanical properties of this material and to assess its suitability for use in boat construction. In this way, the material is tested

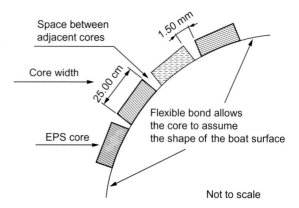

Fig. 10.2 Basic structural arrangement, as per the dimensions in Fig. 10.1, wherein laminated shear webs are placed between the cores.

numerically. When testing numerically, a computational simulation of a prismatic ferry structure is made. The prismatic geometry was chosen just for simplicity.

10.2 Basic core materials

The cellular structure of various woods and foamed plastics makes them among the most commonly used core materials for composites used in hull construction. In general, these core materials should have good shear strength and rigidity, the ability to bond easily, low weight, good resistance to aqueous deterioration, and sufficient crushing strength resistance to withstand loading (Scott, 1996). Cellular structures are available in various sizes and weights and can be constructed from a range of materials including aluminum, fiberglass laminates, and waterproof paper. These materials are generally both light and rigid, but as they cannot support concentrated loads, they require special treatment in order to ensure an effective bond between the core and the faces of the laminate. Consequently, the use of these materials is sometimes restricted to the interiors of boats, such as planing boats.

Softwoods are the most common choice for a wood-based core material, as when hardwoods like balsa are used, they can crack the laminate when they swell and do not bond well to fibers. While softwoods in particular can drape very well over curved boat surfaces, their use is generally avoided below the waterline because of the possibility of rotting, swelling, and degradation. Currently, PVC is the most commonly used foamed plastic and is used mainly because it softens when heated and can be draped over curved surfaces such as those found on boats. This tendency to soften with temperature, however, also makes it unusable in places on boats that are subjected to high temperatures, such as decks. Cross-linking polyurethane with PVC can overcome this issue, but results in the loss of some of the desirable properties of the original material.

EPS can serve as a reasonable replacement for many of the core materials currently used in marine applications as it is very light, less expensive, resistant to fungi, and impermeable. Unfortunately, it is also a somewhat weaker material, and its low shear

Table 10.1 **Mechanical properties of different core materials**

Property	PVC (Renicell 240)	Balsa LD7	Honeycomb PP30–5	EPS 3	EPS 3 with shear webs
Density (kg/m^3)	240	90	100	14	38
Compression strength (MPa)	4	5.4	1.62	0.062	179.26
Compression modulus (MPa)	131	1847.8	72.39	- -	151,684.66
Tensile strength (MPa)	3.30	7	1.21	0.12	206.84
Shear strength (MPa)	2.50	1.6	0.52	0.062	75.84
Shear modulus (MPa)	96.53	96.53	13.79	3.31	- -

and compressive strength makes it susceptible to delamination and damage. However, these limitations can generally be overcome by increasing the core thickness or using shear webs. The shear web concept, under the name lattice webs, is now being adopted for civil infrastructure applications (Fang et al., 2017). Unfortunately, this material is incompatible with polyester resin, although this problem can typically be overcome by binding a thin PVC cover to the central polystyrene core to act as a barrier. Epoxy resin can also be mixed with glass strands to produce shear webs. These various core designs are compared in Table 10.1.

10.3 Composite structure concepts

The design of stiffened composite panels is a key factor in the effectiveness of composite structures. The assessment of such structures in a design environment consequently requires models that can provide a rapid assessment of the reliability of the final construction. Given that stiffeners play an indispensable role in enhancing the strength and stiffness of these structures (Yu et al., 2015), panels for use in ship fabrication have been previously investigated through experimental testing with preform frames under in-plane uniaxial compressive loads (Mouring, 1999).

Typically, sandwich-structured composites contain a web that separates the outer layers. These webs are usually oriented in a specific direction to provide continuous longitudinal support, with added stands providing transverse support. This makes the overall structure highly orthotropic (Romanoff and Varsta, 2007). Loads applied in such structures are supported by internal tension and compression forces in the outer material (Atkinson, 1997), while shear forces are transferred to the core, allowing it to work as a homogenous structure. This means that the materials used must be able to handle significant tension and compression. Fig. 10.1 shows simplified top and side views of a newly proposed structure that overcomes these aforementioned difficulties.

Here, vertical epoxy shear webs make up for EPS's lack of strength and rigidity, and while a core thickness of 15 mm is used in this instance, the availability of EPS in different thicknesses makes the structure readily scalable. Moreover, as the webs are both longitudinal and transverse, there is no need for formal framing, which thereby provides a light and rigid composite material while freeing up internal space. Shear webs can even be used as part of the framing system when thicker cores are present. Of course, if this is the case, then it would also be possible to use thinner faces and webs, such that the faces that cover the core would act like the flanges of an I-beam. As the shear, compressive, and tensile strengths of the faces and webs are far higher than those of the EPS core, the voids between the webs provide a molded surface for the whole assembly and can prevent buckling.

The material can be delivered already bonded on one side, thus allowing it to more effectively drape along the surface of a boat under construction. This is fairly straightforward when using core blocks that form voids between the webs, as shown in Fig. 10.2.

An added advantage of using an epoxy resin in this structure is that, unlike polyester resins, it can elongate as much as or even more than the reinforcement. This helps prevent structural failure prior to reaching the limit of the reinforcement and improves the overall strength of the composite. To clarify, Fig. 10.3 shows an example of a typical fabrication process in the construction of a reduced model of a boat using the material investigated, along with a description of the principal steps.

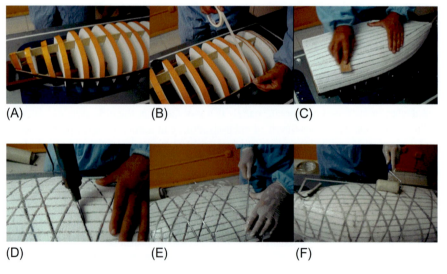

Fig. 10.3 Typical process for fabricating a vessel using the test material. (A) Preparation of a hull with EPS divisions. (b) Placing of double-sided tape. (c) Sanding of the surface to flatten the curvature. (d) Preparation of slits with a soldering iron. (e) Introduction of roving wire in the available slits. (f) Finishing of the outer layer with fabric space and fiberglass.

10.4 Economic viability

10.4.1 Cost

Ultimately, the cost of any new composite must be competitive with that of other materials, even if its performance is significantly greater. The basic structural arrangements of EPS and Divinycell units are shown in Fig. 10.4, while a comparison of the price of these two panel types is given in Table 10.2. These prices were budgeted in Brazil, but are listed in American dollars (USD).

It should be considered that, from a cost analysis perspective, industrial process engineering recommends that the production operations of a product focus on minimizing the total costs, which implies the production in economic batches, represented by the balance between the cost of the stock of raw material, which increases if bought in large quantities, and the cost of the number of manufacturing batches, which generally decreases when larger quantities are produced in a single batch (production cycle). In this context, it is important to note that the costs of producing (CAPEX)

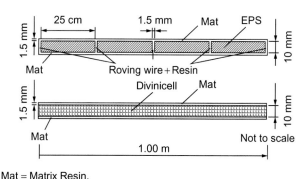

Fig. 10.4 Structural arrangements of the two possible core materials.

Table 10.2 Price breakdown of panels made using EPS or Divinycell as the core material

Item	Panel with EPS core	Panel with Divinycell core
Matrix (300 g/m^2)	7.56	7.56
Roving wire	1.41	0.00
Epoxy resin	84.92	0.00
Polyester resin	0.00	35.28
EPS (1 m^2)	1.00	0.00
Divinycell (1 m^2)	0.00	75.59
Total price	94.89	118.43

and maintaining (OPEX) the internal structure of a marine vessel constitute a large percentage of the lifetime costs of a ship. Vessels designed to have minimal maintenance costs over the entirety of their service life will therefore typically consist of larger members and be more expensive to produce. This means that it is important to design the internal structure of marine vessels in a way that strikes an effective balance between these two competing cost aspects in order to minimize the total monetary cost to the ship owner (Temple and Collette, 2015).

The marine boat structure community is also currently pursuing improved structural performance while simultaneously reducing the costs of construction and life cycle maintenance. In addition to reducing structural weight, efforts have been made over the last 70 years to improve many other properties related to boat design, and composite structures have the very clear potential to help achieve. Of the low-cost options available for fabricating composite ship structures, the most promising are the resin transfer molding and filament winding techniques described by Critchfield et al. (1994).

Historically, the model used for these structures has been based on a plate, as the curvature of ship plating is sufficiently small for individual panels to be approximated as flat plates (Rajendran and Lee, 2009).

10.4.2 Weight comparison with Divinycell structure

For comparison, we chose the Divinycell with the lowest available density (38 kg/m^3). As the case of boat used here is prismatic in all directions, its weight can be calculated for each linear meter of the structure. The thicknesses of the internal and external faces, as well as that of the core, were set to the same value as that of the EPS structure. The only noticeable difference in this structure is that the reinforcements are external to the plate. The weight breakdown of the aforementioned Divinycell plate is provided in Table 10.3, and the values for the EPS plate are provided in Table 10.4. Note that the two structures are very close in weight, although the EPS plate is somewhat heavier. However, the lower density Divinycell structure is not as useful mechanically and is much more expensive.

Table 10.3 **Weight of a Divinycell plate**

Item	Weight
Divinycell plate	14.906 N (1.52 kgf)
External reinforcements in the transverse direction	10.101 N (1.03 kgf)
External reinforcements in the longitudinal direction	10.101 N (1.03 kgf)
Internal/external fiberglass plate faces	211.824 N (21.6 kgf)
Inner lining of reinforcements (Divinycell)	1.177 N (0.12 kgf),
Total weight	248.108 N (25.3 kgf)

Table 10.4 **Weight of the new composite**

Item	Weight
Base polystyrene plate	5.492 N (0.56 kgf)
Inner fiberglass reinforcements	56.486 N (5.76 kgf),
Internal/external fiberglass plate faces	211.824 N (21.6 kgf)
Total weight	273.606 N (27.9 kgf)

10.5 Case study: A vessel structural computational design

10.5.1 Structural arrangement and geometry

The girder of a ship's hull is a large floating structural system made up of plate panels and stiffeners and is subjected to both still-water and wave loads. The static forces acting on it in still-water are created by the longitudinal distributions of weight and buoyancy (Shu and Moan, 2011); however, the structure may collapse in sea conditions if the structural capacity is less than the work load. Consequently, both the working load and hull girder capacity are vital aspects to the safety of a ship (Pei et al., 2015).

To test the composite in this study, we used a rectangular prismatic barge for use in calm, deep waters as the model, using the following main dimensions: an overall length of 20.00 m, a molded breadth of 4.00 m, a molded draught of 2.13 m, a maximum draft of 0.50 m, a displacement at full load of 364.754 kN (41.0 tf), a light draft of 0.023 m, a light displacement of 16.12 kN (1.812 tf), and a load capacity of 348.634 kN (39.188 tf).

This simulated boat has eight compartments, separated by seven watertight bulkheads. In order to simulate both hogging and sagging conditions, we carried out two distinct simulations, in which four separate compartments were alternately loaded, as shown in Fig. 10.5.

The vertical bending moment is of crucial importance for ensuring the survival of vessels in rough seas, and in the case of conventional vessels, it is normally considered that the wave-induced maximum vertical bending moment is experienced in head seas

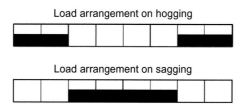

Fig. 10.5 Load arrangement for the two simulations.

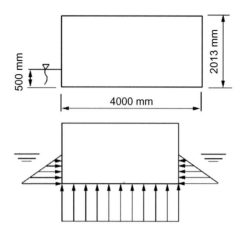

Fig. 10.6 Hydrostatic load acting on the hull.

applied to maximum sagging and hogging (Zhu and Moan, 2014). As implied by the load capacity and molded draft in Fig. 10.5, a load of 8.715 kN/m^2 (0.9796 tf/m^2) is distributed at the bottom of each loaded compartment and hydrostatic pressure is present at the bottom and on the portions in contact with water. In order to maintain static equilibrium, we applied the sum of the gravitational loads distributed over the floor area from bottom to top. To clarify,

Fig. 10.6 shows the locations and directions of the hydrostatic forces that maintain a balance with the applied external forces, and which represent the boundary condition of the model. Thus, these are the only forces acting on the system, which provide the self-equilibrium in each loading condition shown in Fig. 10.5. It is for these loading conditions that we calculated the distributions of the stresses and strains in the ferry structure.

10.5.2 Computational modeling

Any design strategy needs to take into account two important aspects: material selection, and the structural arrangement and scantlings most appropriate for the chosen material (Stenius et al., 2011). As such, engineering design consists of several steps including mathematical modeling, the application of physics, and theoretical and computational analysis (Lee et al., 2003). For the purposes of this study, we computationally elaborated a basic plate (Fig. 10.7) using the finite element method (FEM) with a commercial finite element package (SAP2000, 2015), which has been used in the past to evaluate the design of high-speed marine craft (Townsend et al., 2012). We used a three-dimensional model for the discretization of a plate containing solid elements and built a model by taking into account the actual dimensions and properties of the materials of each separate component of the plate. These included details of the

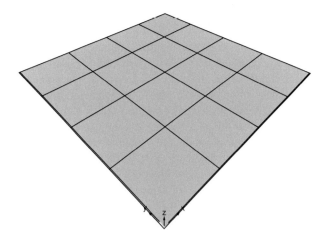

Fig. 10.7 Basic three-dimensional FEM model of the plate.

corner and middle of both models, as shown in Fig. 10.1 and Fig. 10.4. The basic plate of the new composite material served as the basis of the numerical experiment and constituted the main piece covering the entire barge and bulkheads of the model.

We determined the mathematical relationship between the solid elements using an isoparametric formula that includes nine optional, unique bending modes. These unique bending modes significantly improve the bending behavior of the element if the element geometry is rectangular. We used a $2 \times 2 \times 2$ numerical integration scheme to evaluate the stresses in the local coordinates at the integration points, which were then extrapolated to the joints. We estimated the approximate error for the stresses from the difference between the calculated value and other elements attached to the same joint. This, in turn, provided an indication of the accuracy of the calculation that could be used in the selection of a new and more accurate finite element mesh. All six stress and strain components were active for this element (Cook, 1974; Bucalem and Bathe, 2011).

We completed the computational modeling of the vessel described in subsection 10.5.1 using 35,183 eight-node solid modeling elements. The construction process was based on replicating the basic plate as many times as necessary in each of three orthogonal directions (X, Y, and Z) until the boat was fully formed. This process included replication of the frontal part, back, bottom, cover, laterals, and internal divisions. Special care was taken to ensure that all nodes of each plate were connected to each other at all points of the model, which included both the internal nodes of the plate and the union between plates.

A similar technique using FEA has been employed to design what is primarily a glass-reinforced polymer composite with an airfoil-shaped sail and canopy-style configuration known as the composite advanced sail (CAS), which can reduce weight and maintenance costs and improve the load capacity. For modeling, this CAS structure is divided into four separate components to account for variation in material composition (Eamon and Rais-Rohani, 2009).

10.5.3 Computational results and discussion

We classified the hogging and sagging conditions summarized in Fig. 10.5 as COMB1 and COMB2, respectively, with both simulations assuming a draft depth of 0.50 m and considering the self-weight of the structure. The overall displacement of the vessel for both loading conditions indicated a maximum vertical absolute displacement of 22 and 30 mm, respectively, while the maximum relative observed sag was 2.6%. We note that epoxy resin impregnated with fiberglass can elongate by up to about 5%. A maximum horizontal displacement of 66 and 30 mm was obtained for the front and sides (Fig. 10.8) of the deformed structure, respectively.

Table 10.5 summarizes the stress results and Table 10.6 shows the mechanical properties of interest for fiberglass impregnated with epoxy resin (Torabizadeh, 2013). It is important to note that these results were obtained by considering the initial dimensions of the plate, which can be modified and adjusted at any time.

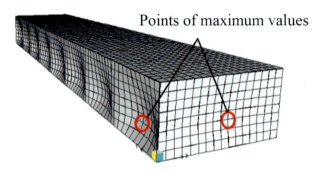

Fig. 10.8 Deformation from pressure (front cover and side).

Table 10.5 **Maximum normal stresses (MPa)**

Condition	Deck	Bottom
COMB1	55 (Tension)	343 (Compression)
COMB2	159 (Compression)	628 (Tension)

Table 10.6 **Mechanical properties of fiberglass complexed with epoxy resin**

Property	Value
Tensile strength	715 MPa
Compressive strength	570 MPa
Shear strength	70 MPa
Poisson's ratio	0.22

10.6 Conclusions

In this study, simulation and comparison showed that the proposed composite is not only feasible for use in boat hull construction, but also quite competitive with existing materials. Even considering the relatively poor mechanical properties of EPS, its use as a simple core in conjunction with shear webs made of epoxy resin improves the overall strength of the proposed composite to the extent that it is competitive with the much higher-quality sandwich-structured Divinycell. Furthermore, as EPS is at worst roughly the same weight as Divinycell, it is a particularly desirable construction material for speedboats, especially given that it is approximately 20% less expensive.

Overall, this new composite appears to provide an excellent alternative for use in light and medium vessels, although further laboratory research has yet to be undertaken to address its other mechanical properties and weight capacity. To address these issues, new computational models will be developed with plate elements that should prove much simpler and quicker to use than solid element models. However, models made with FEM three-dimensional elements will be used for comparison.

References

Atkinson, R., 1997. Innovative uses for sandwich constructions. Reinf. Plast. 41 (2), 30–33.

Bucalem, M.L., Bathe, K.J., 2011. The Mechanics of Solids and Structures–Hierarchical Modeling and the Finite Element Solution. Springer, Germany.

Cerracchioa, P., Gherlonea, M., Di Sciuvaa, M., Tesslerb, A., 2015. A novel approach for displacement and stress monitoring of sandwich structures based on the inverse finite element method. Compos. Struct. 127, 69–76. https://doi.org/10.1016/j.compstruct.2015.02.081.

Chalmers, D.W., 1994. The potential for the use of composite materials in marine structures. Mar. Struct. 7 (2–5), 441–456. https://doi.org/10.1016/0951-8339(94)90034-5.

Cook, R.D., 1974. Concepts and Applications of Finite Element Analysis. John Wiley and Sons, Inc., NJ.

Craugh, L.E., Kwon, Y.W., 2013. Coupled finite element and cellular automata methods for analysis of composite structures with fluid–structure interaction. Compos. Struct. 102, 124–137. https://doi.org/10.1016/j.compstruct.2013.02.021.

Critchfield, M.O., Judy, T.D., Kurzweil, A.D., 1994. Low-cost design and fabrication of composite ship structures. Mar. Struct. 7, 475–494. https://doi.org/10.1016/0951-8339(94)90036-1.

Crupi, V., Epasto, G., Guglielmino, E., 2013. Crupi comparison of aluminum sandwiches for lightweight ship structures: Honeycomb vs. foam. Mar. Struct. 30, 74–96. https://doi.org/10.1016/j.marstruc.2012.11.002.

Davies, P., Choqueuse, D., Bigourdan, B., 1994. Test-finite element correlations for non-woven fibre-reinforced composites and sandwich panels. Mar. Struct. 7, 345–363. https://doi.org/10.1016/0951-8339(94)90030-2.

Eamon, C.D., Rais-Rohani, M., 2009. Integrated reliability and sizing optimization of a large composite structure. Mar. Struct. 22, 315–334. https://doi.org/10.1016/j.marstruc.2008.03.001.

Fang, H., Liu, W., Zou, F., Shi, H., 2017. Composite sandwich structures reinforced with GFRP lattice webs for civil infrastructure applications.Unpublished keynote presentation at The Third China International Congress on Composite Materials, Xiaoshan (Hangzhou) ~ China, 21–23 October 2017. http://boatbuildercentral.com/products.php?cat=10. Consulted March 2015.

Kimpara, I., 1991. Use of advanced composite materials in marine vehicles. Mar. Struct. 4 (2), 117–127. https://doi.org/10.1016/0951-8339(91)90016-5.

Lee, K.W., Chong, T.H., Park, G.J., 2003. Development of a methodology for a simplified finite element model and optimum design. Comput. Struct. 81, 1449–1460. https://doi.org/10.1016/S0045-7949(03)00084-1.

Mouring, S.E., 1999. Buckling and post buckling of composite ship panels stiffened with preform frames. Ocean Eng. 26, 793–803. https://doi.org/10.1016/S0029-8018(98)00025-0.

Mouritz, A.P., Gellert, E., Burchill, P., Challis, K., 2001. Review of advanced composite structures for naval ships and submarines. Compos. Struct. 53 (1), 21–42. https://doi.org/10.1016/S0263-8223(00)00175-6.

Pei, Z., Iijima, K., Fujikubo, M., Tanaka, S., Okazawa, S., Yao, T., 2015. Simulation on progressive collapse behavior of whole ship model under extreme waves using idealized structural unit method. Mar. Struct. 40, 104–133. https://doi.org/10.1016/j.marstruc.2014.11.002.

Rajendran, R., Lee, J.M., 2009. Blast loaded plates. Mar. Struct. 22, 99–127. https://doi.org/10.1016/j.marstruc.2008.04.001.

Romanoff, J., Varsta, P., 2007. Bending response of web-core sandwich plates. Compos. Struct. 81 (2), 292–302. https://doi.org/10.1016/j.compstruct.2006.08.021.

SAP2000, 2015. Integrated Software for Structural Analysis and Design, Analysis Reference Manual. Computer and Structures, Inc., Berkeley, CA.

Scott, R.J., 1996. Fiberglass Boat Design and Construction, second ed. SNAME, NJ.

Shu, Z., Moan, T., 2011. Reliability analysis of a bulk carrier in ultimate limit state under combined global and local loads in the hogging and alternate hold loading condition. Mar. Struct. 24, 1–22. https://doi.org/10.1016/j.marstruc.2010.11.002.

Stenius, I., Rosén, A., Kuttenkeuler, J., 2011. On structural design of energy efficient small high-speed craft. Mar. Struct. 24, 43–59. https://doi.org/10.1016/j.marstruc.2011.01.001.

Temple, D.W., Collette, M.D., 2015. Minimizing lifetime structural costs: Optimizing for production and maintenance under service life uncertainty. Mar. Struct. 40, 60–72. https://doi.org/10.1016/j.marstruc.2014.10.006.

Torabizadeh, M.A., 2013. Tensile, compressive and shear properties of unidirectional glass/Epoxy composites subjected to mechanical loading and low temperature services. Indian J. Eng. Mater. Sci. 20, 299–309.

Townsend, N.C., Coe, T.E., Wilson, P.A., Shenoi, R.A., 2012. High speed marine craft motion mitigation using flexible hull design. Ocean Eng. 42, 126–134. https://doi.org/10.1016/j.oceaneng.2012.01.00.

Yang, N., Das, P.K., Blake, J.I.R., Sobey, A.J., Shenoi, R.A., 2013. The application of reliability methods in the design of tophat stiffened composite panels under in-plane loading. Mar. Struct. 32, 68–83. https://doi.org/10.1016/j.marstruc.2013.03.002.

Yu, Z., Hu, Z., Wang, G., 2015. Plastic mechanism analysis of structural performances for stiffeners on bottom longitudinal web girders during a shoal grounding accident. Mar. Struct. 40, 134–158. https://doi.org/10.1016/j.marstruc.2014.11.001.

Zhu, S., Moan, T., 2014. Nonlinear effects from wave-induced maximum vertical bending moment on a flexible ultra-large containership model in severe head and oblique seas. Mar. Struct. 35, 1–25. https://doi.org/10.1016/j.marstruc.2013.06.007.

Part Three

Applications

Offshore wind turbines

Puyang Zhang
State Key Laboratory of Hydraulic Engineering Simulation and Safety,
Tianjin University, Tianjin, China

Chapter Outline

11.1 Introduction 317
11.2 The load-bearing characteristics of composite bucket foundation 321
    11.2.1 Force transfer mechanism of arc transitional section 323
    11.2.2 Top cover load-bearing mode 328
11.3 Model tests on the bearing capacity of composite bucket foundation 329
    11.3.1 The deformation mechanism and the soil-structure interactions of CBF under horizontal load 330
    11.3.2 Failure envelope 332
11.4 Model tests on the installation of composite bucket foundation 333
    11.4.1 Test 1 336
    11.4.2 Test 2 337
    11.4.3 Comparisons 342
11.5 Conclusions 342
References 343

11.1 Introduction

The foundation of an offshore wind turbine unit consists of a tower frame and a seafloor foundation. In China, the mono pile, three-pile tripod foundation, multipile jacket foundation, and multipile cap foundation are typical foundations (see Fig. 11.1) for offshore wind turbines, which can effectively convert loads and improve the overturning resistance (high-pile cap foundation, 2010; multi-piles foundation, 2013; high socketed pile cap foundation, 2015; Longyuan Zhenhua No. 2, 2012). A research group led by Professors Lian and Ding at Tianjin University and DaoDa Offshore Wind Company innovated a new foundation based on the conventional bucket foundation and, for the first time, proposed a composite bucket foundation (CBF, shown in Fig. 11.2) and one-step installation technique for offshore wind turbines suitable for the marine area offshore of China. This proposed structure effectively converts the extremely large bending moment of the turbine tower to limited tensile and compressive stresses within the foundation structure via a transition section; in addition, the proposed structure effectively solves the deformation compatibility and cracking control problems of steel-concrete structures via prestressing and makes full use of the properties of the

Marine Composites. https://doi.org/10.1016/B978-0-08-102264-1.00011-X
Copyright © 2019 Elsevier Ltd. All rights reserved.

Fig. 11.1 Typical foundations for offshore wind turbines in China. (A) high-pile cap foundation and tower and turbine transportation and installation in Donghai Bridge offshore wind farm (high-pile cap foundation, 2010); (B) multipiles foundation in Longyuan Jiangsu Rudong Offshore (Intertidal Zone) Demonstration Wind Farm (multi-piles foundation, 2013); (C) high socketed pile cap foundation in Fujian Nanridao offshore wind farm (high socketed pile cap foundation, 2015); (D) Longyuan Zhenhua No. 2 (800 tons self-elevating offshore wind vessel) driving 57.5 m single pile with 530 tons (Longyuan Zhenhua No. 2, 2012).

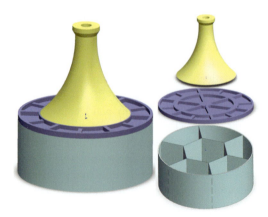

Fig. 11.2 CBF structure.

material (Ding et al., 2013, 2015a,b; Lian et al., 2014, 2012a,b, 2011; Liu et al., 2015a,b; Zhang et al., 2013a,b, 2013c,d, 2014a,b, 2015).

In 2010, the first composite BFB (Bucket Foundation with Bulkheads) for offshore wind turbines was successfully installed at the offshore test facility in Qidong City, China, as shown in Fig. 11.3A. In 2016, two BFBs were also installed in Xiangshui City, China (see Fig. 11.3B). At the same time, a series of tests were conducted on the

(A)

(B)

Fig. 11.3 A prototype with CBF. (A) Concrete skirt for 2.5 MW offshore wind turbine; (B) Steel skirt for 3 MW offshore wind turbine.

prototype, including a structural test under construction conditions, a static test after the installation was complete, a dynamic test after the installation was complete, and a sinking test under simulated large bending-moment working conditions. The development of core techniques for the CBF and the construction of offshore wind turbines address the constraints of large-scale development of offshore wind power in China. The three primary core techniques are as follows: (1) a new prestressed steel strand steel plate concrete composite bucket wind turbine foundation structure system is proposed; the system displays certain force transfer characteristics and stress distributions; (2) the modeling and calculation of the bearing capacity of the ground for the large-scale CBF with the bucket head cover as the main load-bearing component have been established, and the limitations of other foundation structures in soft ground are thereby overcome; (3) the "floating transport-sinking-leveling" construction technique is proposed for the composite bucket offshore wind turbine foundation structure.

Based on the characteristics of the CBF (no pile needs to be driven for the entire structure, there is no impact load, and negative-pressure sinking is unnecessary for installing the CBF), an innovative special construction vessel has been designed to replace the hoisting vessel to install offshore wind turbines. This special construction vessel can be used to provide integrated transportation and one-step installation. Similar to planting trees in the ground, the special construction vessel can be used to "plant" wind turbines in the seabed. The one-step installation technique developed by our team achieves complete wind turbine structure installation. The proposed scheme (see Fig. 11.4) has seven primary aspects: the prefabrication of the foundation, the floating transport of the foundation, the assembling of the wind turbine on land and the testing of the complete wind turbine unit, the floating transport of the complete wind turbine unit, the on-site one-step installation, consignment, and recycling of the unit. Compared with conventional wind turbine installation, the primary advantage of the one-step installation technique is the assembly of the wind turbine on land, which avoids the use of large equipment in the marine environment, reduces the installation difficulty, is easier, improves the installation quality and structural reliability, reduces the risk of damage to the components of the wind turbine resulting from the offshore construction environment and technical limitations to a certain degree, and reduces the investment costs of various construction steps. The one-step installation technique allows for rapid offshore construction. In addition, the technique is safe, economical, and efficient and can allow for the recycling and reuse of the wind turbine and the foundation structure. The technique also supports the standardization and large-scale development of offshore wind power and significantly increases the construction speed of offshore wind farms.

The floating transport of the complete wind turbine unit has peculiarities in terms of the foundation and vessel and differs substantially from conventional vessel transport (see Fig. 11.5). First, the wind turbine unit and the vessel have different floating states. As an air floating structure, the towing of the wind turbine foundation structure exhibits complex motions and is affected by many factors. Second, to accommodate the bucket foundation, the structure of the one-step installation transport vessel also differs from that of a conventional vessel. Last and most importantly, the floating transport of the complete wind turbine unit is no longer that of a single floating body, but instead is the coupled motion of a composite floating body structure, which

Fig. 11.4 One-step installation technique.

exhibits the more complex coupled motion of an air floating structure and a real floating structure.

In contrast, the one-step offshore wind turbine installation and transport technique does not require large offshore hoisting equipment, and thus, to a large degree, simplifies the construction process, shortens the construction period, circumvents the impacts of complex sea conditions and uncertainties, and provides a technical basis for the rapid construction of offshore wind farms.

11.2 The load-bearing characteristics of composite bucket foundation

In order to investigate the bearing capacity of the CBF in a typical soil profile in China, a three-dimensional finite element model for the bearing capacity behavior of CBF is established by using the universal finite element analysis software ABAQUS, as

Fig. 11.5 One-step vessel operation steps (A) Unit loading; (B) Gas floating jacking; (C) Unit transportation; (D) Unit installation; (E) Foundation penetration; (F) suction penetration with adjusting tilt.

shown in Fig. 11.6. The CBF has an outer diameter of 30 m and a clear wall height of 12 m. The finite element method (FEM) model of the foundation with the soil and the CBF are shown in Fig. 11.6A. Fig. 11.6A also shows the typical mesh with approximately 66,000 elements used in the present study. The vertical, horizontal, and rotational displacements at the bottom boundary of the soil were fully fixed, and horizontal displacements were prevented on the side boundary of the soil. Fig. 11.6C shows the seven rooms divided inside the bucket by steel plates, while the prestressed tendons are shown in Fig. 11.6D. The six peripheral rooms have the same proportions with same subdivision plate of 7.5 m long, and the middle one is a little larger. The bottom diameter of the offshore wind-powered 3-MW turbine tower is 4.4 m, and the bucket foundation top cover diameter is 30 m. A proper connection between the tower bucket and the top cover of the bucket foundation can ensure a

Offshore wind turbines

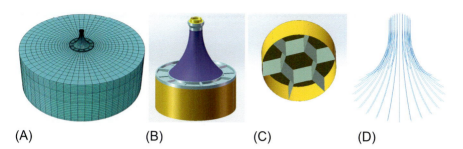

Fig. 11.6 FEM models; (A) 3 MW foundation model with soils; (B) CBF; (C) Subdivision plates inside the bucket; (D) prestressed tendons.

fairly good transfer of the upper structural load to the bucket foundation. In the 3-MW turbine bucket foundation, the distance between the upper edge of the transitional segment to the soil surface is 19.8 m, and the arc portion is 16.8 m.

Using the parameters given in Table 11.1, soil was modeled as an elastic-plastic model based on the Mohr-Coulomb failure criterion. If the CBF is assumed to be an ideal elastoplastic model, Young's modulus is 36 GPa and Poisson's ratio is 0.2 for transmission part made of C60 concrete, while Young's modulus is 210 GPa, the yield strength is 345 MPa, and Poisson's ratio is 0.3 for steel bucket skirt and subdivision plates. For analyses, the foundation and the soils are all simulated by C3D8R elements. The contact pair algorithm in the ABAQUS is employed to simulate the contact features of the interface between the foundation and the soil. Coulomb's friction law is used to estimate the tangential ultimate frictional resistance. The foundation and the soil are stuck together and no slip happens when the shear stress on the contact surface is less than the ultimate frictional resistance. Similarly, slip occurs along the contact surface if the shear stress exceeds the ultimate frictional resistance. The contact in the normal direction of the interface is considered to be hard contact.

11.2.1 Force transfer mechanism of arc transitional section

The ultimate loads of 50-year return period (YRP) on the upper structure of the wind turbine that are transferred to the bottom surface of the tower bucket are the following: horizontal load of 1523 kN, bending-moment load of 109 MNm, and vertical load of 7011 kN. The load includes a safety coefficient of 1.3 and a structure importance coefficient of 1.1 (American Society of Civil Engineers, 2006), which are applied to the surface of the transitional segment of the bucket foundation. Based on the hydrological condition of the wind turbine field and considering the wave and current loads, the horizontal load increased to 2209 kN. To determine the load transfer mode from the upper structure downwards through the arc transition segment, the bending moment, shear force, and axial force data of each horizontal cross section of the transitional segment under the action of the ultimate load of 50YRP were obtained, as shown in Fig. 11.7A. The method for extraction of the cross section internal force is also shown in Fig. 11.7B.

Table 11.1 Soil properties

Soil layers	Thickness (m)	Wet density (g/cm^3)	Poisson's ratio	Compression modulus (MPa)	Cohesion (kPa)	Internal friction angle (°)
Silt	3.9	1.98	0.25	9.04	9.0	31.0
Silty clay	6.8	1.78	0.39	3.13	14.0	12.1
Silty sand	10.8	2.03	0.21	13.06	4.4	33.5
Silty clay	7	1.89	0.37	3.83	21.8	13.7
Silt	1.8	1.98	0.25	15.04	21.4	29.6
Silty sand	4.3	1.96	0.21	13.03	4.7	33.2
Silty clay	6.4	1.96	0.37	5.70	38.5	17.8
Silt	6.1	2.05	0.25	11.22	10.8	29.1

Offshore wind turbines

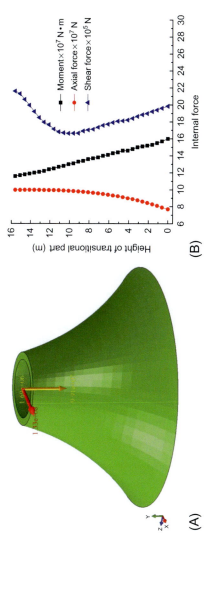

Fig. 11.7 Transitional part and internal force. (A) Transitional segment; (B) The cross section internal force.

The primary characteristic of an offshore turbine combined with a bucket foundation is that the upper structure bears a tremendous bending-moment load. The bending moment of the transition segment cross section gradually increases from the top down, which includes the impact of the horizontal load. The calculated stress results of each cross section are shown in Fig. 11.8A, from which it can be seen that, under a bending-moment load, the tensile and compressive stresses at the upper portion of the transitional segment are rather high, approximately 15.3 MPa. The transitional segment cross section area gradually increases downwards, and the stress gradually reduces; at the bottom of the transitional segment, the tensile and compressive stresses on the two sides were only 0.61 MPa, which means that the large load of the bending moment at the upper structure became relatively low tensile and compressive stresses when transferred to the top surface of the bucket foundation through the transitional segment.

For concrete materials, the tensile stress at the upper portion of the transitional segment is too high, and the requirement is not satisfied. Thus, prestressed steel must be added. The prestressed steel is shown in Fig. 11.8B, which totals 48 bundles, and the cross sectional area of each bundle is 1530×10^{-6} m^2 with a tensile stress of 1320 MPa. The axial force of the transitional segment cross section is primarily produced by the tensile force of the prestressed steel. The maximum axial force of the transitional segment cross section is approximately 1100 MN, which is close to the combined tensile force of the prestressed steel.

The stress value of the prestressed steel under the ultimate load of 50YRP is shown in Fig. 11.8B, in which the maximum value is 1336 MPa, and the minimum value is 1161 MPa. The resultant compressive stress on the concrete is shown in Fig. 11.8C; the distribution profile along the transitional segment height exhibits a similar characteristic as the bending moment-induced cross section tensile and compressive stresses. According to the above calculations, the tensile and compressive stress of each cross section under the combined action of a bending moment and the prestressed steel are the combination of the side compressive stress of the compressed cross section along the load direction and are the tensile stress on the tensioned side of the cross section.

The calculation results are shown in Fig. 11.8C. It can be observed from the figure that the tensile and compressive stresses are greatest at the top portion of the transitional segment; the tensile stress is approximately 2.83 MPa, and the compressive stress is approximately 27.8 MPa. For C60 concrete, both of these values are between the design value and standard value, which is essentially consistent with the numerical calculation results in Fig. 11.8D and E. An appropriate amount of steel is still required in the actual project to satisfy the standard requirements. Near a height of 10 m of the transitional segment, the cross section tensile stress reduces to 0, i.e., under the ultimate load of 50YRP, there is no tensile stress in the concrete under the water surface; this is conducive to corrosion resistance in sea water. The stress value near the bottom of the transitional segment is relatively small; the compressive stress is approximately 1.96 MPa. No stress concentration appears in any cross section throughout the entire transitional segment. The huge bending-moment load at the upper structure of the turbine gradually diffuses downwards through the arc transitional segment with the

Offshore wind turbines 327

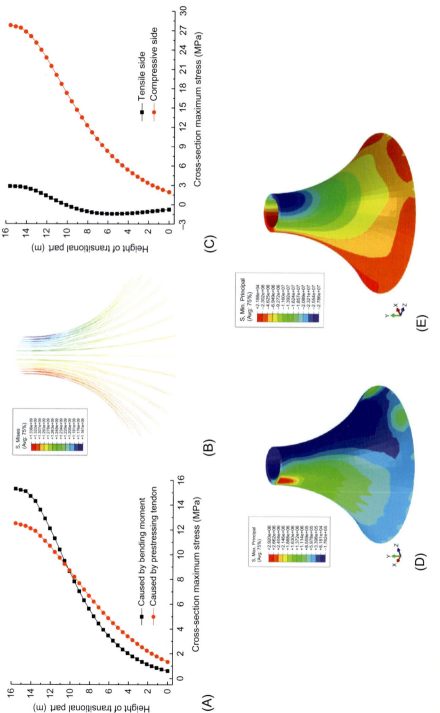

Fig. 11.8 Transitional segment internal stresses under the ultimate load of 50YRP. (A) Cross section stresses with height; (B) Stress value of the prestressed steel; (C) Maximum stress with height; (D) Transitional segment tensile stresses; (E) Transitional segment compressive stresses.

prestressed steel and becomes relatively small tensile and compressive stresses. The force transfer mechanism is correct, and the structural integrity is effectively used, which means the combined bucket foundation has a high degree of safety.

11.2.2 Top cover load-bearing mode

A bucket foundation primarily includes two parts, the top cover and the bucket wall; the load-bearing mode is also generally divided into two modes, the bucket top and bucket wall load bearing. Fig. 11.9A shows the distribution characteristic of the soil pressure on the top cover and the bucket wall when the tilting degree of the bucket foundation under the bending-moment load reaches 3°. The location of the center of rotation is identified based on the finite element calculation result, which is located near and on the left side of the central axis of the bucket foundation in the opposite direction of the load. The degree of impact of the top cover, bucket wall, and bulkhead on the bucket foundation load nearing the capacity under the horizontal and vertical loads and the ultimate load of 50YRP, respectively, is shown in Table 11.2; the location of the rotation center and the soil pressure distribution under the ultimate load of 50YRP are shown in Fig. 11.9B.

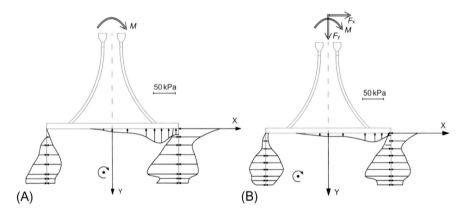

Fig. 11.9 Soil pressure distribution of CBF. (A) The tilting degree 3° of the bucket foundation; (B) Ultimate load of 50YRP.

Table 11.2 **Bearing proportion of different part of CBF**

	Top cover (%)	Skirt (%)	Bulkheads (%)
Bending moment (3‰)	57.9	36.3	5.8
Horizontal load (3‰)	50.6	42.1	7.3
Vertical load (0.1 m)	70.7	17.9	11.4
Ultimate load of 50YRP	59.4	29.8	10.8

Offshore wind turbines

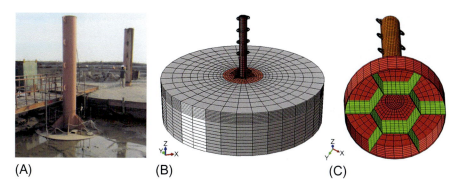

Fig. 11.10 Test and FEM models. (A) Horizontal loading at the height; (B) Overall model of bucket foundation and soil; (C) Layout of rooms inside the bucket foundation.

It can be observed from Table 11.2 that the load-bearing ratio of the top cover under all four types of loads exceeds 50%, which in the case of the ultimate load of 50YRP is closest to reality; in this situation, the top cover bears approximately 60% and the bucket wall and bulkhead bear approximately 30% and 10% of the load, respectively. The configuration of the bulkheads is primarily for the convenience of towing and submerging and can only provide a small safety margin under loading; therefore, it can be determined that the load-bearing mode of the offshore wind turbine combined with a bucket foundation is a "cover-load-bearing type", such that the top cover bears most of the load and is supplemented with bucket wall load bearing.

11.3 Model tests on the bearing capacity of composite bucket foundation

The CBF evaluation was carried out in a large artificially excavated test pool located along the coast of Jiangsu. The soil was placed in the tank by an excavator layer by layer, to a total depth of 3 m. The top soil requires elaborate preparation to make it flat. The soil around the pool is mainly fine sand. The saturated density of the soil is 1806 kg/m^3 and the water content is 33.5%. Its liquidity index and plasticity index are 1.19 and 11.75, respectively. The compressive modulus E_s of the soil is taken as 3.69 MPa for a proper ultimate state, and its cohesion strength c and internal friction angle φ are 3.84 kPa and 7.14 degrees, respectively.

The bucket foundation in the tests has an outer diameter of 3.5 m and a clear wall height of 0.9 m. The seven rooms are divided inside the bucket by bulkheads as shown in Fig. 11.10. The six peripheral rooms have the same proportions, and the middle one is a little larger. A steel tube is connected to the lid and reinforced by six ribbed plates as part of this CBF, and the tube is also used for horizontal loading as part of a wind turbine tower. Lifting lugs are required and attached to the tube for applying horizontal load at the height of 2 m, 3 m, 4 m, and 5 m, as shown in Fig. 11.10A. Meanwhile, in order to have a clearer knowledge on the ultimate bearing capacity of the CBF in

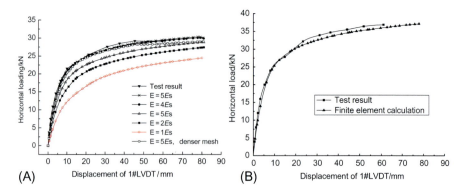

Fig. 11.11 Load-displacement relationship curves of CBF. (A) Horizontal load at the height of 3 m; (B) Horizontal load at the height of 4 m.

saturated muddy clay, a three-dimensional finite element model for the bearing capacity behavior of CBF is established by using the universal finite element analysis software ABAQUS, as shown in Fig. 11.10B and C.

11.3.1 The deformation mechanism and the soil-structure interactions of CBF under horizontal load

Fig. 11.11 shows the load-displacement relationship of the CBF under different loading heights. The curves contain both inflection and extreme points. The inflection points indicated that the plastic failure of the soil took place outside the wall under the action of the corresponding load, but the soil pressure on the lower surface of the lid at the passive zone still had a growth trend for the horizontal restraint of the bulkheads. The horizontal tension loading at the extreme points is taken as the ultimate bearing capacity of the CBF. Under such a condition, the failure mode of the foundation can be determined according to the distribution of the equivalent plastic strain in the soil.

According to the test results and finite element analysis, the calculated positions of instantaneous rotation centers of the CBF are shown in Fig. 11.12. The horizontal load tests at the heights of 3 m and 4 m were conducted twice each. The position of the instantaneous rotation centers moves from the lid center to the bottom of the bulkhead at the loading direction with the increase of horizontal load. The rotation center is approximately $0.8\,L$ below the ground surface around the bucket at the loading direction under ultimate horizontal load, (where L is the depth of the bucket). Its position and the schematic diagram of the CBF rotation are shown in Fig. 11.13; the coordinate of the rotation center is assumed to be $X_0 = \frac{\sqrt{3}}{4}R$, $Z_0 = 0.8\,L$. The soil pressure on the outer wall in the loading direction is shown in Fig. 11.14, along with the test results, the finite element calculation results, and the formulas, all of which are in basic agreement.

Offshore wind turbines 331

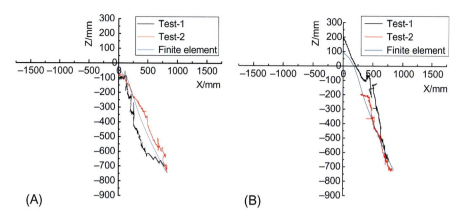

Fig. 11.12 Positions of the instantaneous rotation centers of the CBF. (A) Horizontal load at the height of 3 m; (B) Horizontal load at the height of 4 m.

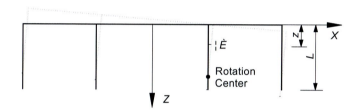

Fig. 11.13 The rotation of CBF.

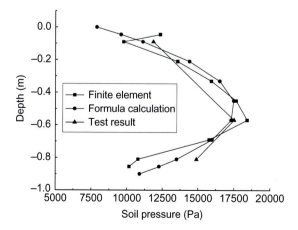

Fig. 11.14 Passive earth pressure on the outer wall.

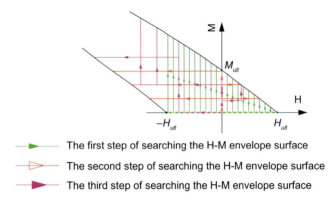

— The first step of searching the H-M envelope surface
▷ The second step of searching the H-M envelope surface
▶ The third step of searching the H-M envelope surface

Fig. 11.15 Steps of searching the H-M envelope surface.

11.3.2 Failure envelope

In offshore wind turbine engineering, the foundations are usually subjected to a huge horizontal load and a huge bending moment for the high tower tube. The composite bearing capacity of the CBF will be clearer for establishing the failure envelope on the basis of these experiments. In order to obtain the failure envelope of the foundation under combined loading, load-controlled analysis and displacement-controlled analysis have been used in numerical simulation. The results show that the yield surfaces determined by the load-controlled and displacement-controlled approaches are consistent.

The failure envelope of the foundation was obtained using the load-displacement control search method, which basically consists of the following steps. 1) A specific horizontal load (H_i, a proportion of H_{ult}) is applied based on the ultimate horizontal bearing capacity H_{ult} and ultimate moment bearing capacity M_{ult}. The load H_i remains unchanged when a radian displacement is applied until the corresponding moment load no longer increases. The failure envelope is obtained as shown in Fig. 11.15. 2) A specific moment load (M_i, a proportion of M_{ult}) is applied based on the ultimate moment bearing capacity M_{ult}. The load M_i remains unchanged when a horizontal displacement is applied until the corresponding horizontal load no longer increases. 3) A lower horizontal load H_i or moment load M_i is applied to the top surface of the CBF, increasing alternately within the failure envelope, and a horizontal displacement or radian displacement is applied at the end of each search. The search method of the three steps is shown in Fig. 11.15, and the failure envelope of H-M is finally shown in Fig. 11.16.

As a qualitative research in Fig. 11.16, there are two possible shapes of the failure envelope based on different extreme methods. For the solid lines in Fig. 11.16, as mentioned in Section 11.4.1, the loading at the extreme points is taken as the ultimate bearing capacity. The solid envelope curves are close to linear in limited spaces and have a larger envelope scope. For the dashed lines, the ultimate bearing capacity is determined by using the tangent intersection method, which is a little conservative. The

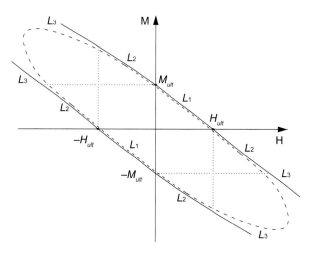

Fig. 11.16 Schematic diagram of the *H-M* envelope surface.

failure envelope of CBF can be divided into three sections: L_1, L_2, and L_3, where L_1 comes into being when H and M are in the same direction and it is independent with the search path, L_2 comes into being when H and M are in the opposite direction and one of them is in the scope of uniaxial ultimate value, L_3 comes into being when H and M are in the opposite direction and none of them are in the scope of uniaxial ultimate value, and L_2 and L_3 require a proper search method. From the results, we can see that the horizontal load and moment have influences that counteract the CBF when they are in opposite directions. At this time, the composite ultimate bearing capacity may exceed the uniaxial ultimate bearing capacity of H_{ult} or M_{ult}.

11.4 Model tests on the installation of composite bucket foundation

To obtain the pressure inside the compartment, soil pressure, and water pore pressure, pressure transmitters, soil pressure cells, and pore water piezometers were fixed on the CBF (see Fig. 11.17). There are 26 soil pressure sensors with a diameter of 0.02 m embedded in the steel plate, with eight in the top cover (see Fig. 11.17A), two at the skirt tip, along the skirt eight toward inside and the remaining eight toward outside (see Fig. 11.17B–D). Meanwhile, there are eight water pore pressure sensors embedded in the skirt wall of the caisson, with four toward inside and the remaining four toward outside, while there are another two sensors fixed on the top cover of the CBF. In addition, there are seven pressure transmitters on the top lid of the bucket for every compartment. The main equipment used in model tests includes the gas/water pump system and tube system and data collection system. The layout scheme and picture of the experimental equipment on test site are illustrated in Fig. 11.18.

In order to investigate the feasibility of tilt adjusting technique, two installation tests were carried out for two different sinking methods. For Test 1, the sinking

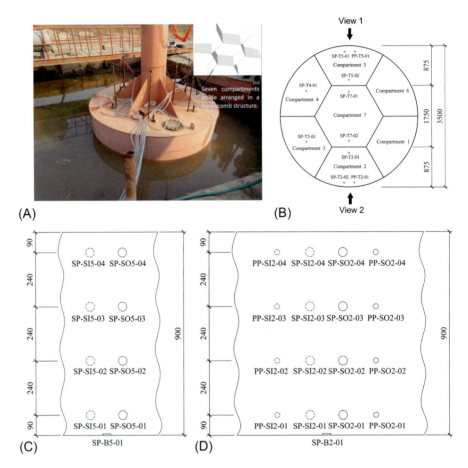

Fig. 11.17 The test model of CBF. (A) Pictures of CBF with sensors; (B) On the top lid of CBF; (C) Along the skirt of compartment 5 (View 1); (D) Along the skirt of compartment 2 (View 2).

process proceeded without interruptions under negative pressure. The process took approximately 20 min from the time when sinking under self-weight began to when the top cover contacted the mud surface. The negative-pressure tubes of seven compartments of the CBF converged to a common tube that was connected to the negative-pressure pump. During the sinking process under negative-pressure, the valves of the negative tubes of all the compartments were controlled to level the foundation in a timely operation. When the foundation started to tilt, the valves of the lower compartments were switched to reduce the suction gradually until they were closed; however, the upper compartments were allowed to continue to sink. After the required levelness of the foundation was achieved, all of the valves were opened so that the entire foundation could sink. This process was repeated until the bucket foundation was completely sunk.

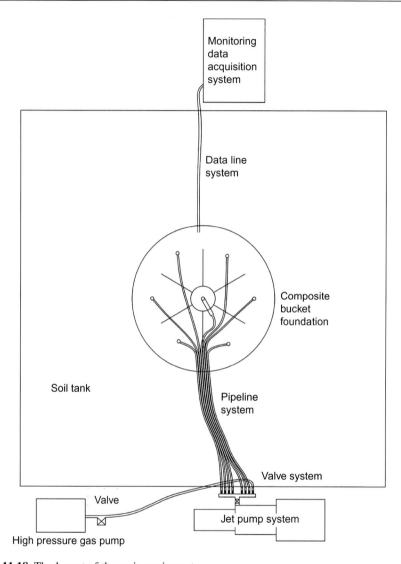

Fig. 11.18 The layout of the main equipment.

By comparison, in Test 2, a negative pressure was produced in the top compartments to level the foundation (the maximum height difference between the two ends was within 1 cm) and, each time the bucket foundation sank 10–15 cm, the negative-pressure production stopped. The inclination of the bucket foundation was calculated using a gradienter, an inclinometer, and a ruler to directly measure the distance between the head cover surface and the water surface. A negative pressure was produced in all of the compartments to allow the bucket foundation to gradually sink into the clay until the head cover contacted the mud surface.

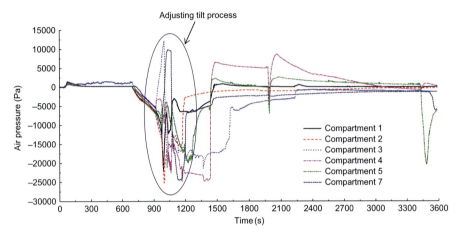

Fig. 11.19 Air pressures in all compartments in Test 1.

11.4.1 Test 1

Fig. 11.19 shows the variations in the air pressure in each compartment during the sinking process under self-weight and the sinking process under the negative pressure of the CBF. In the initial stage of sinking under negative pressure, the negative pressure of each compartment of the bucket foundation increased linearly to approximately −0.01 MPa. As the bucket foundation continued to sink, the mean extreme value of the negative pressure of each compartment was approximately −0.02 MPa. Even though the overall vertical sinking of the bucket foundation was achieved by controlling the valve of each compartment, the negative-pressure value of each compartment could not be accurately controlled under the actual conditions. The valves of some compartments were not opened sufficiently rapidly after they were closed, resulting in a positive pressure in these compartments for a short period of time. The sinking speed of the CBF and the minimum negative pressure necessary for sinking could not be effectively controlled. The problems that arose during this sinking experiment were systematically solved in the subsequent CBF sinking experiments. An improved scheme was also developed for use as a reference for sinking wind turbine foundations in practical applications.

Fig. 11.20 illustrates the variations in the pore pressure in compartment 2 during the penetration process. A top cover pore pressure sensor was used to verify the pressure sensor measurements. The values measured by the two types of sensors were essentially the same. The negative-pressure values measured in compartment 2 were used to plot the variation in the sidewall pore water pressure, showing that the pore water seepage from the negative pressure in the bucket produced an excess pore water pressure on the internal and the external walls of the bucket. The excess pore water pressure at the top of the internal wall was close to the negative pressure and gradually decreased. The excess pore water pressure at the bottom of the external wall was close to the negative pressure inside the bucket and gradually rose to zero at the top of the external wall.

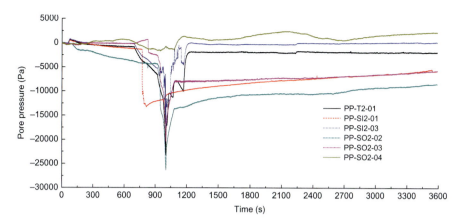

Fig. 11.20 Pore pressures of compartment 2 in Test 1.

Fig. 11.21 shows the variations in the earth pressure during the sinking process of the CBF. The earth pressure of the internal (see Fig. 11.21A) and the external walls (see Fig. 11.21B) of the bucket and the head cover (see Fig. 11.21C) was significantly affected by the negative pressure. When the negative pressure was being produced, the earth pressure of the head cover that was measured by the sensor was essentially the same as the negative-pressure value produced in each compartment. The earth pressure at the internal side of each compartment was negative, all of the earth pressures at the external side of compartment 2 were negative, the negative pressure at the bottom was relatively larger than the other pressures, and the negative-pressure value at the top was relatively smaller than the other pressures, which was in agreement with the distribution of the excess pore water pressure.

11.4.2 Test 2

The sinking process under a negative pressure lasted for a relatively long time in this experiment. In the initial leveling stage, the bucket foundation sank under self-weight for the first 30 cm. A negative pressure was then produced in all of the compartments, and the bucket foundation sank to a further depth of 10 cm. After the foundation was leveled, a power outage occurred. The sinking under negative pressure continued the next day. Real-time monitoring was not conducted for the first 40 cm of sinking because of instabilities in the voltage.

Fig. 11.22 shows the pressure variations in all of the compartments during the last 50 cm of sinking the bucket foundation. A very short time was needed for the bucket foundation to sink 10–15 cm each time. The extreme negative pressure in each compartment was in the −0.01 MPa range during the sinking of the CBF and each compartment exhibited a different extreme negative pressure.

Fig. 11.23 illustrates the variations in the pore pressure in compartment 2 in Test 2. Fig. 11.24 shows the variations in the earth pressure during the sinking process of the

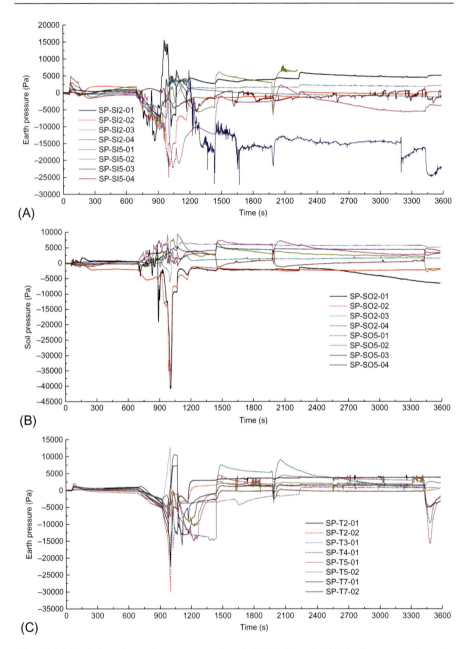

Fig. 11.21 Variations in earth pressure results of CBF in Test 1. (A) Earth pressure along the internal skirt of compartment 2 and 5 in Test 1; (B) Earth pressure along the external skirt of compartment 2 and 5 in Test 1; (C) Earth pressure on the top lid in Test 1;

(Continued)

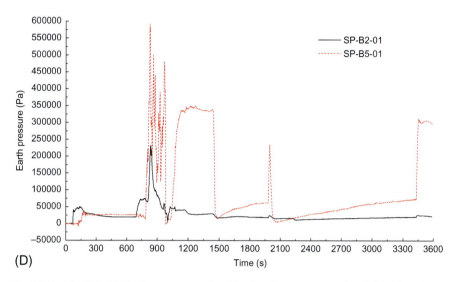

Fig. 11.21, Cont'd (D) Earth pressure at the skirt tip of compartment 2 and 5 in Test 1.

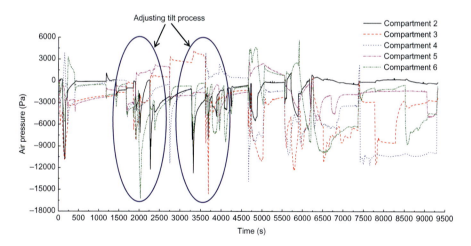

Fig. 11.22 Air pressures in the compartments in Test 2.

CBF. The earth pressure of the internal (see Fig. 11.24A) and the external walls (see Fig. 11.24B) of the bucket and the head lid (see Fig. 11.24C) was significantly affected by the negative pressure, while the Fig. 11.24D shows the variations of the earth pressure at the bottom of the wall of the bucket. The adjusting tilt processes were indicated by the change of pore pressures and earth pressures along the skirt wall. The

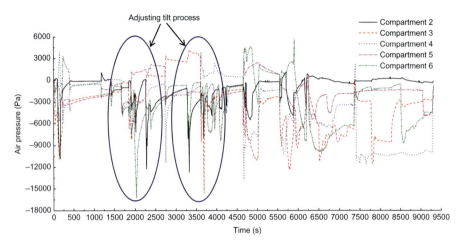

Fig. 11.23 Pore pressures of compartment 2 in Test 2.

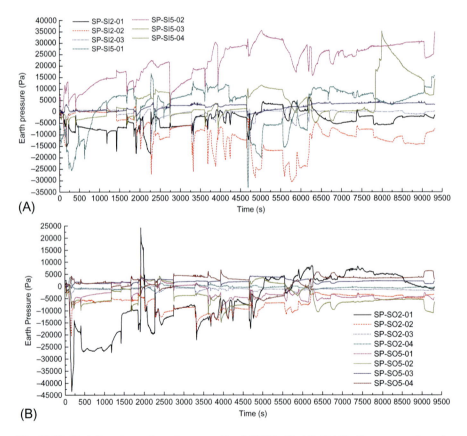

Fig. 11.24 Variations in earth pressure results of CBF in Test 2. (A) Earth pressure along the internal skirt of compartment 2 and 5 in Test 2; (B) Earth pressure along the external skirt of compartment 2 and 5 in Test 2;

(Continued)

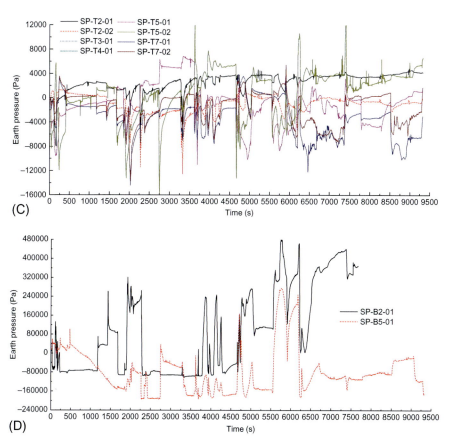

Fig. 11.24, Cont'd (C) Earth pressure on the top lid in Test 2; (D) Earth pressure at the skirt tip of compartment 2 and 5 in Test 2.

foundation was leveled by producing a negative pressure in the top compartments, pumping air into the bottom compartments or using both of the aforementioned methods. After the foundation had stabilized for 20–30 min, the inclination was measured again. If the required inclination (i.e., that the maximum height difference between the two ends should be within 1 cm, a levelness of 0.29%) was not obtained, then the foundation was leveled further. If the requirement was met, then a negative pressure was produced to allow the foundation to sink 10–15 cm. This process of sinking-adjusting-stabilizing-sinking was repeated until the head cover contacted the mud surface. A negative pressure was continuously produced in the top compartments to level the foundation. The production of negative pressure continued for several hours to strengthen the soil mass inside the bucket. The maximum height difference between the two ends was maintained to within 1 cm for the final leveling of the bucket foundation.

11.4.3 Comparisons

For Test 1, during the sinking process under negative pressure, the pore water seepage resulted in the vanishing of the resistance on the sidewall. To balance the gravity of the bucket foundation with the force generated at a certain sinking speed, the end resistance at the bottom of the bucket displayed an increasing trend during the sinking process under negative pressure. After the sinking process under self-weight had stabilized, the earth pressures at the bottom were 0.019 MPa and 0.026 MPa. During the sinking process under self-weight, the bucket foundation tilted to a certain extent, resulting in an uneven distribution of the earth pressure at the bottom. As the sinking depth increased, the maximum earth pressure at the bottom of compartment 2 reached 0.22 MPa, whereas the mean maximum value of compartment 5 was approximately 0.35 MPa.

From the results of Test 1, it was found out that there was a main tilt adjusting process started at about 1000 s in the Test 1. The compartment 3 was lowest among the compartments. Firstly, the positive pressure in compartment 3 was increased to reduce the levelness while increasing the negative pressure of compartment 1, 5, and 6. Then, it was a backup plan operated by increasing the positive pressure in the middle larger room 7 to lift up the whole foundation. Consequently, the penetration process could be restarted by applying different negative pressure for all compartments until the CBF was installed into the designed depth.

In Test 2, there were two main tilt adjusting process. From 500 s to 2000 s, the negative pressures in different compartments were almost the same until the obvious tilt adjusting process appeared around the 2000 s. The pressures in compartment 1 and 6 were negatively increased to reduce the levelness of the CBF, while those in compartment 3 and 4 were positively increased. The next tilt adjusting process happened around the 3500 s by increasing negative pressures of compartment 2 and 3 with adding the positive pressure of compartment 5 and 6. Meanwhile, the corresponding values of pore and earth pressures illustrated the occurrence of tilt and adjusting processes, which show that the compartment 1 and 6 were higher than compartment 3 and 4 around 2000 s and the compartment 2 and 3 were higher than compartment 1 and 6 around 3500 s. The tilt of the CBF to a certain extent would result in an uneven distribution of the earth pressure at the bottom. When the levelness of the CBF was limited, the pressure values obtained from the skirt tip of compartment 2 and 5 were almost the same, for example, from 2500 s to 3500 s.

11.5 Conclusions

The composite bucket foundation (CBF) is a new type of foundation for offshore wind turbines, which can be adapted to the loading characteristics and development needs of offshore wind farms due its special structural form. There are seven rooms divided inside the CBF by steel bulkheads, which are arranged in a honeycomb structure. The six peripheral rooms with the skirt have the same proportions, while the middle orthohexagonal one is a little larger. With the seven-room structure, the CBF has reasonable motion characteristics and towing reliability during the wet tow construction process. Through extensive research into the force transfer characteristics and stress mechanisms of composite structure systems, a CBF structure system with definite force transfer and stress systems has been developed. The large bending moment and horizontal force of the wind turbine tower are transferred to and dispersed into the sea floor soil through a prestressed curved concrete transition section, the head cover of the bucket foundation, the sidewalls of the bucket foundations, and the internal steel compartment plates. The deformation mechanism, the soil-structure interaction, and the ultimate bearing capacity of the CBF are determined depending on the tests. Based on the position of the rotation center, soil pressure and ultimate bearing capacity for the CBF are presented. Moreover, by means of numerical simulation, the envelope curve of ultimate bearing capacity of the CBF is described in H-M load space, clarifying the load-bearing characteristics of the CBF under combined loads.

The penetration depth and levelness determine the required bearing capacity and stability of the CBF for the expensive wind turbines. Specially, the tilt installation tolerance is very limited for offshore wind foundations, which has critical impacts and an insignificant reduction in turbine working behaviors. The installation model tests on a CBF with seven compartments were performed in order to investigate the feasibility of a foundation tilt adjusting technique by applying suction/positive pressure and intermittent pumping among the rooms. The tests represent the tilt adjusting techniques of the CBF. Two sinking strategies were taken in the tests that are the continuous sinking by applying different suction in different compartments without any interruption and the sinking process of sinking-adjusting-stabilizing-sinking by applying negative/positive pressure among the divided rooms of CBF. The results show that increasing the negative pressure in lower side compartment and the positive pressure in higher side compartment can effectively reduce the levelness of the CBF. Meanwhile, lifting the foundation by pumping the positive pressure in the middle larger compartment and applying the different negative pressure in all compartments again are another alternative method when the adjusting tilt process meets difficulties. The on-site tests data are effective for further study on the design and construction of CBF with seven compartments.

References

American Society of Civil Engineers, 2006. Minimum Design Loads for Buildings and Other Structures. ASCE/SEI 7-05.

Ding, H.Y., Lian, J.J., Li, A.D., Zhang, P.Y., 2013. One-step-installation of offshore wind turbine on large-scale bucket-top-bearing bucket foundation. Trans. Tianjin Univ. 19 (3), 188–194.

Ding, H.Y., Liu, Y.G., Zhang, P.Y., Xiong, K.P., 2015a. Influential factors of bucket foundation for offshore wind turbine. Trans. Tianjin Univ. 21 (3), 264–268.

Ding, H.Y., Liu, Y.G., Zhang, P.Y., Le, C.H., 2015b. Model tests on the bearing capacity of wide-shallow composite bucket foundations for offshore wind turbines in clay. Ocean Eng. 103, 114–122.

[Online] high socketed pile cap foundation, 2015. http://fjsdy.com/showgcdt.aspx?cid=262136&id=1114.
[Online] high-pile cap foundation, 2010. http://www.eiicn.com/news/focus/2010/0401/8098.html.
Lian, J.J., Sun, L.Q., Zhang, J.F., Wang, H.J., 2011. Bearing capacity and technical advantages of hybrid bucket foundation of offshore wind turbines. Trans. Tianjin Univ. 17 (2), 132–137.
Lian, J.J., Ding, H.Y., Zhang, P.Y., Yu, R., 2012a. Design of large-scale prestressing bucket foundation for offshore wind turbines. Trans. Tianjin Univ. 18 (2), 79–184.
Lian, J.J., Sun, L.Q., Zhang, J.F., Wang, H.J., 2012b. Bearing capacity and technical advantages of composite bucket foundation of offshore wind turbines. Trans. Tianjin Univ. 17 (2), 132–137.
Lian, J.J., Chen, F., Wang, H.J., 2014. Laboratory tests on soil–skirt interaction and penetration resistance of suction caissons during installation in sand. Ocean Eng. 84, 1–13.
Liu, R., Chen, G.S., Lian, J.J., Ding, H.Y., 2015a. Vertical bearing behaviour of the composite bucket shallow foundation of offshore wind turbines. J. Renew. Sustain. Energy. 7(1). 013123.
Liu, R., Zhou, L., Lian, J.J., Ding, H.Y., 2015b. Behavior of monopile foundations for offshore wind farms in sand. J. Waterw. Port Coast. Ocean. 142(1), 04015010.
[Online] Longyuan Zhenhua No. 2, 2012. http://www.chinaequip.gov.cn/2012-09/02/c_131822296_2.htm.
[Online] multi-piles foundation, 2013. http://news.cableabc.com/enterprise/20131028017773.html.
Zhang, P.Y., Ding, H.Y., Le, C.H., 2013a. Model tests on tilt adjustment techniques for a mooring dolphin platform with three suction caisson foundations in clay. Ocean Eng. 73, 96–105.
Zhang, P.Y., Ding, H.Y., Le, C.H., 2013b. Installation and removal records of field trials for two mooring dolphin platforms with three suction caissons. J. Waterw. Port Coast. Ocean 139 (6), 502–517.
Zhang, P.Y., Ding, H.Y., Le, C.H., 2013c. Motion analysis on integrated transportation technique for offshore wind turbines. J. Renew. Sustain. Energy. 5(5), 053117.
Zhang, P.Y., Ding, H.Y., Le, C.H., 2013d. Hydrodynamic motion of a large prestressed concrete bucket foundation for offshore wind turbines. J. Renew. Sustain. Energy. 5(6), 063126.
Zhang, P.Y., Ding, H.Y., Le, C.H., 2014a. Seismic response of large-scale prestressed concrete bucket foundation for offshore wind turbines. J. Renew. Sustain. Energy. 6(1), 013127.
Zhang, P.Y., Ding, H.Y., Le, C.H., 2014b. Seismic response of large-scale prestressed concrete bucket foundation for offshore wind turbines. J. Renew. Sustain. Energy. 6(1), 013127.
Zhang, P.Y., Han, Y.Q., Ding, H.Y., Zhang, S.Y., 2015. Field experiments on wet tows of an integrated transportation and installation vessel with two bucket foundations for offshore wind turbines. Ocean Eng. 108, 769–777.

Marine renewable energy

12

Ramona B. Barber, Michael R. Motley
Department of Civil and Environmental Engineering, University of Washington, Seattle, WA, United States

Chapter Outline

12.1 Introduction 345
12.2 Bend-twist deformation coupling 347
12.3 General turbine design parameters 350
    12.3.1 Power control system 350
    12.3.2 Site-specific design 351
    12.3.3 Turbulence 353
    12.3.4 Cavitation, vibration, noise 353
12.4 Composite-specific design considerations 354
    12.4.1 Rate dependence 355
    12.4.2 Scaling concerns 356
12.5 Potential performance benefits of composites 358
    12.5.1 Lifetime performance 358
    12.5.2 Effect on power and thrust 359
12.6 Conclusions 359
References 360
Further reading 362

12.1 Introduction

In recent years, global energy concerns have become a driving force in technological advancements, creating a need for clean and renewable energy resources and an increased interest in wind and tidal turbines. While research and development has been prevalent in wind energy for many years, exploration of ocean energy resources is relatively new. Because of its potential for clean, reliable, and predictable energy extraction, tidal energy has become an increasingly attractive option, especially as more nations commit to sustainable goals. Recent studies have shown that the hydrokinetic resources of marine currents alone have the potential to supply a significant fraction of future electricity needs (Ben Elghali et al., 2007; Blunden and Bahaj, 2007). Tidal currents are among the most consistent and reliable of the potential energy sources contained in the ocean. While currents in the open ocean typically move at only a few centimeters per second, when constrained by topography to a narrow channel or strait, the same current gains peak velocities of 2–3 m/s or more

(Fraenkel, 2002). Better still from the standpoint of power generation, the variability of tidal energy is predictable on average, not stochastic like that of wind, wave, or solar power. The magnitude and direction of tidal current velocities are regular and predictable to a high degree of accuracy. Thus, the kinetic energy of the tides is both extremely potent and able to deliver power predictably to a time table, which eases the integration of tidal energy into existing electricity networks (Strategic Initiative for Ocean Energy, 2013a). Yet another advantage is that tidal turbines are similar to wind turbines in operation and design, both extracting energy from the surrounding flow. This provides the nascent technologies of the tidal energy field with decades of applicable research and experimental data to inform new studies.

Though there are many benefits to marine current energy extraction, there are significant engineering challenges. Developments in marine hydrokinetic (MHK) turbine design can be strongly informed by advancements in the closely related fields of wind turbines and marine propellers; however, the technologies are not directly transferable. While tidal turbines are subject to many of the same loading and operating conditions as marine propellers, the variability in blade geometry and system constraints creates differences in optimal designs and requires turbine-specific design and analysis programs. Similarly, marine turbine design can borrow heavily from the wind turbine industry, but tidal turbines cannot be analyzed in the same manner as wind turbines due to the added mass and other dynamic effects of the much denser fluid (Strategic Initiative for Ocean Energy, 2013b). As a result, blade design is a critical factor in the implementation of MHK turbines, as they must withstand the large, dynamic fluid forces inherent to the maritime environment. A tidal turbine has the potential to capture more energy per year than a wind turbine of the same size due to the higher density of the surrounding flow, but this greater yield comes with an increase in drag and hydrodynamic loading (Young et al., 2010). Maintenance needs are considerably harder to address for underwater turbines purposefully placed in locations of extreme current, and the slender blades are potentially vulnerable to damage by marine debris (Bahaj and Myers, 2003). An additional concern specific to marine turbines is fluid cavitation, which can cause performance decay, corrosion, vibration, and fatigue (Kumar and Saini, 2010). All of these obstacles underline the need for a turbine uniquely suited to its harsh environment and able to operate for long periods of time without maintenance.

To that end, most MHK turbine blades are constructed from fiber-reinforced polymer (FRP) composites. Composite materials provide excellent strength-to-weight and stiffness-to-weight ratios, improved fatigue resistance and damping properties, and can be easier to manufacture in complex shapes compared to traditional metallic alloys. Further, the anisotropic nature of these materials can be exploited by hydroelastically tailoring the design to improve performance over the expected operational life, notably in spatially varying or off-design flow conditions (e.g., Barber, 2014). This can create additional complexity in the turbine blade design problem, as changing the material layup of the composite laminate can change the overall hydrodynamic response of the system; however, through proper design, the intrinsic bend-twist deformation coupling behavior of anisotropic composites can be utilized to develop a passive pitch adaptation, in which elastic deformations are tailored to dynamically vary with the loading condition. MHK turbines often experience highly unsteady and nonuniform inflow profiles due to boundary layer effects, free surface waves, currents, turbulence, and interaction between blades and nearby structures; the effective fluid inflow angle thus varies constantly as the blade rotates in a spatially and temporally varying flow. It can, therefore, be valuable to take advantage of the intrinsic bend-twist coupling behavior of anisotropic composites to passively adjust the pitch distribution of the blades in order to maintain an effective angle of attack at each blade section. These fluid-structure interaction designs have the potential to improve system performance by increasing lifetime energy capture, reducing hydrodynamic instabilities, and improving efficiency, load shedding, cavitation behavior, fatigue life, and structural integrity.

While many marine turbines and turbine blades are constructed of composite materials, these materials are not typically used in wave energy applications at this time. Most wave energy converters operate by aiming for resonance with local wave conditions, and thus, are required by design to be highly massive. Because of this, the weight savings made possible by the use of composite materials do not provide the same benefit to wave energy devices as in other applications. Additionally, the technology of wave energy converters is younger than that of tidal turbines, and in many cases, these devices are still in the iterative design stage. The high cost of composite materials compared to traditional metallic alloys is often prohibitive for this stage of development. However, the use of composite materials has much to offer to wave energy converters in terms of strength, corrosion resistance, and fatigue life, and as the technology converges and matures, these devices will likely move toward increased application of marine composites.

12.2 Bend-twist deformation coupling

The behavior of a composite structure is directly related to the materials used and the manner in which the laminate is manufactured. A composite laminate is constructed of a number of plies that work together as a structural system, where a typical ply can range in thickness from 0.125 to 0.175 mm (Gay and Hoa, 2007). The number of plies is dependent upon the structural demand. For structures that require large thicknesses, soft cores such as balsa or synthetic foams can be sandwiched between laminates. Plies of carbon fiber-reinforced polymers are typically constructed as unidirectional composites to reduce cost (Greene, 1999); however, it is possible to design a laminate with directional material properties by varying the angle of the fibers of each lamina. The ply stacking sequence and orientation of the fibers through the thickness are what defines the material response. In general, the longitudinal stiffness (E_x) of a laminate along the axis of the fiber is much higher than its transverse stiffness (E_y) perpendicular to the fiber. Fig. 12.1 shows the resulting laminate longitudinal ($E_{x'}$), transverse ($E_{y'}$), and shear stiffnesses ($G_{xy'}$) as the angle of the material fibers (θ_n) relative to the longitudinal axis of the laminate is varied from 0 to 90°. For both the high and low fiber orientations, there is a distinct strong and weak bending axis; however, as the fiber angle approaches 45°, the bending stiffness becomes relatively low in each direction (as shown in Fig. 12.2).

Fig. 12.1 Transformed longitudinal, transverse, and shear stiffnesses as a function of unidirectional fiber angle for a composite member in flexure (Motley, 2011).

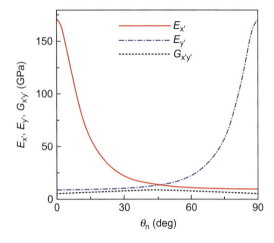

Fig. 12.2 Bend-twist deformation coupling for composite beams under the same load with unidirectional fiber angles of 10° (*left*), 20° (*center*), and 30° (*right*), with respect to the longitudinal axis (Barber, 2017).

Generally, for cantilevered structures similar to marine turbine blades, a symmetric ply layup is used to create a quasi-isotropic material in order to reduce design complexity. Exploiting the anisotropic nature of a composite layup is possible, however, by carefully designing the laminate layup sequence to provide a bend-twist coupling mechanism. The angle at which the fibers of each individual laminate layer are oriented defines both the direction in which the blade will pitch as well as the magnitude of blade deformation; thus, the laminate stacking sequence is a critical design parameter for passive pitch adaptation. In practice, composite blades are composed of tens to hundreds of very fine laminate layers. Each lamina can be defined by a primary fiber angle from the radial axis of the blade. However, modeling each individual layer is impractical due to the high degree of computational expense and provides a more detailed analysis than is appropriate for this stage of research. It has been shown that a multi-layered structure can be modeled using an equivalent unidirectional fiber angle, θ_{eq}, which can be found such that the effective stiffness and degree of bend-twist coupling are approximately equal to that of the full structure (Young, 2008; Liu and Young, 2009; Motley and Young, 2011a, b). The unidirectional fiber angle model needs to only have enough layers in the through thickness to provide the necessary mesh density for convergence, making computation vastly more efficient. The equivalent fiber angle of a multi-layered structure can be found by means of a composite plate analysis; in general, there are many different laminate layup sequences that will result in similar load-deformation behavior and bend-twist coupling and can be modeled with the same θ_{eq}. While this is appropriate to determine the load-deformation characteristics and trends in the stress profile for these linear-elastic structures, detailed analysis of various multilayer models is necessary after optimization to verify structural integrity and blade performance.

Recently, the design of bend-twist coupled composite blades has been proposed to increase the performance and structural integrity of MHK turbines. Tidal turbines stand to benefit from composite materials and bend-twist coupling mechanisms, often referred to as passive control mechanisms, in many of the same ways as marine propellers and wind turbines. The higher stiffness- and strength-to-weight ratios of composites and their resistance to corrosion are all valuable properties to the durability of MHK turbines (Mohan, 2008). Beyene and Peffley (2009) investigated geometrically adaptive composite blades and find flexible turbine blades to be more efficient than standard rigid blades. Nicholls-Lee et al. (2008, 2011, 2013) developed an iterative solving process for bend-twist coupled marine turbine blades and found that they produced a higher power output over a wider range of tip speed ratios when compared to a rigid metallic blade. However, their model produced unrealistically large blade deformations and did not consider site-specific parameters. Murray et al. (2013, 2015) also explored the creation of a low-order methodology for the design of adaptive composite blades.

Previous work by Barber (2014) presented an examination of the capabilities of passive pitch control of MHK turbine blades under both instantaneous and long-term variable amplitude loading with consideration for practical design and operational constraints. Potential performance improvements demonstrated in this work included increased power generation, reduced hydrodynamic instabilities, and improvements in efficiency, load shedding, and structural performance. It was shown that the orientation of the composite fibers in an adaptive blade could be tailored to create turbine blades that adapt in various ways to the local fluid flow.

12.3 General turbine design parameters

12.3.1 Power control system

To fully consider the potential benefits of adaptive composite blades, it is necessary to understand the basics of marine turbine control. Turbine controls generally fall into two categories: active pitch control, in which mechanical actuators in the hub of each turbine blade dynamically adjust the pitch of each blade to match the local flow condition, and active speed control, in which generator torque on the turbine shaft is regulated to manage rotational speed. These control strategies are used to prevent overloading the turbine generator in conditions of excess power capture and also to avoid extreme loads on the blades. Traditional MHK turbines can be either fixed speed-fixed pitch, variable speed-fixed pitch, fixed speed-variable pitch, or variable speed-variable pitch. These categories correspond to different methods and levels of control that affect the performance and power generation of the turbine. Because the output of each turbine is necessarily limited to the rated (maximum) power or torque of the attached generator, some sort of control scheme is generally desirable in order to reduce the load on blades and generator in extreme operating conditions. Controls are also used for various other reasons, including the prevention of cavitation along the blades and, in some locations, minimizing interaction with sensitive marine mammal species.

Active control used to vary blade pitch is generally enforced by adjusting the angle of attack of each blade relative to the fluid inflow. This strategy can be used to decrease generated lift above rated power, therefore slowing the rotation of the turbine, often referred to as an aerodynamic or hydrodynamic braking system. In extreme circumstances, the turbine can be stopped by adjusting the blades such that the leading edge of each blade points directly to the oncoming flow and no lift is generated. On the other hand, controlling the speed of the turbine rotation uses opposite means to achieve the same effect. These variable speed turbines apply generator torque to the rotor shaft in order to regulate the tip speed ratio. Above rated power, variable speed control can effectively increase the angle of attack and cause the flow over the blades to stall. Fluid stall occurs when the angle of attack of the blade becomes too large and the flow over the blade detaches from the top of the hydrofoil and forms turbulent eddies and creates a reduction in generated lift which further reduces rotor speed. Stall-controlled turbines are difficult to stop completely without imposing extreme loads on the generator, and forcing fluid stall increases the hydrodynamic loading on the blades significantly (Whitby and Ugalde-Loo, 2014). However, variable speed turbines have the benefit that a single brake mechanism in the generator is, in most cases, easier to construct and maintain than an active pitch mechanism acting in the hub of each blade. Even simpler, fixed speed-fixed pitch turbines are the easiest of the design varieties to manufacture, requiring fewest active control mechanisms. Generally, a fixed speed-fixed pitch turbine will have the ability to decouple the rotating shaft from the generator driver, allowing the turbine to spin freely in overload conditions to avoid damage to the generator; however, this can result in extremely large blade loads at high velocities. On the opposite end of the spectrum, variable speed-variable pitch systems are the most complex to design, containing both mechanical and aero- or hydrodynamic braking systems, but have the ability to reduce blade loading significantly, optimize performance, or maintain rated power over a range of inflow velocities.

On average, variable speed-variable pitch turbines have the potential to generate the most power; however, the associated higher initial costs and higher maintenance needs required by the active control systems can result in a higher cost per unit of electricity generated compared to simpler system designs (Wood et al., 2010). Relative to wind turbines, active control mechanisms in MHK devices can be more costly and more difficult to control effectively. Most of the proportionally larger cost lies in the challenges of designing mechanisms for and conducting maintenance in the submarine environment. The difficulty in efficient control is due to lower maximum pitching rates and the fact that the active systems cannot react instantaneously to changes in the inflow (Winter, 2011).

In contrast to active control mechanisms, passively controlled rotor blades are able to rapidly adjust to changing flow conditions. This creates the possibility of specifically designing blade geometries that will alter system performance by way of load-dependent deformations. Passive adaptation does not require an active driver to change the blade pitch, but instead creates a nearly instantaneous structural response that is difficult to achieve with an active mechanism because of the high flow variation and excitation frequencies in water. One method of passive control exploits the coupled bend-twist characteristics of composite materials, as discussed above, to create an adaptive pitch mechanism in the turbine blades.

12.3.2 Site-specific design

Because of the importance of velocity profile characteristics in adaptive turbine design, it is essential to inform numerical simulations with realistic inflow data. The state of technology at present indicates that only areas with peak currents of 2 m/s or more are viable sites for tidal turbine implementation. While 2 m/s (3.9 kts) is a very strong current relative to most tidal locations, some sites experience currents that exceed 5 m/s. As thrust increases with the square of velocity and power with velocity cubed, the difference between the extreme loads and available power experienced at a 2 m/s site and a 5 m/s site suggests that optimal technology and design may be highly dependent on location (Bryden and Melville, 2004; Fraenkel, 2007). Researchers at the University of Washington have been collecting baseline data to inform the design, siting, and permitting of a potential pilot scale tidal energy array in Admiralty Inlet in Puget Sound, WA (Polagye and Thomson, 2013). As an example, a four-day sample from a 104-day stationary survey (December 2012–April 2013) using a bottom-mounted acoustic Doppler current profiler is shown in Fig. 12.3. While the velocity and direction of tidal flows are vastly more predictable than wind (Wood et al., 2010), local variations in the flow profile for any given tidal cycle are to be expected due to local bathymetry and boundary effects caused by the sea floor and water surface (Strategic Initiative for Ocean Energy, 2013a). An example of this phenomenon can easily be seen in the figure. The predictable bi-directional nature of the average inflow is evident, with an

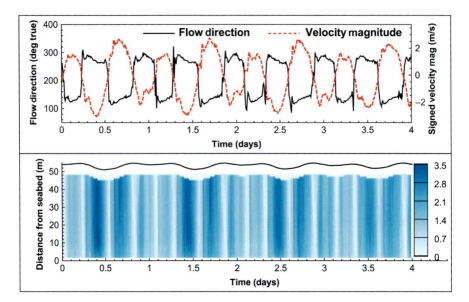

Fig. 12.3 *Top*: Measured flow direction and depth-averaged velocity data from Admiralty Inlet, WA; *Bottom*: Measured velocity magnitude over depth from Admiralty Inlet, WA (Barber, 2017).

approximate 180° change in the flow direction, and variations such as boundary effects from the seabed and water surface can be seen as well. This demonstrates the highly dynamic nature of the instantaneous velocity profile, a profile difficult and impractical to model. For numerical simulations, boundary layer effects are often approximated with a power-law estimate of the flow velocity across the water column. It is important to note that the power law is a general approximation for the mean profile and the model will likely not hold for any instantaneous profile depending on the location of the turbine in the water column and its position in the tidal cycle. Near the seabed especially, vertical shear can become a significant component of the total force on the blade. While the accuracy of the power-law estimate is a function of location in the water column, it is adequate to provide a rough profile estimate that includes a representation of the variable amplitude loading that an MHK turbine blade will experience, allowing for the consideration of potential blade fatigue or other structural instabilities. Marine turbines experience highly nonuniform inflow profiles because of boundary layers, free surface waves, currents, turbulence, and interaction between blades and nearby structures; thus, the effective fluid inflow angle on each blade varies constantly as it rotates in a spatially and temporally varying flow. Due to the constant and nearly instantaneous pitch adaptation inherent to passive control systems, they are ideal for these highly unsteady loading environments where fixed geometries can become suboptimal in off-design conditions, as the intrinsic bend-twist coupling behavior of anisotropic composites will passively adjust the pitch distribution of a blade in order to maintain an effective angle of attack at each blade section.

12.3.3 Turbulence

MHK turbines are typically deployed at energy-dense sites where fluid flow can be extremely complex. In addition to the changes in mean inflow velocity due to tidal flow, boundary layer flow, free surface waves, currents, local bathymetry, and surrounding structures or turbines in an array combine to create highly nonuniform inflows that vary spatially and temporally. These complex flow fields typically contain turbulence at a large array of length and time scales (Thomson et al., 2012). Previous studies have found that the blade loads and fatigue life of MHK turbines can depend strongly on a variety of turbulence parameters. Blackmore et al. (2015, 2016) found that both turbulence intensity and turbulent length scale can have large impacts on blade and rotor loads, while a parametric study conducted by Milne et al. (2010a,b) identified longitudinal turbulence intensity as the dominant factor for blade loading, though integral length scale had a larger influence on extreme and fatigue loads.

Overall, studies agree that both turbulence intensity and integral length scale have a significant impact on turbine loads and lifetime performance. Because of the strength and fatigue life benefits available with the use of marine composites, these materials have much to offer like hydrokinetic turbines operating in complex, turbulent conditions. Studies indicate that additional benefits can be found with the application of adaptive blade technology. The use of composite blades was shown to reduce both mean streamwise blade loading and load variation (Barber et al., 2017b). These results are significant to the lifetime robustness of an MHK, as reductions in these areas would lead to improved fatigue behavior and lower extreme loads, increasing structural reliability of the turbine. Additionally, wake characterization indicated that the use of adaptive blades could accelerate the recovery of the mean wake velocity while not allowing an increase in turbulence intensity downstream of the turbine (Barber et al., 2017b). This introduces the possibility of decreasing the spacing between turbines and could be highly valuable in formulating turbine arrays, providing another potential benefit of the use of adaptive composite blade technology for marine turbines.

12.3.4 Cavitation, vibration, noise

Another potential hazard to the performance and longevity of a hydrokinetic turbine is cavitation. Every marine structure that moves at a high velocity relative to the incident flow is at risk of cavitation. In accordance with Bernoulli's principle, an increase in the velocity of a fluid will cause a decrease in pressure. When the pressure in the flow falls below that of the vapor pressure of the fluid, vapor bubbles form and collapse at a high rate, causing highly localized areas of extreme pressure. Cavitation susceptibility is a major limiting criterion in modern marine turbine design. Research has shown that cavitation contributes to pitting, corrosion, vibration, and fatigue of marine structures (Bahaj and Myers, 2003; Wang et al., 2011). Additionally, the vibration caused by cavitation dramatically increases noise generation, a particular concern of turbine systems located in or near sensitive marine mammal habitats. Cavitation can be a function of high relative mean inflow velocities, but can also be initiated or exacerbated by turbulent fluctuations or wave orbital velocities.

The onset of cavitation also has a drastic effect on the power generation of the system. In general, the lift and drag coefficients of a hydrofoil do not vary as the ambient pressure decreases; the pressures on the upper and lower surfaces decrease in tandem. However, as cavitation develops along the top of the hydrofoil, the pressure on the top surface can no longer decrease because it is limited by the vapor pressure. The pressure on the bottom surface is not restricted, however, and continues to fall, progressively approaching the pressure on the top of the hydrofoil. As the pressure differential across the hydrofoil decreases, there is a corresponding decrease in the lift coefficient. Additionally, the flow separation induced by cavitation changes the shape of the pressure distribution along the hydrofoil and causes an increase in drag. Thus, cavitation causes both a decrease in lift and an increase in drag, limiting the power generation as well as exacerbating the loads on the system.

Though the use of composite materials does not necessarily provide reduced cavitation susceptibility, the inherent strength, resistance to corrosion, and superior fatigue performance gained with the use of marine composites are especially beneficial under cavitating conditions. Additionally, studies have indicated that the use of bend-twist coupled composite material could reduce cavitation on hydrofoils (Gowing et al., 1998). Preliminary results have shown that adaptive composite turbine blades can be used to delay cavitation and decrease cavitation volume compared to their nonadaptive counterparts in spatially varying flows (Barber and Motley, 2016).

Reducing cavitation susceptibility would allow marine turbines to operate more efficiently. The turbine performance and structural system would be more robust in conditions of increased turbulence or significant wave orbital velocities. Additionally, the maximum rotational speed of the turbine rotor is limited by the pressure reduction caused by the speed of the blade tips. In raising that limit, a turbine could successfully operate at higher rotational speeds. This would provide a larger window for the turbine to operate at maximum efficiency. Furthermore, a reduction in cavitation susceptibility would allow the installation of turbine arrays in shallower waters where the ambient pressure in the fluid is lower. This would drastically expand the number of possible sites for tidal power generation as well as moderate the cost of installation and implementation, as power transfer cables could be shorter and the systems would be more accessible for maintenance.

12.4 Composite-specific design considerations

Typical composites are manufactured from some combination of resins, reinforcements, and, in the case of sandwich laminates, soft cores. This combination of materials is common for marine turbine blades, which at full scale are generally constructed of an outer composite shell, interior spars, and a soft core such as structural foam. The most common reinforcement materials used in marine composites are glass and carbon fibers. Glass fibers are used for their cost-effectiveness and workability characteristics and are most commonly found in general marine applications. Carbon fibers and similar specialty reinforcements are used for highly optimized structures. Carbon fibers provide the highest strength- and stiffness-to-weight ratios and are highly water- and corrosion-resistant, resulting in excellent fatigue properties. Typical resins, often referred to as matrix materials, are polyesters, vinyl esters, and epoxies. For standard construction and for applications with lower demand, polyesters supply sufficient structural properties; however, vinyl esters and epoxies provide superior performance at higher costs. For highly engineered structures, glass and carbon fiber-reinforced composites are manufactured using vinyl ester or epoxy resins. Vinyl esters increase corrosion resistance and hydrolytic stability and provide superior impact resistance (Greene, 1999). Epoxies are generally found in highly optimized designs and provide superior fatigue and corrosion resistance. Because of handling concerns, epoxies can be difficult to use for large-scale marine structures such as renewable energy devices (Greene, 1999); however, these laminates, while costly, provide the best structural performance in terms of fatigue, corrosion, and impact resistance, factors which are critical for marine composite structures.

The strength and stiffness of the fibers in a composite material are, by design, much higher than that of the matrix material. Fiber strengths and stiffnesses can be on the order of 50–100 times the strengths and stiffnesses of the matrix material. The mass densities of the materials are much closer, even 1:1, and typically the fibers are slightly denser than the matrix. Because of these large discrepancies, matrix failure is typically seen before fiber failure (Greene, 1999). An additional design consideration for composite materials is that, compared to metallic alloys, composite materials tend to be more susceptible to geometric and material imperfections due to the complex manufacturing process (Potter et al., 2005, 2008). For a composite rotor, random variations due to fiber misalignments, voids, laminate properties, and boundary conditions can add another level of variation in the overall system response, especially when the performance depends on the deformations, which are governed by the load conditions, material response, and fluid-structure interactions.

12.4.1 Rate dependence

Traditionally, nonadaptive marine turbine blades have been designed to optimize performance for a specific or a few flow condition(s). In determining the performance of an MHK turbine, it is common to examine power (C_P) and thrust (C_T) coefficients. Often, deformations are generally assumed to be sufficiently small such that the geometry can be assumed to be fixed, or rigid, enabling the use of nondimensional parameters to characterize operating conditions. The nondimensional tip speed ratio ($\lambda = 2\pi nR/V$, where n is the rotational frequency of the turbine, R is the turbine radius, and V is the inflow velocity at the turbine) is generally used to categorize the operating condition for turbines and represents the ratio of the mean axial inflow velocity to the tangential velocity at the blade tip. Using the tip speed ratio λ allows for comparisons between numerical and experimental models over different ranges of inflow conditions or rotational speeds (i.e., direct comparisons between a rapidly rotating turbine in a high-speed flow and a slowly rotating turbine in a low-speed flow, operating at the same λ). This comparison does not fully capture Reynolds number effects; however, as most full-scale turbines operate in conditions such that the lift and drag coefficients are independent of Reynolds number, this is a reasonable assumption and provides a valuable simplification to the design process.

The performance of adaptive turbine blades, on the other hand, is governed by *both* the material and geometry design, and it is strongly influenced by the load-dependent deformations. As noted above, an adaptive blade is designed to change pitch as a function of the loading induced by the surrounding fluid flow, creating a constantly morphing turbine geometry. Thus, the performance of an adaptive system depends on the magnitudes, not just the ratio, of the inflow speed and rotational velocity, which when combined creates the total dimensional blade loads. As shown in Fig. 12.4, if the nondimensional tip speed ratio is changed by fixing only the velocity and varying the rotational frequency as a function of inflow speed, the predicted performance (power and thrust coefficients) will be different than if the rotational frequency is fixed and the velocity is varied across the same range of tip speed ratios (Motley and Barber, 2014). To that end, when designing an adaptive turbine blade and comparing it with its rigid or nonadaptive counterpart, the full range of expected, site-specific operating conditions must be taken into account for an accurate representation of the turbine's performance over its design life. In most cases, it is most appropriate to define an adaptive turbine performance as a function of inflow velocity as opposed to tip speed ratio, as the power control system will often determine the rotational speed as a function of velocity.

12.4.2 Scaling concerns

Because of the size and expense of a typical MHK turbine, much of the technology is developed through a series of scaled experiments. From a structural mechanics perspective, scaling of a composite material is nontrivial because of the inherent

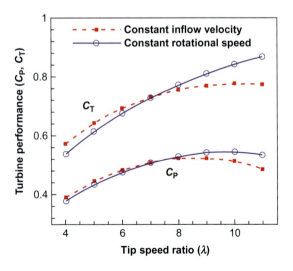

Fig. 12.4 Predicted turbine performance of an adaptive turbine considering changes in tip speed ratio as independent functions of inflow velocity or rotational speed (Barber, 2017).

anisotropic laminate constitutive properties—a scaled down composite is simply constructed from fewer laminae. For a nonadaptive blade, this is sufficient, as the structural engineering concepts associated with small deformations can be fully decoupled from the hydrodynamic scaling issues, which are relatively well-established for nonadaptive rotors. For adaptive blades where the load-dependent deformations are not only nonnegligible but also critical to the overall performance, these structural and hydrodynamic problems cannot be uncoupled and the scaling issues become more complex, as fluid-structure interaction similarity becomes a critical part of the physical test design. A detailed discussion of the scaling requirements for adaptive composite propellers can be found in the literature (Young, 2010; Motley and Young, 2012; Young et al., 2016), but the concepts are generally applicable to adaptive turbine blades and are summarized here and the text below references these works.

Typically, experimental studies of scale-model turbines are conducted in a flume, towing tank, or cavitation tunnel filled with water, where the scaling factors for gravity, fluid density, viscosity, and speed of sound are all approximately 1. Assuming full, three-dimensional geometric similarity, a characteristic length scale $\lambda_D = D_M/D_F$ can be defined, where D_M is the diameter of the model-scale turbine and D_F is the diameter of the full-scale turbine. For similar operating conditions, the scaling ratios of the tip speed ratio and cavitation number must also be 1, which can be achieved through control of the inflow velocity, turbine rotational frequency and, if necessary, pressure inside a cavitation tunnel. Similar fluid-structure interaction response is obtained with scaling ratios for solid density and Poisson's ratio of 1 and effective structural stiffness ratios are then scaled by:

$$\lambda_{E_i} = \lambda_{G_{ij}} = \lambda_n^2 \lambda_D^2$$

where E_i is the Young's modulus in the i-direction, G_{ij} is the shear modulus in the ij-direction, and n is the angular velocity of the rotor.

When maintaining the above similarity conditions, Froude number, Reynolds number, and Mach number lead to conflicting requirements and the physics of the experiment define the scaling approach. Froude number similarity is critical for flow conditions where the gravitational forces are important, such as for slowly rotating marine turbines near the free surface where hydrostatic effects become nonnegligible relative to hydrodynamic forces. For adaptive turbine blades, Froude scaling is the only approach that maintains the four principal forces: solid elastic restoring force, gravitational force, hydrodynamic inertial force, and turbine inertial force. Reynolds number represents the ratio of viscous and inertial forces in the fluid, and for many design conditions, viscous effects tend to be small relative to inertial and gravitational forces. Froude number similarity can be easier to achieve than Reynolds number similarity because Froude scaling allows for slower than full-scale velocity rather than faster than full-scale velocity, which is beyond the scope of many experimental facilities.

Mach number similarity is important for the issue of flow compressibility, which for most turbine tests will be negligible. Mach scaling does offer several advantages, however, specifically for adaptive composite blades. Mach scaling allows the same material to be used for both the model-scale and full-scale turbines, which can be helpful for examination of the stress distributions and failure susceptibility of the blades. Care must be taken, however, in interpreting results related to structural performance, as failure strength can be affected by manufacturing issues and material and geometric uncertainties. Nonetheless, Mach scaling allows for a more comprehensive representation of full-scale structural integrity than Froude or Reynolds scaling for composite blades, where it may not be feasible to find appropriate model-scale materials.

12.5 Potential performance benefits of composites

12.5.1 Lifetime performance

Composite materials are also popular because they can provide improvements in lifetime performance, specifically in terms of material robustness and fatigue-resistant properties. The more costly materials (e.g., carbon fibers, vinyl esters, and epoxies) provide the maximum resistance to material degradation. Sometimes these materials are referred to as corrosion-resistant; however, this is not an accurate representation. These materials are susceptible to thermal and ultraviolet (UV) corrosion. Vinyl esters and epoxies are highly sensitive to UV corrosion if exposed for any significant period of time. Gel coats or UV screens can be applied to reduce this tendency. Along with the UV corrosion goes the thermal corrosion, which these materials can experience even with the addition of a gel coat or UV screen. Resins that are overheated can soften and attempt to cure further, weakening the material. Because vinyl esters and epoxies cure at higher temperatures, UV and thermal corrosion are less prevalent in these materials. Polymeric matrix materials may have free volume between the molecules which allows diffusion and hence tends to permit absorption of water. In extremely cold environments, the water can freeze and cause expansion in the material, creating stress concentrations (Greene, 1999). Water absorption is a factor in the corrosion of the reinforcement, specifically with seawater. It has also been found that immersion in seawater can significantly reduce the impact resistance of glass-fiber-reinforced polymers (Strait et al., 1992).

Fatigue resistance is difficult to quantify for composite materials. It is generally agreed that composite materials outperform metals in fatigue tests; however, this fact rests heavily on the material quality. The inherent uncertainties and the complexity of the failure mechanisms of composite structures also contribute to the relatively slow advancement of knowledge about the fatigue behavior of composite materials. In addition to a detailed description of the load spectrum and history, the fatigue analysis requires a deep knowledge of the material responses (Marin et al., 2007). The anisotropic nature of the material itself can be difficult to measure, and inherent flaws such as fiber misalignments, microcracks, and voids can be expected. Active research is only recently begun to quantify the material response and fatigue behavior when operating in harsh seawater environments. Composite marine turbines tend to operate in harsh environments, where changes in wind, waves, or currents and proximity to the sea floor can have significant impact on the effective inflow. When these uncertainties are coupled with the inexact nature of fatigue life modeling, a deterministic fatigue life estimate is not a reasonable goal; some level of probabilistic analysis that considers the inherent randomness of the fatigue life is required. The fatigue behavior of composite materials can generally be defined by one of three models: fatigue life, residual strength and stiffness, and progressive damage (Degrieck and Van Paepegem, 2001). All three models require extensive experimental testing due to the complex failure mechanisms, anisotropic behavior of fiber-reinforced composites, and dependence on loading conditions. As a result, there are many different fatigue models available in the literature.

12.5.2 Effect on power and thrust

The use of adaptive composite blades can provide numerous performance benefits to a MHK turbine. Depending on the overall design goals, an adaptive composite turbine blade can be designed to increase power consumption, shed additional load, and improve lifetime performance; however, these are often competing goals and require some tradeoff. Blades designed to pitch toward feather decrease angle of attack, effectively reducing lift and drag and the corresponding power and thrust coefficients. The opposite is true for blades designed to pitch toward stall. Generally, it has been shown that the possible increase in power capture from pitch to stall blades is offset by the increasing rate of pitch change that can lead to static or dynamic divergence and system instability. Conversely, designing the blade to pitch to feather has been shown to significantly reduce blade loads while maintaining a relatively close power coefficient relative to a nonadaptive blade, which could lead to longer operational life to offset the costs of reduced energy capture.

12.6 Conclusions

As discussed here, composite FRPs are becoming common materials for the construction of marine renewable energy devices, especially for tidal and current energy conversion. From a research and development perspective, this technology is rather immature, but the increased strength- and stiffness-to-weight ratios, superior corrosion resistance, and ability for tailored design have made composites a popular alternative. Additionally, adaptive tailoring mechanisms are emerging as a potential control mechanism for which composite materials can be applied to improve system performance. It should be noted, however, that there are still significant research questions to fully implement these technologies. Because these devices are relatively new, knowledge related to long-term operation in seawater is not available and has relied on accelerated aging techniques and laboratory fatigue tests. Tests to failure, including fatigue and ultimate strength tests, have also reinforced the relative uncertainty associated with composite material properties, which thus requires large safety factors and overly designed structural systems. As geometries become more complex and technologies allow for the design and manufacture of larger systems, construction, implementation, and maintenance concerns will also need to be addressed. Momentum for using composites for marine renewable energy applications, however, is growing, and it can be expected that many of these questions will be answered in the coming years.

References

Bahaj, A.S., Myers, L.E., 2003. Fundamentals applicable to the utilisation of marine current turbines for energy production. Renew. Energy 28, 2205–2211.

Barber, R.B., 2014. Passive Pitch Control in Marine Hydrokinetic Turbine Blades. Master's thesis, University of Washington.

Barber, R.B., 2017. Adaptive Pitch Composite Blades for Axial-Flow Marine Hydrokinetic Turbines. Ph.D. dissertation, University of Washington.

Barber, R.B., Motley, M.R., 2016. Cavitating response of passively controlled tidal turbines. J. Fluids Struct. 66, 462–475.

Barber, R.B., Hill, C.S., Babuska, P.F., Aliseda, A., Motley, M.R., 2017b. Performance of an adaptive pitch marine hydrokinetic turbine in turbulent inflow. 12th European Wave and Tidal Energy Conference, Cork, Ireland.

Ben Elghali, S.E., Benbouzid, M.E.H., Charpentier, J.F. (Eds.), 2007. Marine tidal currentelectric power generation technology: state of the art and current status.In Electric Machines & Drives Conference. pp. 1407–1412.

Beyene, A., Peffley, J., 2009. Constructual theory, adaptive motion, and their theoretical application to low-speed turbine design. J. Energy Eng. 112–118.

Blackmore, T., Gaurier, B., Myers, L., Germain, G., Bahaj, A.S., 2015. The effect of freestream turbulence on tidal turbines.In European Wave and Tidal Energy Conference, Nantes, France.

Blackmore, T., Myers, L., Bahaj, A.S., 2016. Effects of turbulence on tidal turbines: Implications to performance, blade loads, and condition monitoring. Int. J. Marine Energy 14, 1–26.

Blunden, L.S., Bahaj, A.S., 2007. Tidal energy resource assessment for tidal stream generators. J. Power Energy 221 (2), 137–146.

Bryden, I., Melville, G.T., 2004. Choosing and evaluating sites for tidal current development. J. Power Energy 218, 567–577.

Degrieck, J., Van Paepegem, W., 2001. Fatigue damage Modelling of fibre-reinforced composite materials: Review. ASME Appl. Mech. Rev. 54 (4), 279–300.

Fraenkel, P.L., 2002. Power from marine currents. J. Power Energy 216, 1–14.

Fraenkel, P.L., 2007. Marine current turbines: Pioneering the development of marine kinetic energy converters. Proc. Inst. Mech. Eng. A: J. Power Energy 221, 159–169.

Gay, D., Hoa, S., 2007. Composite Materials: Design and Applications, seconnd ed CRC Press.

Gowing, S., Coffin, P., Dai, C., 1998. Hydrofoil cavitation improvements with elastically coupled composite materials. Proceeding of 25th American Towing Tank Conference, Iowa City, IA.

Greene, E., 1999. Marine Composites, second edition Eric Greene Associates, Inc.

Kumar, P., Saini, R.P., 2010. Study of cavitation in hydro turbines - a review. Renew. Sust. Energ. Rev. 14, 374–383.

Liu, Z., Young, Y.L., 2009. Utilization of bending-twisting coupling effects for performance enhancement of composite marine propellers. J. Fluids Struct. 25, 1102–1116.

Marin, J., Barroso, A., Paris, F., Canas, J., 2007. Study of damage and repair of blades of a 300 kW wind turbine. Energy 33 (7), 1068–1083.

Milne, I.A., Sharma, R.N., Flay, R.G.J., Bickerton, S., 2010a. The role of onset turbulence on tidal turbine blade loads.17th Australasian Fluid Mechanics Conference.

Milne, I.A., Sharman, R.N., Flay, R.G.J., Bickerton, S., 2010b. A preliminary analysis of the effect of the onset ow structure on tidal turbine blade loads. OCEANS 010 IEEE-Sydney, 1–8.

Mohan, M., 2008. The advantages of composite material in marine renewable energy structures. RINA Marine Renewable Energy Conference.

Motley, M.R., 2011. Probabilistic Design and Analysis of Self-Adaptive Composite Marine Structures. Ph.D. Dissertation, Princeton University.

Motley, M.R., Barber, R.B., 2014. Passive control of marine hydrokinetic turbine blades. Compos. Struct. 110, 133–139.

Motley, M.R., Young, Y.L., 2011a. Influence of uncertainties on the response and reliability of self-adaptive composite rotors. Compos. Struct. 94 (1), 114–120.

Motley, M.R., Young, Y.L., 2011b. Performance-based design and analysis of flexible composite propulsors. J. Fluids Struct. 27 (8), 1310–1325.

Motley, M.R., Young, Y.L., 2012. Scaling of the transient hydroelastic response and failure mechanisms of self-adaptive composite marine propellers. Int. J. Rotating Mach.

Murray, R.E., Doman, D.A., Pegg, M.J., Nevalainen, T., Gracie, K., Johnstone, C.M., 2015. Design tool for passively adaptive tidal turbine blades.11th European Wave and Tidal Energy Conference, Nantes, France.

Murray, R.E., Gracie, K., Doman, D.A., Pegg, M.J., Johnstone, C.M., 2013. Design of a passively adaptive rotor blade for optimized performance of a horizontal-axis tidal turbine.10th European Wave and Tidal Energy Conference, Aalborg, Denmark.

Nicholls-Lee, R.F., Boyd, S.W., Turnock, S.R., 2011. A method for analyzing uid structure interactions on a horizontal axis tidal turbine. 9th European Wave and Tidal Energy Conference, United Kingdom.

Nicholls-Lee, R.F., Turnock, S.R., Boyd, S.W., 2008. Performance prediction of a free stream tidal turbine with composite bend-twist coupled blades. 2nd International Conference on Ocean Energy.

Nicholls-Lee, R.F., Turnock, S.R., Boyd, S.W., 2013. Application of bend-twist coupled blades for horizontal axis tidal turbines. Renew. Energy 50, 541–550.

Polagye, B., Thomson, J., 2013. Tidal energy resource characterization: Methodology and field study in admiralty inlet, Puget sound, US. Proc. Inst. Mech. Eng. A: J. Power Energy 227 (3), 352–367.

Potter, K., Campbell, M., Langer, C., Wisnom, M., 2005. The generation of geometrical deformations due to tool/part interaction in the manufacture of composite components. Compos. Part A 36, 301–308.

Potter, K., Khan, B., Wisnom, M., Bell, T., Stevens, J., 2008. Variability, fibre waviness and misalignment in the determination of the properties of composite materials and structures. Compos. Part A 39, 1343–1354.

Strait, L., Karasek, M., Amateau, M., 1992. Effects of seawater immersion on the impact resistance of glass fiber reinforced epoxy composites. J. Compos. Mater. 26, 2118–2133.

Strategic Initiative for Ocean Energy, 2013a. Ocean Energy: State of the Art.

Strategic Initiative for Ocean Energy, 2013b. Ocean Energy Technology: Gaps and Barriers.

Thomson, J., Polagye, B., Durgesh, V., Richmond, M.C., 2012. Measurements of turbulence at two tidal energy sites in Puget sound, WA. IEEE J. Ocean. Eng. 37 (3), 363–374.

Wang, J., Piechna, J., Müller, N., 2011. A novel design and preliminary investigation of composite material marine current turbine. Arch. Mech. Eng. 58 (4), 355–366.

Whitby, B., Ugalde-Loo, C.E., 2014. Performance of pitch and stall regulated tidal stream turbines. IEEE Trans. Sustainable Energy 5 (1), 64–72.
Winter, A.L., 2011. Differences in fundamental design drivers for wind and tidal turbines. OCEANS, 2011 IEEE.
Wood, R.J.K., Bahaj, A.S., Turnock, S.R., Wang, L., Evans, M., 2010. Tribological design constraints of marine renewable energy systems. Phil. Trans. R. Soc. A 368, 4807–4827.
Young, Y.L., 2008. Fluid-structure interaction analysis of flexible composite marine propellers. J. Fluids Struct. 24, 799–818.
Young, Y.L., 2010. Dynamic hydroelastic scaling of self-adaptive composite marine rotors. Journal of Composite Struct. 97–106.
Young, Y.L., Motley, M.R., Yeung, R.W., 2010. Three-dimensional numerical modeling of the transient fluid-structure interaction response of tidal turbines. J. Offshore Mech. Arct. Eng. 132. 011101.
Young, Y.L., Motley, M.R., Barber, R.B., Chae, E.J., Garg, N., 2016. Adaptive composite marine propulsors and turbines: Progress and challenges. Appl. Mech. Rev. 68(6). 060803.

Further reading

Barber, R.B., Hill, C.S., Babuska, P.F., Wiebe, R., Aliseda, A., Motley, M.R., 2017a. Adaptive composites for load control in marine turbine blades. In 36th International Conference on Ocean, Offshore and Arctic Engineering, Trondheim, Norway.

Propulsion and propellers

Y. Hong*, X.D. He*, G.F. Qiao[†], R.G. Wang*
*Science and Technology on Advanced Composites in Special Environment Laboratory, Harbin Institute of Technology, Harbin, China, [†]School of Civil Engineering, Harbin Institute of Technology, Harbin, China

Chapter Outline

13.1 Introduction 363
13.2 The characteristics of composite propeller 364
    13.2.1 The structural characteristic 364
    13.2.2 The working characteristic 365
    13.2.3 The difference between the composite and metal propellers 368
13.3 The calculation and evaluation method of composite propeller 369
    13.3.1 The finite-element method and the PSF-2 program 369
    13.3.2 The coupled 3-D FEM/VLM (vortex-lattice methods) method 371
    13.3.3 The coupled FEM/BEM (boundary element method) method 373
    13.3.4 The coupled FEM/CFD (computational fluid dynamics) method 377
13.4 Performances of composite propeller 381
    13.4.1 The open water performances 381
    13.4.2 The hydro-elastic performance in nonuniform wake 383
    13.4.3 Structural dynamic characteristics 384
    13.4.4 Performance optimization 385
13.5 Conclusions and future trends 386
Reference 387

13.1 Introduction

As one of the most important components of the ship propulsion system, the design and development of the propeller has always been of concern. The traditional marine propeller is made of nickel-aluminum-bronze (NAB) alloy because this material has a very high yield strength. However, in recent years, people have found a lot of problems with the NAB propeller in the process of using. Firstly, the NAB propeller has poor anticavitation ability and is prone to fatigue-induced cracking. Secondly, the NAB propeller is heavy and has relatively poor acoustic damping properties that can lead to noise problem from vibration. At the same time, it is very expensive for NAB to machine into a complex shape, such as propeller. The above problems make the designers assess the feasibility of other materials for fabricating propeller.

In recent years, the rapid development of fiber-reinforced materials brought a new opportunity to the propeller designers. Compared with the metal materials, fiber-reinforced composites have higher specific strength and specific stiffness, lighter weight, good damping performance, and design-ability. These excellent material properties make the fiber-reinforced composites effectively applied to produce the marine propeller because the excellent design-ability and the proper damping property of composites provide the foundation for the marine propeller to reduce weight, noise, and increase the fuel efficiency.

In this chapter, the composite marine propeller, as a new type of propulsion, will be analyzed and discussed in depth. We will separately elaborate the following aspects: the structural and work characteristics, the design and analysis methods, and the performance evaluation of composite propeller.

13.2 The characteristics of composite propeller

13.2.1 The structural characteristic

The propeller is mounted on the tail of the hull and consists of a few blades and a hub. The hub is a truncated cone that is connected to the tail shaft to transfer the load. The propeller blade is a typical curved surface part with a helical surface shape, which is determined by the value of a series of cylindrical sections. All blades are fixed and distributed around the hub, forming overlapping relationships in space. In general, a fairing (that is, a hub cap) is mounted on the rear end of the hub, which forms a streamlined body with the hub to reduce water resistance. The side of the propeller blade seen from the rear of the ship is called the pressure surface, and the other side is called the suction surface. When the propeller is rotating, the edge of the blade first entered the water is called the leading edge, and the other side is called the trailing edge, as shown in Fig. 13.1.

For the traditional metal propeller, the structural characteristics of the propeller are only affected by the material properties, the geometrical characteristics of the blade, and the connection type of the blade and the hub. The geometrical characteristics are determined by the geometrical parameters of the blade, such as diameter, disk ratio,

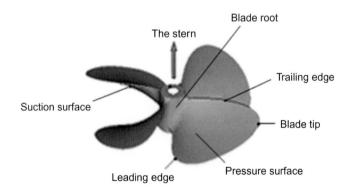

Fig. 13.1 The component names of propeller.

pitch, rake, skew, chord length, etc., and the connection type depending on the design is the fixed pitch propeller or the variable pitch propeller. However, the structural characteristics of the composite propeller are very different from those of the metal propeller.

For the composite marine propeller, the rich material and flexible molding provide a good foundation for the diversification of composite propeller structures. The diversified design is mainly reflected on the following aspects: (a) From the point of view of selecting materials, the blades can be prepared from the carbon fiber-reinforced composites with good comprehensive performance, or the blades also can be prepared by selecting many different types of the fiber-reinforced composite according to the design requirements. (b) From the point of view of designing the internal structure, the propeller blade can be designed as the laminated composite structure (Fig. 13.2) (Lin et al., 2009), or it can also be designed as the sandwich composite structure; (c) from the point of view of determining the connection type, the blade and the hub can be integral molding or can be separately molded and then assembled (Fig. 13.3). When the blade and the hub are molded separately, the composite blade and the metal hub may be used, or the composite material blade and the composite material hub may be used.

13.2.2 The working characteristic

In general, we can regard the propeller blade as a part of the wing at a certain angle of attack and the inflow velocity. As shown in Fig. 13.4, it has a small angle of attack between the wing and the flow direction. At that time, the fluid velocity above the airfoil is greater than that below the airfoil. The cross section a–b is greater than the cross section a′–b′, the pressure P_2 above the wing is less than the static pressure P_1, the cross section b–c is less than the cross section b″–c′, the pressure P_3 below the wing is greater than P_1, so it forms a pressure difference between the upper and lower

Fig. 13.2 The laminate molding of composite propeller blade.

Fig. 13.3 The assembled composite propeller.

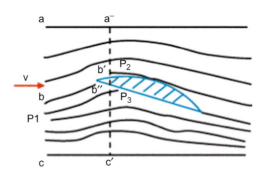

Fig. 13.4 The working principle of the propeller.

surfaces of the wing. This pressure difference, together with the frictional force generated when the fluid flows through the wing, synthesizes a total fluid force R. Divide the total fluid force R into two components: a component X parallel to the direction of fluid flow, to prevent the forward movement of the wing, which is called resistance; another component Y perpendicular to the direction of fluid flow, called lift, which forms the propeller thrust.

After understanding the working principle of the propeller, we further discuss the force and torque acting on the propeller blades. It should be noted that the propeller generates the thrust by fluctuating water, due to the relationship between the force and the reaction force, the flow is also affected by the propeller to obtain an additional velocity (commonly referred to as the axial-induced velocity) that is opposite to the thrust direction. At the same time, the circumferential induction velocity is also obtained in the rotation direction of the propeller. Therefore, the flow around the propeller should be discussed, in addition to considering the forward and rotation speeds of the propeller, but also the axial induction speed and circumferential induction speed should also be considered.

Fig. 13.5 shows the fluid velocity polygon of the blade element at any radius. It can be seen that the relative velocity acting on the blade element is the result of the synthesis of undisturbed flow velocity V_p, circumferential velocity u, axial-induced velocity c_{u1}, and circumferential-induced velocity c_{u2}. The synthetic speed acts on the blade element with a certain angle of attack. According to the working principle of the propeller, we can see that the fluid force will be generated on the blade element, that is, the lifting force dY and the resistance dX.

Fig. 13.6 shows the force polygon of the blade element at any radius. According to the force polygon, the thrust $dP (dP = dY_T - dX_T = dY \cos \beta - dX \sin \beta)$ and rotation resistance $dQ (dQ = dY_Q + dX_Q = dY \sin \beta + dX \cos \beta)$ of the blade element can be obtained. Integrating from the hub to the blade tip in the radial direction, and multiplied by the number of blades, the total thrust and torque of propeller can be obtained.

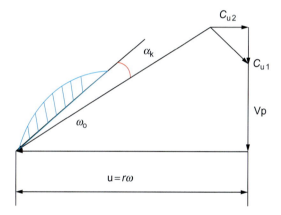

Fig. 13.5 The speed polygon of blade element.

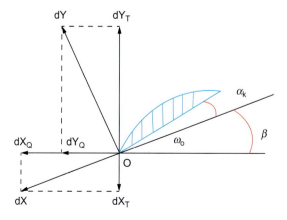

Fig. 13.6 The force polygon of the blade element.

As far as the composite propeller be concerned, the principle of hydrodynamic action is the same as that of metal propellers. In the same inflow case, the initial thrusts and torques generated by the composite and metal propellers are the same exactly, but because the composite material has good elasticity. The deformation of the blade element at any radius is larger, as shown in Fig. 13.7, resulting in the effective angle of attack decreasing and the hydrodynamic characteristics of composite propeller also changing.

13.2.3 The difference between the composite and metal propellers

Based on the above analysis, the great difference in the structure and working characteristics is exhibited between the composite propeller and the metal propeller, which leads to the need to consider more details in the design and analysis methods of composite propeller and provides a good opportunity to improve the performance of the propeller.

In general, metal propellers are designed primarily around their functional requirements (hydrodynamic, cavitation, noise). By changing the profile of the propeller blades and controlling the inflow around the propeller, the hydrodynamic design requirements under the design conditions are achieved, while meeting the design objectives of high efficiency and low noise. However, metal propellers are often limited by the nature of the metal material. In order to meet the requirements of propulsion efficiency at the design condition, metal propellers often have to reduce the propulsion efficiency at other operating speed. At the same time, because of the larger density and the poor damping performance of metal, the performance of the metal propeller to further improve is also restricted to a large extent.

In the case of composite propellers, the nature of the composites determines the research difference between the composite propeller and the metal propeller, and it also shows the potential advantages of the composite propeller:

(1) Compared with the traditional metal propeller, the deformation of the composite propeller blades is much larger, so the inflow angle on the blade element at any time will change greatly, which causes the thrust of the blade to change greatly. This situation shows that

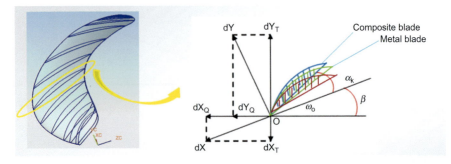

Fig. 13.7 Comparing the force polygon of the composite and metal blade element at any radius.

the composite propeller has a great hydro-elasticity, so the design of composite propeller must take into account the effect of fluid-solid coupling.
(2) The hydro-elasticity can be effectively utilized in the design of composite propeller, to achieve the maximum operating efficiency of the propeller in multiple working conditions.
(3) Considering the structural complexity of the composite propeller, the strength of the composite propeller must be analyzed and evaluated.

13.3 The calculation and evaluation method of composite propeller

In recent years, a considerable amount of research has been directed toward the calculation and evaluation of composite marine propellers for their excellent advantage. In this part, we will discuss the classification of the numerical calculation of composite propellers according to the literature; each classification is discussed as the representative of the method used by authors in the literature. These discussions only refer to the author's own usage and preferences in the simulation method and do not represent any general consensus about the superiority of the method. The literature regarding these themes is extensive, and that which is referred to here is not exhaustive.

13.3.1 The finite-element method and the PSF-2 program

Lin developed an earlier approach based on the finite-element method (FEM) and the PSF-2 program for analyzing a thick-shell composite blade (Lin, 1991a,b). In order to obtain the blade loading under the hypothetical operating conditions, the pressure coefficient difference was predicted by using the PSF-2 program, which is a software program developed by Kerwin and Lee (1978) for analyzing the flow fields of a propeller in steady and noncavitating flows. As shown in Fig. 13.8 (Lin, 1991a), the calculated cross sectional pressure load differences are applied to the pressure side of the blade.

In author's research, the propeller was considered as the partial-composite structure. The cross sectional material arrangement of this composite propeller is shown in the Fig. 13.9 (Lin, 1991a). The thick-laminate technique was adopted to account for 3-D features of composite blade. Considering that composite materials are layers of orthotropic materials, the skin composite layer and shear-web were regarded as composed by a numbers of sublaminate. Effective moduli were derived by properly integrating the constituent lamina properties through the thickness of the sublaminate.

Two different kinds of elements were used for constructing the finite-element model of the composite blade, as shown in Fig. 13.10 (Lin, 1991a), where 15-node curved solid elements were used in the high curvature leading edge, the remaining was 20-node elements. The nodal points on the intersection of blade and hub are set as the fixed boundary condition. The MPC (multipoint constraints) was employed on the dividing juncture between isotropic and composite zones to ensure structural connectivity.

The stress and elastic deflections were calculated using the commercial finite-element software ABAQUS and compared with that of a geometrically identical isotropic metal propeller blades.

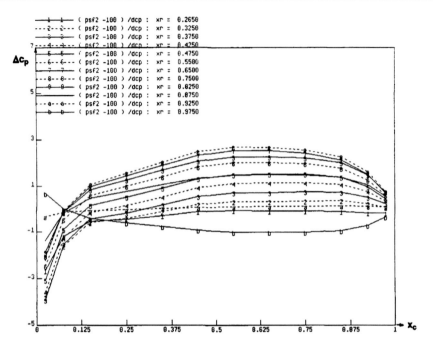

Fig. 13.8 Pressure difference ($\triangle C_p$) versus chord-wise station (x_c) for different stations (PSF-2 code).

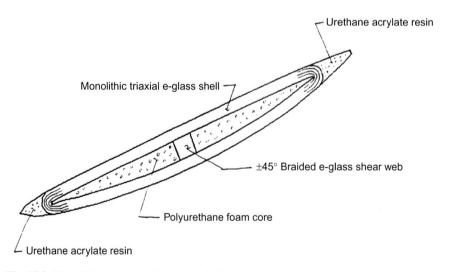

Fig. 13.9 Material arrangement in composite blade.

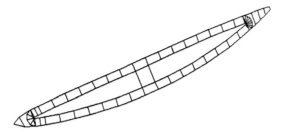

Fig. 13.10 Cross sections of finite-element model of composite blade.

13.3.2 The coupled 3-D FEM/VLM (vortex-lattice methods) method

In 1996, a coupled 3-D FEM/VLM method was presented by Lin and his partners (Lin and Lin, 1996, 1997) for the hydro-elastic analysis of composite marine propellers in steady, noncavitating flows. The PSF-2, a noncavitating vortex-lattice method (VLM), was used to assess the strength of the vortex of a marine propeller; the FEM was used to calculate the structural displacements of a marine propeller.

13.3.2.1 The PSF-2 program based on lifting surface theory (Lin and Lin, 2005)

According to the lifting surface theory, by assuming that the blade is sufficiently thin, the blades can be expressed by a distribution of vortices and sources on the mean camber surface of each blade, as well as via trailing vortices into the wake.

Considering the solid boundary conditions at the control point of each lattice element, the strength of a vortex or a source can be determined by

$$n_i \cdot V_{T,i} = 0 \quad i = 1, 2, L, \ (N \times M) \tag{13.1}$$

where, n_i, $V_{T,i}$ represents the normal vector and the total velocity of control point i, respectively. $V_{T,i}$ includes the inflow velocity V_∞, the rotational velocity V_Ω, and the velocity V_q induced by the vortex or source.

Because the strength of the sources representing blade thickness can be predetermined through a strip-wise application of thin-wing theory at each radius, Eq. (13.1) is converted into a system of simultaneous equations that specify the horseshoe vortex strength of the blade, and it can be expressed as:

$$\sum_{j=1}^{(N \times M)} A_{ij}\Gamma_i + n_i \cdot \left(V_\infty + V_\Omega V_q^{Source}\right)_i = 0 \quad i = 1, 2, L, \ (N \times M) \tag{13.2}$$

where A_{ij} denotes a coefficient of influence, which is the normal velocity at control point i resulting from a unit strength of horseshoe vortex j. The total number of unknowns (Γ) equal the number of control points ($N \times M$) on the blade. This equation can be expressed in matrix form as follows:

$$G\Gamma = -n \cdot \left(V_\infty + V_\Omega + V_q^{Source}\right) \tag{13.3}$$

According to the Bernoulli equation, the pressure at a point in the steady flow condition can be expressed by

$$P - P_\infty = -\rho\left[V_\infty \cdot V_\Omega + \frac{1}{2}V_\Omega^2 + \left(V_\infty + V_\Omega + \frac{1}{2}V_q\right) \cdot V_q\right] \tag{13.4}$$

In which $V_q \cdot V_q$ is a quadratic term obtained from $V_{q,\,i-1} \cdot V_{q,\,i}$ for linearization, and the subscripts i and i-1 denote the iteration number. So, the pressure difference between the pressure side and the suction side of the blade can be calculated by

$$\Delta P = \rho\left[\left(V_\infty + V_\Omega + \frac{1}{2}V_{q,i-1}\right) \cdot V_{q,i}\bigg|_{back} \left(V_\infty + V_\Omega + \frac{1}{2}V_{q,i-1}\right) \cdot V_{q,i}\bigg|_{face}\right] \tag{13.5}$$

where $V_{q,\,i}$ represents the velocity induced by the vortex and source singularity at iteration i, and the term V_q can be expressed as:

$$V_q = \sum_m \sum_n u_{mn}^{vortex}\Gamma_{mn}^{vortex} + \sum_m \sum_n u_{mn}^{vortex}\Gamma_{mn}^{vortex} \tag{13.6}$$

where, u_{mn} denotes the velocity induced by a vortex or source of unit strength calculated based on the Biot-Savart law. Considering the strength of the source, Γ_{mn}^{source} has already been determined by the strip-wise application of the thin-wing theory at each radius; there is only the need to determine the vortex strengths Γ_{mn}^{vortex}. Eq. (13.5) then becomes

$$\Delta P = \rho \sum_m \sum_n \left[\frac{\left(V_\infty + V_\Omega + \frac{1}{2}V_{q,i-1}\right) \cdot u_{mn}^{Vortex}\bigg|_{back}}{\left(V_\infty + V_\Omega + \frac{1}{2}V_{q,i-1}\right) \cdot u_{mn}^{Vortex}\bigg|_{face}}\right] \cdot \Gamma_{mn}^{Vortex} \tag{13.7}$$

$$= \{AC\}^T\{\Gamma_{mn}^{Vortex}\}$$

13.3.2.2 Coupled 3-D FEM/VLM method

For constructing the FEM of composite blade, a 3-D degenerated shell element with five degrees of freedom was applied. The equilibrium equation for a geometrically nonlinear analysis can be expressed as:

$$\{[K_l] + [K_0] + [K_g]\}\{u\} = \{F_{ext}\} + \{F_I\} \tag{13.8}$$

where $[K_l]$, $[K_0]$, $[K_g]$ denote the linear stiffness matrix, the initial displacement matrix, and the geometric stiffness matrix, respectively; $\{u\}$ represents the node displacement vector; F_I is the fluid force; $\{F_{ext}\}$ is the external forces excluding the fluid force.

The fluid forces on the blades can be expressed as

$$F_I = \int_A N^T n \cdot P dA = \int_A N^T n \cdot N dA \cdot \{\Delta P\} = [B]\{\Gamma^{vortex}\} \qquad (13.9)$$

where $\{\Delta P\}$ is the matrix of pressure difference calculated using Eq. (13.7); N is the interpolation matrix; dA denotes the interaction area between the fluid and the structure; n represents the outward normal vector; and $[B]$ is the coupled matrix.

Substituting Eq. (13.9) into Eq. (13.8), leads to

$$\{[K_l]+[K_0]+[K_g]\}\{u\} - [B]\{\Gamma^{vortex}\} = \{F_{ext}\} \qquad (13.10)$$

The fluid and solid governing equations were fully coupled in Eq. (13.10), which can be solved using a Newton-Raphson procedure. In the structural solution part, there are three convergence criteria, namely the blade displacement, the external forces, and the external work. In the fluid solution part, there are two convergence criteria, namely the wake velocities just behind the trailing edge and the singularity strength.

Subsequently, the researchers completed a series of extended research work based on the coupled FEM/VLM method. Combined with the genetic algorithm, Lin discussed the effects of stacking sequence on the hydro-elastic behavior of composite propeller blades (Lee and Lin, 2004). Combined with a 3-D stress evaluation procedure and the Hashin material failure criterion (Hashin, 1980), Lin assessed the strength evaluation of composite marine propellers, the 3-D equilibrium equations were employed to resolve the transverse shear stresses, the least squares method was applied to smooth the stress distributions (Lin and Lin, 2005). Most recently, Lin and Tsai (2008) and Lin and Lee (2009) discussed the free vibration characteristics of composite propeller and the test results of optimization design, respectively.

13.3.3 The coupled FEM/BEM (boundary element method) method

In contrast to the work of Lin and Lee et al., a coupled FEM/BEM (boundary element method) was applied by Young (2008) to investigate the fluid-structure interaction of flexible composite propeller in subcavitating and cavitating flows. The 3D BEM is a low order potential-based fluid solver developed by the hydrodynamic group at The University of Texas at Austin. Different from VLM, BEM did not discretize the mean camber surface, but discretize the actual blade surface, thus taking into account the effect of thickness loading coupling, which makes BEM better capture the flow details. Moreover, the BEM is able to account for the effects of flow unsteadiness, complex cavitation or ventilation patterns, nonzero blade trailing edge thickness, and varying blade submergence, all of which provide the basis for fully exploring the hydro-elastic performance of composite propellers.

13.3.3.1 The governing equation of BEM (Young, 2007)

Consider a composite propeller subjected to a general effective inflow wake $\vec{q}_E$ and rotated at a constant angular velocity ω. To avoid the need for a moving mesh, a rotating blade-fixed coordinate system (x, y, z), as shown in Fig. 13.11, is used to solve the fluid problem. The inflow velocity can be determined as follows

$$q_{in}(x, t) = q_E(x_s, r_s, \theta_s) - \Omega \times x \tag{13.11}$$

where, the inflow velocity $\vec{q}_{in}$ is defined with respect to the blade-fixed coordinate system, the effective inflow velocity $\vec{q}_E$ is defined with respect to the ship-fixed coordinate system (x_s, y_s, z_s) with the origin located at the center of the hub. r_s, θ_s, θ, and $\vec{\Omega}$ can be further expressed as

$$r_2 = \sqrt{y^2_s + z^2_s} = \sqrt{y^2 + z^2}$$
$$\theta_s = \arctan(z_s/y_s) = \theta - \omega t, \quad \theta = \arctan(z/y) \tag{13.12}$$
$$\Omega^{ur} = [-\omega, 0, 0]^T$$

Assuming the flow is incompressible and inviscid, the total velocity can be expressed as

$$\vec{v} = \vec{q}_{in} + \vec{\nabla} \Phi \tag{13.13}$$

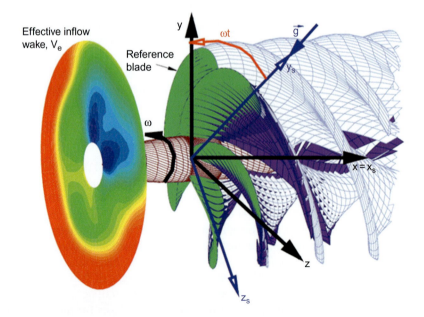

Fig. 13.11 The rotating blade-fixed coordinate system (Young, 2008).

where Φ is the perturbation velocity potential corresponding to the propeller-induced flow field.

The momentum equation can be expressed as follows:

$$\partial \vec{v}/\partial t + \left(\vec{v}\cdot\vec{\nabla}\right)\vec{v} = -\vec{\nabla}(P_t/\rho) + \vec{g} - \vec{\Omega}\times\left(\vec{\Omega}\times\vec{x}\right) - 2\vec{\Omega}\times\vec{v} \qquad (13.14)$$

where the terms $\vec{\Omega}\times\left(\vec{\Omega}\times\vec{x}\right)$ and $\vec{\Omega}\times\vec{v}$ denote the centrifugal acceleration and Coriolis acceleration, respectively. P_t, ρ, and g represent the total pressure, the fluid density, and amplitude of the gravitational acceleration, respectively.

The continuity equation can be expressed as follows:

$$\left(\vec{\nabla}\cdot\vec{v}\right) = 0 \qquad (13.15)$$

Substituting Eqs. (13.12), (13.13), (13.15) into Eq. (13.14), the momentum equation can be rewritten as follows:

$$\partial\vec{q}_E/\partial t - \vec{v}\times\left(\vec{\nabla}\times\vec{q}_E\right) = -\vec{\nabla}H \qquad (13.16)$$

where

$$H = \frac{\partial\Phi}{\partial t} + \frac{P_t}{\rho} + \frac{1}{2}|\vec{v}| - \vec{g}\cdot\vec{x} - \frac{1}{2}\left(\vec{\Phi}\times\vec{x}\right)^2 \qquad (13.17)$$

Assuming $\partial\vec{q}_E/\partial t = 0$ (steady effective wake), integration of Eq. (13.16) between two points on the same streamline yields the following equation for the absolute total pressure P_t:

$$P_t = P_o + \rho\left[\frac{1}{2}|\vec{q}_{in}|^2 - \frac{\partial\Phi}{\partial t} - \frac{1}{2}|\vec{v}|^2\right] \qquad (13.18)$$

where $P_o = P_{atm} + \rho g d_s$ is the absolute hydrostatic pressure at $\vec{x}$, P_{atm} is the atmospheric pressure, d_s is the submerged depth of point $\vec{x}$ from the free surface.

13.3.3.2 Coupled 3-D FEM/BEM method

Considering the effect of fluid-structure interaction, the propeller-induced perturbation potential Φ can be defined as

$$\Phi = \phi + \varphi \qquad (13.19)$$

where ϕ, φ represent the perturbation potential due to large rigid-blade rotation and small elastic blade deformation, respectively.

Correspondingly, in conjunction with Eqs. (13.13), (13.18), the total pressure P_t can be decomposed into two parts

$$P_t = P_r + P_v$$
$$P_r = P_o + \rho \left[\frac{1}{2} |\vec{q}_{in}|^2 - \frac{\partial \phi}{\partial t} - \frac{1}{2} |\vec{v}_{tr}|^2 \right] \quad (13.20)$$
$$P_v = \rho \left[-\frac{\partial \varphi}{\partial t} - \vec{q}_{in} \cdot \nabla \varphi \right]$$

where P_r, P_v denote the hydrodynamic pressure due to rigid-blade rotation and elastic blade deformation, respectively. $\vec{v}_{tr}$ represents the total velocity without considering the effect of elastic blade deformation and can be expressed as

$$\vec{v}_{tr} = \vec{p}_{in} + \nabla \phi \quad (13.21)$$

In this research, a low-order BEM is used to solve the boundary-value problems for ϕ and φ. The detailed formula derivation and explanation can be found in literature (Young, 2008).

Hence, the inertial and dissipative hydrodynamic force due to elastic blade motion can be expressed as

$$\{F_{hv}\} = -[M_H]\{\ddot{u}\} - [C_H]\{\dot{u}\} \quad (13.22)$$

where $[M_H], [C_H]$ represent the hydrodynamic added mass matrix and the hydrodynamic added damping matrix, respectively.

$$[M_H] = \rho \int [N]^T [H][T] ds$$
$$[C_H] = \rho \int [N]^T [V_{in} \cdot \nabla H][T] ds \quad (13.23)$$

The coupling equation of motion for the composite blade can be expressed as:

$$([M] + [M_H])\{\ddot{u}\} + ([C] + [C_H])\{\dot{u}\} + [K]\{u\} = \{F_{ce}\} + \{F_{co}\} + \{F_h\} \quad (13.24)$$

where $\{\ddot{u}\}, \{\dot{u}\}$, and $\{u\}$ denote the local nodal acceleration, velocity, and displacement vectors, respectively. $[M], [C], [K]$ are the global mass, damping, and stiffness matrices, respectively. $\{F_{ce}\}, \{F_{co}\}, \{F_h\}$ are the centrifugal force, Coriolis force, and hydrodynamic force, respectively. There terms can be further expressed as

$$[M] = \int \rho_s [N]^T [N] dV$$
$$[C] = \int c[N]^T [N] dV \quad (13.25)$$
$$[K] = \int [B]^T [D][B] dV$$

$$\{F_{ce}\} = \int \rho_s [N]^T \{f_{ce}\} dV, \quad f_{ce} = -\Omega \times (\Omega \times (x+u))$$

$$\{F_{co}\} = \int \rho_s [N]^T \{f_{co}\} dV, \quad f_{co} = -2\Omega \times \dot{u} \tag{13.26}$$

$$\{F_h\} = \int [N]^T \{P_r\} dS$$

where, the variables ρ_s and c represent the mass density and frequency independent mass damping of the blade, respectively. The matrices [N], [B], and [D] are the displacement interpolation matrix, the strain-displacement matrix, and the material constitutive matrix, respectively.

According to the above analysis, a stacking 3-D quadrilateral solid element in the thickness direction is used to construct the finite-element model of composite blade. The BEM is used to compute the hydrodynamic pressures due to rigid-blade rotation, as well as the hydrodynamic added mass and damping matrices. The commercial FEM software ABAQUS/Standard is used to calculate the blade deformation by solving Eq. (13.24). In order to consider the interaction between the large elastic displacement and the pressure distribution, iterations are implemented between BEM and FEM solvers until the solution (thrust coefficient, propeller efficiency, and maximum displacement) converges.

13.3.4 The coupled FEM/CFD (computational fluid dynamics) method

A 3-D FEM/CFD coupled method is proposed by He and Hong (He et al., 2012) to analyze the hydro-elastic performances of composite marine propeller in steady and unsteady flows. Unlike the VLM and BEM based on fluid inviscid assumptions, the viscous of the flow is considered in the CFD method. In the coupled FEM/CFD method, the CFD method based on Reynolds Averaged Navier-Stokes (RANS) equation is used for the fluid analysis in place of conventional methods based on potential theory. At the same time, a nonconforming layered solid element (Taylor and Beresford, 1976) is used for constructing the finite-element model of composite propeller blades for the structure analysis.

13.3.4.1 RANS method

The mass and momentum conservation equations are written as follows.
Continuity equation:

$$\frac{\partial \rho}{\partial t} + \frac{\partial}{\partial x_i}(\rho u_i) = 0 \tag{13.27}$$

Navier-Stokes equation:

$$\frac{\partial}{\partial t}(\rho u_i) + \frac{\partial}{\partial x_j}(\rho u_i u_j) = -\frac{\partial p}{\partial x_i} + \frac{\partial}{\partial x_j}\left(\mu \frac{\partial u_i}{\partial x_j} - \rho \overrightarrow{u_i' u_j'}\right) + S_i \quad (13.28)$$

where $\vec{u}$ is the velocity vector in the Cartesian coordinate system, p is the static pressure, μ is the molecular viscosity, $-\overline{\rho u_i' u_j'}$ is the Reynolds stresses.

$-\overline{\rho u_i' u_j'}$ has to be modeled to close the momentum equation and it can be expressed as follows

$$-\overline{\rho u_i' u_j'} = \mu_t \left(\frac{\partial u_i}{\partial x_j} + \frac{\partial u_j}{\partial x_i}\right) - \frac{2}{3}\left(\rho k + \mu_t \frac{\partial u_i}{\partial x_i}\right) \delta_{ij} \quad (13.29)$$

where δ_{ij} is the Kronecker delta, μ_t and k represent the turbulent viscosity and the turbulent kinetic energy, respectively.

RANS equations are closed with Boussinesq's eddy-viscosity assumption and two transport equations for solving the turbulence velocity and the turbulence time scale.

In author's research, a $k - \omega$-based SST model is chosen as the turbulence model (Menter, 1994). The SST model is an improvement of the baseline $k - \omega$ model, and it takes into account the transport of the turbulent shear stress by limiting the eddy viscosity ν_t with the following equation

$$\nu_t = \frac{a_1 k}{\max(a_1 w, SF_2)} \quad (13.30)$$

where $\nu_t = \mu_t/\rho$ and $\alpha_1 = 5/9$, S represents an invariant measure of the strain rate, ω represents the turbulent frequency. F_2 is a blending function, which restricts the limiter to the wall layer and can be calculated by

$$F_2 = \tanh\left(\arg_2^2\right) \quad (13.31)$$

where

$$\arg_2 = \max\left(\frac{2\sqrt{k}}{\beta' \omega y}, \frac{500\nu}{y^2 \omega}\right) \quad (13.32)$$

where $\beta' = 0.09$, y is the distance to the nearest wall, ν is the kinematic viscosity.

Contrasting with other two-equation turbulence models, the SST $k - \omega$ model has good performance for wall-bounded boundary layer flows, and it can highly accurately predict the onset and the amount of flow separation under adverse pressure gradients by including the transport effects into the eddy viscosity (Qing, 2002).

13.3.4.2 Steady fluid-structural coupling

The research of the open water performance of composite propeller is regarded as a steady fluid-solid coupling problem under uniform flow. In that case, the composite blades are subjected to centrifugal force and hydrodynamic pressure load, resulting in bending and torsional deformation. Considering that the deformation behavior of the blade is a geometric nonlinear behavior, the equilibrium equation of the finite-element model can be expressed as:

$$\{[K_l]+[K_0]+[K_g]-[K_r]-[K_r]\}\{u\}=\{F_r\}+\{F_h\} \qquad (13.33)$$

where $[K_l]$ is the linear stiffness matrix; $[K_0]$ is the initial displacement matrix; $[K_g]$ is the geometric stiffness matrix; $[K_r]$ is the rotation stiffness matrix; $\{u\}$ is the node displacement vector in the local coordinates; $\{F_h\}$ is the fluid force; $\{F_r\}$ is the centrifugal load. These matrices are defined as follows:

$$\begin{aligned}
[K_l] &= \int B^T D B_L dV \\
[K_g] &= \int G^T \begin{bmatrix} \sigma_x & \tau_{xy} \\ \tau_{xy} & \sigma_y \end{bmatrix} G dV \\
[K_0] &= \int (B^T D B_L + B_L^T D B_L + B_L^T D B) dV \\
[K_r] &= \rho \int [N]^T [A][N] dV
\end{aligned} \qquad (13.34)$$

where B is the linear strain-displacement matrix; B_L is the nonlinear strain-displacement transformation matrix; D is the material matrix; G is the coordinate transformation matrix; $[N]$ is the displacement interpolation function matrix, which depends on the element types used in the structural calculation; $[A]$ is the angular velocity matrix; ρ is the density of composites.

The equation is constructed based on the full-Lagrangian integral and is solved using the Newton-Raphson method.

At the same time, the fluid force acting on the blade surface can be obtained by using the computational fluid dynamics analysis method based on the RANS equation:

$$F_h = \int_A [N]^T n P dA \qquad (13.35)$$

where dA is the interface between the fluid and the blade; n is external normal vector; P is the total pressure acting on the blade and can be divided into two parts:

$$P = P_r + P_v \qquad (13.36)$$

where P_r is the hydrodynamic pressure due to the rotation of the blade; P_v is the hydrodynamic pressure due to the elastic deformation of the blade.

Substituting Eqs. (13.35) and (13.36) into Eq. (13.33) yields the steady fluid-solid coupling equations of the composite propeller under open water conditions:

$$\{[K_l]+[K_0]+[K_g]-[K_r]\}\{u\}-\int_A [N]^T n\{P_v\} dA = \{F_r\} + \int_A [N]^T n\{P_r\} dA \qquad (13.37)$$

13.3.4.3 Transient fluid-structural coupling

At the same time, the hydro-elastic behaviors of the composite propeller during the underwater operation are studied and analyzed. The interaction between the elastic force, the inertial force, and the hydrodynamic force of the composite propeller is analyzed by using the hydro-elastic theory. In this process, the hydrodynamic force acts on the composite propeller; the magnitude of the force depends on the displacement, velocity, and acceleration of the propeller vibration. At the same time, the hydrodynamic force changes the displacement, velocity, and acceleration of the composite propeller. The physical properties of this interaction exhibit the coupling phenomenon in terms of inertia, damping, and elasticity between the water and the propeller. By the inertial coupling, the propeller produces the attached mass; by the damping coupling, the propeller produces the attached damping; by the elastic coupling, the propeller produces the attached stiffness. Their magnitude depends on the flow conditions and the boundary conditions of the water and the propeller; the solution is more complex.

To analyze the hydro-elastic behavior of the composite propeller, a reference blade is chosen to construct the finite-element model. The dynamic equation of the composite blade is expressed as

$$[M]\{\ddot{U}\} + [C]\{\dot{U}\} + [K]\{U\} = \{F_r\} + \{F_h\} \tag{13.38}$$

where $[M]$, $[C]$, and $[K]$ are the mass, damping and total stiffness matrices, respectively. $\{\ddot{U}\}, \{\dot{U}\}$, and $\{U\}$ are the acceleration, velocity, and displacement vectors, respectively. On the right side of Eq. (13.38), $\{F_r\}$ is the centrifugal force, $\{F_h\}$ is the generalized fluid force and it is the value that changes over time. According to its specific physical meaning, $\{F_h\}$ can be considered as three parts

$$F_h = F_h^{(1)} + F_h^{(2)} + F_h^{(3)} \tag{13.39}$$

where $F_h^{(1)}$, $F_h^{(2)}$, and $F_h^{(3)}$ are fluid force, radiation force, and restoring force, respectively. $F_h^{(2)}$ and $F_h^{(3)}$ can be further expressed as

$$F_h^{(2)} = -([\overline{A}]\{\ddot{U}\} + [\overline{C}]\{\dot{U}\}) \tag{13.40}$$

$$F_h^{(2)} = -\lfloor \overline{B} \rfloor \{U\} \tag{13.41}$$

It can be seen from Eqs. (13.40) and (13.41) that the matrix $[\overline{A}]$, $[\overline{C}]$ and $[\overline{B}]$ describe the corresponding parts of hydrodynamic force which are, respectively, in phase with the acceleration, velocity, and displacement of the blade vibrating in water. So they are called added mass, added damping, and added stiffness matrixes, respectively.

Substituting Eqs. (13.39)–(13.41) into Eq. (13.38), the hydro-elastic vibrating equation of the composite blade can be expressed as

$$([M] + [A])\{\ddot{U}\} + ([C] + \lfloor \overline{C} \rfloor)\{\dot{U}\} + ([K] + \lfloor \overline{B} \rfloor)\{U\} = F_h^{(1)} + F_r \tag{13.42}$$

Based on the above theory, the hydrodynamic characteristics of the composite propeller can be solved by applying the CFD method based on RANS equation. It is worth noting that the hydrodynamic cannot be solved directly for incompressible fluids in CFD solver, but utilizing a Semi-Implicit Method for Pressure Linked Equations (SIMPLE) algorithm for completing the coupling of pressure and velocity and solving the equations of incompressible fluids (Van Doormal and Raithby, 1984). Then, the hydrodynamic is calculated and transferred to the FEM Solver. The fluid-structural interaction of the composite marine propeller is calculated by using the commercial FEM/CFD software ANSYS/ANSYS CFX based on the fluid model and the finite-element model. The nonmatching meshes can be supported by the CFX GGI technology (ANSYS, 2009). The conservative interpolation scheme is used when sending flows (Force), and the profile preserving interpolation scheme is used when sending fluxes (Displacement). The Newton-Raphson numerical procedure is applied to solve the coupling Eq. (13.42).

13.4 Performances of composite propeller

13.4.1 The open water performances

It is well-known that open water performance reflects the performance characteristics of propellers operating in uniform flow. Because the composite propeller must take into account the steady fluid structural coupling problem in uniform flow, the study of the open water performance specifically refers to the studies of the structural static and hydrodynamic performances in uniform flow. Considering the effect of cavitation phenomena on propeller performance, at the same time, many researchers also compared the performance of the subcavitation and cavitation composite propellers.

By analyzing the parameters such as deformation, principal stress, and shear stress of the composite blade, the structural static performance of the composite propeller is discussed and analyzed. On the other hand, the hydrodynamic performance of the composite propeller is usually evaluated using a series of nondimensional parameters, namely, thrust coefficient, torque coefficient, and efficiency. These parameters can be expressed as follows:

$$\text{Thrust coefficient } K_T = \frac{T}{\rho n^2 D^4} \tag{13.43}$$

$$\text{Torque coefficient } K_Q = \frac{Q}{\rho n^2 D^5} \tag{13.44}$$

$$\text{Efficiency } \eta = \frac{K_T}{K_Q} \cdot \frac{J}{2\pi} \quad J = \frac{V_a}{nD} \tag{13.45}$$

where T, Q denote the thrust and the torque, respectively. D is the propeller diameter, n is the propeller rotational speed, ρ is the density of fluid, J is the advance coefficient, V_a is the speed of advance.

According to the existing research, the open water performances of the composite propeller are mainly exhibited in the following aspects:

(1) Composite propeller blade has larger deflection, which is a new behavioral characteristic compared with the metal propeller blade.

When the composite propeller blades were composed of a lower-modulus FRP materials, the maximum deflection at the tip of the composite blade is an order of magnitude larger than that of the metal blade (Lin, 1991a). In order to solve this problem, an effective method is to use a higher modulus FRP materials or a hybrid composite material, in order to improve the integral stiffness of the composite propeller. Another method is that the initial composite propeller blade was designed to have a higher pitch near the tip (Young, 2008). When the composite propeller is subjected to the hydrodynamic force, the pitch angle of the blade tip is reduced due to the elastic bending and torsional coupling effect of the orthotropic composite; the deformation of the composite propeller will be effectively controlled. Around this behavioral characteristic of the composite propeller, many researchers have carried out a series of extended studies (Young, 2008; Liu and Young, 2009). The geometry parameters (pitch, rake, skew, etc.) were used to investigate the deformation and complete the reverse design engineering (Lee et al., 2015).

(2) The strength design and evaluation of composite propeller are more complex than that of the metal propeller.

A composite propeller, referred in literature (Lee et al., 2014), is composed of a lower-strength glass-fiber-reinforced composite; the in-plane stress and inter-laminar shear stress of the composite propeller almost closed to the ultimate stress levels of the material, which means that the composite blade had reached the failure point of the design condition. In that case, the composite blade would be limited to operation at speeds well below the design value used in the analysis.

According to the above example, it is not difficult to find the significant difference of the strength evaluation method between the composite and metal propellers. The conventional metal propeller is evaluated by Von-Mises stress; however, the composite propeller is completely different. Due to the diversity of the structure (laminated or sandwich) and material anisotropy, each element has its unique fiber orientation and strengths are significantly influenced by fiber directions. Therefore, the strength of composite propeller should be evaluated by stress components (fiber direction, fiber transverse direction, shear…) or by failure criteria for composite (Tsai-Hill, Hashin's, etc.). Recently, a number of studies have been done on the strength evaluation of composite propellers (Lin and Lin, 2005).

(3) In the case of geometric equivalent design, the hydrodynamic performance of the composite propeller is lower than that of the metal propeller at the design operating condition. However, the design-ability of the composites can improve the overall hydrodynamic performance of the composite propeller, including the design operating condition and the off-design conditions.

Just as the discussion of Motley in literature (Motley et al., 2009), the pitch angle of the rigid propeller is fixed and does not vary with advance coefficients; in this case, the rigid propeller often only optimizes the hydrodynamic performance for the design conditions to maintain the maximum propulsion efficiency; the cost of doing so is that the hydrodynamic performance for the nondesign conditions is sacrificed. Different from the rigid propeller, the composite propeller can optimize pitch angle by utilizing the bending-torsional coupling characteristic and making it a function of advance coefficients. The result is that the composite propeller will achieve a satisfactory propulsion efficiency over the entire advance range. This result was also applicable to full-scale composite propeller (Das and Kapuria, 2016).

13.4.2 The hydro-elastic performance in nonuniform wake

In addition to open water performance, the hydro-elastic performance in nonuniform flow is also the focus of composite propeller research. Because the propeller and the hull are an integral system, the hydrodynamic forces of propeller and hull influence each other. When the hull sails and produces nonuniform wake in the surrounding, the propeller runs in the nonuniform wake. Thus, the hydro-elastic analysis is needed.

Because the structural characteristics and the working characteristics of the composite propeller are much different from those of the metal propeller, the hydro-elastic behaviors of the composite blade are much more complex than those of the metal blade. The comprehensive hydro-elastic analysis is greatly needed for guaranteeing the hydrodynamic and strength demands of composite propeller, maximizing the efficiency in a wide range of operating conditions, and at the same time, controlling the high vibration that is caused by harmonic loading on the blade due to an unsteady hydrodynamic environment and highly flexible rotating blades.

The work of many researchers has confirmed this phenomenon. He and Hong (He et al., 2012) discussed the hydro-elastic performance of the coupled-composite blade in nonuniform wake. A series of fluid and structural parameters during a rotation in nonuniform flow were analyzed and evaluated for the composite blade and the metal blade, including the hydrodynamic load coefficients, the tip node displacement, the maximum Von-Mises stress, and the harmonic loads. At the same time, Liu and Young (2009) analyzed the key blade harmonics, the shaft harmonics, the hydrodynamic axial force, and bending moment coefficients in the spatially varying (nonuniform) wake. Mulcahy (Mulcahy et al., 2010) presented a hydro-elastic tailoring method for flexible composite propeller blades to adapt it to changes of flow conditions due to rotation in a spatially varying wake.

According to the analysis results, it is not difficult to find the following characteristics of the composite propeller in nonuniform wake:

(1) The hydro-elastic effects of the composite propeller were larger than those of the metal propeller, mainly reflected on that the oscillation characteristics of all hydro-elastic performance of the composite blade were significantly higher than those of the metal blade.
(2) The vibratory hub loads of the composite blade were much smaller than those of the metal blade.
(3) The composite propeller in a spatially varying (nonuniform) wake can be designed to produce a higher efficiency than the metal propeller at both design and nondesign conditions.

13.4.3 Structural dynamic characteristics

As the typical underwater dynamic machine, the dynamic characteristic of composite propellers is an important issue that must be paid attention to in the research. On the one hand, it is necessary to fully understand the underwater wet mode characteristics (natural frequency, modal shape, etc.) of the composite propeller blades, which provides the basis for the dynamic frequency modulation design and hydro-elastic analysis of the composite propeller. On the other hand, the design of composite propeller must meet its structural and functional requirements, and as much as possible to increase its structural damping, in order to effectively control the vibration of composite propeller. For the composite propeller, due to the internal structure of the composite blade being anisotropic, the solution of the problem will be more theoretical and engineering practical significance.

In recent decades, the underwater dynamic characteristics of composite propeller have attracted much attention of researchers. Young (2008) predicted the natural frequencies and the mode shapes of a relatively rigid composite propeller and a relatively flexible composite propeller in air and in water by using the coupled BEM–FEM solver and compared with the measured natural frequencies. Motley (Motley and Young, 2011) analyzed and compared the natural frequencies and the mode shapes of the same size NAB propeller and carbon-fiber propeller. Lin (Lin and Tsai, 2008) discussed the natural frequencies and the mode shapes of a MAB (Manganese aluminum bronze) blade and symmetrical composite blades by using the FEM. He and Hong (Lee et al., 2014) analyzed the natural frequencies in air and in water of a metal blade and composite blades and calculated the ratio of the dry and wet natural frequencies.

Based on the above discussion and analysis, it shows several notable features of the underwater dynamic characteristics of composite propeller:

(1) The predicted fundamental frequencies in water agree reasonably well with the measured fundamental frequency.
(2) The mode shapes are almost the same in air and water.
(3) Due to the effect of the added mass, the natural frequency in the water is usually lower than the natural frequency in the air. For composite propeller, the natural frequencies are approximately 50% lower in water (Young, 2008). Considering $\omega \propto \sqrt{1/M}$, the effects of the added mass on the vibration characteristics of the composite blade exceed that on those of the metal blade (Lin and Tsai, 2008; Hong et al., 2011).
(4) Different constituent materials have an important effect on the natural frequency of the composite propeller. The more soft the material, the lower the frequency (the natural frequencies of the relatively flexible composite propeller are slightly lower than the relatively rigid composite propeller (Young, 2008)). Similarly, the natural frequencies of the CFRP propeller are significantly lower than those of the NAB propeller because of the decreased effective stiffness (Liu and Young, 2009).

After discussing the underwater modal characteristics of the composite propeller, the structural damping, as another significant structural dynamic parameter, also raises the concerns of the researcher. When a composite blade is subjected to forced vibrations at frequencies near resonance, small exciting forces will induce large vibratory inertia forces so that high-amplitude vibrations and severe stresses can be caused.

A sufficient damping source is available to decrease the stress induced. Compared with metal materials, composite materials have good material damping performance due to the viscoelastic characteristics of the polymer matrix. Nowadays, the key is how to predict and design the structure damping of the composite propeller, so that the vibration of the composite propeller can be effectively controlled.

Recently, Hong (Hong et al., 2012) proposed a hybrid method for calculating the structural damping of composite propeller. This method evaluated the structural damping of composite propeller by completing the statistical prediction of high-frequency material damping, the numerical calculation of modal damping, and the conceptual modification of Rayleigh damping, in turn. The research found

(1) The contribution of different materials to structural damping is different. The carbon/epoxy composites show a higher mass proportional damping capacity, and the glass/epoxy composites exhibit a higher stiffness proportional damping capacity.
(2) In addition to the type of material, the stacking sequence of composite materials also has a great influence on the mass and stiffness proportional damping.
(3) The effect of the stiffness proportional damping on the structural damping capacity of the composite blade is greater than that of the mass proportional damping.
(4) The structural damping of the composite blade exhibits excellent design-ability. The dynamic responses of the composite blade can be improved by optimizing design of the structural damping.

13.4.4 Performance optimization

The design of the composite propeller involves many aspects: including structure, fluid, and material. On one hand, the designers hoped that the composite propeller meets its functional requirements: higher propulsion efficiency, lower cavitation probability, and lower noise. On the other hand, the structural demands, including lighter weight, higher strength, and lower vibration, also are the important aspects needed to consider.

In order to meet the many design requirements of composite propeller, the researchers have presented a series of optimization methods and optimization schemes of composite propeller. Lee (Lee and Lin, 2004) utilized the genetic algorithm, a robust tool for solving problems that involve many discrete variables, to determine the best stacking sequence of the composite propeller. Combined with the predeformed design, two optimization requirements were satisfied for composite propeller, including to reduce the range over which the torque varies and improve the cooperation between the propeller and the engine. Young et al. (Pluciński et al., n.d.; Young et al., 2010) firstly optimized the efficiency of the composite marine propeller for many inflow velocities by using a genetic algorithm method; subsequently, a first-order reliability-based design and optimization methodology is proposed to optimize the design parameters and discuss the effect of uncertainties on material and load parameters. He and Hong (He et al., 2012) completed the optimization of the hydrodynamic load coefficients (per blade) and the vibratory hub loads by considering the ply angles and stacking sequences as the design variables; an improved low-vibration composite propeller design was presented. Manudha et al. (Herath et al., 2015) developed an optimization scheme using the Cell-based Smoothed Finite Element Method (CS-FEM) combined with a Genetic Algorithm (GA); this work was implemented under a variety of parameter settings. Recommendations were presented for achieving an ideally passive pitch varying propeller. Next, Manudha et al. (Herath et al., 2014) also presented a layup optimization algorithm combining NonUniform Rational B-Splines (NURBS)-based FEM with real-coded Genetic Algorithm (GA) for composite marine propeller. The hygrothermal effects were investigated to enable the optimization of nonsymmetric layups. At the same time, the multiobjective, multimaterial, and multiple layer thickness optimization were further discussed.

Based on the above research results, the composite propeller has achieved a comprehensive performance optimization, including: efficiency, weight, strength, vibration characteristics, etc., which provides a good foundation for further improved design of composite propeller.

13.5 Conclusions and future trends

As a new type of ship propulsion device, in recent years, composite propeller has become a hot spot of research in many countries. Composite materials possess excellent unique properties: anisotropy, high specific strength, high specific stiffness, and superior damping properties, which provided a new opportunity for composite propellers to replace conventional metal propellers, and at the same time, sparked a new exploration of composite propeller in terms of design ideas and performance evaluation.

In recent years, a considerable amount of research has been directed toward the composite marine propellers for their excellent advantage. In addition to the above-mentioned various research methods, performance characteristics and optimization, and evaluations, researchers have also done a lot of work on the preparation and applications (Mouritz et al., 2001; Anon, 2003; Yeo et al., 2014), test evaluation (Lin et al., 2009; Paik et al., 2013; Kumar and Wurm, 2015), scale effects (Young, 2010), and cavitation performance (Yamatogi et al., 2009; Hsiao and Chahine, 2015) of composite propeller. According to the existing research results, it has been found that the hydrodynamic performances of the composite propeller were slightly reduced by using the design schemes of the metal propeller. However, the hydrodynamic performances of the composite propeller in the design and off-design conditions were improved by utilizing the predeformation and structural optimization method. On the other hand, compared with the metal propeller, the weight of composite propeller greatly reduced, which effectively controlled the shaft vibration. At the same time, the unique high structural damping performance and design-ability of composite propeller also provided a good basis for vibration control.

Future work will be close to the practical application and is committed to solving the propeller-excited vibration, noise, anticavitation corrosion, and other aspects of content.

References

Anon, 2003. World's largest composite propeller successfully complete sea trials. Mater. Des. 2, 16.
ANSYS, 2009. ANSYS Version 12.0 Documentation.
Das, H.N., Kapuria, S., 2016. On the use of bend–twist coupling in full-scale composite marine propellers for improving hydrodynamic performance. J. Fluids Struct. 61, 132–153.
Hashin, Z., 1980. Failure criteria for unidirectional fiber composites. J. Appl. Mech. 47, 329–334.
He, X.D., Hong, Y., Wang, R.G., 2012. Hydroelastic optimisation of a composite marine propeller in a non-uniform wake. Ocean Eng. 39, 14–23.
Herath, M.T., Natarajan, S., Prusty, G.B., John, N.S., 2014. Smoothed finite element and genetic algorithm based optimization for shape adaptive composite marine propellers. Compos. Struct. 109, 189–197.
Herath, M.T., Natarajan, S., Prusty, G.B., John, N.S., 2015. Iso-geometric analysis and genetic algorithm for shape-adaptive composite marine propellers. Comput. Methods Appl. Mech. Eng. 284, 835–860.
Hong, Y., He, X.D., Wang, R.G., 2011. Dynamic responses of composite marine propeller in spatially wake. Polym. Polym. Compos. 19 (4&5), 405–412.
Hong, Y., He, X.D., Wang, R.G., 2012. Vibration and damping analysis of a composite blade. Mater. Des. 34, 98–105.
Hsiao, C.T., Chahine, G.L., 2015. Dynamic response of a composite propeller blade subjected to shock and bubble pressure loading. J. Fluids Struct. 54, 760–783.
Kerwin, J.E., Lee, C.S., 1978. Prediction of steady and unsteady marine propeller performance by numerical lifting-surface theory. Trans. SNAME 86, 218–253.
Kumar, J., Wurm, F.H., 2015. Bi-directional fluid–structure interaction for large deformation of layered composite propeller blades. J. Fluids Struct. 57, 32–48.
Lee, Y.J., Lin, C.C., 2004. Optimized design of composite propeller. Mech. Adv. Mater. Struct. 11, 17–30.
Lee, H., Song, M.C., Suh, J.C., Chang, B.J., 2014. Hydro-elastic analysis of marine propellers based on a BEM-FEM coupled FSI algorithm. Int. J. Nav. Archit. Ocean Eng. 6, 562–577.
Lee, H., Song, M.C., Suh, J.C., Cha, M.C., Chang, B.J., June 2015. A numerical study on the hydro-elastic behavior of composite marine propeller.Fourth International Symposium on Marine Propulsors, Austin, Texas, USA.
Lin, G., 1991a. Comparative Stress-Deflection Analyses of a Thick-Shell Composite Propeller Blade. Technical Report, David Taylor Research Center, DTRC/SHD-1373-01.
Lin, G., 1991b. Three-dimensional stress analyses of a fiber-reinforced composite thruster blade. Symposium on Propellers/Shafting. Society of Naval Architects and Marine Engineers, Virginia Beach, VA, USA.
Lin, H.J., Lin, J.J., 1996. Nonlinear hydroelastic behavior of propellers using a finite-element method and lifting surface theory. J. Mar. Sci. Technol. 1, 114–124.
Lin, H.J., Lin, J.J., 1997. Effect of stacking sequence on the hydroelastic behavior of composite propeller blades. Eleventh International Conference on Composite Materials. Australian Composite Structures Society, Gold Coast, Australia.
Lin, H.J., Lin, J.J., 2005. Strength evaluation of a composite marine propeller blade. J. Reinf. Plast. Compos. 24, 1791–1807.
Lin, H.J., Tsai, J.F., 2008. Analysis of underwater free vibrations of a composite propeller blade. J. Reinf. Plast. Compos. 27, 447–458.

Lin, C.C., Lee, Y.J., Hung, C.S., 2009. Optimization and experiment of composite marine propellers. Compos. Struct. 89, 206–215.
Liu, Z.K., Young, Y.L., 2009. Utilization of bend–twist coupling for performance enhancement of composite marine propellers. J. Fluids Struct. 25, 1102–1116.
Menter, F.R., 1994. Two-equation eddy-viscosity turbulence models for engineering applications. AIAA-J. 32, 1598–1605.
Motley, M.R., Young, Y.L., 2011. Performance-based design and analysis of flexible composite propulsors. J. Fluids Struct. 27, 1310–1325.
Motley, M.R., Liu, Z., Young, Y.L., 2009. Utilizing fluid-structure interactions to improve energy efficiency of composite marine propellers in spatially varying wake. Compos. Struct. 90, 304–313.
Mouritz, A.P., Gellert, E., Burchill, P., Challis, K., 2001. Review of advanced composite structures for naval ships and submarines. Compos. Struct. 53, 21–41.
Mulcahy, N.L., Prusty, B.G., Gardiner, C.P., 2010. Hydroelastic tailoring of flexible composite propellers. Ships Offshore Struct. 5 (4), 359–370.
Paik, B.G., Kim, G.D., Kim, K.Y., Seol, H.S., Hyun, B.S., Lee, S.G., Jung, Y.R., 2013. Investigation on the performance characteristics of the flexible propellers. Ocean Eng. 73, 139–148.
Pluciński M.M., Young Y.L., Liu Z.K., n.d. Optimization of a self-twisting composite marine propeller using genetic algorithms. 16th International Conference on Composite Materials.
Qing, L.D., 2002. Validation of RANS predictions of open water performance of a highly skewed propeller with experiments.Conference of Global Chinese Scholars on Hydrodynamics.
Taylor, R.L., Beresford, P.J., 1976. A non-conforming element for stress analysis. Int. J. Numer. Methods Eng. 10, 1211–1219.
Van Doormal, J.P., Raithby, G.G., 1984. Enhancement of the SIMPLE method for predicting incompressible fluid flows. Numer. Heat Transfer 7, 147–163.
Yamatogi, T., Murayama, H., Uzawa, K., Kageyama, K., Watanabe, N., 2009. Study on cavitation erosion of composite materials for marine propeller.17th International Conference on Composite Materials. Japan.
Yeo, K.B., Leow, W.J., Choong, W.H., Tamiri, F.M., 2014. Hand lay-up GFRP composite marine propeller blade. J. Appl. Sci. 14 (22), 3077–3082.
Young, Y.L., 2007. Time-dependent hydroelastic analysis of cavitating propulsors. J. Fluids Struct. 23, 269–295.
Young, Y.L., 2008. Fluid-structure interaction analysis of flexible composite marine propellers. J. Fluids Struct. 24, 799–818.
Young, Y.L., 2010. Dynamic hydroelastic scaling of self-adaptive composite marine rotors. Compos. Struct. 92, 97–106.
Young, Y.L., Baker, J.W., Motley, M.R., 2010. Reliability-based design and optimization of adaptive marine structures. Compos. Struct. 92, 244–253.

Offloading marine hoses: Computational and experimental analyses

Maikson L.P. Tonatto*,†, Pedro Barrionuevo Roese‡, Volnei Tita§, Maria M.C. Forte*, Sandro C. Amico*
*Post-Graduation Program in Mining, Metallurgical and Materials Engineering, Federal University of Rio Grande do Sul (UFRGS), Porto Alegre, Brazil, †Centre for Innovation and Technology in Composite Materials (CIT^eC), University of São João Del Rei, São João Del Rei, Brazil, ‡Petrobras E&P, Rio de Janeiro, Brasil, §Department of Aeronautical Engineering, São Carlos School of Engineering, University of São Paulo, São Carlos, Brazil

Chapter Outline

14.1 Introduction 391
    14.1.1 Flexible pipe in the offshore industry 391
    14.1.2 Bonded flexible pipe: offloading marine hoses 392
    14.1.3 Components of an offshore marine hose 393
14.2 Types of models 400
    14.2.1 Constitutive models for hyperelastic materials 400
    14.2.2 Failure models for composite materials 401
    14.2.3 Mechanical behavior of flexible pipe 403
    14.2.4 Hydrodynamic models 404
14.3 Offloading hoses: computational and experimental analyses 406
    14.3.1 Strength analysis 406
    14.3.2 Stiffness analysis 410
14.4 Concluding remarks 413
Acknowledgments 413
References 414

List of symbols

$[F_i]$	second order stress tensor
$[F_{ij}]$	four order stress tensor
A_r	cross section area of rebar
C_A	additional mass coefficient

Marine Composites. https://doi.org/10.1016/B978-0-08-102264-1.00014-5
Copyright © 2019 Elsevier Ltd. All rights reserved.

C_d	drag coefficient
C_m	inertia coefficient
E_1	longitudinal elastic modulus
E_2	transverse elastic modulus
E_3	transverse-thickness elastic modulus
f_i, f_{ij}	Tsai-Wu coefficients
F	Morison's force
F_{12}	in-plane shear strength
F_{1c}	longitudinal compressive strength
F_{1t}	longitudinal tensile strength
F_{23}	interlaminar shear strength in the 2–3 plane
F_{2c}	transverse compressive strength
F_{2t}	transverse tensile strength
G_{12}	in-plane shear modulus
G_{23}	interlaminar shear modulus in the 2–3 plane
G_{31}	interlaminar shear modulus in the 3–1 plane
G_i-	Tsai-Hill coefficients
I_1, I_2, I_3	strain invariants
J	volume elastic ratio
K	bulk modulus
S	projected area
$\dot{u}$	acceleration of the particle
u	speed of the particle
V	volume displaced
W	strain energy absorbed per unit volume
α	factor that determines the contribution of the shear stress in the fiber failure under tension
γ_{12}	in-plane shear strain
γ_{31}	interlaminar shear strain in the 3–1 plane
γ_{23}	interlaminar shear strain in the 2–3 plane
ε_1	longitudinal normal strain
ε_2	transverse normal strain
ε_3	transverse-thickness tensile strain
θ	orientation angle of the rebar
$\lambda_1, \lambda_2, \lambda_3$	elongation ratios
λ_m	statistical coefficient from micromechanics analyses
μ	initial shear modulus
ν_{ij}	Poisson's ratio in ij plane
ρ	fluid density
σ_1	longitudinal stress component
σ_2	transverse stress component
σ_3	transverse-thickness tensile stress component
σ_{ij}	stress components in tensor notation
τ_{12}	in-plane shear stress component
τ_{31}	interlaminar shear stress component in the 3–1 plane
τ_{23}	interlaminar shear stress component in the 2–3 plane

14.1 Introduction

Offshore offloading hoses have played an important role in the oil industry for the last 40 years, being extensively used for fluid transfer operations in a dynamic environment. Starting from rough, handcrafted parts using the same materials available for automobile tires, the offshore discharge hoses have evolved through the 1980s and 1990s, incorporating new technology in their manufacturing, engineering, and testing.

Due to their intricate geometry and the complex mechanical behavior of the materials used in their construction, offshore offloading hoses pose a good challenge regarding computational modeling. Predicting the overall mechanical behavior of the component before its construction is only possible with refined computational finite element analysis (FEA). The numerical models, in turn, require detailed mechanical characterization of the materials to provide accurate input for nonlinear materials' constitutive laws. In this chapter, the whole process of testing materials, building the FEA model, and testing an offshore discharge hose is reviewed in detail.

14.1.1 Flexible pipe in the offshore industry

As the oil industry moved to the offshore environment, driven by the need to explore new oil reserves, many technological challenges appeared. One particular component that had to be fully reengineered to comply with the offshore environment was the piping used to transport crude oil across different installations. Differently from onshore installations, where piping is stationary, the offshore environment exposes the piping to a high level of bending (Yang et al., 2015). To comply with that, the flexible pipe solution was developed making the pipe compliable with the movements imposed by waves and wind, instead of trying to resist them.

However, to achieve that, some of the pipe strength needed to be sacrificed, and this balance must be considered for every application. Riser flowlines bring oil from a subsea oil well to a floating production platform. Normally, most of their length is fully submerged, crossing the sea in the vertical position. This configuration minimizes movement imposed by the environment, but requires strength since the top section needs to support the weight of the suspended pipe below it. On the other hand, marine discharge hoses, which are used to transfer oil from a production platform to a crude carrier, float parallel to the sea surface, being fully subjected to the dynamics of the waves and wind, yielding large deformations. However, these hoses only need enough strength to support the operating pressure and the loads due to their own inertia when forced to move by the environment.

For applications like the riser flowlines aforementioned, the type of construction called "unbonded flexible pipe" is more suited. This kind of pipe, depicted in Fig. 14.1, is usually limited to small diameters, typically <7 in. (175 mm), and is built

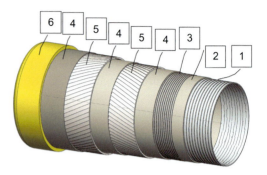

Fig. 14.1 3D image of flexible riser: (1) Stainless steel carcass, (2) liner, (3) carbon steel pressure armor (4) antiwear, (5) carbon steel tensile armors, (6) polymer external sheath.

by "braiding" steel wires into layers which are separated by antifriction tapes. These pipes have the ability to allow relative movement among the layers and display high tensile capacity, burst, and compressive strength, together with enough flexibility to allow bending of the structure to form a catenary curve, from a vertical point in the platform to the horizontal seabed.

On the other hand, applications such as marine offloading hoses require bonded flexible pipes, where reinforcement layers are embedded into a flexible matrix and overlaid to produce the pipe. In this case, all reinforcement layers end up consolidated together (usually by rubber vulcanization) at the end of the manufacturing process. This kind of pipe displays very high flexibility and allows large diameters (>20″ or 508 mm ID), although operating pressures and collapse and tensile strength are much smaller than those of unbonded pipes.

14.1.2 Bonded flexible pipe: Offloading marine hoses

Offshore offloading hoses are found in a variety of construction and shapes. Their main application is in systems called CALM—Catenary Anchor Leg Mooring, SALM—Single Anchor Leg Mooring, EPS—Early Production System, CMBM—Conventional Multi Buoy Mooring, and FPSO—floating production storage offloading.

A hose line is built by assembling together different hose sections (11.5 m each), as shown in Fig. 14.2 (Dunlop Oil and Marine Ltd., 2006), because the loads vary in certain positions (Tanaka and Saito, 1981). For instance, for a monobuoy system, there is

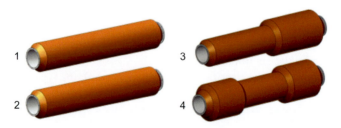

Fig. 14.2 Hoses sections used in an assembled line: mainline (1), tail hose (2), first off buoy (3), and barbell tanker rail (4).

the mainline (1), the tail hose (2), the first off buoy (3), and the barbell tanker rail (4). The first off buoy and tail hoses are subjected to higher loads and, because of that, they receive extra reinforcement layers.

Normalization of transferring operations was first established in the Oil Companies International Marine Forum—OCIMF, in 1991. In this document, the main operational parameters were defined as: (a) Working pressure of 1034 or 2068 kPa, (b) collapse resistance of −85 kPa, (c) minimum bending radius within four to six times the internal diameter, and (d) working temperature of −20°C to 92°C. Basic testing of materials and acceptance tests were also defined. In 2002, the American Petroleum Institute (API) released the 17K standard, which extended the tests and encompassed all kinds of bonded flexible pipes. Besides, API 17K also demanded from the flexible pipe sector the use of computational analysis for product design. Later, in 2009, OCIMF standard was revised, incorporating new test procedures for prototypes, such as section crushing test and fatigue testing followed by destructive pressure test (OCIMF 2009, 2009).

Although these hoses have evolved technologically, most structures are still manufactured employing materials and construction design from the 1970s (Dunlop Oil and Marine Ltd., 2006; Trelleborg, 2008). With the development of new materials, rubber products started being reinforced with high strength fibers in applications where weight and strength are vital. However, actual advances in oil transportation necessarily involve comprehensive understanding of the behavior of new materials to allow design innovations.

Regarding hose failure mechanisms, as discussed by Zandiyeh (2006) and Lassen et al. (2014), they may be grouped into: (a) manufacturing, stock, and transport and (b) service. External damage occurs when the hose is in contact with the tanker or kink failure is caused by overloading in bending and crushing. The main failures found in service are presented in Table 14.1.

14.1.3 Components of an offshore marine hose

Although offshore hoses existed before the 1960s, and great technological advance was incorporated in its manufacturing, the basic components of the hoses have changed little since then. Most technological advances were incorporated in the field of materials, project engineering, manufacturing, and testing.

Table 14.1 **Failure modes traditionally found in hoses**

Loading mode	Failure criteria
High internal pressure	Rupture of steel cables near nipple end
High external pressure	Plastic collapse of spiral due to radial compression load
High tensile	Bucking of steel spiral due to radial compression load
Tensile and bending	Nipple end deformation, spiral ovalization
Reeling	High local stresses at hose transition to nipple
	Ovalization of spiral in the hose mid-section

These hoses can be of the floating or submerged type, with single or double carcass, but in this chapter, focus will be given to the floating, double carcass offshore discharge hose class 300. For the latter, the basic structural components are shown in Fig. 14.3, namely, the first carcass and the end fitting, which is composed of a nipple and a flange. The nonstructural components comprise the second carcass, whose main role is to avoid oil spill in case the primary carcass fails, the flotation elements, and the cover.

A closer look into the first carcass, in Fig. 14.4, reveals that it is composed of: (a) liner, (b) sets of reinforcing plies, and (c) wire helix or spiral. Since the first carcass and the end fittings are the load-bearing components, these structures and their components are further detailed below.

Fig. 14.3 3D image of the double carcass hose: (a) flange, (b) nipple, (c) inner (first) carcass, (d) outer (second) carcass, (e) floaters, and (f) cover.

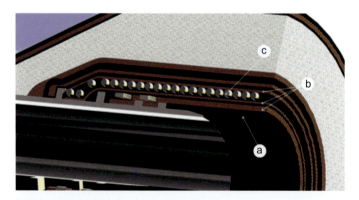

Fig. 14.4 Typical components of the first carcass: (a) liner, (b) sets of plies, and (c) wire helix.

14.1.3.1 End fitting

One of the most critical parts in an offloading hose is the transition between the main carcass and the end fitting. Anchoring of the first carcass onto the end fitting is critical to prevent failure by detachment when the load is applied. Also, abrupt stiffness transition in this area may result in stress concentration.

The end fitting of an offloading hose usually consists of a flange and a nipple. The former is manufactured from forged carbon steel and allows connectivity between hose sections to produce a continuous line. In most constructions, the flange is connected to a nipple that is integrated into it. The flexible carcasses are connected on the nipple. Drilling of the flange holes follows ANSI 16.5 standards (Dunlop Oil and Marine Ltd., 2006) and the flanges are assembled using nuts and bolts.

The nipple consists of a section of steel pipe manufactured separately from the flange, and circumferentially welded to it. Good anchoring of the carcass to the nipple is obtained by chemical adhesion and mechanical joining. For that, the nipple has "ribs" welded onto its surface to act as anchoring points for the reinforcing plies of the carcasses. The plies are laid over these ribs and then tied to them with steel wire around the nipple's circumference.

14.1.3.2 Liner and elastomeric body

The liner is the innermost layer of an offshore discharge hose and consists of an unreinforced layer of oil-resistant rubber. Its primary function is to prevent contact between the fluid being transported (crude oil) and the structural components of the hose.

Chemical resistance of the rubber employed in the liner depends mainly on the aromatic content of the oil to be transported, since these compounds are prone to degrade the polymeric materials of the hose. The working temperature also affects selection of a specific type of rubber. Permeability to gaseous substances present in the crude oil is also an important feature of rubbers. Excessive gas permeation through the liner layer, combined with pressure variations observed during offloading operations, may lead to a phenomenon known as "blistering", i.e., gas bubbles accumulated inside the elastomeric material increase in volume when a reduction in pressure occurs.

Styrene butadiene copolymer rubber is often chosen due to abrasion, tensile, and hysteresis characteristics. Nitrile rubber (NBR), a butadiene and acrylonitrile copolymer of polar nature, is widely used to operate in contact with oil due to its resistance to many hydrocarbons (Dick, 2001). In particular, API 17B (2002) recommends the use of NBR rubber, hydrogenated nitrile rubber (HNBR), or polychloroprene (CR) for the liner (API 17B, 2002). Table 14.2 shows their main characteristics. The resistance of NBR rubber to aromatics is increased by increasing its acrylonitrile content, which, however, reduces its mechanical properties.

Six different types of experiments can be used to characterize the mechanical properties of the elastomer: uniaxial tension, uniaxial compression, equibiaxial tension, pure shear, simple shear, and volumetric tests (MSC.Software, 2010). Marczak et al. (2006) reported that most elastomers display nonlinear tensile vs strain behavior.

Table 14.2 **Main elastomers used in flexible pipes**

	Fragilization temperature (°C)	Maximum continuous operating temperature (°C)	Characteristics
NBR	−20 to 40	125	• Properties depend on the acrylonitrile content; • Good resistance to hydrocarbons; • Good mechanical properties; • Good water resistance and heat resistance; • Low ozone and UV resistance.
HNBR	−40 to 50	150	• Good resistance to hydrocarbons; • High mechanical properties; • High water and weather resistance.
CR	−30 to 40	100	• Good resistance to hydrocarbons; • Good mechanical properties; • High weather resistance.

Shear and biaxial tensile tests yield curves with higher modulus than those obtained from uniaxial tensile test. The compression test exhibits a much more aggressive gradient when compared with shear, biaxial tensile and uniaxial tensile tests.

14.1.3.3 Reinforcing plies

The reinforcing plies are the main load-bearing component of offloading hoses. They are responsible for providing strength and stiffness under internal pressure, tensile, bending, and twisting loads. The reinforcing plies consist of twisted polymeric cords, similar to those used in automobile tires, embedded in a rubber matrix. As in a classic composite laminate, the reinforcing cords provide strength and stiffness in their direction, while the elastomeric matrix provides flexibility and a continuous medium for stress transfer.

For the reinforcing plies to actually work as a composite material, adhesion between the cords and the rubber matrix is mandatory. Using cords instead of fibers or wires brings an advantage because the uneven surface of the twisted cord provides mechanical anchoring. Chemical adhesion is also necessary and is usually achieved by dip coating, a process where the cords are dipped into a reactive mixture that will add functional groups to the cord surface, which will, in turn, promote their chemical bonding to the rubber in the subsequent vulcanization process.

In the manufacturing of the offloading hose, reinforcing plies are applied in opposite angles over a metallic mandrel, producing a tubular, laminated structure. The angle of the cords with respect to the hose axis typically varies from 30 to 50°, depending on the desired balance between axial and tangential stiffness and strength. Lower angles will favor axial and bending stiffness, whereas wider angles will favor tangential stiffness and strength. Angle variation will also greatly impact load distribution between the plies and the wire helix when, for instance, the hose is subjected to internal pressure.

Polyamide (Nylon), polyaramid (e.g., Kevlar, Nomex, and Twaron), polyester (Dacron), steel wire, and glass (E-glass) cords are common reinforcements. The polyamide and polyester are traditionally used in offloading hoses due to their high elongation at break, even though they display low strength. Polyaramid cords have higher strength and modulus compared with polyamide and polyester cords, as reported by Hahn (2000) (Table 14.3). In particular for polyaramid cords, care must be taken when they are subjected to compression loads. Studies in the literature show that these cords have decreased strength when subjected to compression fatigue. The main mechanism of failure is caused by microkinking (kink bands) due to buckling of the cords (Tonatto et al., 2017a).

In order to obtain the final cords, the fibers are subjected to twisting in two steps, the fibers are first twisted to produce the yarns and later the yarns are twisted to produce the cord. The most common types of twist are "S" and "Z," according to the direction of rotation (Lambillotte, 1989). Greater twisting degree improves

Table 14.3 **Properties of common fibers**

Properties	Polyaramid	Polyamide	Polyester	Carbon fiber	E-glass fiber
Density (g/cm^3)	1.44	1.14	1.38	1.78	2.58
Decomposition temperature (°C)	450	–	–	3700	–
Melting temperature (°C)	–	255	260	–	825
Tensile strength (MPa)	2800	950	1160	3400	2000
Elastic modulus (GPa)	80	6	14	238	73
Elongation at break (%)	3.3	20	13.5	1.4	2.0

properties such as elongation at break and compressive fatigue strength, but decreases tensile strength and modulus. According to Onbilger and Gopez (2008), the twist multiplier (TM) may be calculated according to Eq. (14.1):

$$TM = 0.0137 \cdot TPI \cdot \sqrt{DL} \qquad (14.1)$$

where TPI is the number of cord twists, in turns/in; DL is the linear density of cords, in denier.

Kovac and Kersker (1964) studied the potential of using polyester cords in tires, which received wide market acceptance due to their excellent elongation at break and ease of production. In order to increase their low tensile strength, Papero et al. (Papero et al., 1967) studied hybrid constructions. The 100% polyamide cord had 6.1 gf/dm fracture toughness and 31% elongation at break, while the 100% polyester had 5 gf/dm and 18%, respectively, and the hybrid (30% polyester/70% polyamide) displayed a better balance of properties, i.e., 6.4 gf/dm fracture toughness and 27% elongation at break. Another work, by Onbilger and Gopez (2008), also focused on hybrid constructions, and the use of 2 yarns of Kevlar (1500 denier each) and 1 yarn of nylon (1260 denier) increased maximum elongation at break to 7.6% and residual fatigue strength to 48%.

14.1.3.4 Wire helix

The wire helix, or spiral, generates an imbalance in the construction, thus making the mechanical behavior of the hose even more complex. The main contribution of the wire helix is to provide the hose with stability when bent to a small radius. In the absence of a supportive structure, the circular section would easily become an ellipse and ultimately lead to the collapse of the section for large deformations. This collapse is highly undesirable during an oil transferring operation, causing a pressure spike that could cause damage to the equipment or to the hose line.

Bonded together with the reinforcing plies, the wire helix supports crushing loads originated by bending of the hose, even when the hose is bent to a curving radius as small as 2.5 m. The wire helix also shares, with the reinforcing plies, some of the tangential loads due to the internal pressure. Its angle is close to 90° with respect to the hose axis, thus aligned with these tangential forces.

Another mechanical effect due to the presence of the wire helix is the coupling between axial and torsional loads. When twisted, the hose will stretch or shorten, depending on the direction of the twist, and when stretched, it will tend to rotate.

Although some designs have attempted different approaches, the wire helix in most offloading hoses consists of a steel wire with a circular section, e.g., 12.7 mm in diameter, which is wrapped around the metallic mandrel during manufacturing (Ernest, 1948) and mechanically anchored to the nipple. A typical pitch of the helix is 40 mm, and the space among the wires is filled with blocks of rubber so that, after vulcanization, there are no voids in the hose body.

The wire helix is traditionally manufactured using steel. But, high performance composite materials have been already mentioned for this structural application

Table 14.4 **Mechanical properties of steel SAE 1045 (Cabezas and Celentano, 2002) and carbon fiber/epoxy composite (Composite materials Handbook, 1999)**

Properties	SAE 1045	Carbon fiber T300/epoxy 976
Density (g/cm^3)	7.85	1.60
Longitudinal modulus (GPa)	222	135
Transversal modulus (GPa)	222	9.2
Poisson's ratio	0.30	0.32
Tensile longitudinal strength (MPa)	450	1454
Compressive longitudinal strength (MPa)	450	1296
Tensile transversal strength (MPa)	450	39
Compressive transversal strength (MPa)	450	206

(Jansen, 2016). Carbon fiber is vastly used in structural polymeric composites due to comparable elastic modulus, very high strength, and very low density compared to steel, as can be seen in Table 14.4.

Pultrusion and filament winding are two widely used manufacturing processes for composites due to characteristics such as reliable fiber positioning, high fiber volume fraction, low void content, and process automation. Both processes are able to produce spiral- or ring-shaped parts, as shown in Fig. 14.5, for a composite spiral manufactured by radial pultrusion (Jansen, 2016).

Another possible construction for this load-bearing component is to use rings instead of a spiral. For instance, Mahfuz et al. (1999) evaluated rings, extracted from cylindrical composite tubes, subjected to radial compression to determine failure modes. It was verified that the rings with 90° fiber angles presented greater resistance to radial compression and that the main failure mode was by delamination.

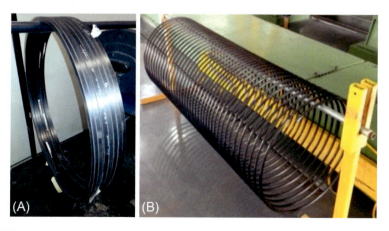

Fig. 14.5 Composite spiral (A) and rings (B) prototypes.

The main failure mode in radial compression of laminates oriented in the axial (0°) direction produced by filament winding is delamination between the layers of rings. Chen et al. (2007) manufactured and analyzed new materials with good adhesion to improve interlaminar strength of rings made by filament winding. Samyn et al. (2007) analyzed the stress field in short beam testing of straight, curved convex and concave profiles and compared it to bearing ring samples. It was concluded that conventional short beam tests represented well the stress field in comparison with bearing ring.

14.2 Types of models

14.2.1 Constitutive models for hyperelastic materials

Parameters that represent nonlinear behavior in hyperelastic models are not easily determined. Thus, several constitutive theories for large elastic strain have been developed based on strain energy functions (Gent, 2001), such as Arruda-Boyce, Marlow, Mooney-Rivlin, Neo-Hookean, Ogden, polynomial, Van der Waals, and Yeoh. These models can be divided into phenomenological and micromechanical models. Those of the first group are more commonly used and are formulated considering the general behavior of material coupons during mechanical testing. The others are based on physical phenomena in micro-scale level, using statistical approaches to include complex micro-mechanism effects in the constitutive model.

A general treatment of stress-strain relationship for solid rubbers was developed by Rivlin (1956), assuming that the material is isotropic with an elastic response and incompressible in volume. Elastic properties of rubbers can be related to the strain energy based on the strain invariants I_1, I_2, and I_3, thus their elastic behavior is mathematically described based on continuum mechanics. The invariants can be obtained by the following equations:

$$I_1 = \lambda_1^2 + \lambda_2^2 + \lambda_3^2 \tag{14.2}$$

$$I_2 = \lambda_1^2\lambda_2^2 + \lambda_2^2\lambda_3^2 + \lambda_3^2\lambda_1^2 \tag{14.3}$$

$$I_3 = \lambda_1^2\lambda_2^2\lambda_3^2 \tag{14.4}$$

where λ_1, λ_2, and λ_3 are the main elongation ratios, defined as the ratio between elongated and non-elongated length of edges for a small cubic element.

The λ_i values are given by $(1 + \varepsilon_i)$, where ε_i is the corresponding principal strain and the values of I_i becomes zero in the nonelongated state when $\lambda_1 = \lambda_2 = \lambda_3 = 1$. Besides, for an incompressible material, I_3 is equal to "1," so only I_1 and I_2 need to be evaluated. Therefore, the strain energy density W can be written as shown in Eq. (14.5):

$$W = \sum_{i+j=1}^{\infty} C_{ij}(I_1 - 3)^i (I_2 - 3)^j + \frac{1}{D}(J - 1)^2 \tag{14.5}$$

where D = inverse of two times the bulk modulus, $K = 2/D$ ($D = 0$ for incompressible materials); J = volumetric elastic ratio.

Arruda-Boyce's model is based on Eq. (14.6), where the parameters can be more easily determined:

$$W = \mu \sum_{i=1}^{5} \frac{C_i}{\lambda_m^{2i-2}} \left(\bar{I}_1^i - 3^i\right) + \frac{1}{D}\left[\frac{J^2-1}{2} - \ln(J)\right] \quad (14.6)$$

where

$$C_1 = \frac{1}{2}; C_2 = \frac{1}{20}; C_3 = \frac{11}{1050}; C_4 = \frac{19}{7000}; C_5 = \frac{519}{673750} \quad (14.7)$$

and μ = initial shear modulus; λ_m = elongation at which the stress-strain curve changes significantly.

Marlow's model considers that the material is subjected to axial tension for each value of uniaxial deformation, $\epsilon = \lambda - 1$. The nominal uniaxial tensile load is called $T(\epsilon)$. To obtain the corresponding strain energy, given the value of the first invariant $\hat{I}$, the stress-strain curve is integrated from "zero" to $\left[\lambda_T(\hat{I}) - 1\right]$ strain. Since $\hat{I}$ is arbitrary, the strain energy density W is determined assuming that $W(3) = 0$ as shown in Eq. (14.8):

$$W(I) = \int_0^{\lambda_T(\hat{I})-1} T(\epsilon) d\epsilon \quad (14.8)$$

where $\lambda_T(\hat{I})$ is the uniaxial elongation.

Thus, the stress-strain curve used in the integral, which reproduces the strain energy density obtained using experiments, can be applied to determine the model parameters for that material. Commonly, experimental data is required to identify model parameters by using, for example, an optimization process. In these cases, Eq. (14.8) becomes the general constitutive relation for incompressible materials in which the strain energy is a function of the first invariant (Marlow, 2003).

14.2.2 Failure models for composite materials

Two methods are usually employed to identify failure in composite materials. The first is defined as *first ply failure* and determines that the laminate fails when the first layer fails, disregarding alternative loading paths in the other layers of the laminate and producing more conservative predictions. Also, it only requires one failure criterion. Alternatively, the *last ply failure* concept is more complex and considers that the laminate fails only when the last layer fails, because the loading is redistributed between layers as long as any remaining layer exists (Ribeiro et al., 2013).

There are several failure criteria for composite materials. Some of them do not distinguish between matrix and fiber failure, such as Tsai-Hill, and Tsai-Wu. In these cases, failure initiation is assumed to indicate simultaneous failure of fiber and matrix. Another group of failure criteria distinguishes between matrix and fiber failure, such as Hashin, Christensen (Christensen, 1998), and Puck (Puck and Schurmann, 2001). Some of these criteria are briefly presented below.

14.2.2.1 Tsai-Hill failure criterion (Kaw, 2006)

Tsai-Hill failure criterion is one of the oldest criteria used for transversely isotropic composites. This failure criterion can be applied to a triaxial stress state according to Eq. (14.9):

$$(G_2+G_3)\sigma_1^2 + (G_1+G_3)\sigma_2^2 + (G_1+G_2)\sigma_3^2 - 2G_3\sigma_1\sigma_2 - 2G_2\sigma_1\sigma_3$$
$$-2G_1\sigma_2\sigma_3 + 2G_4\tau_{23}^2 + 2G_5\tau_{13}^2 + 2G_6\tau_{12}^2 = 1 \quad (14.9)$$

where G_1, G_2, G_3, G_4, G_5, and G_6 coefficients can be found in Kaw (2006).

14.2.2.2 Tsai-Wu failure criterion (Tsai and Wu, 1970)

Tsai-Wu criterion assumes that failure of the ply can be predicted according to Eq. (14.10):

$$f(\sigma_k) = [F_i][\sigma_i] + [F_{ij}][\sigma_i][\sigma_j] = 1 \quad (14.10)$$

where i, j, and $k = 1, 2, \ldots, 0.6$; $[F_i]$ and $[F_{ij}]$ are the second and fourth order strength tensors, respectively.

For transversely isotropic materials, it is possible to write this criterion according to Eq. (14.11):

$$f_1\sigma_1 + f_2\sigma_2 + f_3\sigma_3 + f_{11}\sigma_1^2 + f_{22}\sigma_2^2 + f_{33}\sigma_3^2 + f_{44}\tau_{23}^2 + f_{55}\tau_{13}^2$$
$$+ f_{66}\tau_{12}^2 + 2f_{12}\sigma_1\sigma_2 + 2f_{13}\sigma_1\sigma_3 + 2f_{23}\sigma_2\sigma_3 = 1 \quad (14.11)$$

where $f_1, f_2, f_3, f_{11}, f_{22}, f_{33}, f_{44}, f_{55}, f_{66}, f_{12}, f_{13}$, and f_{23} coefficients can be found in (Tsai and Wu, 1970).

This is a simple criterion based on a quadratic approximation, where the stresses in different directions can be combined and treated in a unified way. However, fiber compression failure mode is not accurately evaluated.

14.2.2.3 Hashin failure criterion (Hashin, 1980)

Unlike the previous two criteria, Hashin failure criterion is formulated considering four different failure modes, identified below:

- If $\sigma_1 \geq 0$, fiber failure under tension is calculated by Eq. (14.12):

$$\left(\frac{\sigma_1}{F_{1t}}\right)^2 + \alpha\left(\frac{\tau_{12}^2 + \tau_{13}^2}{F_{12}^2}\right) = 1 \qquad (14.12)$$

- If $\sigma_1 < 0$, fiber failure under compression is calculated by Eq. (14.13):

$$\left(\frac{\sigma_1}{F_{1c}}\right)^2 = 1 \qquad (14.13)$$

- If $\sigma_2 + \sigma_3 \geq 0$, polymer matrix failure under tension is calculated by Eq. (14.14):

$$\frac{(\sigma_2 + \sigma_3)^2}{F_{2t}^2} + \frac{\tau_{23}^2 - \sigma_2\sigma_3}{F_{23}^2} + \frac{\tau_{12}^2 + \tau_{13}^2}{F_{12}^2} = 1 \qquad (14.14)$$

- If $\sigma_2 + \sigma_3 < 0$, polymer matrix failure under compression is calculated by Eq. (14.15):

$$\left[\left(\frac{F_{2c}}{2F_{23}}\right)^2 - 1\right]\left(\frac{\sigma_2 + \sigma_3}{F_{2c}}\right) + \frac{(\sigma_2 + \sigma_3)^2}{4F_{23}^2} + \frac{\tau_{23}^2 - \sigma_2\sigma_3}{F_{23}^2} + \frac{\tau_{12}^2 + \tau_{13}^2}{F_{12}^2} = 1 \qquad (14.15)$$

where the α factor determines the contribution of shear stress to fiber failure under tension. If $\alpha = 0$, Hashin criterion shows fiber failure under tension similar to the maximum stress criterion. Fiber failure under compression is not influenced by shear stress, although it is known that shear stress influences compression failure in the polymer matrix (Barbero, 2013).

In all criteria, it is possible to detect initial failure at particular points of the laminate. Thus, if a numerical approach is used, for example finite element method (FEM), it may be identified which integration point fails in the model and applies a specific damage evolution law in order to perform stiffness degradation of the ply of the finite element to which this point of failure belongs. One way to execute this process consists of using the multicontinuum criterion (Hansen and Garnich, 1994; Mayes and Hansen, 2003), which performs progressive failure analysis of matrix and fibers considering independent failure modes. However, there are different alternatives to carry out progressive analyses, combining different failure criteria and degradation laws (Tita, 2016).

14.2.3 Mechanical behavior of flexible pipe

There are different methods in the literature to evaluate the behavior of flexible pipes. Alfano et al. (2009) developed a constituent model for flexible risers and a procedure for identifying input parameters using a multiscale approach. It was formulated based on the Euler-Bernoulli beam model, with the addition of pressure terms to the generalized stress to account for internal and external pressures. A nonlinear relationship between stresses and strains in the beam was used. Sævik (2011) compared experimental results to analytical predictions for flexible pipes. However, those models can only be used to unbonded hoses.

On the other hand, Lotveit and Often (1990) developed an analytical methodology adapted from FEM for evaluating unbonded and bonded flexible pipes. Pidaparti (1997) used a 3D finite element model to investigate the mechanical behavior of two layers of a cord/rubber laminate under combined tensile and torsional loading. More recent studies (Cho et al., 2013; Saeed et al., 2014; Rafiee, 2013) show similar methods that predict the behavior of structures based on the stiffness matrix, which is obtained using homogenized orthotropic elastic properties. Thus, the orthotropic constitutive relation is expressed by using three elastic moduli, E_1, E_2, and E_3, six Poisson coefficients, ν_{ij}, and three shear moduli, G_{23}, G_{31}, and G_{12}, as in:

$$\begin{bmatrix} \epsilon_1 \\ \epsilon_2 \\ \epsilon_3 \\ \gamma_{23} \\ \gamma_{31} \\ \gamma_{12} \end{bmatrix} = \begin{bmatrix} \dfrac{1}{E_1} & \dfrac{-\nu_{21}}{E_2} & \dfrac{-\nu_{31}}{E_3} & 0 & 0 & 0 \\ \dfrac{-\nu_{12}}{E_1} & \dfrac{1}{E_2} & \dfrac{-\nu_{32}}{E_3} & 0 & 0 & 0 \\ \dfrac{-\nu_{13}}{E_1} & \dfrac{-\nu_{23}}{E_2} & \dfrac{1}{E_3} & 0 & 0 & 0 \\ 0 & 0 & 0 & \dfrac{1}{G_{12}} & 0 & 0 \\ 0 & 0 & 0 & 0 & \dfrac{1}{G_{23}} & 0 \\ 0 & 0 & 0 & 0 & 0 & \dfrac{1}{G_{31}} \end{bmatrix} \begin{bmatrix} \sigma_1 \\ \sigma_2 \\ \sigma_3 \\ \tau_{23} \\ \tau_{31} \\ \tau_{12} \end{bmatrix} \quad (14.16)$$

It is important to bear in mind that the hyperelastic behavior of cords and rubber makes the elastic properties variable according to the stress level. There are different ways to model this effect, one of them consists of using specific finite element models. Kondé et al. (2013) carried out a comparative analysis to verify the behavior of aircraft tires composed of rubber laminates. They compared the constitutive relation of Eq. (14.16) with rebar elements, which perform a purely hyperelastic analysis, and they found good accuracy in the results. These rebar components may be defined using the particular cross section A_r of each rebar component, the spacing between two consecutive cords, and the orientation angle θ of the rebar cord in the local frame. More details about rebar elements and calculation methodology may be found in ABAQUS/Standard (2010).

14.2.4 Hydrodynamic models

Gong et al. (2014) developed a finite element model via Orcaflex software to perform a dynamic analysis of an offshore installation. They investigated the effects of dynamic laws, including wave surfaces, ocean currents, movements, and contact between structures. The analysis resulted in the detailed dynamic behavior of the pipe including deflection, axial tension, and bending moment. The authors focused on the comparison between static and dynamic loads, verifying significant differences between them and stressing the need to include dynamic effects when analyzing submarine structures.

Takafuji and Martins (2011) evaluated the fatigue life of a steel riser using Orcaflex software, considering 76 different irregular waves found on the Brazilian coast. Fatigue life was calculated by using the S-N (stress vs. n cycles) curve of steel and Miner's law for cumulative damage. The authors obtained the fatigue life along the riser line for different conditions. Svoren (2013) and Grov (2014) performed similar global analyzes to determine the fatigue life of a flexible riser. The complexity of this component required not only the global model, but also a local model for a more refined evaluation of the stress distribution. For that, the authors coupled their global model, which was developed with SIMLA and/or RIFLEX software, with the local model, which was built with BFlex software.

For the hydrodynamic analysis, it is important to determine the weighted average height of the largest waves and the peak period of the spectrum. These data represent the wave spectra. It is also important to define the wave type, Jonswap (Joint North Sea Wave Project), which is defined by the statistical distribution of the entire state of the sea and is one of the most common. The effects of currents must also be taken into consideration in the design of offshore structures due to: (a) Large displacement and slow drift of anchored platforms; (b) The current at the beginning to the drag and forces of support of submerged structures; (c) Vibrations of slender structural elements and movements of large structures; (d) The interaction between strong currents and waves affects height and period of the wave.

The current profile can be defined by the speed variation as a function of sea depth. However, for floating hoses, this variation can be incorporated into the calculation in a simplified way, with only the surface current condition, composed of the speed and direction to be modeled. Statistical distribution information of currents and speed profile are available for most areas of the world and they must be known prior to oil exploration (Det Norske Veritas, 2010).

Another important effect refers to wind loads, which act on the normal direction in relation to the surfaces of structures, and are generally time-dependent due to speed variations. Frictional loads due to the tangential drag must also be considered when a large surface is swept by the wind. The resulting structure response due to wind loads is an overlap of the static and dynamic response due to near-natural frequency excitation. The dynamic effects may be the resonance response due to wind turbulence and vortex motion.

Some structures commonly found in an offshore assembly do not interfere with the wave behavior, such as jackets, risers, mooring lines, submersible equipment, and hoses. The Morison formulation (Morison, 1953) has been used in several studies dedicated to find the loads involved in these structures. It assumes that the forces can be calculated from important parameters of the flow at the body surface such as pressure, speed, and acceleration. In the Morison Eq. (14.17), the first term considers the inertial portion, related to the accelerations of the fluid, and the second term shows the drag portion related to the square of the relative speed between the body and the fluid.

$$F = C_m \cdot \rho \cdot V \cdot \dot{u} + C_d \cdot \frac{1}{2} \cdot \rho \cdot S \cdot u \cdot |u| \tag{14.17}$$

where F is the Morison force; C_m is the inertia coefficient, calculated by $C_m = 1 + C_A$; C_A is the additional mass coefficient; ρ is the density of the fluid; V is the volume displaced; $\dot{u}$ is the particle acceleration; C_d is the drag coefficient; S is the projected area; u is the particle speed. Dimensionless coefficients C_m, C_A, and C_d are obtained based on experimental tests. Tonatto et al. (2018) developed a multiscale analyses. Firstly, a finite element model of the traditional system was built to perform a macroscale analysis of global hydrodynamics loads in the hose line using Orcaflex software. Then, a nonlinear finite element model was developed for mesoscale analysis using Abaqus software. The global hydrodynamics loads were included in the model using Morison Eq. (14.17).

14.3 Offloading hoses: Computational and experimental analyses

14.3.1 Strength analysis

The use of computational simulation by finite element analysis (FEA) to solve specific problems of offloading hoses can be quite laborious and complex. The use of FEA to hoses extends from the design to the prediction of operating conditions and life of the structure. For the design of the hose, the numeric models are used to predict local response of the hose, in terms of stress and strain, when under a particular type of loading. Thus, using a reliable model, the hose performance can be optimized in terms of materials used in its construction at considerably lower cost compared to experiments.

Tonatto et al. (2017b) conducted several burst pressure tests of hoses to evaluate strength of the carcasses. The original dimensions of the hoses were length around 10.7 m (35 ft) and internal diameter of 500 mm (20 in.). Two prototypes were manufactured, [$\pm 45_{20}/90/\pm 45_{10}/\pm 45_{16}$] and [$\pm 45_{16}/90/\pm 45_{6}/\pm 45_{14}$], using polyester cords in the inner carcass and polyamide cords in the outer carcass. Other prototypes were manufactured using hybrid (polyaramid and polyamide) cords following [$\pm 36_{6}/90/\pm 36_{6}/\pm 50_{8}$].

The produced prototype hoses were submitted to burst testing following the methodology available in OCIMF 2009 (see Fig. 14.6A). A pump was used to apply pressure into the hose, using ambient temperature water, and pressure was monitored with a manometer. The hoses were coupled at their ends by flanges, which allowed free displacement as shown in Fig 14.6B.

FEM were developed using non-dedicated commercial software. Complex nonlinear calculations for the contact between the hyperelastic rubber, which was modelled with Arruda-Boyce's, and the polyester, polyamide, and hybrid reinforcement cords, modelled using Marlow's theory, have been implemented in the models. The stress vs. strain experimental curves have been used to identify the parameters of these hyperelastic models.

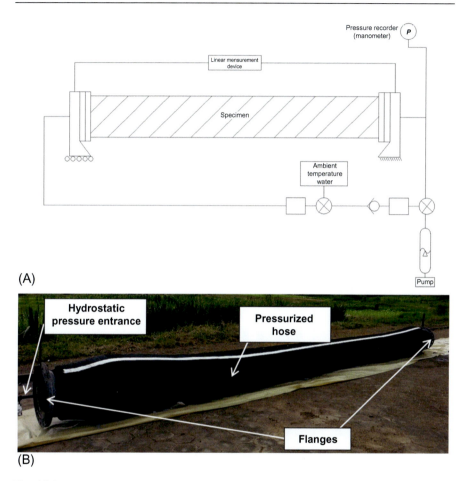

Fig. 14.6 Burst test (Tonatto et al., 2017b): (A) Scheme and (B) hose during the test.

An axisymmetric model using FE software Abaqus was employed to predict failure under internal pressure of the central section of the hose (500 mm long). The hose configuration includes the inner and outer carcasses, where the former is comprised of layers of reinforcement below and above the spiral and the latter only has reinforcing plies. During burst testing, the inner carcass is pressurized until rupture, losing its structural role after that. After that, the fluid enters the cavity between the two carcasses and the outer carcass eventually pressurizes until failure.

Embedded elements technique was used to implement the embedding of the reinforcing plies and the spiral in the rubber. Regarding the boundary conditions, the nodes located at $x = 0$ were restrained so that movement can only occur in the y axis. The nodes located at $x = 500$ were connected to the reference point (RP) located at $x = 500$, $y = 0$, using MPC constraint.

Table 14.5 shows the burst pressure predictions given by the axisymmetric model along with the experimental results. The numerical results were obtained monitoring

Table 14.5 Maximum load in reinforcement bars, axial displacement, and burst pressure (Tonatto et al., 2017b)

Specimen	Carcass	Finite element modeling					Experimental	
		Maximum load in rebar (N)	Stress in spiral (MPa)	Radial displacement at burst (mm)	Burst pressure (MPa)	Burst pressure (MPa)	Difference[a] (%)
[±45$_{20}$/90/ ±45$_{10}$/±45$_{16}$]	Inner	199.9	967	16.5	14.6	14.1	3.5
	Outer	501.0	–	169.7	5.8	5.4	7.4
[±45$_{16}$/90/ ±45$_6$/±45$_{14}$]	Inner	199.9	849	21.9	12.2	12.2	0.1
	Outer	501.0	–	168.5	5.2	5.1	1.9
[±36$_6$/90/ ±36$_6$/±50$_8$]	Inner	360.7	1030	17.9	14.8	16.0	7.5
	Outer	449.0	–	74.8	5.4	–	–

[a]Difference = |(Burst pressure_FEM)−(Burst pressure_Exp)|/(Burst pressure_Exp).

the maximum load in rebar or stress in spiral. The model was highly refined and showed good convergence with the experimental results, with a variation ranging from 0.1% to 7.5% for burst pressure. Such variations are highly acceptable considering the manual manufacturing of the hose and measurement uncertainties during testing, as well as the hypotheses adopted in the computational model. In addition, the results obtained with the numerical model and the experiments were well above the minimum limit specified by OCIMF, which is 10.5 MPa.

If an ordinary hose is bent until collapse (kink) of the circular section, the steel spiral within the hose undergoes plastic deformation and obstructs the section. However, recent advances in the processing of carbon fiber composites have allowed high-quality curved profiles for selected applications. In that context, in another study by Tonatto et al. (2016), a crush test was performed to compare different load-bearer components steel spiral with composite spiral or rings.

Among the load-bearer components analyzed, a spiral (average diameter of 560 mm) of circular cross section (diameter 12.7 mm and pitch 36 mm) was manufactured with epoxy resin and carbon fiber (CF) by curved pultrusion (named here E/PU/S). A rigid square plate was used to apply uniaxial compression loading on the top of the sample, leading to crushing, as shown in Fig. 14.7.

A FE model was used for progressive damage analysis of the complex failure behavior of carbon fiber composite spirals and ring. As for the boundary conditions,

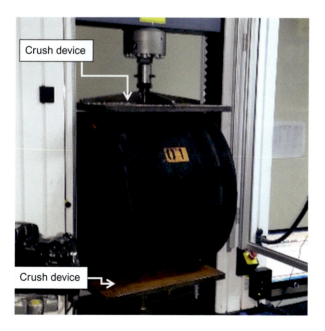

Fig. 14.7 Photograph of the crushing test.
Adapted from Tonatto, M.L., Forte, M.M., Tita, V., Amico, S.C., 2016. Progressive damage modeling of spiral and ring composite structures for offloading hoses. Mater. Des. 108, 374–382.

the hose section was fixed in the lower support using a RP (rigid plate) and, in the upper support using another RP, where displacements and rotations were fixed in all directions except for the 2-direction (y-direction), where a displacement of 150 mm was gradually applied.

Fig. 14.8 shows the numerical and experimental results of the crushing of the hose section with the composite component. Good correlation between experimental and numerical curves is seen and deviation in the maximum load is only around 0.5%, with an associated difference in displacement of ca. 0.8%. By using the damage model, it was possible to verify that damage was severe. The high stiffness of the composite produced with curved pultrusion leads to more severe localized damage in the part because the applied load does not distribute much to the surroundings, leading to composite failure.

Well-defined load peaks are observed in the numerical curves, caused by failure of the spirals as shown in Fig. 14.8. The first peak corresponds to failure in the upper region of the spiral and the second failure in the lower region in a very similar load level of the first one. The third failure occurs when one of the side regions fails, being immediately followed by failure of the other side.

14.3.2 Stiffness analysis

A 3D model of the $[\pm 45_{20}/90/\pm 45_{10}/\pm 45_{16}]$ stacking sequence was developed to obtain bending, torsional, and tensile stiffnesses of the hose. Similar parameters to that of the previously described axisymmetric model were used. Symmetry was considered

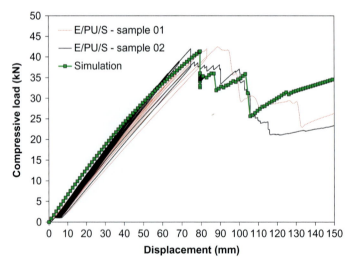

Fig. 14.8 Compressive load vs displacement curve (numerical and experimental data) for the hose with E/PU/S spiral.
Adapted from Tonatto, M.L., Forte, M.M., Tita, V., Amico, S.C., 2016. Progressive damage modeling of spiral and ring composite structures for offloading hoses. Mater. Des. 108, 374–382.

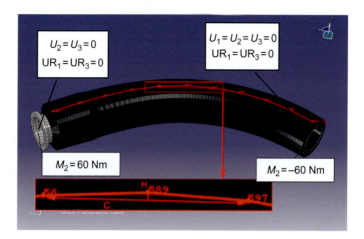

Fig. 14.9 3D numerical model used in the stiffness analysis.

with respect to the transverse center axis for lower computational cost. The reinforcement cords were modeled using rebar elements, as two-dimensional surface of revolution, and an individual cylinder was modeled for each reinforcement layer using 4 nodes quadrilateral elements (SFM3D4). The spiral was modeled using one-dimensional beam elements (B31) and its circular section with a radius of 7.5 mm. The elastomeric body was modeled using 8 nodes hexahedral elements (C3D8RH) and the nipple-flange was modeled as a rigid body.

Bending stiffness analysis was performed by imposing a bending loading on the hose structure. The purpose of the bending test is to determine radius of the curvature as a function of bending load, using for that node displacements on the surface of the hose. Fig. 14.9 shows an image generated by simulation using the developed numerical model.

Radius of curvature and bending moment values were used to calculate flexural stiffness (EI) along the length of the hose for a moment (M_2) of 60 kN m, according to the OCIMF Guide 2009. Fig. 14.10 shows bending stiffness values along the hose length and it is observed that bending stiffness is around 230 kN m^2 in the central region and increases towards the ends of the hose due to the very high stiffness of the nipple-flange.

Fig. 14.11 shows a 3D image of the numerical model used to obtain torsional stiffness. A torque of 50 kN m was applied at the control point in the nipple-flange and, at the other end, all directions and rotations were fixed. The boundary conditions and loads were chosen for replicating the test conditions and the numerical torsional moment versus angular deformation curves were produced using the variable UR1 at the control point where moment was applied.

Fig. 14.12 shows the experimentally obtained and the numerical curves for clockwise and counterclockwise twisting. Good approximation between experimental and numerical results is seen, that is, the simulation was able to adequately predict the torsional behavior of the hose, including the distinct response for each torsional direction, referring to the opening or closing of the spiral during twisting.

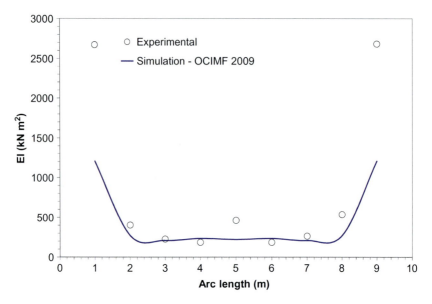

Fig. 14.10 Bending stiffness vs arc length curve (numerical and experimental results) for a bending moment of 60 kN m.

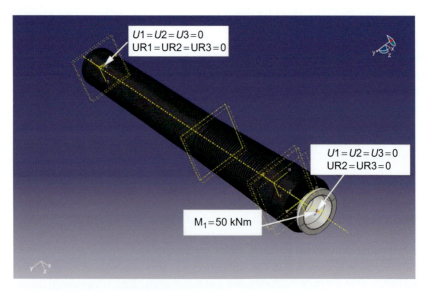

Fig. 14.11 3D numerical model used for torsional stiffness analyses.

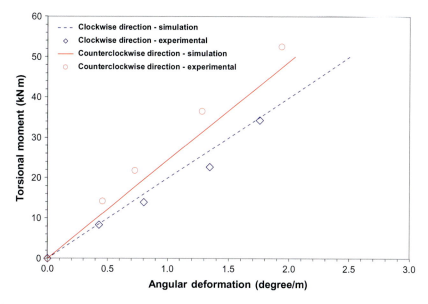

Fig. 14.12 Torsional moment vs angular deformation curves (numerical and experimental results).

14.4 Concluding remarks

In this chapter, advanced topics related to marine hoses manufactured were briefly presented, including the main components and their respective characteristics (roles, parameters, and mechanical behavior). The use of advanced polymers and composite materials was also introduced. Appropriate numerical analysis on commercial software and actual experimental tests of marine hoses comprised burst analyses of carcasses, radial compressive loading of central sections, and stiffness evaluations. Commonly used numerical methods applied to these hoses do not include high strain and displacements, or nonlinear effects (material, geometry, and contact), in addition to high anisotropy, failing to provide an adequate response of complex bonded hose structures. Despite that, the numerical method presented, which included hyperelastic models and failure evaluation and damage in composites, was able to predict with good accuracy the offloading marine hose behavior.

Acknowledgments

The authors gratefully acknowledge Petrobras (contract number: 2000.0067652.11.9) for the financial support. They also acknowledge the Coordination for the Improvement of Higher Education Personnel (CAPES) (Process no. PROEX-048/1666504) and the financial support of the National Council for Scientific and Technological Development (CNPq).

References

ABAQUS/Standard, 2010. Theory Manual and Example Problems Manual, Release 6.9.
Alfano, G., Bahtui, A., Bahai, H., 2009. Numerical derivation of constitutive models for unbonded flexible risers. Int. J. Mech. Sci. 51, 295–304.
API 17B, 2002. Recommended Practice for Flexible Pipe. American Petroleum Institute, Washington.
Barbero, E.J., 2013. Finite Element Analysis of composite material using Abaqus. CRC Press.
Cabezas, E.E., Celentano, D.J., 2002. Experimental and numerical analysis of the tensile test using sheet specimens. Universidade de Santiago de Chile, Santiago.
Chen, W., Yu, Y., Li, P., Wang, C., 2007. Effect of new epoxy matrix for T800 carbon fiber/epoxy filament wound composites. Compos. Sci. Technol. 67, 2261–2270.
Cho, J., Jee, Y., Kim, W., Han, S., Lee, S., 2013. Homogenization of braided fabric composite for reliable large deformation analysis of reinforced rubber hose. Compos Part B-Eng 53, 112–120.
Christensen, R.M., 1998. The numbers of elastic properties and failure parameters for fiber composites. J. Eng. Mater. Technol. 120, 110–113.
Composite Materials Handbook, 1999. Polymer Matrix Composites Materials Properties. vol. 2. Department of Defense, USA.
Det Norske Veritas, 2010. DNV-RP-C205—Environmental condictions and environmental loads. (Norway).
Dick, J.S., 2001. Dick, J.S. (Ed.), Rubber Technology: Compounding and Testing for Performance. Hanser, Munchen.
Dunlop Oil and Marine Ltd, 2006. Offshore Hose manual.
Ernest, S. (1948, November 9). Patent No. US 2661026 A.
Gent, A.N., 2001. Engineering with Rubber: How to Design Rubber Components, second ed. Hanser, Kempten.
Gong, S., Xu, P., Bao, S., Zhong, W., He, N., Yan, H., 2014. Numerical modelling on dynamic behaviour of deepwater S-lay pipeline. Ocean Eng. 88, 393–408.
Grov, T.A., 2014. Fatigue of Flexible Risers Considering Alternative Constitutive Models. Norwegian University of Science and Technology, Trondheim.
Hahn, C., 2000. Characteristics of p-aramid fibers in friction and sealing materials. J. Ind. Text. 30 (2), 146–165.
Hansen, A.C., Garnich, M.R., 1994. A multicontinuum theory for structural analysis of composite material systems. Compos. Eng. 5, 1091–1103.
Hashin, Z., 1980. Fatigue failure criteria for unidirectional fiber composites. J. Appl. Mech. 47, 329–334.
Jansen, K., 2016. Mass production of curved profiles for car bodies—process and machines. Automotive Composite Conference and Exhibitionpp. 1–13.
Kaw, A.K., 2006. Mechanisms of Composite Materials, ninth ed. CRC Press.
Kondé, A., Rosu, I., Lebon, F., Brardo, O., Devésa, B., 2013. On the modeling of aircraft tire. Aerosp. Sci. Technol. 27, 67–75.
Kovac, F.J., Kersker, T.M., 1964. The development of the polyester. Text. Res. J. 69–79.
Lambillotte, B. D. (1989, Janeiro). Fabric Reinforcements for Rubber. J. Ind. Text., 18, 162–179.
Lassen, T., Lem, A.I., Imingen, G., 2014. Load response and finite element modelling of bonded offshore loading hoses. In: Proceedings of the International Conference on Offshore Mechanics and Arctic Engineering—OMAE, Paper 23545. pp. 1–17.

Lotveit, S.A., Often, O., 1990. Increased reliability through a unified analysis tool for bonded and non-bonded pipes. Adv. Subsea Pipeline Eng. Tech. *24* (N-1364), 81–110.

Mahfuz, H., Pal, A., Rahman, M., & Jeelani, S. (1999). Failure Analysis of Thick Composite Rings Under Diametral Compression. 12th International Conference on Composite Materials (ICCM-12), pp. 9.

Marczak, R., Gheller Jr., J., Hoss, L., 2006. Caracterização de elastômeros para simulação numérica. SENAI, São Leopoldo.

Marlow, R.S., 2003. A general first-invariant hyperleastic constitutive model. In: Busfield, J., Muhr, A. (Eds.), Constitutive Models for Rubber III. pp. 157–160.

Mayes, J.S., Hansen, A.C., 2003. Composite laminate failure analysis using multicontinuum theory. Compos. Sci. Technol. 64, 379–394.

Morison, J.R., 1953. The Force Distribution Exerted by Surface Waves on Piles. University of California, California.

MSC. Software, 2010. Nonlinear Finite Element Analysis of Elastomers. Santa Ana.

OCIMF 2009, 2009. Guide to Manufacturing and Purchasing Hoses for Offshore Moorings. Oil Companies International Marine Forum, London.

Onbilger, D.G., Gopez, F., 2008. Aramid yarn as a tensile menber in products. In: Rubber & Plastic News.pp. 14–16 Fevereiro 25.

Papero, P.V., Kubu, E., Roldan, L., 1967. Fundamental Property Considerations in Tailoring a New Fiber. Text. Res. J. 37, 823–833 Outubro.

Pidaparti, R., 1997. Analysis of cord-rubber composite laminates under combined tension and torsion loading. Compos. Part B Eng. 28B, 433–438.

Puck, A., Schurmann, H., 2001. Failure analysis of FRP laminates by means of physically based phenomenological models. Compos. Sci. Technol. 62, 1633–1662.

Rafiee, R., 2013. Experimental and theoretical investigations on the failure of filament wound GRP Pipes. Compos Part B-Eng 45, 257–267.

Ribeiro, M.L., Vandepitte, D., Tita, V., 2013. Damage model and progressive failure analyses for filament wound composite laminates. Appl. Compos. Mater. 20, 975–992.

Rivlin, R., 1956. Large elastic deformation. In: Eirich, F.R. (Ed.), In: Rheology, Theory and Applications, vol. 1. Academic Press.

Saeed, N., Ronagh, H., Virk, A., 2014. Composite repair of pipelines, considering the effect of live pressure-analytical and numerical models with respect to ISO/TS 24817 and ASME PCC-2. Compos. Part B Eng. 58, 605–610.

Sævik, S., 2011. Theoretical and experimental studies of stresses in flexible pipes. Comput. Struct. 89, 2273–2291.

Samyn, P., Schepdael, V., JS, L., Paepegem, V., Baets, D., Degrieck, J., 2007. Short-beam-shear testing of carbon fibre/epoxy ring segments with variable cross-sectional geometry as a representative selection criterion for full-scale delamination. ASTM J. Test. Eval. 35, 310–320.

Svoren, D.F., 2013. Fatigue Analysis of Flexible Risers. Norwegian University of Science and Technology, Trondheim.

Takafuji, F.C., Martins, C.A., 2011. Comparison of different approaches for fatigue damage accumulation in steel risers.30th International Conference on Ocean, Offshore and Arctic Engineeringp. 9.

Tanaka, M., & Saito, H. (1981, March 31). Patent No. US 4259553 A.

Tita, V., 2016. Composite structures design and analysis. In: Dynamics of Smart Systems and Structures. first ed. Springer International Publishing, pp. 217–263.

Tonatto, M.L., Forte, M.M., Tita, V., Amico, S.C., 2016. Progressive damage modeling of spiral and ring composite structures for offloading hoses. Mater. Des. 108, 374–382.

Tonatto, M.L., Forte, M.M., Amico, S., 2017a. Compressive-tensile fatigue behavior of cords/rubber composites. Polym. Test. 61, 185–190.

Tonatto, M.L., Tita, V., Araujo, R.T., Forte, M.M., Amico, S.C., 2017b. Parametric analysis of an offloading hose under internal pressure via computational modeling. Mar. Struct. 51, 174–187.

Tonatto, M.L.P., Tita, V., Forte, M.M.C., Amico, S.C., 2018. Multi-scale analyses of a floating marine hose with hybrid polyaramid/polyamide reinforcement cords. Mar. Struct. 60, 279–292.

Trelleborg, A.B., 2008. Kleline Hoses. Clermont Ferrand: (company catlalogue).

Tsai, S.W., Wu, E.M., 1970. A General theory of strength for anisotropic materials. J. Compos. Mater. 5, 58–80.

Yang, X., Saevik, S., Sun, L., 2015. Numerical analysis of buckling failure in flexible pipe tensile armor wires. Ocean Eng. 108, 594–605.

Zandiyeh, A.R., 2006. Fatigue-life prediction in offshore marine hoses. Oilfield Engineering with Polymers 2006, 1–12 , March.

Modern yacht rig design

Hasso Hoffmeister
DNV GL, Hamburg, Germany

Chapter Outline

15.1 Introduction 418
15.2 "Why" is a rig? 419
    15.2.1 Aerodynamics: slenderness 419
    15.2.2 Means of adjustment: stiffness 419
    15.2.3 Ease of handling: complexity 420
    15.2.4 Robustness: strength 420
15.3 Modern rig configurations 421
15.4 Selected design considerations 422
    15.4.1 Spreader sweep 422
    15.4.2 Rigging angles 423
    15.4.3 Pretension 424
    15.4.4 Stability 426
    15.4.5 Mast section 426
    15.4.6 Number of spreaders in a rig 427
    15.4.7 Diamond jumper arrangements 428
    15.4.8 Headstay arrangements 428
    15.4.9 Backstay arrangements 430
    15.4.10 Size and size effects 431
15.5 Why weight savings? 432
15.6 Material selection 432
    15.6.1 Comparison 433
    15.6.2 Composites mechanical properties 435
15.7 Rig analysis technologies 437
15.8 Statics and dynamics 438
15.9 Rig loads 439
    15.9.1 Monohulls 441
    15.9.2 Multihulls 441
    15.9.3 Developing a load matrix 442
15.10 Design criteria; safety margins, reserve factors 442
    15.10.1 Strength 442
    15.10.2 Standing rigging 443
    15.10.3 Mast and spars 444
    15.10.4 Stiffness 444
15.11 Future trends 445
    15.11.1 Fast monohulls 445
    15.11.2 Wing rigs 447
References 448
Further reading 448

15.1 Introduction

All sailors know WHAT a rig of a sailing yacht is. A yacht rig is dominant. A yacht rig is conspicuous. Even when the rig of a sailing yacht, its "main engine", is turned off, it cannot hide. A rig stands out. A rig is also a representing figurehead: "Look, I am a racing yacht!", another: "Look, I am here to defy all storms" (Fig. 15.1).

To aid the nonsailing readers, a glossary of sailing terminologies has been included after the references in this chapter. Every sailor knows a rig consists mainly of spars and stays. In former times, when materials were not as advanced, the sail plan had to be portioned in many smaller scale sails, all of which needed to be set and reefed by many hands. This required entering the rig in all weather conditions, often with more than a dozen crew aloft, with all the associated risks.

Today, modern material and modern technology have made it possible to travel faster—and safer. Sails are easier to handle, with less crew. Earlier, rigs used to be made from raw steel, wood, and wire ropes and sails made from cotton, but today rigs and sails are predominantly made from carbon and carbon composites.

In this chapter, we are looking into modern rig and rigging technology partly from the perspective of a classification society involved in the verification of rig designs for more than a hundred years. The invention of carbon composites prompted the industry to ask DNVGL to develop a new design standard; this was after a major rig failure of an early carbon fiber rig in 1994. DNVGL has since been entrusted to certify the design of most of the modern carbon sailing yacht rigs on the Superyacht market.

Fig. 15.1 Rig of a modern Ketch.

This chapter aims to provide an overview of modern rig design on an intermediate level. It considers the methodologies and analysis technologies used, assuming reasonable but not excessive budgets are available for optimization. This will usually be the case in most superyacht projects, where those yachts are predominantly used for pure cruising, and occasionally for (mostly coastal) racing.

Having accomplished the activities to develop a standard with its associated methodologies, we would like to assist in answering the question: "WHY is a rig?"

15.2 "Why" is a rig?

One first "why" is easy to answer: A rig is there to hold the sails. But having a closer look, it is also quite complex; this unique function is at the same time associated with a great variety of aspects which need to be considered for making the system work as efficiently and as reliably as possible. Sails are simply used to generate aerodynamic driving forces to pull a yacht through the water. Still, there are not only the aerodynamics playing a role as with aircraft, but we also have to deal with the fact that we are working at the boundary layer between two fluids; air and water. Sail design and rig design always need to go hand in hand with the conditions, but also restrictions, imposed by the hydrodynamics. Both aero and hydro systems should work together well to achieve the best results. In general terms, both systems should match in their physical efficiency, otherwise the system (either one) won't work well.

In order to understand the reasons for certain rig configurations, designs, and construction, the underlying principles will be illuminated. Not only the technological principles themselves are decisive, but also how the available techniques can be approached and solved. Even though today highly sophisticated design tools are available, great experience and understanding is required on what is essential. There are a few traps and some "don'ts"; things to consider and/or advised to be avoided from the beginning. This reduction makes the exploration of the design space slightly less complex.

A sailing yacht rig has four underlying functions:

15.2.1 Aerodynamics: Slenderness

The rigs' only objective is to act as a carrier system for the aerodynamically effective sails, providing a "shape platform". The rig should ideally be as inconspicuous as possible, to generate as little parasitic aerodynamic drag as possible. The mast and all standing rigging elements are ideally as slim as possible.

15.2.2 Means of adjustment: Stiffness

Considering that the rig elements shall be aerodynamically slim, they shall still provide a certain stiffness, so that the sails can work efficiently. A mast is designed to be intentionally bent to adjust the aerodynamic shape of the sails, for different wind

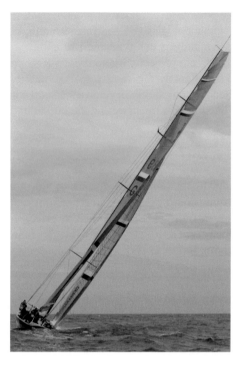

Fig. 15.2 Rig as an "aerodynamic shape platform."

conditions and heading angles and to depower a sail when the wind is strong. Sailmakers and rig designers must work hand in hand, so that the sails/rig package works together as a unit (Fig. 15.2).

15.2.3 Ease of handling: Complexity

Considering the two aforementioned aspects, most often a compromise needs to be made with regard to the ease of handling. A rig needs to be controllable by the number of crew available and possibly also be designed to the capabilities and experience of the crew. A racing yacht with high manning provides more possibilities to control/adjust the rig to the aerodynamic objectives than a shorthanded owner-driven cruising yacht, where the crew might prefer a more simplistic rig, yet making the handling easier but restricting the possibilities for trimming/adjustments.

15.2.4 Robustness: Strength

Fulfilling the above-mentioned demands, a rig has to stand up for structural and operational safety, in all designated and foreseeable weather and sea-conditions. Various different safety margins were established over time. All of these have to accommodate the sometimes-great variance of structural loads which a rig experiences; depending on individual trim, pertinent dynamics occurring during service, the diversity of structural design, and the detail of how these structures were analyzed. Further, a certain conservatism shall be exercised covering some margin of human error or mishandling.

15.3 Modern rig configurations

A modern rig must accommodate the requirements from a modern sail plan. Sails with high aerodynamic efficiency are possible today only with modern materials. From these, what we call modern "Bermudian" rig style has developed, flying reasonably high aspect ratio sails. The latter are most effective for covering the need for the highest versatility of upwind/downwind sailing.

To achieve the desired sail shape, low stretch sail fabric is necessary as well as stiff enough sail battens to span a plane, often with highly efficient square top mainsails. The sloop style rig, featuring one main sail accompanied with one headsail, has proved to be very efficient in the physical environment of a sailing yacht. Compromises are only made in order to size down the sail plan or if upwind performance is not the first priority. Sail plans are then divided up to reduce sizes; this can lead to adding another mast to the rig. Ketch or Schooner rigs are more complex in handling and design than sloop rigs. Sometimes, masts are interconnected so that trimming both masts in a perfect way is very difficult. Also, a connected rig can interfere with the global hull girder stiffness; so designing the rig for hull stiffness sometimes needs to be considered. Having both masts separated avoids both downsides, but requires more longitudinal separation, so that each mast can individually be supported (Fig. 15.3).

A rig system comprises two different character structural members: Beams and Trusses. Beams are mainly there to carry compressive and bending loads; trusses are there to carry tensile loads only. A clever assembly of beams and trusses fulfills all the four demands mentioned above. Masts, spreaders, and booms act as beams; standing rigging acts as truss elements. Combined, this assembly acts as a framework or a spaceframe, providing the required strength and stiffness and provide a good compromise to minimize the aerodynamic drag they generate.

This holds true for most modern stayed rig systems, whereas an unstayed rig solely relies on one beam (the mast) in bending. The latter, yet rare application, is realized by free standing wing masts or rigs like the so-called Dynarig, as described by Pekins et al. (2004), which is not further pursued in this chapter.

Rigging elements supporting a mast are typically distinguished between "lateral" rigging (shrouds) and "longitudinal" rigging (stays). The lateral rigging is mainly designed to support the mast in transverse direction, opposing the side forces generated by the sails. Longitudinal rigging is to support the rig framework and also to carry sails. There is a high amount of interaction between these two systems, particularly when the supporting spreaders are angled (swept).

Fig. 15.3 Modern rig configurations, a sloop (one mast) and a ketch (two masts with the forward being higher than the aft).

15.4 Selected design considerations

Among the full diversity of design criteria for a sailing yacht rig, there are some obvious key aspects which should be considered as primary:

15.4.1 Spreader sweep

Spreaders, often multiple sets, are the beam elements designed to span a framework of vertical shrouds, diagonal shrouds, and the mast. When the spreaders have 0° sweep angle (Fig. 15.4) (all spreaders on both sides are in one plane with the mast; pointing only sideways), this lateral framework is quasi-independent from the effects of the longitudinal rigging and longitudinal bend. We call this system an "in-line spreader"

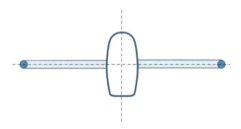

Fig. 15.4 "In-line"-rig with great trim versatility.

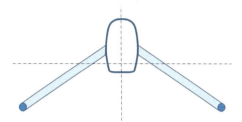

Fig. 15.5 "Swept spreader"-rig with great global bending stiffness.

rig. This rig type is used, when there is a requirement for a large range of mast bends in order to control the shape of the mainsail.

When the spreaders (and the associated lateral rigging) are being swept back by a certain angle (Fig. 15.5), the system of shrouds and mast is no longer in line. The system now comprises a 3D spaceframe, where any longitudinal bending required to influence the shape of the mainsail is harder to achieve, because this system is obviously stiffer in longitudinal direction. On the other hand, masts can be designed to be slightly slimmer in this 3D spaceframe, because the mast alone does not have to provide the bending stiffness; rather the whole spaceframe system.

A rig system needs to be bent longitudinally, to adjust the sail aerodynamics. There is a certain spreader sweep angle range within which this works well. Apart from the in-line spreader situation, sweep angles below ca. 4° are to be avoided because the framework would be too "flat" and there is the danger of overloading the shrouds in tension and the mast in resulting compression, when subjected to bending. Sweep angles above 25 degree seem to be ineffective.

On today's rigs, we often see sweep angles around 15 degree with modern racing yachts and around 20 degree for cruising yachts.

15.4.2 Rigging angles

Cap shroud angles (see Fig. 15.6) usually comprise an attachment angle of around 10 degree to the mast, which, together with the system of vertical and diagonal shrouds and spreaders, provide a good compromise on the headsail sheeting angle, for headsails designed to overlap the shroud base. Modern racing yachts can often realize much smaller headsail sheeting angles; so small that these sails cannot overlap the shroud base. Wider rigging angles (see Fig. 15.6) are obviously structurally more efficient, but much wider lateral rigging envelope can also be problematic, because the rigging weights are moved far out and this could lead to the excitation of torsional instability of the rig system.

Attachment angle of longitudinal rigging is often driven by the mast height/hull length relation and often comprises angles in the 20 degree range.

On multihulls, which offer the possibility of a much wider shroud base, the rig system configuration is often different from those seen on monohulls. The mast is often supported by only three rigging elements, two shrouds, and a headstay. To stiffen up the long freestanding portion of the mast, a self-supporting system of diamond shrouds is often used.

Fig. 15.6 Cap shroud angle, shroud base width, and jib leech profile for a narrow and a wide shroud base rig.

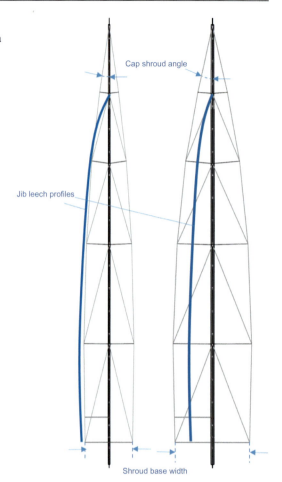

15.4.3 Pretension

On the mentioned rigs comprising multiple spreader sets, the side shrouds system provides the lateral strength and stiffness to the rig. Especially when the rig comprises swept spreader envelope spanning the spaceframe, it is of utmost importance to maintain tension on the leeward cap shrouds when the rig is loaded by wind forces; otherwise the rig system loses its 3D spaceframe stiffness and provides only 2D spaceframe stiffness with a much lower stiffness and strength, which can compromise the rigs' integrity, as schematically shown in Fig. 15.7.

The reason for maintaining some reserve tension on the leeward rigging is that as long as a rigging element is under tension, it contributes as a spring in the spaceframe system. While a rig is heeled over, the tension on the windward side increases, while the tension on the leeward side reduces, leaving the mast compression theoretically constant.

Modern yacht rig design

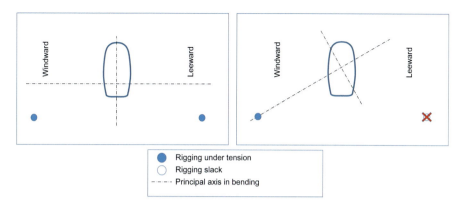

Fig. 15.7 Shift of bending principal axis and drop in global stiffness under lack of sufficient rig pretension.

Thus, windward and leeward rigging under tension provides two springs supporting the mast; once the leeward shroud slackens, there is only one spring left active and the sideways deflections over-proportionally increase with increasing load. This effect is illustrated in the schematic Fig. 15.8. The orange curve representing both windward and leeward shrouds tight, thus is active across a certain heel range and has

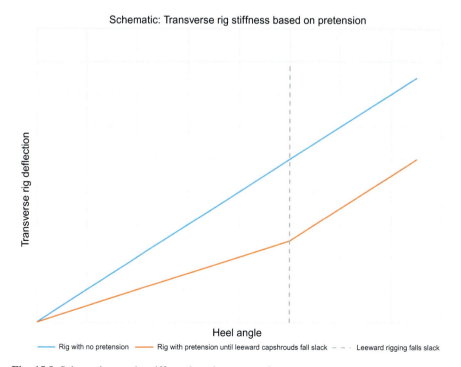

Fig. 15.8 Schematics on rig stiffness based on pretension.

higher sideways stiffness compared to the untensioned rig. Once the pretensioned rig exceeds the point where the leeward rigging falls slack, the stiffness (slope of the deflection graph) is similar to that of the untensioned rig. Rigs relying on the spaceframe stiffness shall, in general, operate below the point where leeward cap-shrouds fall slack (in Fig. 15.8 indicated by a dashed line).

Setting a rig under pretension is realized by pumping the mast up at its step (base), most often using hydraulic rams. This procedure requires the shrouds to be adjusted in a way that the final tune sequence provides perfect settings and often a very slight positive bend of the mast (convex bend with the mainsail luff; ca. 0.5% of mainsail luff length "P").

15.4.4 Stability

Besides the pure minimum strength requirements for endurance and structural reliability, the stiffness of a rig system is not only important for providing an appropriate shape platform for the sails, but also is essential to provide dimensional stability and safety against a collapse in buckling. Prone to buckling are obviously beam elements of the rig which are subjected to compressive forces. The compression load within the mast column is not a constant, but increases from top to heel of the mast (see Fig. 15.19). In transverse direction, the mast is supported at each spreader attachment; those act as transverse supports. In longitudinal direction, the mast is supported by the longitudinal stay. These supports are not in general totally fixed, because they imply a certain stiffness; for a longitudinal stay support, the stiffness of this support obviously depends on the attached stay stiffness. The less stiffness the stay has, the softer this support point acts. This occurs, for example, in the Buckling mode shown in Figs. 15.22 and 15.23.

In the longitudinal direction, the buckling structure is comprised of the mast including its transverse spaceframe. Consequently, a mast with higher spreader angles is stiffer in longitudinal direction, as the rig is acting as a framework.

15.4.5 Mast section

The selection of an appropriate mast section depends on several different aspects and, obviously, an optimization loop will have to be completed to find the best solution:

(a) Minimize section size for minimizing aerodynamic drag

A mast section of a nonrotating, stayed mast needs to be as small as possible to minimize aerodynamic interference with the sails. The highest aerodynamic efficiency will be required for apparent wind angles from ahead, but need to provide a good compromise for beam winds.

(b) Optimize section shape for minimizing aerodynamic drag

A common compromise of a mast section shape is a "bullet"-shaped section, often with depth/width ratios of around 2. For longer sections, large areas of flow detachments can occur on the leeward side (Fig. 15.9), reducing the aerodynamic efficiency of the whole system.

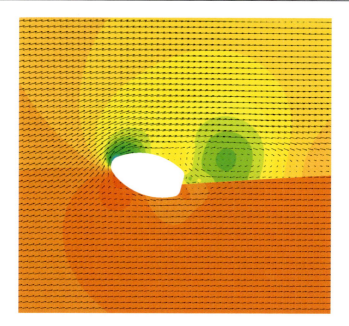

Fig. 15.9 CFD Plot showing pressure and wind speed around mast section. Courtesy of Yacht Research Unit Kiel.

(c) Maximize section shape for maximizing section stiffness at minimum structural weight

The least material invested to provide high stiffness is that featured by a large section size. However, there are limitations to this optimization. A certain minimum wall thickness of the mast section has to be provided in order to avoid shell buckling. Also, a mast needs to accommodate local strength attachments and reinforcements.

Obviously the three main criteria for section design are partly contradictory. High profiled racing syndicates often will invest in a tailor-made optimized section shape development which will be the "optimized compromise."

Most mast manufacturers for cruising rigs will choose from an available set of shapes and sizes as a best of available compromises. This is because carbon masts will be built in negative moulds or over mandrels, both of which are comparatively costly to produce.

15.4.6 Number of spreaders in a rig

The question of how many spreader sets are needed in a rig is not so easy to answer. It appears that, for taller rigs, more spreader sets are common; but why is this so?

In fact, the decision on the correct number of support points is influenced by different aspects, which all interact with each other.

A designer will start his mast design with mast section shape, which has low magnitude of aerodynamic interference, but as light as possible. This optimization loop has to incorporate the design criteria from the previous paragraphs. Having chosen a section, the transverse stiffness accommodated by both the geometric shape and the material invested, together with the local mast compressive load, will define a minimum support distance in terms of buckling length. The latter parameters are directly linked; the higher the compressive forces, the shorter the support distance shall be or the stiffer the section.

Rigs with narrow chainplate beam consequently suffer higher compressive forces in the mast (and tensile forces in the shroud). A narrow rig envelope can be forced by a hull shape or by the requirement to have a narrow sheeting angle for an overlapping headsail.

The result of the section optimization with regard to stability, strength, and the occurring mast compression often results in typically three spreader sets for yachts in the 50–100 ft. (15–30 m) range and goes up to five when yachts are longer than 150 ft. (46 m).

In the early 2000s, the Team New Zealand for the 30th America's Cup invented the so-called Millennium-Rig. This rig featured a reduced number of spreader sets while still adding transverse supports. This was featured by crossing diagonals. This idea was followed up by almost all America's Cup syndicates when designing their IACC (International Americas Cup Class) yacht rigs. This design feature, however, remained unique and did not divert into any other application. One reason for this is that those rigs were pure in-line spreader rigs. They had a great versatility to mainsail trim and also the whole mast could be cranked forward in order to provide better downwind performance. These rigs were difficult to operate and the more recent development of designing lighter and faster boats diminished the requirement for pure downwind sailing; instead, gennakers fixed on bowsprits came in fashion, as were nonoverlapping upwind headsails. Today, most performance cruising yachts and also racing yachts feature rigs with swept spreaders, offering a light and stable solution with minimum drag (Fig. 15.10).

15.4.7 Diamond jumper arrangements

Diamond and jumper arrangements are systems which are self-contained. They often serve to stiffen/reinforce a region of the mast which is not supported by shrouds/stays. Sometimes, they are adjustable and serve as an additional trimming device. Often they are seen on multihull rigs, where the shroud base is wide and the spreader-less rig is held by only one set of shrouds.

15.4.8 Headstay arrangements

In addition to a mainsail, with its luff fixed to the mast, a modern sail plan of a performance-oriented sailing yacht typically features at least one, out of a choice of many, headsails. The larger the yacht, the more difficult the handling of these large sails gets. This is a reason why most large yachts are equipped with more than one headstay, each around which a headsail of different sizes is furled, ready for easy deployment. Some of these additional headstays are permanently rigged and additionally support the mast in longitudinal direction, while some may be temporary options; those will often be furled on a high strength composite stay, which all together can be

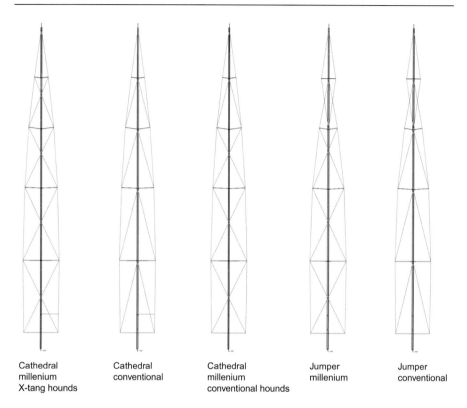

Fig. 15.10 Different shroud arrangement alternatives for an in-line spreader rig in early development phase.

stowed below deck when not in use. These are usually staysails set in addition to a gennaker or as a stormsail or so-called Code sails set in front of the outermost permanent headstay, for light upwind conditions.

All of these stays on which sails are hoisted require a certain amount of tension. This is to avoid excessive transverse sag imposed by the wind force. Most upwind headsails are shaped so that they develop their flying shape in a sideways stay sag of between 1% and 3% related to its length for normal conditions.

The transverse loading of these headstays by wind force induces axial forces which analytically can be estimated using the formulae for catenary sag. A simplification of this complicated differential equation gives reasonably exact results:

$$F_T = \frac{F_{Lat}}{8 \cdot s}$$

F_T = tensile force resulting from sagged catenary due to a constant lateral line load.
F_{Lat} = cumulated line load acting laterally on the stay.
s = transverse sag, expressed in fractions of the stays length.

The challenge is to find proper tension settings for the headstays when they are used in different combinations and conditions. The best possible solution to maintain minimum tension on unemployed stays while not exceeding maximum target values on employed ones is to engage length adjustment equipment. Releasing the so-called real time adjusters provides more sag to a sail, thus reducing the tension; or the other way around. In certain situations, it is required to have some reserve tension on the stays so that they are active in stabilizing the rig; this can be achieved then by shortening the stay. These adjusters work by means of hydraulic cylinders (Fig. 15.11).

15.4.9 Backstay arrangements

As headstays, the aft rigging is designed to support the rig in longitudinal direction. Usually, the effective attachment angles for longitudinal rigging are significantly greater than for transverse rigging; this is simply due to the fact that a yacht is offering more length than breadth and thus the chainplate base is larger. An even more important purpose than simply supporting the mast in longitudinal direction is to counteract the forces generated in the headstays by the phenomenon of the catenary, as already described in Section 15.4.8.

Often, a permanent, yet adjustable, backstay is mounted on a short aft-swept masthead crane. This measure provides space to accommodate the mainsail headboard and, at the same time, allows some convex roach in the mainsail leech.

Fig. 15.11 Illustration of headstay sag "s."

Modern yacht rig design

For rather slender rigs with the need for a lot of room for mainsail shape adjustment through different mastbends, there is the necessity for one or more further longitudinal supports. Those will be put into effect as running backstays ("runners" or "checkstays"), which have to be de −/employed when changing tack. Runners are also required for counteracting sail flying headstays which are mounted in addition to the main headstay.

Often, the offset between the attachment points of counteracting stays is used to achieve desired action, for example to bend the mast when the loads increase, as schematically indicated in Fig. 15.12.

In order to reduce weight and windage, runners attached at different points of the rig are combined by leading them together; while still offering full adjustment range.

15.4.10 Size and size effects

The expression "PanMax" has originally been established in the merchant shipping business. It expresses a maximum size of seagoing ships, particularly container vessels, to be able to pass the Panama Canal locks before they were increased in size in June 2016. However, the maximum air draft for passing the Panama Canal is still 205 ft. (62.5 m). Most of the Superyachts have to include this limitation in their design plan, should they intend to travel through the canal. However many of the very big Sailing yachts are "PostPanmax" (Fig. 15.13)

Modern yacht rigs on large sailing yachts have reached sizes exceeding the above limit by far. At the time of writing this chapter, the tallest carbon mast is 108 m above waterline (Sailing Yacht "A").

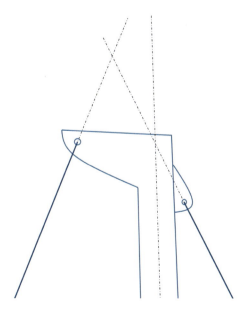

Fig. 15.12 Typical geometric relations on a masthead.

Fig. 15.13 Men working at diagonal tang during mast stepping of a 86 m LoA Ketch "Aquijo."

One of the most noteworthy size effect is that when hoisting a person up the mast, they must be connected to a downhaul line; otherwise, the weight of the hoist line might at some stage exceed the weight of the person, leading to inadvertent upwards acceleration.

15.5 Why weight savings?

It is not up to the author to quantify the weight savings in a rig, but it is obvious that, with the large vertical erection of a rig, a multiple quantity of counter ballast is required in order to balance the rig or keep the righting moment constant. In other words: weight savings in the rig can lead to:

— Improved performance due to higher available righting moment
— Reduced ballast leading to a reduction in overall displacement
— Reduction of inertia effects, higher agility
— A combination thereof

The big milestone in weight savings has been marked with the invention of carbon masts, weighing roughly only half of an aluminum mast of similar strength and stiffness. The second huge step was made with the invention of fiber-based standing rigging (Pemberton and Graham-Jones, 2016). As here steel is replaced by carbon, the absolute weight savings are in a very similar order to those of the mast alternatives.

15.6 Material selection

Masts were built in solid wood or in riveted steel plates a hundred years ago; standing rigging was made from galvanized wire ropes or even natural fiber ropes earlier on. Over the course of time, hollow aluminum profiles found their way in the mid-30s of the last century, pioneered as usual in high profile racing events such as the Americas Cup. The use of aluminum started as riveted plate assembly. Later, after extruding of aluminum alloy was invented and was getting more common, the use of this option gave durable and light spars, which were at the same time affordable and easy to maintain. This holds valid until today. However, the choice of materials used in sailing yacht rigs has significantly broadened over the last century. The quest for more competitiveness, thus lighter rigs, pushed the use of carbon fiber as a structural material. Starting with first commercial rig applications in the early 1990s, they have found full acknowledgement alongside the aluminum spar production since.

Almost a decade later than the introduction of carbon fiber for masts and spars, another replacement process started. This was when fiber rigging was invented to replace steel rigging. Also, today, both technologies exist alongside; the much higher investment cost has, however, hampered a total replacement of steel rigging.

15.6.1 Comparison

If mechanical properties of composites are compared to those of steel (standing rigging) and aluminum (masts and spars), there is the risk of not being objective. As the materials are so vastly different, the base of a comparison must be clearly defined; one should compare typical structural functional elements rather than the materials themselves. However, even that is difficult, because the structural setup and the way a structural element works are different. Additionally, the physical natures of metals and composites require different safety margins in design. Where with some technical applications weight is not critical and it might be alright to look at absolute values of stiffness or strength, often optimizing strength and stiffness includes a look at the structural weight; thus a view at the **specific** properties might be more revealing. If weight of structures is critical, the benefits of composite materials arise, simply because structures can be much lighter for same strength and stiffness; however, at greater labor and material cost.

Fig. 15.14 compares the tensile stiffness in different ways. The absolute values are compared to the specific ones, the latter relating to the material's density. This is done to pinpoint how effective composite materials are when weight matters. Further, more related to design purposes, the ultimate (Fig. 15.15) and the allowable strength values (Fig. 15.16), derived from DNVGL Rules for typical structural hull components, are compared using the same method (absolute vs specific).

Since a large bulk of mast material is designed for stiffness (the bare mast tube), the weight advantage of a carbon mast can be directly estimated from the specific stiffness differences, of course depending on the carbon fiber choice made. A weight of a carbon mast will be roughly 60%–70% of an aluminum mast.

Carbon standing rigging will approximately be only ca. 25% of the weight of stainless steel Nitronic 50 rigging.

Both savings individually amount to similar net savings in the overall weight balance of a rig.

Besides strength and stiffness, metals and composites differ widely when aspects such as degradation modes, robustness, or maintenance are looked into.

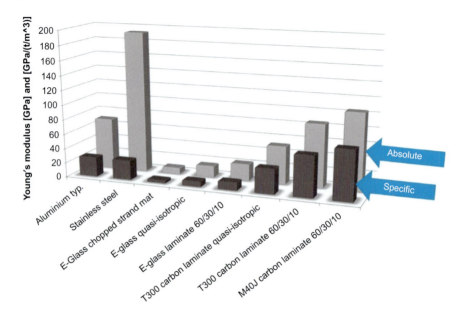

Fig. 15.14 Material comparison on absolute and specific tensile stiffness.

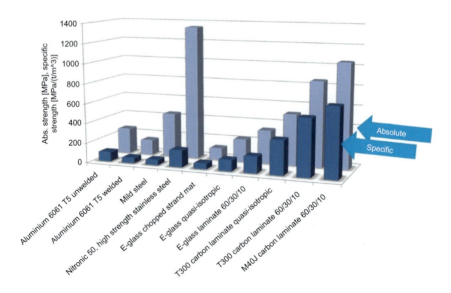

Fig. 15.15 Material comparison on absolute and specific tensile strength.

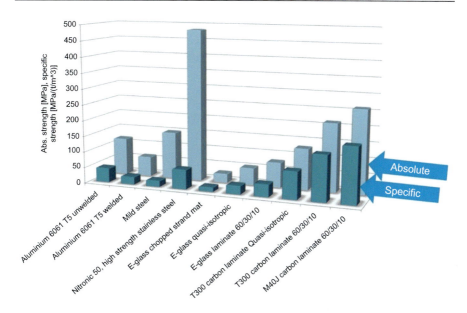

Fig. 15.16 Material comparison on absolute and specific allowable tensile strength; "design values."

15.6.2 Composites mechanical properties

Mechanical properties of composites can be tailored for given load paths by aligning fibers in force directions. For pure tensile elements (e.g., standing rigging), all fibers can be aligned in one direction, providing maximum tensile strength and stiffness. For most other marine components, it is common to find a compromise, where fibers are laid in multiple directions, with or without prioritizing one main direction. Often, off-main-axis properties must fulfill particular structural functions also. Different fiber types and fabric types can be utilized to suit particular needs. If no particular directional properties shall be obtained, fibers can be aligned such that the in-plane properties are similar in all directions ("quasi-isotropic"). To achieve this, the same numbers of fibers need to be arranged, at least in three different directions ("triaxial", 33.3% 0°, 33.3% 60°, and 33.3% 120°). Most often, however, marine structure laminates are arranged such that four different basic fiber directions are allocated, 0°, ±45°, and 90°. For a mast or other longitudinal elements, the 0° direction is generally representing the mast's longitudinal axis. For example, typically, a carbon mast has the following fiber fractions in those directions: 60% 0°, 30% ±45°, 10% 90°.

The off-axis stiffness and strength properties for non-quasi-isotropic composites vary from the given values. This variance increases as more emphasis is given on single fiber directions. Fig. 15.17 illustrates the distribution of the in-plane tensile stiffness of carbon laminates for different orientation fractions.

The through-thickness properties of composites add another dimension to the complex behaviors. Where metallic materials in general can be assessed as completely isotropic in

436　Marine Composites

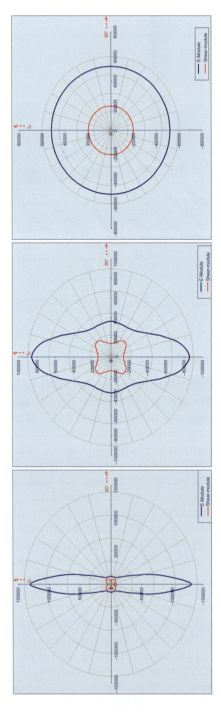

Fig. 15.17 In-plane tensile and shear stiffness of different carbon laminates (at different scale).

all directions, through-thickness stiffness and particularly through-thickness strengths of composites need to be considered when designing composites. At the same time, there are some more advantages associated with the possibilities of tailoring mechanical properties: by having freedom in the layup sequence of composites, properties such as the plate-bending characteristics can be influenced; this can be an important factor when looking at the local shell buckling stability, but also for local strength applications.

Carbon fibers themselves are quite good electrical conductors. Having an electrical conductivity of ca. 2–14 × MS/m (depending on the fiber type), this is comparable to Copper (roughly 58), Aluminum (35), and Nitronic50 stainless steel (14), but within a polymer matrix, the conductivity compared to metals is no more than 1/1000. This property makes carbon fiber laminates worth protecting from lightning strikes, where the metal counterparts can deal with this by themselves. With the low electrical conductivity and the associated high electrical resistance, there is not enough sectional area to conduct a lightning current to ground. The lightning energy, therefore, resolves by heat development, boiling, or combusting of the polymer resin, leading to the lack of integrity of the composite. Since rigs are quite exposed, an appropriate protection of carbon masts and rigging is important.

15.7   Rig analysis technologies

The global view of analyzing a framework consisting of beams and truss elements suggests the use of FEA ("Finite Element Analysis"). Large bend and deflection values occurring in a rig automatically prohibit the use of trigonometrical equations to analyze the structural response; it can lead to drastic over−/underestimation of structural loadings. So does (geometrically) linear FEA analysis; it will only deliver similar results. Consequently, it is highly recommendable to use the geometrically nonlinear option when performing FEA.

The latter condition will automatically be fulfilled, when the structural FEA is complemented by CFD ("Computational Fluid Dynamics") to make up a complete FSI ("Fluid Structure Interaction") with structural rig elements and sail membrane elements in their flying shape under the influence of an air stream or a pressure array (Fig. 15.18) (Fallow, 2016).

Fig. 15.18 Comparison of sail in FSI and on-the water for a superyacht jib (©North Sails).

For every numerical model, there is freedom about how detailed a model is created. If a model is kept simplistic, not all areas of the model will show their explicit behavior. On the other hand, if a mast is modelled as a simple beam only, no influence on local stress, such as concentrations around a local load introduction point, will occur. However, a simpler model can be very efficient in studying the global behavior of the rig. The results of design variations can be easier identified using simpler models. The characteristics best developed using a simplified, thus efficient, model are:

- Shroud and stay arrangements
- Effects of spreader sweep angle and rig envelope (spreader lengths)
- Bend characteristics and optimization to meet sail designer's targets
- Standing rigging choices and arrangements
- Stability developments
- Geometric variants

Thus, for all available common techniques to design a rig for appropriate operational performance and structural reliability, the first step is to develop the global behavior of a rig; by determining the tensile forces in standing rigging elements and forces and moments in the mast and spreader elements (Fig. 15.19).

At a later stage, these results are employed to design the (local) structural details such as shroud attachments, spreaders, etc. Often, analytical methods are used to design these typical details; or detailed FEA can be established, if budgets allow or stress state is too complex to allow for analytical approaches.

15.8 Statics and dynamics

Sailing at a constant speed and constant wind in calm water can be considered as quasi-static condition for the structural behavior of the rig. However, besides the fact that long-term weather is never a constant and also short-term weather will occur as gusty wind in speed and direction, dynamics will add once a yacht is subjected to

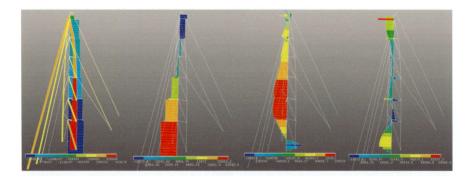

Fig. 15.19 Four out of six global loads in a rig; Axial Forces, Torque, Longitudinal bending moment, Transverse bending moment.

motions in irregular seaways. There are three linear dynamic motions, occurring as translations, surge, sway, and heave. And there are three rotations, yaw, pitch, and roll. Of these six degrees of freedom, only two are usually primarily relevant to rig design; these are the pitching and the rolling motions.

There is not much of a rolling motion observed, when the yacht has sails set. This is because the sails are usually spanned perpendicular to the rolling direction and this dampens this motion significantly. However, the hull gets excited by seaways, so there is a dynamic rolling moment excited, but dampened, by the rig.

The pitching motion is also excited by seaways, particularly significant when the yacht is crossing waves. These motions are hardly damped by the sails, so that pitching motions can develop and thus inertia effects are excited on the rig. It has been shown that the pitching motions are particularly exaggerated, when the yacht is facing waves of a wavelength in the same dimension as the yacht length. The motions are even more drastic when the yacht is carrying no sails and running under engine.

Pitching occurs as a nonharmonic rotational acceleration. The center of rotation is not a fixed geometrical location, but estimated to be located around the half-length waterline. The oscillation around this point is generating high inertia forces in the rig, particularly because the weights are far away from the axis of rotation.

Having determined the frequency of encounter between a yacht and a wave, both with individual speed of travel and with estimating a pitch angle amplitude, it is possible to estimate a rotational acceleration with simple equations.

Maximum rotational angular acceleration; $\dot{\omega}_e$ [rad/s^2]:

$$\dot{\omega}_e = \theta \cdot \omega_e^2$$

where

θ = maximum pitch amplitude [rad]
ω_e = angular frequency of encounter between wave and yacht [rad/s]

This rotational acceleration can be fed into a FEA, if the dynamic behavior of a rig is sought to be critical. In case of lack of real simulations, this value needs to be chosen in a conservative way, as it includes simplifications. Particularly critical are cases in which, due to accelerated motions, rigging falls slack and retightens in a short term. These shock loadings can cause damage on rigs.

Typically, however, rig structural analysis techniques feature only a quasi-static approach. The margins required to cover dynamic effects are usually included in the pertinent safety margins for loads and material usage.

15.9 Rig loads

The only purpose of a rig is to assist sails in generating driving forces to propel the yacht. These driving forces are generated by aerodynamic pressure differentials.

The aerodynamic forces generated by a sail can be divided up depending on the coordinate system of interest, as indicated in Fig. 15.20:

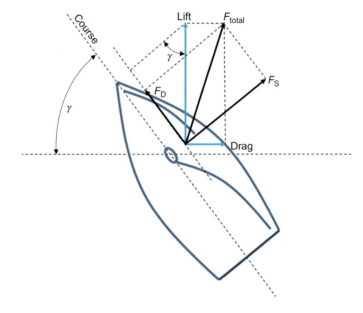

Fig. 15.20 Simplified aerodynamic forces acting on a sailing yacht rig.

- Lift force "Lift" is the portion of the total aerodynamic force which is pointing perpendicular to the apparent wind direction
- Drag force "Drag" is the portion of the total aerodynamic force pointing in the same direction as the apparent wind
- Both of the above combine to the total aerodynamic force "F_T", which can be divided up in:
- Driving force "F_D"; this is the portion of the total aerodynamic force pointing in the heading direction of the yacht
- Side force "F_S"; this is the portion of the aerodynamic force pointing perpendicular to the heading direction

Generally, in upwind conditions, the driving force is generated by aerodynamic lift, and in downwind conditions, by aerodynamic drag. The intersection between these two principles shifts further towards downwind, the more efficient the aero—(rig and sails) and the hydro—(hull) platforms are; modern lightweight racing yachts achieve their leeway VMG (velocity made good) purely by aerodynamic lift effects; they are using gennakers or so-called code sails working as lift producing means of propulsion. In an extreme, seen with current foiling catamarans or with land—or ice-yachts, the upwind sail plan does not at all differ from the downwind sail plan, as the apparent wind always comes from ahead; they virtually do not jibe downwind to change the tack, but perform tacks.

The design of a rig is predominantly subject to the magnitude of the side force generated by the sails. In comparison, the driving force is relatively small. Even in downwind conditions characterized by drag effects (spinnaker set), the rig design is less critical. This phenomenon is amplified by the fact that a rig in transverse direction has a much smaller envelope with tighter rigging angles and smaller chainplate width than in longitudinal direction.

For high budget yachts and rig projects, a full FSI analysis will often be carried out in which the pressure fields and the air flow are implicit.

For all other techniques not enhancing fluid and structure interaction, the aerodynamic loadings on a rig must be idealized.

In the following paragraphs, this simplified method is further pursued; analyzing or designing a rig utilizing a structural FE-model with no sail membrane elements, loaded with explicit forces, analytically developed from an anticipated pressure field. This method requires to split the approach in two categories:

15.9.1 Monohulls

With monohulls, the aerodynamic side force occurs as a heeling force, causing the yacht to heel away from the wind. This withstanding, the hull is equipped with hydrostatic stability from the hull buoyancy distribution plus the fraction of stability caused by the weight and location of the ballast keel. Together, the ability to withstand the heeling moment caused by aerodynamic effects is characterized as the yacht's righting moment. An equilibrium between heeling and righting is achieved at a certain heeling angle of the yacht.

As the aerodynamics work best when the heeling angles are low, modern performance sailing yachts usually limit their heel angle to maximum of 20–30° of heel.

Except when budgets allow, the approach to find the basic rig design load is to use the yacht's righting moment. This is because, as already mentioned, it is in equilibrium with the aerodynamic heeling forces, which would be more complicated to determine.

For monohull yachts with more than one mast, the above-described methodology is a little more complicated; this is because the available righting moment can be "generated" by either one of the masts; which in turn would mean that each mast has to be designed to the design righting moment. In reality, the righting moment is shared between the masts; in different fractions, depending on the sail plan. A conservative share factor will be attributed to each rig, should the share fractions not be exactly realized (in maneuvering situations).

15.9.2 Multihulls

Unfortunately, the simplistic approach of using the righting moment as a starting parameter for the rig load determination fails, because multihulls have excessive righting moment available, due to their large beam. It would lead to heavily over-dimensioned rig structures, if the same approach as for monohulls was followed.

Instead, the rig is being designed using forces determined from the apparent wind. The problem with a multihull is that it does not heel away when loaded. So, there will always be safety margin on top of average design wind speed, to accommodate the possibility of gusts.

15.9.3 Developing a load matrix

When developing a load matrix for feeding into an FEA study, the distinction between external and internal loads acting on or within the rig must be made. External loads are those directly excited by the pressure effects of the wind, while internal loads are those which are secondarily excited. The latter are:

- Those implied by missing membrane sail elements in the FEA. Membrane forces in sails, acting in the plane of a sail and generated by spanning a required flying shape for a soft sail, can reach quite significant magnitudes. Here, again, the catenary formula can serve to estimate the membrane forces. With appropriately chosen sag values for different sails, it is possible to estimate the leech and corner loads of a sail, with the aid of semi-empirical coefficients. Those will be used to apply to the rig model as individual force values.
- Internal loads excited by pretensioning the rig. As mentioned earlier, a certain state of pretension needs to be put to the rig to provide sufficient stiffness and thus stability.
- Self-weight. Especially, when a rig heels over, additional internal forces occur purely implied by the self-weight of the rig and equipment aloft. This effect is not necessarily intuitive, but can be made more transparent when it is assumed that a yacht has zero righting moment across the heeling angle range. No aerodynamic forces are necessary to heel the boat over, but the self-weight of the rig alone causes internal forces; higher tensile forces in the "windward" rigging and higher compression in the mast column. The self-weight effect can cause quite significant internal forces, especially with rigs getting larger. It is recommended to not neglect this effect with yachts longer than ca. 24 m. There are different methods to cope with this effect: Either the FEA can be performed using densities and thus weights attributed to the model elements and adding an appropriate transverse acceleration; or simply by adding representative weight forces applied to the rig model.

A set of load cases will need to be developed featuring varying external wind loads and including the above-mentioned internal forces. Each load case is usually characterized by a different sail setting (and wind speed). As most of these load cases will be developed for what is called "Maximum safe heeling angle," they are all similar in their basic design load parameter, the Righting Moment. However, they all differ in where the loads will be applied (reefed mainsail or different choice and/or size of headsails) and obviously how big their magnitude is (smaller sail produces higher forces to achieve the same heeling/righting moment).

15.10 Design criteria; safety margins, reserve factors

15.10.1 Strength

The structure of a rig must sustain the loads it experiences; and that over time, considering that it will be handled by human beings, not necessarily always in a way which is typically intentionally included in the design space. A usual design space of a sailing yacht would typically include a handling per good seamanship. Even though this benchmark seems to be a bit unseizable, particular semiempirical safety margins and reserve factors have developed over time.

Structural integrity is essential and can't be put in a nutshell better than in the quote from John H. Illingworth in his pioneering book "Offshore":

"... every part and member must be designed not only to fulfil its particular function, but to fulfil it with the design and efficiency of the neighboring members in view and with the character and duty of the boat as a whole in mind."

This particularly holds true for a rig design because there is not often functional redundancy within the assembly of structural elements of a rig, because they are connected in series. As an example, here is a series of structural elements within a piece of standing rigging from bottom up:

Chain plate—clevis pin—toggle—turnbuckle—shroud termination—shroud—shroud termination—clevis pin—lug plate—mast tube.

In the design approach described so far, the load cases defined represent a quasi-static sailing scenario; the reserve factors have to include some allowance for dynamic behavior of the rig. Even though dynamic studies can be undertaken, the array of results may not seem to justify the effort associated to determine these effects.

Structural assemblies are all subjected to physical forces. These forces need to be categorized because they differentiate in nature; they could be purely static and fairly constant (e.g., mainsail halyard) or not; sometimes they include shock loadings (e.g., main sheet in a gybe). The nature of forces is addressed by attributing load factors to them as to represent a so-called design-load.

Further, safety margins are required to compensate for the method of analysis. For the design of details such as a pin/lug connection, simplified engineering methodologies are often used.

The variance of material mechanical properties requires an attribution as well as the production methods.

For some particular items, standard safety margins even cover a fatigue life (standing rigging "cold heads" of Nitronic50 rigging).

15.10.2 Standing rigging

There are generally two solutions for standing rigging for modern rigs:

The long-time proven solution of Nitronic50 [N-50] stainless steel standing rigging. The material comprises very high tensile strength of over 1000 MPa from a cold drawing process. Terminations are most often realized as a "cold-heading" process, where the rod is hydraulically squeezed in the form of an inverted mushroom head, making it possible to fix with structural infrastructural equipment; eyes, turnbuckles, etc. These "cold heads" are at the same time the weak point, as stress concentrations can lead to fatigue failure, if not renewed at intervals. Typically, the overall safety margin of the nominal static breaking load of Nitronic50 rod over the maximum quasi-static working load determined across sufficient number of representative load cases is 2.5 for lateral rigging and 2.0 for longitudinal rigging.

The second, yet more modern solution for staying a rig, is the use of carbon rigging. Often particularly developed for its purpose, it comprises one or a multiple of unidirectional carbon fiber strands, bedded in an epoxy compound. Different manufacturers choose different end terminations and thus their individual structural characteristics cannot be generalized. Therefore, DNVGL has developed a qualification scheme for these products, in which the nominal static tensile strength is not the reference for its structural function, but the "endurance strength". This characteristic is developed in a fatiguing test over 100,000 load cycles. In a residual strength test, the cable must stand twice the intended SWL (safe working load). Following this qualification, the safety margin is actually implied.

15.10.3 Mast and spars

There is a vast diversity of structural solutions within a rig. Spreaders need to be fixed but articulate, standing rigging tangs need to be fixed, sheaves must be installed on roller axles, reinforcements may be required for cut-outs, etc.

Aluminum construction details are typically designed with a surplus of 10% over its yield strength using the design loads.

When designing in composites, a very efficient way of verifying scantlings is to use the simplistic strain criteria. This maximum strain criterion has the advantage of covering all load directions within a laminate to an appropriate degree. The allowable axial strain in carbon laminates in DNVGL Rules is 0.25% for plain laminates, and 0.35%, when they are laterally clamped. The allowable in-plane shear strain is 0.35%, and 0.45%, when laminate is clamped, for example in way of pin bores.

15.10.4 Stiffness

As mentioned already, a rig is there to provide the structural platform for the sails. In order to work efficiently, the sails require some adjustments in shape, depending on the ambient conditions. Therefore, a rig requires a certain flexibility (Fig. 15.21). The resistance against this flexibility is called structural stiffness.

Allowing for the required shape modifications of the sails, there is still a limit to their absolute magnitude. Coming back to the rigging attachment angles of attack: when the flexibility of the standing rigging elements is too great, the framework angles will change and lead to undesirable overloading and uncontrollable deformations.

The whole rig framework is subjected to high compressive load and needs to withstand this load without a buckling collapse. Indeed, it is intended to stay away from this failure mode using quite a significant margin, in order to avoid scenarios where the rig tends to react "indifferent". Using a buckling design criterion has proved practical to evaluate the rig's minimum stiffness. The method of evaluating system stiffness by applying numerical buckling calculations is using the determination of the Eigenvalue.

In the numerical software, two options exist for the solution of an Eigenmode problem; the simpler version is that of a geometrically linear method. This unfortunately ignores the prebend of the rig and uses some other simplifications. But the simplicity of the analysis using a proven safety margin is reason enough that this method will often be preferred over the more complicated nonlinear analysis. The margins used for stiffness of a rig over its linear buckling failure is 2.6–3.1.

Modern yacht rig design

Fig. 15.21 Mast bend as a design target from sail designers.

The buckling often typically occurs as transverse buckling between spreader support points ("local" buckling, Fig. 15.22) and longitudinal buckling of the whole rig framework ("global" buckling, Fig. 15.23).

One further stiffness criteria to address is that of the mast tube itself (Fig. 15.24). The side shells of the mast tube are often of low wall thickness and low curvature. Particularly in proximity to local load introduction points, the sufficient stability of the mast wall needs to be assured. The laminate layup can be optimized for local bending stiffness without compromising the global compressive stiffness or strength of the mast tube. When the mast sections are very large or, for example, in spreader shells, sometimes sandwich cores are used to stiffen the laminate avoiding stability problems. For a safety margin on wall buckling, a value of 2.5 is usually used.

15.11 Future trends

There are a couple of future trends to identify, each in different disciplines:

15.11.1 Fast Monohulls

One area of great development is the design of very fast monohulls, which are assisted by hydrofoils. Faster boats exclude the necessity to sail dead downwind. Due to their high boat speed, the apparent wind angle hardly exceeds 100°, in moderate reaching conditions. The consequential sail choice will require different sail wardrobe; excluding spinnakers and "running" gennakers completely and putting more emphasis on

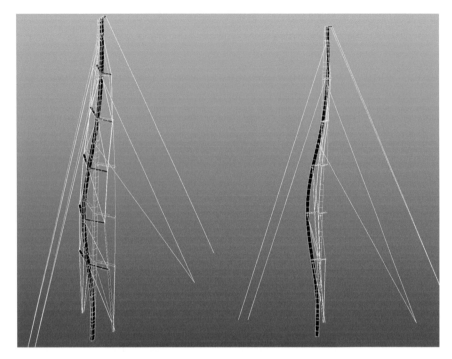

Fig. 15.22 Eigenmode occurring as "transverse buckling."

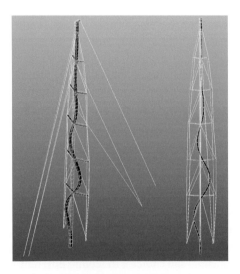

Fig. 15.23 Eigenmode occurring as "longitudinal buckling."

Modern yacht rig design

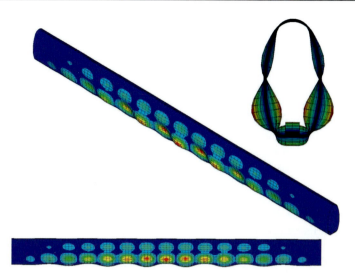

Fig. 15.24 Thin wall buckling mode of a mast section under mainly compressive and bending loads.

lower camber reaching sails being operated in a much lower AWA (apparent wind angle). Reducing windage on yachts like the IMOCA 60's becomes more and more important as a consequence of the high speeds.

The high speeds also make it difficult to adhere to the static righting moment of the yacht as a basic design parameter. This is because the highly fast and dynamic motion of the yachts bring different aspects of stability into the game. Canting keel fins produce lift, particularly when canting axis is tilted upwards, foils produce lift and thus righting moment, and last but not the least, the planning ability of beamy hulls brings up dynamic stability into the righting moment equation.

15.11.2 Wing rigs

The idea of installing a 3-dimensional foil with variable camber, equipped with flaps, has recently been realized on Americas Cup Catamarans. The idea is not totally new, but has passed tentative prototyping efforts done every now and then during the last decades and become a proven technology.

As the hulls on which these wingrigs are installed are highly efficient platforms (using foiling technology and hulls become air-borne), these yachts can reach up to 45kn boat speed in not more than 15kn wind. Even though the America's Cup Teams were given one-design wingrigs, a lot was learnt about the effects associated with lift and particularly with drag. The high apparent wind speeds experienced on these boats forced the designers to work on reducing aerodynamic resistance on rig and platform (and crew).

Wingrig trim operations were as important. Adjusting the camber and also the twist of the wingsail, the desired trim adjustments are needed to be done in shortest time to cope with the wind variability. Using modern sensor technologies, it is not unimaginable that the most efficient wingrig will be one which is automated.

Wingrigs, at least in their rigid foil form, will probably be reserved for top-end grand prix sports like the America's Cup; they cannot be reefed, nor can they be dropped, which makes them impractical for cruising yachts, but currently being observe as a technology transfer to merchant shipping.

Within the initiatives to counter the effects of global warming, shipping is forced to turn "greener" by legislation. The attempts to reduce green-house gas emissions is not only being realized by the use of alternative fuels, such as LNG, but also by the assistance of "sail propulsion systems" (Fritz, 2013). The installation of wingrigs is, besides the installation of "Flettner"-Rotors and "Dynarigs," one of the more popular projects. Certainly, the capital expenditures for such rigs will have to be significantly lower than for an installation on a yacht, thus bringing up the necessity to construct in more classical materials, like steel, aluminum, and polyester sail cloth.

References

Fallow, B., 2016. Computational fluid dynamics (CFD) sail shape optimization, chapter. In: Graham-Jones, J., Summerscales, J. (Eds.), Marine Applications of Advanced Fiber-Reinforced Composites. Woodhead Pulblishing, p. 13.

Fritz, F., 2013. Application of an automated kite system for ship propulsion and power generation. In: Ahrens, U., Diehl, M., Schmehl, R. (Eds.), Chapter Within: Airborne Wind Energy. Springer.

Pemberton, R., Graham-Jones, J., 2016. Applications of composite materials to yacht rigging, chapter 12. In: Grahame-Jones, J., Summerscales, J. (Eds.), Marine Applications of Advanced Fiber-reinforced Composites. Woodhead Pulblishing.

Perkins, T., Dijksta, G., Roberts, D., 2004. The maltese falcon: the realization. In: The Proceedings of HISWA International Symposium on Yacht Design and Yacht Construction 2004.

Further reading

DNVGL "Guidelines for Design and Construction of Large Modern Yacht Rigs", Ed. 12–2016
Illingworth, J.H., 1958. Offshore. p. 133.
Marchaj, C.A., 1979. Aero-Hydrodynamics of Sailing.
Yacht Research Unit des F&E-Zentrums der FH Kiel GmbH

Glossary

Backstays a part of the rigging of a sailing vessel which supports the strain on the upper mast. It runs from the top of the mast to the aft of the vessel. Typically, a backstay would run along the centerline of the vessel, and on most modern yachts, is adjustable. See also runners.

Bermudian the most common type of sailing rig on modern yachts, consisting of a mainsail and a fore sail. The main sail is only supported by the mast and a boom; no upper spars are required.

Boom a spar which is attached to the mast, to support the lower edge (foot) of a sail (typically the mainsail).
Bowsprit a spar which projects forward of the bow, to which foresails (e.g., jib, gennaker) are attached.
Cap shrouds the shrouds which attach to the top of the mast.
Chainplate base chainplates are the attachment points on the hull for the shrouds, and chainplate base refers to the width between the chainplates on either side.
Checkstay a part of the rigging of a sailing vessel which supports the strain on the lower part of the mast. It is similar to a backstay or runners, but is attached to the mast lower down. See also runners.
Downwind a boat is said to be sailing downwind, when it is sailing in a direction which is greater than 90° to the true wind direction.
Furling refers to the process by which sails are rolled up, to store them.
Fennakers are foresails which are a combination of a genoa and a spinnaker. They are intended for downwind sailing, but are attached to the bow (in a similar manner to a genoa) and only one point of the sail is adjusted for changes in angle.
Genoa a larger version of a jib, a sail which attaches to the bow of the boat and then extends beyond the shrouds. It is intended for upwind sailing.
Gybe a manoeuver when sailing downwind whereby the vessel changes direction from sailing at less than 180° to the true wind direction to sailing at an angle greater than 180 degrees (measured in the same direction). This manoeuver causes the sails to move from one side of the boat to the other and, in strong conditions, can cause quite violent motions.
Halyard the ropes which are used to hoist sails.
Headboard the solid piece of material (typically either metal or carbon fiber) which is attached to the highest point of a fabric sail, and through which halyards are attached.
Headsail the term used for any sail which is set in front of the mast.
Heel the term used to describe the transverse inclination a vessel may adopt relative to its upright state.
Jib a sail which attaches to the bow of the boat and then does not extend beyond the shrouds. It is intended for upwind sailing.
Jumpers the separate stays used in conjunction with struts to support the uppermost section of a mast, when the shrouds do not terminate at the highest point of the mast.
Ketch a two-masted yacht with a mizzenmast stepped forward of the rudder which is smaller than the foremast.
Leech the trailing edge of a sail.
Leeward a term denoting the direction at sea in relation to the wind, i.e., downwind (see also windward).
Luff the leading edge of a sail.
Reef to reduce the amount of sail area presented to the wind.
Righting moment the moment caused by the offset between the centre of gravity and the centre of buoyancy, which looks to resist, or right, the vessel when heeling.
Roach when viewed in a profile view, roach is that part of sail area is exceeding a straight line leech; the leech then has a convex shape, providing more sail area for the same luff and foot dimensions.
Runners are the shortened name for a running backstay. These are adjustable backstays which are fixed in pairs, each being attached towards the outboard sides of the vessel and being tensioned alternately on either tack. This allows a larger sail area to be carried, without becoming tangled with a centerline backstay.

Sail aspect ratio the aerodynamic efficiency of a foil or a sail depends on a large number of different characteristics; one of them is the sail area geometric aspect ratio, where the cord of the sail is divided by the span.
Scantlings a nautical construction term for a set of standard dimensions for parts of a structure.
Schooner a sailing yacht with two or more masts, typically with the foremast smaller than the mainmast.
Shrouds the standing (nonmoving) rigging of a sailing vessel which gives the mast its lateral support.
Sloop A yacht with a single mast and only one upwind headsail.
Spars generic term for tubes and poles which are used to support a sail (e.g., masts, booms, spinnaker poles).
Spreaders struts which are attached to a mast in pairs transversely, with the purpose of increasing the angle which the shrouds make to the mast.
Stays a part of the standing rigging of a sailing vessel which gives the mast its lateral support.
Staysail a triangular sail, predominantly intended for upwind sailing, which is typically attached to an inner forestay (behind the main forestay).
Stormsail the smallest of a yacht's upwind triangular headsails.
Upwind a boat is said to be sailing upwind, when it is sailing in a direction which is less than 90° to the true wind direction.
Windward a term denoting the direction at sea in relation to the wind, i.e., upwind (see also leeward).

Composite materials for mooring applications: Manufacturing, material characterization, and design

16

Eduardo A.W. de Menezes*, Laís V. da Silva†, Filipe P. Geiger*, Rogério J. Marczak*, Sandro C. Amico†
*Post-Graduation Program in Mechanical Engineering, UFRGS, Porto Alegre, Brazil,
†Post-Graduation Program in Mining, Metallurgical and Materials Engineering, Federal University of Rio Grande do Sul (UFRGS), Porto Alegre, Brazil

Chapter Outline

16.1 Introduction 452
16.2 Design of composite cables 456
16.3 Mathematical modeling of cables with linearized kinematics 459
16.4 Manufacturing of composite cables 463
16.5 Mechanical characterization and aging of composite cables 466
    16.5.1 Tensile tests 466
    16.5.2 Tensile-tensile fatigue tests 471
    16.5.3 Flexural tests 473
    16.5.4 Cyclic bending tests 474
    16.5.5 Aging behavior 475
16.6 Finite element modeling of composite cables 476
    16.6.1 Model length 478
    16.6.2 Contact between wires 478
    16.6.3 Mechanical properties 480
    16.6.4 FEM models for tensile and bending stresses 484
16.7 Concluding remarks 485
Acknowledgments 486
References 486

List of symbols

1   direction parallel to fiber orientation
2   direction transverse to the fiber orientation (in-plane)

3	direction transverse to the fiber orientation (through-thickness)
A	cross section area
b	binormal vector
E	Young's modulus
G	shear modulus
H	generic vector
I	moment of inertia
J	polar moment of inertia
k	stiffness
m	distributed moment
M	applied moment
n	normal vector
N	number of wires or applied load
p	cable pitch
q	distributed load
r	position vector
R	wire radius
Re	relaxation
S	distance along a wire curve
t	tangent vector
u	displacement vector
α	thermal expansion coefficient or Helix angle
β	lay angle
γ	shear strain
ε	normal strain
ε_u	ultimate tensile strain
θ	angle between S and the $x_2 x_3$ plane
κ	curvature
μ	friction coefficient
ν	Poisson's ratio
ρ	specific weight
σ	normal stress
σ_u	ultimate tensile stress
τ	shear stress or torsion
ϕ	rotation vector

16.1 Introduction

The use of wire ropes, cords, and cables dates back to the stone age, for instance to build hammocks (Sima and Watson, 1961) and suspension bridges (Kawada, 2010). Some mural paintings found in Egypt (2000 BC) also show the civilization applying camel hair ropes in many different ways (Verreet, 2002). However, their manufacturing changed considerably after introducing the helical structure in the 19th century, giving origin to new architectures, such as seale, spiral, stranded, etc. Nevertheless, a common property of these structural elements is their ability to support relatively high tensile stress in comparison to other loadings (torsion, shear, bending) (Costello, 1990).

Fig. 16.1 shows the nomenclature commonly adopted to designate the different cable constituents, where the pitch (p) is the length of a complete twist around the cable core, and nominal diameter is the equivalent diameter of a circumscribed circle that encloses all the wires. According to ISO 17893 (ISO 17893, 2004), the format $1 \times N_1$ is used to designate single-strand cables, where N_1 is the total number of wires (including the core) and $N_2 \times N_3$ format is adopted for multistrand cables, where N_2 is the number of strands (excluding the core) and N_3 is the number of wires in each strand. To illustrate, the 1×10 and 6×19 constructions are shown in Fig. 16.2.

There are also different nomenclatures concerning the wires twisting direction, as depicted in Fig. 16.3, where in the Regular (or Ordinary) lay, the wires and strands are twisted in opposite directions, while in the Lang lay construction, they are twisted in the same direction. In the Alternate lay, adjacent strands are twisted in opposite directions. When a strand twisting direction makes an "S" shape, it is called Left hand construction, and when it is a "Z" shape it is called Right hand. Relative to the strand

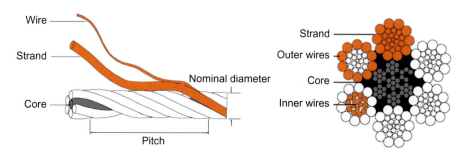

Fig. 16.1 Nomenclature adopted for cable components (Shaw's Wire Ropes, n.d.).

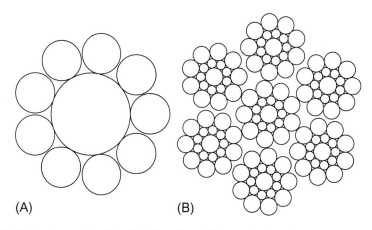

Fig. 16.2 Single-layered 1×10 cable (A) and a multilayered and multistrand 6×19 cable (B).

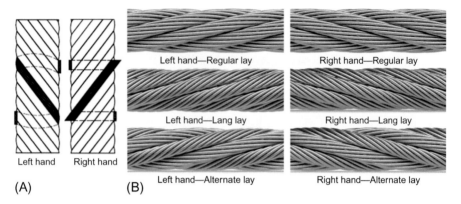

Fig. 16.3 Different cable constructions and their nomenclature considering strand twisting direction (ISO 17893, 2004) (A) and relative direction between wires and strands (B).

construction of a cable, the most applied types are Single lay, Seale construction (all layers with the same wire number), Warrington construction (outer layer with two times more wires than inner layer, intercalating between large and small wires), and Filler construction (outer layer with two times more wires than inner layer, with filler wires in the interstices between them) (ISO 17893, 2004), illustrated in Fig. 16.4.

Wire ropes display an important role today, being commonly applied in mine hoisting, electrical conductors, stowage of satellites, roofs, bridges, and also in anchorage systems of Semisubmersible and TLPs platforms. Due to recent advances in composite materials, there is a growing trend of replacing steel with carbon-fiber reinforced polymer (CFRP) in power cables (Santos et al., 2015), in cable-suspended bridges (Meier et al., 2015), due to the large span allowed by the weight reduction, and also in the offshore industry (Ghoreishi et al., 2007a). Unlike fixed structures, the cost of cable-moored platforms does not dramatically increase as a function of water depth (Adrezin and Benaroya, 1999); however, for deep-water applications, the large self-weight of steel cables makes their usage prohibitive (Ghoreishi et al., 2007a), increasing sagging effect and winch tensions (Harrop and Summerscales, 1989). In addition to their lightweight, composite cables show high specific strength and stiffness, outstanding fatigue behavior (compared to steel cables) (Meier, 2012), non-magnetic properties, low thermal expansion coefficient, corrosion resistance, along with good relaxation and creep behavior (Xie et al., 2014). A comparison between four different CFRP cables and a steel cable is displayed in Table 16.1, while Fig. 16.5 shows a comparison among composite, metallic, and polymeric cables focusing on specific strength and modulus (ratio between the property and the specific weight) (Jackson et al., 2005).

One important drawback of composite materials is its relative high price per unit weight. However, a comprehensive comparison of the usage of cables in two different

Composite materials for mooring applications

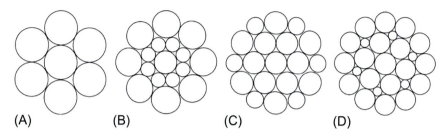

Fig. 16.4 (A) Single lay, (B) Seale, (C) Warrington, and (D) Filler strand constructions (ISO 17893, 2004).

Table 16.1 Mechanical properties of steel and CFRP cables (Xie et al., 2014)

	ρ (kg/m^3)	σ_u (MPa)	E (GPa)	Re (%)	ε_u (%)	α (10^{-6}/°C)
CFRP cable 1	1.63	2140	137	0.3	1.6	0.6
CFRP cable 2	1.63	2550	147	0.3	1.6	0.68
CFRP cable 3	1.63	2022	137	0.3	2	0.6
CFRP cable 4	1.63	2421	159	0.3	1.7	0.6
Steel cable	7.85	1570	196	<2.5	>4	12

ρ, specific weight; σ_u, ultimate tensile stress; E, Young modulus; Re, Relaxation; ε_u, ultimate tensile strain; α, coefficient of thermal expansion.

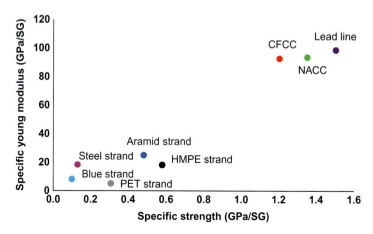

Fig. 16.5 Comparison between cables manufactured using different materials, where CFCC, Leadline, and NACC are CFRP cables.

Table 16.2 Comparison of final constructing cost for structures with steel or CFRP cables (Liu et al., 2016)

Material	Cable net façade		Spoked wheel cable roof	
	Steel	CFRP	Steel	CFRP
ρ (kg/m^3)	7850	1500	7850	1500
Unit price (US$/kg)	10	55	10	55
Amount used (m^3)	0.27	0.18	35.66	31.74
Weight (kg)	2090	291	279,931	47,366
Cost (10^3 US$)	20.9	15 (72%)	2799	2605 (93%)

structures, with CFRP or steel, shows that the composite material presented lower overall cost, as reported in Table 16.2, where similar mechanical properties were achieved with both materials.

This chapter addresses below common techniques used to manufacture composite cables, highlighting the differences and challenges in relation to steel or polymeric cables. CFRP cable testing procedures and results are also included, covering important features for its application by the offshore industry, such as aging, thermal aspects, and static and dynamic mechanical behavior. An introduction to design with these materials is also discussed, focusing on the behavior of some complex geometry structural members, reviewing analytical and numerical tools to predict their mechanical behavior, and aiming also to reduce the need for experimentation.

16.2 Design of composite cables

The design process of composite cables and wire ropes is a complex and challenging task, since the geometry of these structural elements may differ regarding constructive architecture, twisting direction, helix angle, and rods diameter. Some designing requirements cannot be achieved simultaneously, creating conflicting objectives for the designer to deal with, e.g., parameters that improve tensile stress behavior usually decrease the cable flexibility. Since experimental data for composite cables testing is sparse when compared to available results found in the literature for steel cables, care should be taken during the design process of these components in order to validate the usual assumptions.

In order to achieve some flexibility, the external wires of a cable are usually twisted around its core, consisting of a helical geometry and forming a helix angle α between external wires and the plane perpendicular to the cable and a lay angle β (complementary to α) between external wires and cable core. Manufacturers usually inform the twisting through the cable pitch p related to the helix angle according to:

$$p = 2\pi(R_1 + R_2)\tan(\alpha) \qquad (16.1)$$

valid for a single-layered cable, where R_1 is the core radius and R_2 is the external wire radius.

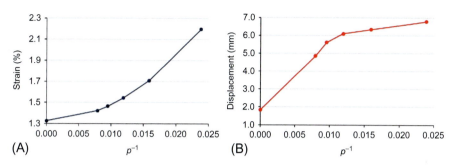

Fig. 16.6 Response of a 1 × 10 cable for five different pitches under tensile (A) (with strain measured at the core) and bending (B) (with vertical displacement measured at the middle cross section) where a zero value for p^{-1} indicates that rods are straight (Menezes et al., 2017a).

As can be seen in Fig. 16.6, where numerical results for a CFRP cable are plotted as function of the pitch, it impacts tensile and bending behavior in different ways. Regarding tensile behavior (Fig. 16.6A), for higher pitches the stress is more homogenously distributed along the wires, improving its tensile stress, while for low pitch the core supports most of the applied load. As for the bending behavior (Fig. 16.6B), flexibility is often necessary in order to twist the cable around sheaves or drums which can be achieved by reducing the cable pitch. During bending deformations, a low pitch will also concentrate stresses in the cable core, but the stress in each outer wire is a function of its distance from the neutral axis and along the length, as observed in Fig. 16.7 (Chen et al., 2017). Bending stresses also corroborate to crack initiation, reducing the cable lifetime. Therefore, higher helix angles provide some stress uniformity, with the drawback of a poor fatigue life, especially when running over sheaves (Feyrer and Ropes, 2015).

Unlike straight beams, cable stiffness has a coupling term between tensile and torsion about its axis, which is highly sensitive to the cable pitch (Utting and Jones, 1987a), decreasing as the pitch increases. Torsion stiffness is also affected by the wires' twisting directions. A regular lay configuration, where the cable wires and strands are twisted in opposite directions, creates a balanced torque along the cable when tensile loading is applied, while the lang lay configuration (were all wires are twisted in the same direction) yields a higher torque, making work with them difficult (Sloan et al., n.d.). By applying flexure or tensile loadings on a cable, external wires move, changing their helix angle. In the case of lang lay configurations, they move in the same direction, while for regular lay the direction is opposite, increasing stresses on the contact region (Feyrer and Ropes, 2015) and lowering cable lifetime. Both configurations, however, show a very similar static tensile stress behavior (Luz et al., n.d.). The drawback of regular lay configuration stands on its lower fatigue life (Evans et al., 2001).

Relative to the cable architecture, by splitting big wires into small ones, it is possible to improve its bending behavior, allowing the cable to be stored in small diameter

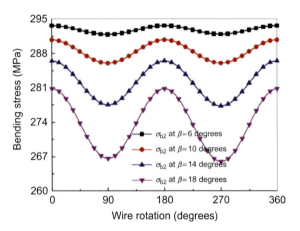

Fig. 16.7 Stress distribution in outer wires along a complete twist around the core for four different lay angles (Chen et al., 2017).

Table 16.3 Suggested and minimum ratio between sheave diameter (D) and cable nominal diameter (d) for four different cable configurations

Cable type	Suggested D/d	Minimum D/d
6 × 7	72	42
6 × 21	45	30
6 × 25	39	26
6 × 42	21	14

Considered strands are single lay for the 6 × 7 cable and filler construction for the others (I. Hanes Supply of SC, n.d.).

sheaves or drums, for instance. It also improves its fatigue life, since the strain level will be lower when bent. With large diameter wires, however, the tensile behavior is improved and abrasion damage or wear effect is reduced. The importance of wire numbers is exemplified by a wire rope manufacturer data, shown in Table 16.3, where the ratio between sheave diameter (D) and cable nominal diameter (d) was evaluated for each configuration.

Despite the outstanding fatigue behavior of CFRP cables, since composite materials fail in relative low strains when compared to steel and polymers, the design process must allow for bending conditions during its application to prevent failure by cyclic bending over drums or sheaves. Another important feature of composite cables is the friction forces between wires. Many cables are protected from UV radiation with polymeric coatings that may be damaged if excessive contact forces are applied.

Designing process involves application, material selection, and geometry. A cable safety factor and fatigue life will rely on this initial design. Therefore, it is important to know all involved variables and measure their impact on the cable stresses.

16.3 Mathematical modeling of cables with linearized kinematics

Possibly, the best way to construct the mathematical model suitable for cables is to describe them as curved beams in three-dimensional space. In order to define a local coordinate system where the equilibrium equations will be attached to, it is necessary to determine the Frenet-Serret triad. A concise explanation will be given here, for more information see Do Carmo (1976), Pressley (2010), and Kreyszig (1968).

Considering a generic curve defined by a vector written as

$$\mathbf{r}(S) = [x_1(S) \quad x_2(S) \quad x_3(S)] \tag{16.2}$$

where $x_1(S)$, $x_2(S)$, $x_3(S)$ are a one-dimensional triplet in which the differential calculus can be used and S is the distance along the curve. Therefore, it is possible to parameterize the geometry through one single parameter. With this representation, three vectors and the respective derivatives can be determined.

The first derivative of $\mathbf{r}(S)$ results in a vector pointed to the tangent direction of the curve, denoted from now on as the tangent vector $\mathbf{t}(S)$. The second derivative of the parameterization represents the normal vector, $\mathbf{n}(S)$, pointing to the center of the arc formed for two close points, or in other words, to the change of $\mathbf{t}(S)$. One more entity is needed to complete the local system; knowing that the last two vectors are orthogonal and unitary, the cross product can be performed. Hence, the result is also a vector with the same properties, the binormal vector $\mathbf{b}(S)$.

Since the tangent vector is unitary, its derivative is orthogonal to itself; however, this rate has no information toward its norm, so a new parameter, called curvature κ, is expressed as

$$\kappa(S) = \left| \frac{d(\mathbf{t}(s))}{dS} \right| \tag{16.3}$$

and is used to represent not only the tangent vector direction change, but also to indicate the curvature radius of the curve at S. Similarly, the torsion τ, represented as

$$\frac{d(\mathbf{b}(S))}{dS} = -\tau \mathbf{n}(S) \tag{16.4}$$

indicates the tendency of the tangent vector to deviate from the osculating plane just as κ defines the variation from the rectifying plane (Fig. 16.8). Therefore, the geometry and the tendencies of any smooth curve in space can be studied by the application of the above definitions. The Frenet-Serret triad is then defined and can be written as

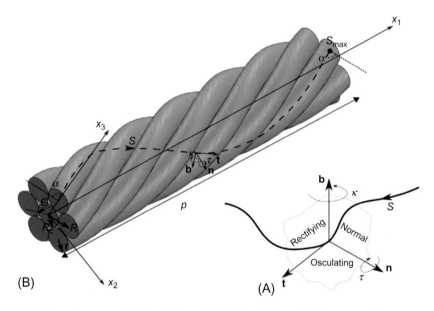

Fig. 16.8 Generic curve with Frenet-Serret triad (A) and its application on a cable coordinates (B).

$$\begin{bmatrix} \dfrac{d(\mathbf{t})}{dS} \\ \dfrac{d(\mathbf{n})}{dS} \\ \dfrac{d(\mathbf{b})}{dS} \end{bmatrix} = \begin{bmatrix} 0 & \kappa & 0 \\ -\kappa & 0 & \tau \\ 0 & -\tau & 0 \end{bmatrix} \begin{bmatrix} \mathbf{t} \\ \mathbf{n} \\ \mathbf{b} \end{bmatrix}. \quad (16.5)$$

From the properties aforementioned, the strand helix can be parametrized as:

$$x_1 = R\cos(\theta); \quad x_2 = R\sin(\theta); \quad x_3 = S\sin(\alpha); \quad \theta = \frac{S\cos(\alpha)}{R} \quad (16.6)$$

where R corresponds to the radius, S is the position along the curve and α denotes the helix angle, i.e., the angle formed by the helix and the $x_2 x_3$ plane, and θ, the angle between S and the $x_2 x_3$ plane, as shown in Fig. 16.8, where one revolution is represented. Since α is constant, in this particular case, the tangent vector inclination is also unchanged and the angle is also equal to α.

The helix representation is defined by a local system which is not aligned to the global coordinates (indicated by x_2, x_3, and x_1 passing through the center core). Results obtained in either systems can be used in a different canonical entity. Therefore, since **T** (transformation matrix) is formed by the local quantities, the multiplication of it by any effect performs the transformation from local to global as

$$\mathbf{H}^{global} = \mathbf{T}\mathbf{H}^{local} \tag{16.7}$$

where **H** is a generic vector and **T** is defined, by normal, tangent, and binormal vectors, as

$$\mathbf{T} = \begin{bmatrix} -\sin\left(\dfrac{S\cos(\alpha)}{R}\right)\cos(\alpha) & \cos\left(\dfrac{S\cos(\alpha)}{R}\right)\cos(\alpha) & \sin(\alpha) \\ \cos\left(\dfrac{S\cos(\alpha)}{R}\right) & \sin\left(\dfrac{S\cos(\alpha)}{R}\right) & 0 \\ -\sin\left(\dfrac{S\cos(\alpha)}{R}\right)\sin(\alpha) & \cos\left(\dfrac{S\cos(\alpha)}{R}\right)\sin(\alpha) & -\cos(\alpha) \end{bmatrix}. \tag{16.8}$$

Parameters κ and τ are defined calculating the norm of the vectors $\mathbf{t}'$ and $\mathbf{b}'$ so that,

$$\kappa = \frac{-\cos^2(\alpha)}{R}; \quad \tau = \frac{\sin(\alpha)\cos(\alpha)}{R} \tag{16.9}$$

The negative sign of the curvature was used to position the global system in the center of the helix. One turn of the helix sets the maximum distance from the beginning to ending point as a function of S. The maximum position is determined when $\theta = 2\pi$, for one pitch, so

$$S_{max} = \frac{2\pi R}{\cos(\alpha)} \tag{16.10}$$

With the geometry and the local coordinate system set, it is possible to develop the equilibrium equations to determine the mechanical response of a long thin rod from the spatial beam theory, more from the equation development is found in Love (2011). The equilibrium equations for such theory can be obtained considering the equilibrium of an infinitesimal portion of the beam. In Fig. 16.9, a generic beam segment is shown with the Frenet-Serret triad represented in the centroidal line with the curvilinear coordinate S along the neutral axis. The cable is subjected to the generic loadings N and M, representing concentrated forces and moments, together with q and m, denoting distributed forces and moments.

The equilibrium of a differential element of cable dS is verified by the sum of forces ($\Sigma N = 0$) and moments ($\Sigma M = 0$):

$$\frac{dN(S)}{dS} + q(S) = 0 \tag{16.11}$$

Assuming, in the Frenet-Serrat basis, that the loads produce effects in different directions, the transformation matrix is used to decompose these components, so that N and q are expressed as

$$N(S) = N_1\mathbf{t} + N_2\mathbf{n} + N_3\mathbf{b}; \quad q(S) = q_1\mathbf{t} + q_2\mathbf{n} + q_3\mathbf{b} \tag{16.12}$$

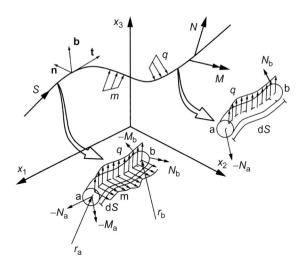

Fig. 16.9 Generic cable showing loading and geometry.

therefore, the local system of equilibrium equations for each direction of the triad can be written as

$$\frac{dN_1}{dS} - \kappa N_2 + q_1 = 0$$
$$\frac{dN_2}{dS} + \kappa N_1 - \tau N_3 + q_2 = 0 \qquad (16.13)$$
$$\frac{dN_3}{dS} + \tau N_2 + q_3 = 0$$

The equilibrium equations for the moments are obtained using the same approach. The sum of the moments is performed, noticing that the forces also produce moments with the brace **r**. Decomposing using the same methodology aforementioned, one can write the system of equilibrium equation for moments as

$$\frac{dM_1}{dS} - \kappa M_2 + m_1 = 0$$
$$\frac{dM_2}{dS} + \kappa M_1 - \tau M_3 - N_3 + m_2 = 0 \qquad (16.14)$$
$$\frac{dM_3}{dS} + \tau M_2 + N_2 + m_3 = 0$$

Now the geometry parameters can be used to determine stresses and the constitutive model is applied to obtain strains of the structure, defining as follows: cable cross section area (A); moment of inertia (I), and polar moment of inertia (J). A good approximation to represent the mechanical behavior of a cable is as a transversally isotropic material, i.e., with the material properties in the axial direction different than the transversal directions. From the Hooke's law, the constitutive relations are written as

$$\varepsilon_1 = \frac{N_1}{E_1 A}; \quad \rho_1 = \frac{M_1 r}{G_{23} J};$$

$$\varepsilon_2 = \frac{-M_2 r}{E_2 I}; \quad \rho_2 = \frac{N_2}{G_{12} A}; \quad (16.15)$$

$$\varepsilon_3 = \frac{-M_3 r}{E_2 I}; \quad \rho_3 = \frac{N_3}{G_{12} A}.$$

In order to obtain the kinematic response, the Principle of Virtual Work (PVW) is applied considering infinitesimal displacements, expressed as

$$\int_V \sigma \delta \varepsilon \, dV = \int_V \mathbf{b} \delta \mathbf{u} \, dV + \int_S \mathbf{t} \delta \mathbf{u} \, dS \quad (16.16)$$

After some mathematical manipulation, the expressions for displacements (**u**) and rotations (**ϕ**) are obtained as

$$\varepsilon_1 = \frac{du_1}{dS} - \kappa u_2; \quad \rho_1 = \frac{d\phi_1}{dS} - \kappa \phi_2;$$

$$\varepsilon_2 = \frac{du_2}{dS} + \kappa u_1 - \tau u_3 + \phi_3; \quad \rho_2 = \frac{d\phi_2}{dS} + \kappa \phi_1 - \tau \phi_3; \quad (16.17)$$

$$\varepsilon_3 = \frac{du_3}{dS} + \tau u_2 + \phi_2; \quad \rho_3 = \frac{d\phi_3}{dS} + \tau \phi_2.$$

With these expressions, one can analyze mechanically a cable as a transversally isotropic beam, either straight or curved. The solution of the systems of equations provides the required information to determine its behavior independently of the shape. All equations obtained are coupled and must be solved simultaneously to provide displacement solutions.

The boundary conditions (BC), the material model (or constitutive relations), and the geometry are integrated to provide results. Solving the first two systems (Eqs. 16.13, 16.14) produce six integration constants related to the particular set of the BC used. This step is by no means easy to carry out, making the use of numerical methods mandatory for the solution of general cases. All available closed-form analytical solutions have limitations to some degree regarding the simplifications applied. References Costello (1990), Ramsey (1990), Osteergard et al. (2012), and Labrosse et al. (2000) are examples and can be used to provide important information regarding particular problems.

16.4 Manufacturing of composite cables

According to a recent review from Liu et al. (2015), the existing CFRP cables can be mainly classified into four types, as shown in Fig. 16.10. Fig. 16.10A shows the CFRP cable in the form of lamella, which is produced by traditional pultrusion such as the CFRP rod shown in Fig. 16.10C. The CFRP cable in Fig. 16.10D is in the form of wire

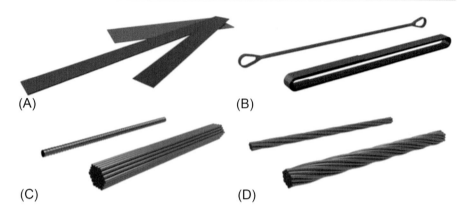

Fig. 16.10 Four main types of CFRP cables (Liu et al., 2015). (A) CFRP lamella, (B) CFRP strip-loop, (C) CFRP rod, and (D) CFRP wire-stand.

rope, which is fabricated by modified pultrusion where CFRP rods are twisted into a helix; finally, the CFRP cable in the form of strip-loop (Fig. 16.10B) is fabricated by winding a continuous CFRP strip on two pins (Liu et al., 2015). Normally, the CFRP cables are made with 60% of volume fraction to assure 60%–70% of the tensile strength of the carbon fibers (1970–3200 MPa (Meier, 2012)). These fibers also have low coefficient of linear expansion, on the order of 0.2×10^{-6} m/m/°C, and high fatigue strength, although with disadvantages like low impact resistance, high electrical conductivity, and high cost (Roddenberry et al., 2014).

The interest in CFRP cables has mostly been for applications in civil engineering, including large companies such as Leadline (Mitsubishi) (Nanni et al., 1996; Zhang et al., 2001) and Tokyo Rope (Motoyama et al., 2002), which have been investing in the replacement of steel cables in the civil sector for more than 15 years. Currently, Tokyo Rope has a commercial cable called Carbon Fiber Composite Cables (CFCC), which is already patented in 10 countries. The CFCC cable is a stranded cable made of several individual rods, e.g., 7, 19, or 37 twisted carbon rods, with nominal diameters varying from 5 to 40 mm (0.2–1.6 in.) (Ali et al., 2015).

Fig. 16.11 presents a summary of the manufacturing process of the CFCC cables from the Tokyo Rope's patent (Ushijima, 2010). In the first step, called layering step, the rod (also known as wires) production, several prepreg rovings (impregnated with a thermosetting resin) are fed from bobbins to a twisting machine and twisted at a predetermined pitch to obtain a composite element (the rods shown in Fig. 16.11A).

In the lapping step, while a plurality of rods is fed out, the synthetic fiber yarns are wound on the periphery of the rods by the lapping machine, as shown in Fig. 16.11B. The purpose of the yarns is to protect the fibers from UV radiation and mechanical abrasion, and also to improve their adhesion with concrete (Roddenberry et al., 2014). Fig. 16.11C illustrates the production of a 1×7 composite cable, where seven bobbins are twisting together at a predetermined pitch (100–200 mm) by the closing machine, forming the 1×7 composite cable. In the curing step, a heat treatment for the curing process is performed at 120–135°C.

Composite materials for mooring applications

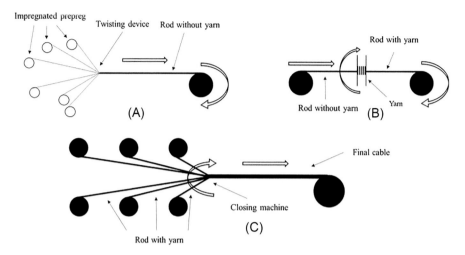

Fig. 16.11 Main steps in CFRP cable production: (A) layering step, (B) lapping step, and (C) production of a composite cable by twisting the rod.

In fact, according to the referred patent, the process to manufacture the cables can be carried out via two routes, both with a single curing step, as described below (Ushijima, 2010):

(i) The cable is formed in a process that a strand having a single twist structure with polymer resin being uncured is fabricated through a layering step, lapping step, and primary closing step, then a plurality of strands with uncured resin are twisted together into a cable in a secondary closing step, and finally the whole is cured in a curing step.

(ii) The cable is formed in a process in which a strand having a single twist structure with resin being cured is fabricated through the layering step, lapping step, primary closing step, and curing step, then a plurality of the strands with cured resin are twisted together into a cable in the secondary closing step.

For cables composed of a core strand in the center, the following manufacturing method can be performed. According to this, since resin of a strand to be core strand has been cured, the synthetic resin base inclusion can be easily applied; in addition, since the core strand has stiffness through curing of the resin, operation of bundling side strands and twisting them together can be smoothly performed.

During manufacturing, the twist direction of the strands is opposite to the twist direction of the cable, reducing rotation and minimizing shape distortion. When this range is low, below 2°, although high tensile strength is obtained, shape deformation occurs, which leads to difficulties in handling due to bending stress caused by the difference in diameter between the strands inside and outside the cable. Consequently, the torsion characteristics of the cable will not be adequate, and when the cable is exposed in a distorted situation to a twist in the opposite direction to that of the direction of the cable, the elements will be spaced from one another, leading to early breakage. In a range limit above 12°, tensile strength will be reduced since the composite does not have high resistance against bending, shearing, and torsion. The optimal range of twist degree to apply for these cables is within 2–8°. Another parameter to be taken into consideration is the pitch (p) in the twisting step, and limiting the twist angle will increase tensile strength without damage or shape deformation and enable a twisting step to be easily carried out using an existing twisting machine.

In composites cables, when the strands are in contact (for instance, when flexure stress is applied), rods are damaged due to a rubbing action or lateral pressure between the rods themselves, consequently sufficient strength cannot be exhibited. However, if there are inclusions in the cable, the contact between the core and the strands is reduced; in addition, since the existence of inclusion possibly expands the core strand, the contact between the side strands themselves is also reduced via such an expansion action, with a consequent reduction in tensile strength because the internal wear (twist abrasion) can be reduced.

Moreover, when the cable is inserted into a hole or cylinder and filler (such as cement or resin) is poured into a space between the periphery of the cable and the hole or cylinder to obtain an anchoring portion, the filler flowing inside of the cables along the longitudinal direction can be prevented, through the gaps between the strands, but even the contact pressure is reduced and the tensile strength can also be reduced.

Problems with fillers led to a development of a filler for these composite cables, where the inclusion preferably has soft synthetic resin so that softness of the cable is not lost. A unification of the inclusion and the strands (from the core) is made using a resin extruder and extruding melted resin around the strand, which is passing through the machine forming a covering layer on the periphery of the strand. Furthermore, the inclusion may be a filament member made of thermoplastic synthetic resin independent of the strand, as shown in Fig. 16.12.

Another type of commercial rod is the Leadline CFRP tendon, which is produced with different surface patterns: smooth, surface-indented, or ribbed. These rods are usually pultruded with pitch-based carbon fiber (manufactured from resin precursor fibers after stabilization treatment, carbonization, and a final heat treatment), and most rods are commonly anchored using a modified wedge anchorage system (Nanni et al., 1996). Technora carbon fiber rod is also similar, with spirally wound rods that are pultruded and impregnated with a vinyl ester resin. Carbon stress rods are also made by pultrusion, then epoxy-impregnated, and usually coated with sand to increase their anchorage when embedded into concrete (Schmidt et al., 2012).

16.5 Mechanical characterization and aging of composite cables

16.5.1 Tensile tests

Prior to the actual testing of the cables, choosing the adequate cable termination is of uttermost importance. Since terminations are submitted to the same loading conditions applied to the cable, they must be accurately designed to support them. These terminations are mostly used for steel or synthetic ropes, and there is not much open literature about the development of terminations for composite cables.

Composite materials for mooring applications

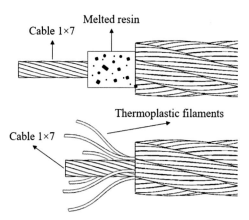

Fig. 16.12 Different types of inclusion.

Fig. 16.13 Eye splice made by burying the rope into the hollow center of the body. Rope is 12-strand HMPE with four strands per carrier (McKenna et al., 2014).

The three most traditional types of terminations are:

(i) Eye splices: In these terminations (Fig. 16.13), the cable performs a loop at the end, assisted or not by a thimble (a metallic or polymeric part that helps to keep the cable twisted). This termination is the simplest and easily performed, being applied when the cable is not submitted to high stresses. It is not employed for composite cables due to their high bending stiffness, which does not allow their bending over a small radius.

(ii) Wire rope clamps: These terminations use a metallic clamp (clip), to pressurize the cable against itself, as depicted in Fig. 16.14, commonly aided by a thimble that helps producing the loop. Usually employed for metallic cables to connect two cables.

(iii) Sockets: These terminations use a metallic component coupled at the cable end, as illustrated in Fig. 16.15. The coupling between socket and cable is usually performed by inserting a resin. Relative twisting and elongation between cable and socket is constrained, nevertheless, this system is versatile and applied for metallic and synthetic fiber cables.

The introduction of composites cables required the development of anchorage systems that allow the rods (subelements) to be anchored to the block. One of the most recent examples of anchoring system is that from Tokyo Rope company, used for the

Fig. 16.14 Wire rope clamp type of termination (Krasucki, 1993).

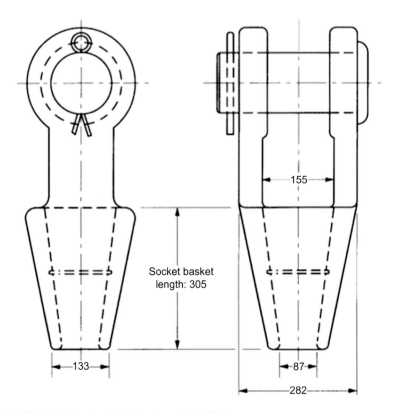

Fig. 16.15 Socket termination (Brandon et al., 2001).

building and construction sector, mainly cable-stayed bridges. The method is described as resin filling and die cast. The resin filling method bonds the cable to a steel socket (like a tube) with conical internal cavity. Epoxy formulations were previously used to fill the cavity and transfer shear stress loads to the anchoring assembly, but nowadays an expansive material (e.g., cement), which showed greater efficiency in ground anchorage systems, is usually applied (see Fig. 16.16). The die cast method attaches the cable to a steel tube by means of a molten and die-molded alloy. Steel wedge clamps the steel tubes similarly to steel tendon systems (Dolan et al., 2001).

Research on carbon fiber composites cables normally aims the construction industry (Noisternig, 2000; Motoyama et al., 2002), for example in cable-stayed bridges, where these materials have obtained greater acceptance especially due to their high

Composite materials for mooring applications

Fig. 16.16 Cables with CFRP rods used in civil engineering structures (Rohleder et al., 2008).

chemical resistance in aggressive environments called salty breezes. In 2000, Noisternig (2000) developed a project on CFRP-stay cables or tendons with a suitable anchorage developed by DYWICARB company. The researchers carefully studied the table of requirements for CFRP-stay cables or tendons, including the standardization and characterization of the material of the CFRP-elements, which can roughly deviate from the recommendations for steel (Noisternig, 2000). The cable was manufactured with wires arranged in parallel over the whole length and with a polyethylene (PE) or polypropylene (PP) coating along its free length, for protection against ultraviolet radiation and wind erosion. Considering the load bearing capacity of the complete stay cable/tendon, a distinctive anchorage system was applied to test the cables, a casing system where the CFRP-wires were inserted into a conically shaped steel socket, which is fixed on the structure over a steel ring nut (Fig. 16.17). The tensile behavior of three different configurations were studied and predicted (see Table 16.4) and 95% of the theoretical tensile strength was reached for DYWICARB systems with 7 and 19 CFRP-wires. Nevertheless, the main fracture occurred near the free length of the cable (close to the anchorage).

Motoyama et al. (2002) also studied the tensile behavior of CFRP cables composed of seven rods, in which six strands were twisted spirally around a core rod, as shown in Fig. 16.18. Anchorage was performed using an expansive cement in the end tabs. All the strands were found to fracture simultaneously and, for high-loading speeds, the samples started breaking near the ends, revealing the effect of test conditions on the results (Motoyama et al., 2002). For tensile tests at high strain rates (up to 10/s), higher ultimate strain and lower breaking load ($\sim$10%) were observed.

Different groups around the world have already done some research focusing on the application of composites cables for deep-water mooring application. Harrop and Summerscales (1989) tested Korvite aramid tendon armored cable in tension. The tendons then showed compressive buckling failure, attributed to shock waves during relaxation on failure. The Freyssinet company, Soficar company, and Doris Engineering company proposed composite tendons of different configurations for the anchoring of TLP (Tension-Leg Platform), from simple structure cables, such

Fig. 16.17 Cable anchorage system (Noisternig, 2000).

Table 16.4 Failure loads of DYWICARB stay cables/tendons with 7, 19 and 91 CFRP-wires (Noisternig, 2000)

Number of CFRP-wires	Failure load (kN)
7	370
19	1020
91	3600

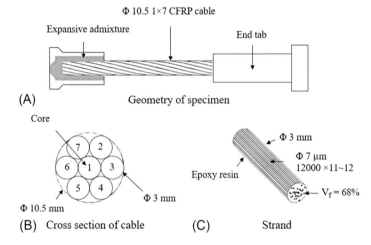

Fig. 16.18 1 × 7 CFRP cables (Motoyama et al., 2002). (A) Geometry of specimen, (B) Cross section of cable, and (C) Strand.

Composite materials for mooring applications 471

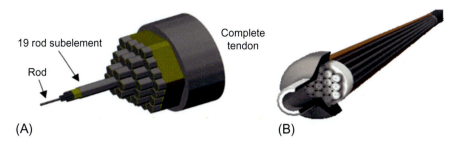

Fig. 16.19 Composite tendon example with 37 subelements (A) and the cable termination (B) (Sparks et al., 2003).

as 13-rod subelement, to complex ones, with 60-rod subelement rods (Sparks et al., 2003) (see Fig. 16.19). Individual rods were analyzed first to determine the guaranteed minimum ultimate tensile stress. The tendon cables were designed for a guaranteed minimum ultimate stress of 2400 MPa; however, they reached 2800 MPa, with an elastic modulus of 160 GPa, being independent on the cable's diameter (Sparks et al., 2003).

Jackson et al. (2005) investigated three prototypes and carried out quasistatic monotonic tensile tests, reaching 96.7% of the designed load (180 ton). Similarly, Botker and Johannessen (2002) developed a prototype tether consisting of 781 Ø6 mm rods bundled into 13 strands, each of them containing 31 or 85 rods. The tether termination, metal end fitting using resin potted cones, was successfully employed in onshore applications and was also tested for offshore loading levels. The tensile strength reached around 1500 MPa and the overall results showed that composite risers and tethers could represent economically and technically feasible TLP solutions for water depths as large as 3000 m.

Menezes et al. (2017b) investigated the tensile behavior of double-layered CFRP cables, commercial 1×19 CFRP cables (12.5 mm diameter)—Lang lay configuration (all rods twisted in the same direction), for improved fatigue behavior, and a Seale construction (same wire number in all layers). The tensile tests were performed according to the JSCE E-531-1995 standard (Test method for tensile properties of continuous fiber reinforcing materials) (JSCE E-531, 1995), with a test speed of 100 MPa/min (see Fig. 16.20). For a tensile loading, the numerical model was able to predict the cable ultimate tensile load with an error of 4.1%, and a difference of only 0.4% was found for the breaking load, where the cables reached 153.9 and 160.4 kN on the numerical model and experimental test, respectively.

16.5.2 Tensile-tensile fatigue tests

In order to fully analyze the potential of CFRP cables to substitute steel wire ropes, it is also necessary to study their fatigue behavior in wider in-service conditions. Fatigue loading may be split into two categories (although certain applications combine the two): pure tensile stress, where the applied load is based on the cable MBL (maximum breaking load), commonly referred as tensile-tensile test, and a static tensile loading

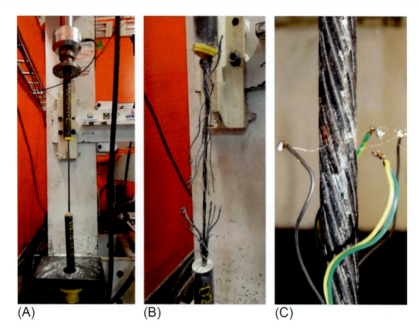

Fig. 16.20 CFRP cable tensile testing: (A) before testing, (B) after rupture, and (C) detail of the strain gage attached to the external rod (Menezes et al., 2017b).

combined with cyclic bending (bending-over-sheave or bending fatigue). Applications involving tensile-tensile fatigue loading include in deck hangers on suspension bridges, in radio mast guy ropes, or in anchor lines, where the Acoustic Emission technique could be used for in-service monitoring of fiber cables (Harrop and Summerscales, 1989).

Due to fatigue problems, steel cables for TLPs are usually designed to keep a period below 4 s and avoid springing and ringing fatigue problems (Odru and Geffroy, 2002), issues that are minimized when CFRP cables are used due to their outstanding fatigue behavior. In 2002, Botker and Johannessen (2002) studied the fatigue behavior of risers and tethers made of carbon fiber composites and found that, in a range of stress amplitude of 700 MPa, their tethers peaked one million cycles below the MBL. Odru and Geffroy (2002) investigated the fatigue behavior of CFRP cables composed of 19 rods under hexagonal grouping, showing that they could support up to 2.1 million cycles with 45% of their MBL under a dynamic amplitude of 400 and 600 MPa, without showing any detectable defect (Odru and Geffroy, 2002). The outstanding fatigue resistance results of composite tethers imply that the natural period in heave, pitch, and roll could be increased relative to conventional TLPs (Botker and Johannessen, 2002).

Sparks et al. (2003) tested 4.7 m long tendons (19-rod subelements) in tensile fatigue using steel anchorages and reached 2.2 million cycles for the cables in two different dynamic range, 400 and 600 MPa (17% and 25% of MBL, respectively). In the three cases tested, these loads were combined with a superimposed axial

Table 16.5 **Fatigue test results of 19-rod subelement**

Sample	Stress range (MPa)	Failure tension (kN)	Ultimate stress (MPa)	Young modulus (GPa)
1	400	1580	2944	162
2	600	1510	2813	150
3	600	1480	2806	150

tension, chosen to yield a maximum stress (static plus dynamic) equal to 45% of MBL and no failure or damage was noticed during the tests. The samples were then loaded under tensile stress until failure, and the results are reported in Sparks et al. (2003). No deterioration was observed after these very severe fatigue tests compared to subelements that were not fatigue cycled, i.e., no significant reduction was found in ultimate strength of the 19-rod subelements, or their anchorages, due to fatigue (Table 16.5).

A partnership between two major companies, DeepSea and Petrobras, performed a dynamic study on composites cables, reported by Jackson et al. (2006). The cable was analogous to spiral strand wire rope, where the mooring line concept is based on a central core consisting of a parallel bundle of seven carbon fiber rods (six twisted around a central rod). The end termination comprised of a socket like those used for steel ropes. The fatigue test was performed in a range of tensile stresses between 10% and 40% of the MBL for 140,000 cycles, following the API specifications for steel ropes. After the fatigue test, the cables passed for a tensile test to measure the residual strength that it was about 205 ton (above the initial breaking load of 180 ton).

Following these studies for offshore applications, Silva et al. (2016a) investigated the tensile-tensile fatigue behavior of 1×7 CFRP cables (nominal diameter 10.5 mm). In order to simulate the variable loads experienced by mooring lines, the test was conducted according to the procedure suggested by Petrobras Technical specifications (PTS) (Petrobras, 2009), which were elaborated based on the requirements of steel cables for mooring. The cables were tested in a range of stress level from 0% to 50% of the MBL, with a test speed of 20% MBL/min, for 12,000 cycles (Fig. 16.21). The average residual strength was around 86% of the original value, showing a higher fatigue strength than steel cables.

16.5.3 Flexural tests

The bending over sheave/drum problem is of great concern when dealing with cables and wire ropes in the offshore industry, since they must be stored for transportation, inspection, or maintenance. When applied in anchorage systems, they also experience flexural stress due to the platform lateral motion and must be stiff enough to resist to it, since the upwelling effect can severally damage such platforms (Niedzwecki and Huston, 1992). Nevertheless, research addressing bending properties of cables and wire ropes are limited (Chen et al., 2015).

Fig. 16.21 1 × 7 cable on the test machine and view of the anchored area (Silva et al., 2016a).

Common tests applied to characterize flexural behavior of composite materials include three-point and four-point bending (ISO 14125, 1998; ASTM D6272-10, 2010). The main advantage of four-point bending is the constant curvature radius produced in the load span due to the constant bending moment, which is an important feature when characterizing cables. While the support span distance must be long enough to prevent shear deformation, a high value could produce sagging (catenary) effect. Fig. 16.22 shows a four-point bending test performed in a 1 × 10 CFRP cable and the load vs. vertical displacement curve.

The cable bending stiffness is directly dependent on the boundary conditions applied. When the cable wire ends are free, they show distinct relative displacement, whereas, if bonded together, they show higher stiffness. Another feature that greatly increases bending stiffness, especially for small diameter cables, is the prestress between wires (Chen et al., 2015), whose updated stiffness value was mathematically modeled by Jolicoeur and Cardou (Jolicoeur and Cardou, 1996).

Another test method available for characterizing flexural behavior is the bending over a variable diameter sheave. This procedure allows a direct measure of strain in function of curvature radius, whose behavior is essentially linear for the experiments conducted by Cutler (Cutler and Knapp, 2010). It is also easier to apply a curvature radius and a tensile loading in this procedure, combining both tensile and flexure stress in one test.

16.5.4 Cyclic bending tests

The cyclic bending is also called bending-over-sheave fatigue, commonly used for steel and synthetic cables, and it refers to the repeated bending under constant tensile load (Chaplin and Potts, 1991). The behavior under bending is crucial for mooring, where tension fluctuations due to motions induced by environmental loadings on the floating structure may be the dominant source of fatigue stresses (Chaplin and Potts, 1991), being tensile and/or bending loadings (Girón et al., 2014).

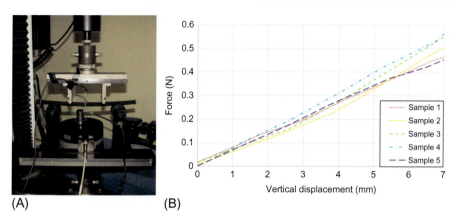

Fig. 16.22 (A) Four-point bending test on a CFRP cable and (B) the force applied at the cable in function of its vertical displacement (Menezes et al., 2016).

Recently, Silva et al. (2016a) have reported the bending behavior of a commercial 1 × 7 CFRP cable. The cyclic bending tests followed Petrobras's Technical Specifications originally based on steel (Chaplin and Potts, 1991) and synthetic cable theories (Gilmore and Thomas, 2009) and later adapted to composites cables. The bending tests were performed in the machine shown in Fig. 16.23 following ABNT standard (ABNT NBR 16272:2014 EN, 2014), with displacement speed of 0.4 m/s and cable transverse length over the sheave of 2 m. A nylon sheave was used to reduce friction, and the relation between sheaves diameters D/d was 45.7, which is relatively high compared with the D/d used for steel ropes and synthetic ropes (usually between 18 (Ridge et al., 2001) and 20 (Davies et al., 2011)). The results were justified and the CFRP cables, under such a critical loading, resisted for 1500 cycles, with a residual strength of 40% of the MLB. Similarly, a cyclic bending test method with the application of a similar tensile stress degree was reported by Jackson for composites cables in 2006 (Jackson et al., 2006), in which the results regarding the bending tests were not presented. Most bending test methods were based on cyclic bending tests for steel wire ropes (Ridge et al., 2001).

16.5.5 Aging behavior

An additional concern about CFRP cables is their aging behavior under aggressive environments. Indeed, several studies are available regarding the general aging of composite materials, but only a few address structures such as CFRP cables in aggressive environments for mooring applications.

Several studies have been conducted to evaluate durability of FRP used as reinforcement in concrete structures, most of them highlight the short-term performance of GFRP bars. Recently, Ali et al. (2015) reported a complex study of commercial CFCC cables aged under marine and alkaline environments. The CFCC tendons were evaluated with and without sustained load in alkaline solution for up to 7000 h. The tendons showed minimal losses in tensile strength (around 10%) for the maximum

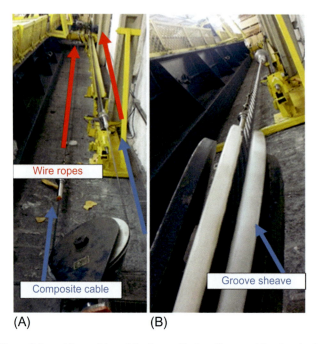

Fig. 16.23 View of the cable positioned in the cyclic bending machine just before testing (A) and image of the grooved sheave (B) (Silva et al., 2016a).

time of 7000 h, as illustrated in Table 16.6. Furthermore, the tensile-strength retention obtained for a service life of 75 and 150 years at mean annual temperatures (MATs) of 10°C and 50°C corresponds to 83% and 80%, and 78% and 75%, respectively.

Similarly, a significant finding was reported by Silva et al. (2016b), where it has shown that seawater, distilled water, and UV radiation have distinct effect on CFRP pultruded rods. These rods lost up to 18% of their tensile strength, showing a strong effect of the environment that deserves more detailed studies.

16.6 Finite element modeling of composite cables

The Finite Element Method (FEM) is a numerical tool widely applied to predict the behavior of wire ropes and cables, showing good correlation with experimental data for steel (Ghoreishi et al., 2007b; Erdomnez and Imrak, 2011) and also for composite materials (Menezes et al., 2017a,b). Many different analytical models also attempted to accurately predict wire ropes and cable behavior over the last decades (Costello, 1990; Elata and Eshkenazy, 2004; Xiang et al., 2015; Utting and Jones, 1987b; Usabiaga and Pagalday, 2008), most of them based on thin rods Love's theory (Love, 2011). Such models usually deviate significantly from experimental data or from numerical results (Ghoreishi et al., 2007b; Jiang, 2011; Jiang and Warby, 2008), because of the many assumptions considered, being the usual ones (Menezes et al., 2018; Cardou and Jolicoeur, 1997; Ghoreishi et al., 2007b):

Table 16.6 Tensile test results of the control and conditioned CFCC specimens (Ali et al., 2015)

Time of immersion (h)	Temperature (°C)	Tensile strength (MPa) Avg.	COV (%)	Retention (%)	Modulus of elasticity (GPa) Avg.	COV (%)	Retention (%)	Ultimate strain (%) Avg.	COV (%)	Retention (%)
0	22	3154	230	100	151.0	0.61	100	2.08	1.71	100
1000	60	3077	2.97	97.6	150.0	0.91	99.3	2.05	2.06	98.6
3000	60	3067	2.99	97.2	148.1	1.09	98.1	2.07	3.68	99.5
5000	60	30,146	3.74	95.6	147.0	0.48	97.3	2.05	3.27	98.6
7000	60	2928	0.35	92.8	147.1	0.22	97.4	1.99	0.54	95.7

(i) Wire contraction and Poisson's effect are neglected;
(ii) Bending and twisting stiffnesses, which are important for high values of α (helix angle), are neglected;
(iii) The change in β (lay angle) is underestimated, leading to discrepant values above 20°. As can be seen in Fig. 16.24, where a wire rope had their stiffness values plotted in function of its lay angle, showing an increasing deviation between the numerical model (red line) and eight different analytical solutions as the lay angle value gets higher (Ghoreishi et al., 2007b);
(iv) Gaps between wires, which are especially important in multilayered cables, are not included;
(v) Plastic or viscoelastic behavior is neglected;
(vi) The material is considered isotropic, which is particularly not suited when modeling composite cables;
(vii) Limited loading and boundary conditions;
(viii) Hypothesis of long length;
(ix) Contact forces due to normal and tangential contacts are neglected;
(x) Surface interaction between wires is either considered smooth ($\mu = 0$) or rough ($\mu = \infty$).

Hypotheses (i) and (ii) are not present in new models, and some recent works also include friction (Argatov, 2011), or transversally isotropic behavior, for cables under tension (Pan, 1992) or bending (Crossley et al., 2003). Nevertheless, most of these assumptions are still present in any analytical model, whereas a numerical model through FEM can easily ignore hypotheses (i)–(vii). By inserting a local coordinate system, as depicted in Fig. 16.25, one can set an orthotropic behavior and allow the external wires mechanical properties to change as they twist around the cable core.

16.6.1 Model length

In order to avoid end effects and allow for wire accommodation in the numerical models, the cable should not be too short. However, it cannot be too long either due to the required computational effort. Indeed, this length must be investigated for every different geometry, but a complete wire twist around the core (1 pitch) is generally considered (Fekr et al., 1999; Menezes et al., 2017a). As can be seen in Fig. 16.26, where a 1 × 10 cable was submitted to tensile stress, a length equal to its pitch (42.16 mm) requires short computational time and yet yield an accurate strain value (measured at the core).

16.6.2 Contact between wires

The consideration of normal and tangential contacts between adjacent wires and between external wires and the core is fundamental to achieve accurate behavior. However, these contact conditions add nonlinearities to the problem, increasing computational time. There are basically two methods for evaluating contact problems in Finite Element Analysis (FEA), Lagrange multipliers and the Penalty method (Khoei and Mousavi, 2010). In the Lagrange multipliers method, nonpenetration conditions are enforced by incorporating constraints forces, while in the latter, penetrations are

Composite materials for mooring applications

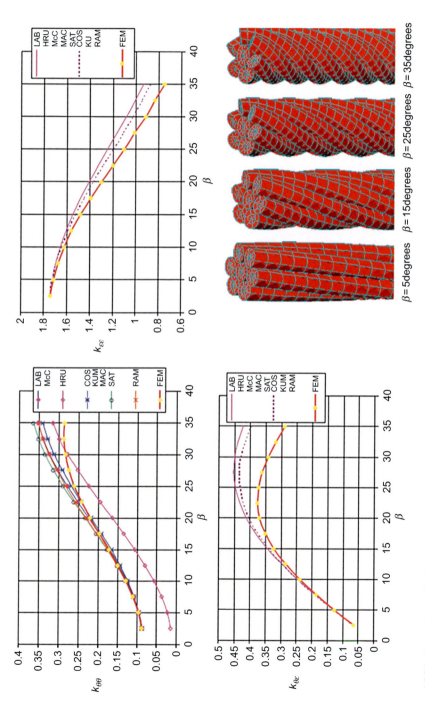

Fig. 16.24 Comparison of a 1 × 7 cable stiffness matrix k for different lay angles, using FEM model and different analytical solutions (Ghoreishi et al., 2007b).

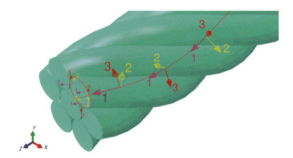

Fig. 16.25 Local coordinate system (1-2-3) and global coordinate system (X-Y-Z) for a 1 × 7 cable.

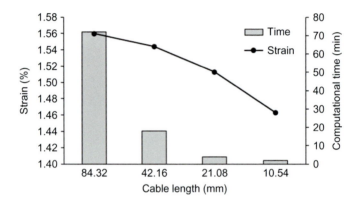

Fig. 16.26 Comparison of computational time and strain value for 1 × 10 cables of different lengths (Menezes et al., 2017a).

introduced and the stiffness matrix is updated to separate the boundaries. Both methods allow for an adjustable contact stiffness, where high value leads to small penetrations but difficult convergence, whereas low stiffness speeds up the convergence process but allow significant penetrations, as shown in Fig. 16.27.

Regarding the friction coefficient, it has been shown to have only a small influence if it is above 0.3 (Jolicoeur and Cardou, 1991) (the CFRP value, for instance, has been reported to be ca. 0.6 (Schön, 2004)) for tensile and for bending stresses (Jiang and Warby, 2008). Fig. 16.28 shows, for five different cases, a small effect for different μ values, and Fig. 16.29 shows the contact stresses on a CFRP cable modeled through FEM (for $\mu = 0.6$).

16.6.3 Mechanical properties

For an elastic transversally isotropic composite material, its stiffness matrix may be expressed in function of five independent engineering constants, for instance: E_1, E_2, ν_{12}, G_{12}, and G_{23}; where: Direction 1 is parallel to the fiber orientation, 2 and 3 are perpendicular to it; E is the Young's modulus; G is the shear modulus; ν is the Poisson ratio.

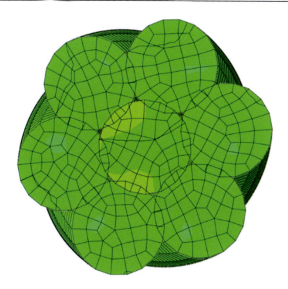

Fig. 16.27 A cable submitted to tensile stress showing significant penetration between wires.

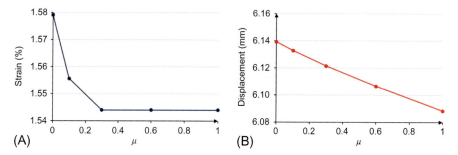

Fig. 16.28 Influence of the friction coefficient in a 1 × 10 CFRP cable under tensile stress (A) with longitudinal strain measured at the core, and bending stress (B) with vertical displacement measured at the middle section.

Although the cable behavior under tensile and bending stresses is mostly influenced by the E_1 value, which is relatively easily measured in individual wires, G_{12} also significantly influences the cable response. The other engineering constants display a less significant role (Menezes et al., 2017a) and may as well be approximated through micromechanics theory (Chamis, 1983; Halpin and Kardos, 1976), invariant-based theory (Tsai and Melo, 2014), or numerically with the Representative Volume Element (RVE) technique (Aliabadi, 2015). Table 16.7 reports some properties estimated for a CFRP evaluated from different ways, showing great differences among some of them, showing that, except for E_1 and ν_{12}, different theories may significantly vary from each other.

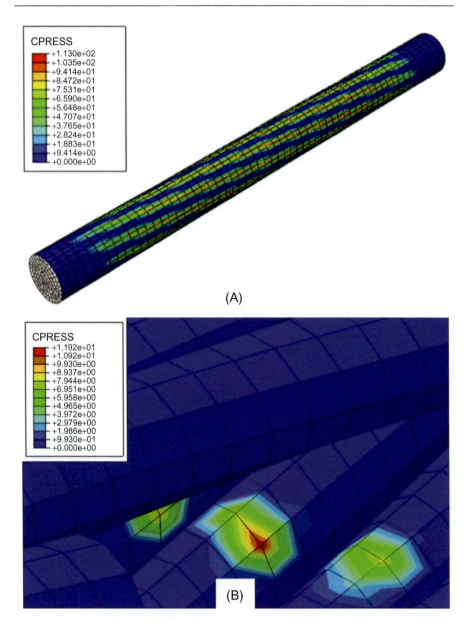

Fig. 16.29 Contact stresses applied by the external wires (peak value of 108.7 MPa) on the core of a CFRP cable under tensile stress (A), and contact stresses on the external wires touching the core (peak value of 11.9 MPa) of a CFRP cable under bending stress (B) (Menezes et al., 2017a).

Table 16.7 Mechanical properties of a carbon/epoxy composite estimated through five different methods

	Carbon T300[a]	Epoxy MY750[a]	ROM	Halpin-Tsai (Halpin and Kardos, 1976)	Chamis (1983)	RVE	Invariant-based (Tsai and Melo, 2014)[b]
E_1 (MPa)	231,000	3350	139,900	139,900	139,900	139,300	139,900
E_2 (MPa)	15,000	3350	8197	16,930	8410	7843	7919
G_{12} (MPa)	15,000	1240	2758	4011	4284	3803	5441
G_{23} (MPa)	7000	1240	2449	3453	3420	2960	2640
ν_{12}	0.200	0.350	0.260	0.260	0.260	0.255	0.328
ν_{23}	0.200	0.350	0.673	1.451	0.229	0.380	0.500

[a]Properties extracted from Kaddour (Kaddour and Hinton, 2004).
[b]The value of E_1 was applied as an input.

16.6.4 FEM models for tensile and bending stresses

Although FEM results may incorporate intrinsic errors, such as element distortion, truncation, mesh errors, and mismatch between the real problem order and the element order, this method is capable of simulating the behavior of cables and wire ropes with relative good accuracy, even for multilayered cables, where analytical models are even more limited (Elata and Eshkenazy, 2004).

A comparison between experimental data, FEM numerical results, and a classical analytical solution (Costello, 1990) is shown in Fig. 16.30A for a 1 × 19 CFRP cable, whose cross section is illustrated in Fig. 16.30B. Force and strain data were acquired, respectively, with a load cell and a strain-gauge (acquisition rate of 10 Hz) positioned on the external wire to avoid crushing. Considering that pure tensile loading was applied, the failure criterion adopted was maximum strain. Since the core is the critical component in such condition, the failure is assumed to occur when the strain value at the core reaches the value obtained from experimental tensile stress tests performed individually at the core (17,600 μ, i.e., 1.76% strain). In fact, similar ultimate strain values were found in tensile tests of an isolated core and in the cable (Motoyama et al., 2002). Fig. 16.30 shows that the numerical model accurately predicted the failure load based on the core strain, since at this moment the external wires strain in both numerical and experimental model shows a good correlation. Numerical model was also able to predict the load distribution with great accuracy in comparison with the analytical solution (Costello, 1990).

Relative to flexural tests, due to excess resin or prestressed wires due to the manufacturing process, the cables may pass from a rough condition (no slippage between wires) to a smoother condition along the test, causing the moment of inertia to greatly decrease, showing a nonlinear response (Menezes et al., 2017b). This behavior is illustrated in Fig. 16.31, for the same CFRP cable of Fig. 16.30B, in which experimental data was again compared with FEM and analytical solutions. Since the

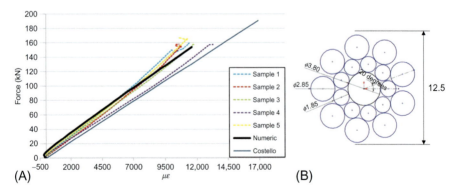

Fig. 16.30 Comparison between experimental data, and numerical and analytical solutions for a 1 × 19 CFRP cable (strain-gauges positioned at external wires to avoid smashing during the test) (A) and the cable cross section (dimensions in mm) (B) (Menezes et al., 2017b).

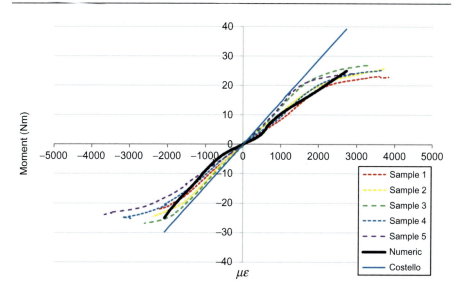

Fig. 16.31 Four-point bending results of a 1 × 19 CFRP cable (Menezes et al., 2017b).

numerical model did not capture the slippage transition, it only meets the experimental data before relative slippage starts and the cable acquires a nonlinear behavior.

16.7 Concluding remarks

Manufacturing techniques, anchorage systems, and characterization of composite cables were reviewed in this chapter. After exploring their potential applications and suitability in substituting traditional materials, an introduction about the design process of cables and wire ropes was presented, aiming to cover strong and weak points of different geometries and constructions regarding their tensile, bending, and fatigue behavior.

Analytical models were briefly discussed, along with their assumptions and simplifications, mentioning also closed-form solutions to be used in early design stages. Also, FEM numerical models showing a good agreement with experimental data for tensile and bending were presented, being able to dismiss costly and time-consuming experimental tests in some cases.

Difficulties in manufacturing CFRP cables were also exposed, and the strategies adopted by main manufacturers were described, showing that current technology is able to produce high quality materials. Characterization tests performed in commercial CFRP cables of different geometries and constructions showed an outstanding performance in tensile-tensile fatigue and static tensile tests. However, due to their high stiffness, their bending is weakened. Aging tests performed in harsh environments pointed a slow degradation in mechanical properties.

As a final statement, due to the aforementioned characteristics, the authors consider CFRP cables good candidates for substituting steel and polymeric cables in mooring lines of deep-water offshore structures.

Acknowledgments

The authors gratefully acknowledge Petrobras for the financial support. They also acknowledge the Coordination for the Improvement of Higher Education Personnel (CAPES) and the National Council for Scientific and Technological Development (CNPq).

References

ABNT NBR 16272:2014 EN, 2014. Steel Wire Ropes—Determination of Bending Fatigue Strength. ABNT NBR 16272:2014 EN.

Adrezin, R., Benaroya, H., 1999. Response of a tension leg platform to stochastic wave forces. Probab. Eng. Mech. 14, 3–17.

Ali, A.H., Mohamed, H.M., ElSafty, A., Benmokrane, B., 2015. Long-term durability testing of tokyo rope carbon cables.ICCM20 International Conference on Composite Materials.

Aliabadi, M.H., 2015. Woven Composites. Imperial College Press, London.

Argatov, I., 2011. Response of a wire rope strand to axial and torsional loads: Asymptotic modeling of the effect of interwire contact deformations. Int. J. Solids Struct. 48, 1413–1423.

ASTM D6272-10, 2010. Standard Test Method for Flexural Properties of Unreinforced and Reinforced Plastics and Electrical Insulating Materials by Four-Point Bending. ASTM D6272-10.

Botker, S., Johannessen, T.B., 2002. Composite risers and tethers: the future for deep water TLPs.Offshore Technology Conference.

Brandon, J.E., Chaplin, C.R., Ridge, I.M.L., 2001. Analysis of a resin socket termination for a wire rope.OIPEEC Round Table Conference.

Cardou, A., Jolicoeur, C., 1997. Mechanical models of helical strands. Appl. Mech. Rev. 50 (1), 1–14.

Carmo, M.P., 1976. Differential Geometry of Curves and Surfaces. Prentice Hall, Englewood Cliffs.

Chamis, C.C., 1983. Simplified Composite Micromechanics Equations for Hygral, Thermal and Mechanical Properties. NASA Technical Memorandum.

Chaplin, C.R., Potts, A.E., 1991. A Critical Review of Wire Rope Endurance Research. Offshore Technology Report HMSO.

Chen, Z., Yu, Y., Wang, X., Wu, X., Liu, H., 2015. Experimental research on bending performance of structural cable. Constr. Build. Mater. 96, 279–288.

Chen, Y., Meng, F., Gong, X., 2017. Study on performance of bended spiral strand with interwire frictional contact. Int. J. Mech. Sci. 128–129, 499–511.

Costello, G.A., 1990. Theory of Wire Rope, first ed. Springer-Verlag, New York.

Crossley, J.A., Spencer, A.J.M., England, A.H., 2003. Analytical solutions for bending and flexure of helically reinforced cylinders. Int. J. Solids Struct. 40, 777–806.

Cutler, K.C., Knapp, R.H., 2010. Cable bend testing over a variable diameter sheave.Proceedings of the Twentieth (2010) International Offshore and Polar Engineering Conference, Beijing, China.

Davies, P., Reaud, Y., Dussud, L., Woerther, P., 2011. Mechanical behaviour of HMPE and aramid fibre ropes for deep sea handling operations. Ocean Eng. 38 (17–18), 2208–2214.

Dolan, C.W., Hamilton, H.R., Bakis, C.E., Nanni, A., 2001. Design Recommendations for Concrete Structures Prestressed With FRP Tendons. FHWA Contract DTFH61-96-C-00019.

Elata, D., Eshkenazy, M.W.R., 2004. The mechanical behavior of a wire rope with an independent wire rope core. Int. J. Solids Struct. 41, 1157–1172.

Erdomnez, C., Imrak, C., 2011. A finite element model for independent wire rope core with double helical geometry subjected to axial loads. Sadhana 36, 995–1008.

Evans, J.J., Ridge, I.M.L., Chaplin, C.R., 2001. Wire failures in ropes and their influence on local wire strain behaviour in tension–tension fatigue. J. Strain Anal. Eng. Des. 36 (2), 236–244.

Fekr, M.R., McClure, G., Farzaneh, M., 1999. Application of ADINA to stress analysis of an optical ground wire. Comput. Struct. 72, 301–316.

Feyrer, K., Ropes, W., 2015. Tension, Endurance, Reliability, second ed. Springer, New York.

Ghoreishi, S.R., Cartraud, P., Davies, P., Messager, T., 2007a. Anallytical modelling of synthetic fiber ropes subjected to axial loads. Part I: a new continuum model for multilayered fibrous structures. Int. J. Solid Struct. 44, 2924–2942.

Ghoreishi, S.R., Messager, T., Cartraud, P., Davies, P., 2007b. Validity and limitations of linear analytical models for steel wire strands under axial loading, using a 3D FE model. Int. J. Mech. Sci. 49, 1251–1261.

Gilmore, J., Thomas, R., 2009. Deepwater synthetic lowering and lifting with enhanced cyclic bend fatigue resistance.DOT Conference.

Girón, A.R.C., Correa, F.N., Hernández, A.O.V., Jacob, B.P., 2014. An integrated methodology for the design of mooring systems and risers. Mar. Struct. 39, 395–423.

Halpin, J.C., Kardos, J.L., 1976. The Halpin-Tsai equations: a review article. Polym. Eng. Sci. 16, 344–352.

Harrop, I., Summerscales, J., 1989. Acousting emission testing of the structural integrity of multicore cable. Br. J. NDT. 31(7).

I. Hanes Supply of SC, n.d., "Wire Rope". Available from: http://www.hanessupply.com/content/pdfs/wireRope101.pdf (Accessed 27 March 2018).

ISO 14125, 1998. Fibre-Reinforced Plastic Composites Determination of Flexural Properties. ISO 14125.

ISO 17893, 2004. Steel Wire Ropes—Vocabulary, Designation and Classification. ISO 17893.

Jackson, D., Shepheard, B., Kebadze, E., Teles, R., Rossi, R., Gonçalves, R.C., 2005. Carbon fibre composites as stay cables for MODU applications.Offshore Technology Conference.

Jackson, D., Dixon, M., Shepheard, B., Kebadze, E., Teles, R., Rossi, R., Gonçalves, R.C., 2006. The development of a carbon fibre mooring line for MODU and permanent mooring in deep and ultra-deepwater.Offshore Technology Conference, Houston, USA.

Jiang, W., 2011. A concise finite element model for pure bending analysis of simple wire strand. Int. J. Mech. Sci. 54, 69–73.

Jiang, W.G., Warby, J.L.H.M.K., 2008. Statically indeterminate contacts in axially loaded wire strand. Eur. J. Mech. A/Solids 27, 69–78.

Jolicoeur, C., Cardou, A., 1991. A numerical comparison of current mathematical models of twisted wire cables under axisymmetric loads. J. Energy Resour. Technol. 113 (4), 241–249.

Jolicoeur, C., Cardou, A., 1996. Semicontinuous mathematical model for bending of multilayered wire strands. J. Eng. Mech. 122, 643–650.

JSCE-E 531, 1995. Test Method for Tensile Properties of Continuous Fiber Reinforcing Materials. JSCE-E 531.

Kaddour, A.S., Hinton, M.J., 2004. Instructions to Contributors of the Second World-Wide Failure Exercise. WWFE-II: Part A.

Krasucki, F., Miskiewicz, K., Wojaczek, A., Fraczek, S., 1993. Electromagnetic Compatibility in Underground Mining: Selected Problems, first ed. Elsevier, New York.

Kawada, T., 2010. History of Modern Suspension Bridge: Solving the Dilemma Between Economy and Stiffness. ASCE Publications, Reston, VA.

Khoei, A.R., Mousavi, S.M.T., 2010. Modeling of large deformation—large sliding contact via the penalty X-FEM technique. Comput. Mater. Sci. 48, 471–480.

Kreyszig, E., 1968. Introduction to Differential Geometry and Riemannian Geometry. vol. 16. University of Toronto Press, Toronto.

Labrosse, M., Nawrocki, A., Conway, T., 2000. Frictional dissipation in axially loaded simple straight strands. J. Eng. Mech. 126 (6), 641–646.

Liu, Y., Zwingmann, B., Schlaich, M., 2015. Carbon fiber reinforced polymer for cable structures—a review. Polymers 7, 2078–2099.

Liu, Y., Zwingmann, B., Schlaich, M., 2016. Advantages of using CFRP cables in orthogonally loaded cable structures. Mater. Sci. 3 (3), 862–880.

Love, A.E.H.A., 2011. Treatise on the Mathematical Theory of Elasticity, fourth ed. Dover Publications, New York.

Luz, F.F., Silva, L.V., Cimini, C.A., Amico, S.C., n.d. Analysis of bending and tensile behavior of CFRP stranded and spiral cables using a numeric model. In: Proc. 20th International Conference on Composite Materials, Copenhagen, Denmark.

McKenna, H.A., Hearle, J.W.S., O'Hear, N., 2014. Handbook of Fibre Rope Technology, first ed. CRC Press, Boca Raton.

Meier, U., 2012. Carbon fiber reinforced polymer cables: Why? Why not? What if? Arab. J. Sci. Eng. 37, 399–411.

Meier, U., Brönnimann, R., Anderegg, P., 2015. Long Term Reliability of CFRPs in Bridge Engineering. ICCM20.

Menezes, E.A.W., Silva, L.V., Silva, V.F., Marczak, R.J., Amico, S.C., 2016. Numerical modelling of carbon fiber reinforced polymer cables.Brazilian Conference on Composite Materials, BCCM3.

Menezes, E.A.W., Silva, L.V., Marczak, R.J., Amico, S.C., 2017a. Numerical model updating applied to the simulation of carbon fiber–reinforced polymer cables under bending and tensile stress. J. Strain Anal. Eng. Des. 52 (6), 356–364.

Menezes, E.A.W., Silva, L.V., Luz, F.F., Jr, C.A.C., Amico, S.C., 2017b. Numerical and experimental analysis of the tensile and bending behavior of CFRP cables. Polym. Polym. Compos. 25 (9), 643–650.

Menezes, E.A.W., Silva, L.V., Luz, F.F., Jr., C.A.C., Amico, S.C., 2018. Metallic and composite cables: a brief review. Int. J. Comput. Aided Eng. 10 (1–2), 179–191.

Motoyama, H., Ohta, T., Ohno, T., Moriya, K., 2002. Dynamic Tensile Properties of CFRP Cables Subjected to High-Speed Loads. Memoirs of the Faculty of Engineering, Kyushu University, vol. 62. pp. 113–127.

Nanni, A., Bakis, C.E., O'Neil, E.F., Dixon, T.O., 1996. Short-term sustained loading of FRP tendon-anchor systems. Constr. Build. Mater. 10 (4), 255–266.

Niedzwecki, J.M., Huston, J.R., 1992. Wave interaction with tension leg platforms. Ocean Eng. 19 (1), 21–37.

Noisternig, J.F., 2000. Carbon fibre composites as stay cables for bridges. Appl. Compos. Mater. 7, 139–150.

Odru, P., Geffroy, R., 2002. Carbon fiber tethers key for ultra-deepwater production. Offshore 62 (8), 141–144.

Osteergard, N.H., Lyckegaard, A., Andreasen, J.H., 2012. A method for prediction of the equilibrium state of a long and slender wire on a frictionless toroid applied for analysis of flexible pipe structures. Eng. Struct. 34, 709–716.

Pan, N., 1992. Development of a constitutive theory for short fiber yarns: mechanics of stample yarn without slippage effect. Text. Res. J. 62, 749–765.

Petrobras, 2009. Carbon Fiber Reinforced Plastic Rope for Offshore Stationkeeping. ET-3000.00-6651-962-PEB-004.

Pressley, A.N., 2010. Elementary Differential Geometry, second ed. Springer, London.

Ramsey, H., 1990. Analysis of interwire friction in multilayered cables under uniform extension and twisting. Int. J. Mech. Sci. 32 (8), 709–716.

Ridge, I.M.L., Chaplin, C.R., Zheng, J., 2001. Effect of degradation and impaired quality on wire rope bending over sheave fatigue endurance. Eng. Fail. Anal. 8, 173–187.

Roddenberry, M., Mtenga, P., Joshi, K., 2014. Investigation of Carbon Fiber Composite Cables (CFCC) in Prestressed Concrete Piles. Final Report, Contract Number BDK83-977-17.

Rohleder, W.J., Tang, B., Doe, T.A., Grace, N.F., Burgess, C.J., 2008. Carbon fiber-reinforced polymer strand application on cable-stayed bridge, penobscot narrows, maine. Transp. Res. Rec. 2050, 169–176.

Santos, T.F.A., Vasconcelos, G.C., Souza, W.A.d., Costa, M.L., Botelho, E.C., 2015. Suitability of carbon fiber-reinforced polymers as power cable cores: Galvanic corrosion and thermal stability evaluation. Mater. Des. 56, 780–788.

Schmidt, J.W., Bennitz, A., Täljsten, B., Goltermann, P., Pedersen, H., 2012. Mechanical anchorage of FRP tendons—a literature review. Constr. Build. Mater. 32, 110–121.

Schön, J., 2004. Coefficient of friction and wear of a carbon fiber epoxy matrix composite. Wear 257, 395–407.

Shaw's Wire Ropes, n.d. Wire Rope Information. Available from: http://www.wireropes.co.nz/wire-rope/wire-information.htm (Accessed 6 October 2017).

Silva, L.V., Menezes, E.A.W., Cimini, C.A., Amico, S.C., 2016a. Fatigue behavior of CFRP cables under tensile and bending loads for offshore applications.CAMX—The Composite and Advanced Materials Expo.

Silva, L.V., Silva, F.W., Tarpani, J.R.A.S.C., 2016b. Ageing effect on the tensile behavior of pultruded CFRP rods. Mater. Des. 110, 245–254.

Sima, Q., Watson, B., 1961. Records of the Grand Historian of China. Columbia University Press, New York.

F. Sloan, R. Nye and T. Liggett, "Improving bend-over-sheave fatigue in fiber ropes," in Oceans 2003 Proceedings, San Diego, USA, 2003.

Sparks, C., Zivanovic, I., Luyckx, J., Hudson, W., 2003. Carbon fiber composite tendons for deepwater tension leg platforms.Offshore Technology Conference, Houston, USA.

Tsai, S.W., Melo, J.D.D., 2014. An invariant-based theory of composites. Compos. Sci. Technol. 100, 237–243.

Usabiaga, H., Pagalday, J.M., 2008. Analytical procedure for modelling recursively and wire by wire stranded ropes subjected to traction and torsion loads. Int. J. Solids Struct. 45, 5503–5520.

Ushijima, K., 2010. Cable Made of High Strength Fiber Composite Material. United States of America Patent US 7650742B2.

Utting, W.S., Jones, N., 1987a. The response of wire rope strands to axial tensile loads—part I. Experimental results and theoretical predictions. Int. J. Mech. Sci. 29 (9), 605–619.

Utting, W.S., Jones, N., 1987b. The response of wire rope strand to axial tensile loads—part II. Comparison of experimental results and theoretical predictions. Int. J. Mech. Sci. 29, 621–636.

Verreet, R., 2002. A Short History of Wire Rope, first ed. PR Gmbh, Aachen.
Xiang, L., Wang, H.Y., Chen, Y., Guan, Y.J., Wang, Y.L., Dai, L.H., 2015. Modeling of multi-strand wire ropes subjected to axial tension and torsion loads. Int. J. Solids Struct. 58, 233–246.
Xie, X., Li, X., Shen, Y., 2014. Static and dynamic characteristics of a long-span cable-stayed bridge with CFRP cables. Materials 7, 4854–4877.
Zhang, B., Benmokrane, B., Chennouf, A., Mukhopadhyaya, P., El-Safty, A., 2001. Tensile behavior of FRP tendons for prestressed ground anchors. J. Compos. Constr. 5 (2), 85–93.

Index

Note: Page numbers followed by *f* indicate figures, and *t* indicate tables.

A

ABAQUS software, 321–323
Accelerated aging techniques, 103–106
  mechanical properties, 104–106
  moisture absorption, 103, 104*f*, 106–111
Acrylic polymer, 43
Active control mechanisms, MHK devices, 350–351
Active fire protection (AFP), 122–123
Adaptive composite blades
  benefits, 350, 353
  cavitation, 354
  design, 349
  scaling requirements, 356–358
  use, 359
Adhesive bonding, 189
Admiralty Inlet, in Puget Sound, 351–352, 352*f*
Advanced polymer composites
  advantages, 117–118
  conventional materials, 117
Aerodynamic braking system, 350–351
Aerodynamic forces, 439–440, 440*f*
Aerodynamic shape platform, 419, 420*f*
Aging behavior, composite cables, 475–476
Aging procedures, 97–98
Air pressures, 337, 339*f*
Aluminum alloys, 117
Angle of attack, 272–273, 365–366, 368
  fluid stall, 350–351
  incidental flow, 281
Anionic Polyamide 6 (A-PA6), 32, 38
ANSYS Fluent 12.1, 273, 279
ANSYS Mechanical 12.1, 274–275, 279, 286–287
Aramid fibers, 14, 15*t*
Arkema, 32, 38
Artificial neural networks (ANN), 66–67, 75–78, 76*f*
ASTM E119, 129
Autoclave processes, 35–36, 35*f*
Automated tape placement, 36–38, 36*f*
Average smoke production rate (SPRav), 130–131
Axial-induced velocity, 366–367

B

Backstay arrangements, rig design, 430–431
Back-up generators, 247–248
Bearing capacity, 321–323, 329–333
Behavioral constraints (BC), 178
BEM. *See* Boundary element method (BEM)
Bending moment, 317–319
  longitudinal and transverse, 438*f*
  transition segment, 326
  vertical, 309–310
Bending-over-sheave fatigue, 474
Bending stress, 484–485
Bend-twist coupling, 272–273
  composite propeller, 279–280
  deformation coupling, 347–349, 348*f*
  SMAHC propeller, 287–292, 294
"Bermudian" rig style, 421
Bernoulli equation, 372
BFB. *See* Bucket Foundation with Bulkheads (BFB)
BFlex software, 405
Bilinear cohesive model, 74–75
Biopolymers, 12–13
Bio-sourced thermoplastics, 47
Biot-Savart law, 372
Bismaleimide (BMI) resins, 8
Blistering, 23
Boat construction. *See* Marine boat construction
Boltzmann gas constant, 148–149
Boundary element method (BEM), 373–377, 384
Bragg grating sensors, 70, 70*f*
Braids, 20

Breakthrough in European Ship and Shipbuilding Technologies (BESST), 161–162
Brinson model, 277
Bucket Foundation with Bulkheads (BFB), 319–320
Buckling mode, 426, 445
 longitudinal, 446f
 mast section, 447f
 transverse, 446f
Burning period, 124
Burst testing, 406–407, 407f, 408t

C

Candlewick effect, 152–154
Cap shroud angles, 423, 424f
Carbon-based nanopaper, 153
Carbon/epoxy composite, 483t
Carbon fiber, 91, 354–355
 acrylic composite, 38f
 composites, 142
 materials selection, 14–15, 16t
 PEEK, 40–41, 40f, 41t
 polyamide 6, 41–43
 reinforced acrylic composites, 43
Carbon Fiber Composite Cables (CFCC), 455f, 464
 aging behavior, 475–476
 tensile test results, 477t
Carbon fiber-reinforced plastic (CFRP), 56–57
 aging behavior, 475
 bending behavior, 457, 457f, 475
 blast pressure profile and deflection, 59f
 composites, 67
 experimental data, 484f
 fatigue behavior, 458, 472–473
 final constructing cost for structures, 456t
 form of lamella, 463–464
 four-point bending result, 485f
 geometry and cross section, 470f
 mechanical properties, 455t
 numerical and analytical solutions, 484f
 numerical results, 457, 457f
 production, 465f
 subsurface grooves, 68f
 tensile stress, 481–482f
 tensile testing, 469, 471–472, 472f

testing procedures and results, 456
 types, 464f
Carbon laminates
 axial strain, 444
 in-plane tensile stiffness, 435, 436f
Carbon-silica reaction, 134–135
Cartesian coordinate system, 273, 378
Cavitation
 erosion, 24
 MHK turbine, 353–354
CBF. See Composite bucket foundation (CBF)
Cellular materials, 188
CFX GGI technology, 381
Classical laminate analysis (CLA), 111
Cleavamine®, 5
Closed-cell PVC foams, 191, 191f
Coats-Redfern method, 150
Code of Federal Regulation, 132
Code sails, 428–429, 440
Cohesive damage model, 74–75
Cold-heading process, 443
Commercial shipbuilding, 86–87
Commingled yarns, 34f
Composite action, 142
Composite advanced sail (CAS), 311
Composite bucket foundation (CBF), 317–319
 air pressures, 337, 339f
 bearing capacity, 321–323, 329–333
 earth pressure results, 337–341, 338–341f
 equipment layout, 335f
 failure envelope, 332–333, 332–333f
 FEM models, 321–323, 323f
 under horizontal load, 330–331, 330–331f
 installation, 333–342
 load-bearing characteristics, 321–323
  cover-load-bearing type, 328–329, 328t
  force transfer mechanism, transitional section, 323–328, 325f, 327f
 load-displacement relationship curves, 330, 330f
 pore pressures, 336–341, 340f
 prototype with, 319f
 rotation, 330, 331f
 soil pressure distribution, 328, 328f
 structure, 318f
 test model, 334f
 vessel operation steps, 320–321, 322f

Index

Composite cables, 452–456
  aging behavior, 475–476
  anchorage system, 467–469, 470f
  components, 453f
  cyclic bending tests, 474–475, 476f
  design process, 456–458
  finite element method, 476–478, 479f
    bending stresses, 484–485
    contact between wires, 478–480, 481–482f
    mechanical properties, 480–483, 483t
    model length, 478, 480f
    tensile stresses, 484–485
  flexural tests, 473–474, 475f
  inclusions, 466, 467f
  Lang lay construction, 453–454, 454f, 457, 471
  manufacturing, 455f, 463–466
  mathematical model with linearized kinematics, 459–463
  mechanical properties, 455t
  single-layered and multilayered, 453f
  stress distribution in outer wires, 458f
  tensile-tensile fatigue tests, 471–473, 474f
  tensile tests, 466–471, 472f, 477t
  terminations, 466–467
  types, 455f
Composite foams. *See* Syntactic foams
Composite materials, 86, 346–347
  advantage, 272
  design consideration, 354–358
  drawback, 454–456
  effect on power and thrust, 359
  fiber reinforced, 272–273
  fire reaction, 136–141
  lifetime performance, 358–359
  mechanical properties, 435–437
  use, 347
Composite propeller, 279–282, 281f
  advantages, 368–369
  assembled, 366f
  coupled 3-D FEM/VLM method, 371–373
  coupled FEM/BEM, 373–377
  coupled FEM/CFD, 377
    RANS equation, 377–378
    steady fluid-structural coupling, 378–379
    transient fluid-structural coupling, 380–381

deformation, 274–276
design of thickness, 281–282, 283f
finite-element method, 369–370, 371f
geometrical characteristics, 364–365
geometry, 279–280, 281f
hydro-elastic performance, 382–383
inter-laminar shear stresses, 382
laminate molding, 365f
material properties, 282t
metal *vs.*, 368–369
open water performances, 283f, 381–383
performance optimization, 385–386
PSF-2 program, 369–370, 370f
structural characteristics, 364–365
structural dynamic characteristics, 384–385
working characteristics, 365–368, 366f
Composite structures, 118–126, 305–306
  average crack density *vs.* laminate stress curve, 167f
  constrained optimization problem, 174–175
  general objective and methodology, 162–163
  material characterization, 170–174
  material safety factors, 163–170
  operational limit, 163–170, 164f
  strength reliability analyses, 163, 169–170
  structural design exploration, 174–181
  structural reliability analyses, 164–165
Compression moulding, 36, 36f
Compressive properties
  polyvinyl chloride foams, 192–198
  syntactic foams, 206–209
Computational fluid dynamic (CFD) method, 148–151, 294, 377
  mast section, 427f
  RANS equation, 377–378
  rig analysis technologies, 437
  steady fluid-structural coupling, 378–379
  transient fluid-structural coupling, 380–381
Cone calorimeter test ISO 5660, 127–128
Connectors, 248
Consolidation, thermoplastic matrix composites, 35
Constituent-level predictive methods, 111
Constitutive models
  for hyperelastic materials, 400–401
  shape memory alloy, 277

Continuity equation, 375, 377–378
Continuous aligned fibers, 91
Continuous fibers, 18–19
Cooling, thermoplastic matrix composites, 35
Coordinate system
 Cartesian coordinate system, 273, 378
 laminate, 275–276
 local, 459, 461, 478, 480f
 rotating blade-fixed, 374, 374f
Core materials, 21–22
Corporate social responsibility (CSR), 24–25
Corrosion-resistant materials, 358
Coulomb's friction law, 323
Coupled spiral inductors (CSI), 67
Cowoven fabrics, 34f
Crushing test, 393, 409f
Cyclic bending tests, 474–475, 476f
Cyclic butylene terephthalate (CBT), 12, 38
Cyclic fatigue damage, 63f, 78–79
Cyclics™, 32, 38

D

Damage assessment
 artificial neural networks, 75–78
 cohesive damage model, 74–75
 E-glass fiber/vinylester composite, 71f, 73t
 environmental effects, 62–64
 finite element modeling, 71–72, 71f
 Hashin damage model, 72–73
 impact loading, 60–62
 impulsive loading, 59–60
 laminated composite sandwich panel, 60f
 micro-buckling, 61f
 modeling cyclic fatigue damage, 78–79
 nondestructive damage detection
  experimental methods, 64–68
  in situ damage detection, 69–70
 numerical and theoretical modeling, 70–79
 sandwich composite panel, 61f
 scanning electron micrographs, 62–63, 62f
 USS Fitzgerald warship's collision, 55, 56f
Darcy's law, 38, 231
Decay period, 125
Deformation analysis
 active control, 273
 bend-twist deformation coupling, 347–349, 348f
 composite propeller, 274–276

 propeller blade under operating conditions, 292–295, 293t
Degradation, of marine composites, 23–24
Degree of crystallinity, 32, 39
Delamination, 74
Diamond jumper arrangements, rig design, 428
Diffusion, 23, 107
Digital image correlation (DIC), 59–60, 170–174
Diglycidyl ether of bisphenol A (DGEBA), 5, 6t
Divinycell units
 panels with, 307t
 structural arrangements, 307
 weight, 308, 308t
DNVGL, 418, 443–444
Durability testing
 accelerated aging techniques, 103–106
 accelerated moisture absorption, 106–111
 constituent-level predictive methods, 111
 epoxy/glass laminate, 89f
 loading and durability requirements, 87–89
 material selection, 90–93
 saturated specimens, 98–103
 sea water conditioning techniques, 93–98
Dynarig, 421
DYWICARB company, 468–469, 470t

E

Earth pressure
 during CBF sinking process, 337–341, 338–341f
 passive, on outer wall, 331f
Economic viability
 cost, 307–308
 weight comparison, Divinycell structure, 308, 308t
E-CR glass, 91
Efficient material utilization, 162
E-glass, 91
 balsa wood sandwiches, 188
 fiber/vinylester composite, 71f, 73t
 noncrimp fabrics, 63
 polyester composite, 56
 vinyl ester composites, 133
Elastomers, 395–396, 396t
Elium™, 32, 38, 43

Index

Environmental effects
  damage assessment, 62–64
  recycling and, 46–48
  testing, 95–96
Environmental scanning electron microscope (ESEM), 135
Environmental stress concentration, 24
Epoxide resin (Ep), 5
Epoxy resins, 90
Epoxy-vinyl ester resins, 90
EPS. *See* Expanded polystyrene (EPS)
Equilibrium equations, 461
  finite-element model, 378–379
  geometrically nonlinear analysis, 372–373
  moments, 462
  three-dimensional, 373
European network for Lightweight Applications at Sea (E-LASS), 86
Exotherm temperature, 238–239, 239f
Expanded polystyrene (EPS), 308
  mechanical properties, 302–305, 305t
  panels with, 307, 307t
Eye splice termination, 467, 467f

F

Fatigue behavior
  CFRP cable, 458, 472–473
  composite materials, 358–359
Fatigue testing, 102
FEA. *See* Finite element analysis (FEA)
FEM. *See* Finite element method (FEM)
Fiber Bragg grating (FBG) strain sensors, 69
Fiberglass, boat construction, 302–303
  characteristics, 302
  mechanical properties, 312t
Fiber-matrix interface, 16–18
Fiber-matrix interfacial failure, 89
Fiber-reinforced composites, 141, 363–364
Fiber-reinforced plastics (FRP), 56, 58
  composites, 62–63
  disadvantages, 58
  laminates, 169–170
  preimpregnated, 166
  strength reliability analyses, 182
Fiber-reinforced polymer composites (FRPCs), 115–118, 346–347
Fiber selection, durability testing, 91
FIBERSHIP projects, 86
Fickian absorption, 89
Fickian diffusion, 106–111
  application, 108–109, 109f
  limitations, 110–111
Filament winding, 399
Filler construction, 453–454, 455f
Film stacking, 34f
Finite element analysis (FEA), 311
  offloading hoses, 406–410
  rig analysis technologies, 437, 442
Finite element method (FEM), 71–72, 274–275
  and BEM, 373–377, 384
  and CFD, 377–381
  composite bucket foundation, 321–323, 323f
  composite cables, 476–485, 479f
  composite propeller, 369–370, 371f
  three-dimensional plate, 310–311, 311f
  and vortex-lattice method, 371–373
Finite volume method, 273
Fire curves
  designing, 125–126
  hydrocarbon fire curve, 126
  standard and hydrocarbon, 125f
  standard fire, 126
Fire growth index (FGI), 153f
Fire growth rate index (FIGRAta), 130
Fire performance
  advanced polymer composites
    advantages, 117–118
    conventional materials, 117
  composite fire reaction
    computational fluid dynamic models, 148–151
    Janssens' method, 148
    kinetic parameters of resin, 152t
    Quintiere's fire growth model, 147–148
  composite structures, 118–126
  Deepwater Horizon offshore platform, 118, 122f
  enhancement, 152–155
  epoxy/glass fiber composite, 155t
  Indian submarine, 118, 121f
  structural performance
    polymer sandwich composites, 142–143
    postfire mechanical properties, 143–145
    single skin laminates, 141–142
  US Navy submarine, 118, 121f

Fire reaction
  composite materials, 136–141
  pyrolysis reaction, 133–136
Fire requirements, durability testing, 88
Fire safety
  development periods, 123–125, 123f
  engineering, 120–122
  and protection requirements, 131–132
    international maritime organization codes, 131–132
    US Naval structures, 132t
  strategies, 122–123, 123f
Fire testing
  Cone calorimeter test ISO 5660, 127–128
  ISO 1182 for noncombustible materials, 129
  room corner test ISO 9705, 126–127
First order shear deformation theory (FSDT), 273, 276
First-Ply Failure (FPF), 166
Fixed speed-fixed pitch turbine, 350–351
Flame retardant, 152–154, 155t
Flashover period, 124
Flax fiber, 47, 48f
Flexible pipe
  bonded, 392–393, 392f
  mechanical behavior, 403–404
  offshore industry, 391–392
  unbonded, 391–392
Flexural strength, 98
Flexural tests, composite cables, 473–474
Floating transport-sinking-leveling construction technique, 319–320
Flow mesh, 243
Fluid-structure interaction (FSI), 279
  rig analysis technologies, 437, 437f, 441
  steady, 378–379
  transient, 380–381
Foam core sandwich structures, 188–190
Force transfer mechanism, transitional section, 323–328, 325f
Fractional free volume (FFV), 23
Frenet-Serret triad, 459–461, 460f, 462f
Froude number, 357
FRP. See Fiber-reinforced plastics (FRP)
FSDT. See First order shear deformation theory (FSDT)
FSI. See Fluid-structure interaction (FSI)
Full-scale room corner test, 126–127
Fully developed region, 125

G

Galvanic corrosion, 24
Gel time, resin infusion, 238, 265
Genetic algorithm (GA), 373, 385–386
Glass fiber
  durability testing, 91
  materials selection, 16, 17t, 18–19
  polypropylene repair, 46
  reinforced acrylic composites, 43
Glass fiber-reinforced composite, 136f, 137
Glass fiber-reinforced plastic (GFRP), 56–57
  composite panels, 130f
  with epoxy matrix, 154
  residual damage, 59f
Glass microballoons (GMBs), 203–206, 206–207f
Glass transition temperature, 4, 6t
Governing equation, boundary element method, 374–375
Graphite epoxy composites, 279–281, 283–285
Gravity, resin infusion, 257–258
Growth period, 124

H

Hand layup, 189
Hand mixing, resin infusion, 258, 259f
Hand tools, resin infusion, 249, 250f
Hardwoods, 304
Hashin damage model
  fiber/matrix failure, 72–73
  laminated composite panels, 70–71
Hashin failure criterion, 402–403
Headstay arrangements, rig design, 428–430, 430f
Heating, thermoplastic matrix composites, 34
Heat release rate (HRR), 137–140, 154f
Heat release rate average (HRRav), 130
HexPly® Prepreg Technology, 21, 21t
High-Speed Code requirements, 131–132
Hollow glass microspheres (HGMs), 203–205
Hollow particles, 203–206
Hooke's law, 271–272, 462–463
Hull construction, 301, 304, 310f
Hybrid weave, 34f
Hydrocarbon fire curve, 126

Index

Hydrodynamic braking system, 350–351
Hydro-elasticity method, 369
   composite propeller, 380
   in nonuniform flow, 383
Hydrostatic pressure
   loading, 31–32
   test on carbon/PEEK, 46f
   underwater structures, 32–33, 42, 45

I

Ignition, 124, 137
Impact loading, 60–62
Impact properties
   finite element analysis, 213
   polyvinyl chloride foams, 198–201
   syntactic foams, 209–214
Impact testing, 101–102
Incipient period, 124
Inclusions, in cable, 466, 467f
Infiltration techniques, 189–190
Infrared (IR) camera, 69–70
Infusion, 38
In-line spreader rig, 422–423, 422f, 428, 429f
In-service monitoring, 58
In situ damage detection, 69–70
Insulation, 129
Integrity, 129
Interlaminar fracture, 74–75
Interlaminar shear stress (ILSS), 98, 101f
   calculation, 276
   composite propeller, 382
International maritime organization (IMO) codes, 131–132
ISO 1182, for noncombustible materials, 129
Isoparametric formula, 311
Iterative solving process, 349

J

Janssens' method, 148
Joint North Sea Wave Project, 405
JSCE E-531-1995 standard, 471
Just-in-time (JIT), 21

K

Ketch rig, 418f, 421
Kevlar®, 14
Kissinger method, 150
Knitted fabrics, 19–20

L

Lagrange multipliers method, 478–480
Laminates
   calculation method, 162
   carbon/epoxy cross-ply, 168–169, 169f
   composite, 60
   FRP, 169–170
   glass/epoxy cross-ply, 168–169, 168f
   stiffness, 165f
   ULS, 166
Lang lay construction, 453–454, 454f, 457, 471
Large eddy simulations (LES) combustion, 148–149
Leading edge, 364
Leadline (Mitsubishi), 455f, 464, 466
Leak detectors, 248–249, 249f
Life cycle analysis (LCA), 48, 161
Life cycle assessment (LCA), 24–25
Life cycle considerations, 24–25
Life cycle cost analyses (LCCA), 161
Life cycle costing (LCC), 24–25
Lifting surface theory, 371–372
Lightweight Constructions at Sea (LÄSS), 161–162
Lightweight design, 163
Linear elasticity, 275
Liquid composite moulding (LCM) process, 12
Load-bearing capacity, 129–130
Load-bearing characteristics, 321–329
Load Cases (LC), 175, 178f
Load-displacement relationship curves, 330f
Loading conditions, 190
Loads in rig, 438f, 439–442

M

Machine mixing, 259–260
Mach number, 357–358
Manufacturing processes, 34–38
   durability testing, 92
   sea water conditioning techniques, 97
Marine boat construction, 301–304
   cellular structure, 304
   composite structure concepts, 305–306
   computational simulation, 303–304
   core materials, 304–305, 305t

Marine boat construction *(Continued)*
  economic viability, 307–308
  fiberglass, 302–303
    characteristics, 302
    mechanical properties, 312*t*
    structural arrangement, 303–304*f*
    vessel structural computational design, 309–312
Marine fouling, 24
Marine-grade resins, 90
Marine hydrokinetic (MHK) turbine, 346, 351–353, 359
  bend-twist deformation coupling, 347–349, 348*f*
  composite-specific design considerations, 354–358
  design parameters
    cavitation, vibration, noise, 353–354
    power control system, 350–351
    site-specific design, 351–352
    turbulence, 353
  FRP composites, 346–347
  performance benefits of composites, 358–359
  pitch control, 349–350
  scaling concerns, 356–357
Marine propeller, 279. *See also* Propeller
Marine-sourced materials, 25
Mast section, 426–427, 427*f*, 447*f*
Materials
  characterization, 170–174
  options, 32–33
  safety factors, 163–170
Matrix, 4–13
  cracking, 66, 66*f*
  materials, 354–355
  thermoplastic polymers, 8–13
  thermosetting resins, 5–8
Maximum safe heeling angle, 442
Mean squared error (MSE), 77–78
Metallic propellers, 272–273
Metal propeller, composite *vs.*, 368–369
Methacrylic resins, 8
Meyer contact law, 200–201
MHK turbine. *See* Marine hydrokinetic (MHK) turbine
Microcracking, 89
Microplastics, 25
Millennium-Rig, 428

Mine counter measures vessel (MCMV), 56–57, 117
Mohr-Coulomb failure criterion, 323
Moisture absorption
  accelerated aging techniques, 103, 104*f*
  and degradation, 88–89
Moisture effects
  PVC foams, 201–203
  syntactic foams, 214–216
Momentum equation, 375
Monohulls, 441, 445–447
Morison formulation, 405–406
Moulding compounds, 21
Multihulls, 423, 441
Multipoint constraints (MPC), 369

N

Natural materials, 188
Navier-Stokes equation, 273, 378
Newton-Raphson numerical procedure, 373, 381
Nickel-aluminum-bronze (NAB), 363, 384
Nitinol fibers, 273, 296
  SMAHC layer, 285, 287–292, 288–291*t*
Nitrile rubber (NBR), 395
Noncombustible materials, ISO 1182 for, 129
Noncrimp fabrics (NCF), 19–20, 63, 172–173, 173*f*
Nondestructive damage detection
  experimental methods, 64–68
  matrix cracking, 66, 66*f*
  in situ damage detection, 69–70
Non-Fickian effects, 23, 94

O

Offshore fire, 116
Offshore offloading hoses
  bending stiffness *vs.* arc length curve, 412*f*
  bonded flexible pipe, 392–393
  components, 393–400
  compressive load *vs.* displacement curve, 410*f*
  computational analysis, 406–412
  constitutive models, 400–401
  double carcass hose, 394*f*
  elastomers, 395–396, 396*t*
  end fitting, 395
  experimental analysis, 406–412

failure models, 393t, 401–403
first carcass, 394f
flexible pipe, 391–392
hydrodynamic models, 404–406
liner and elastomeric body, 395–396
mechanical behavior of flexible pipe, 403–404
oil industry, 391
reinforcing plies, 396–398
steel SAE 1045, 399t
stiffness analysis, 410–412, 411f
strength analysis, 406–410
torsional moment vs. angular deformation curves, 413f
torsional stiffness analyses, 412f
Offshore wind turbines
BFB, 319–320
in China, 317–320, 318f
foundation (see Composite bucket foundation (CBF))
one-step installation technique, 320–321, 321f
Oil Companies International Marine Forum (OCIMF), 393
Open water performances
composite propeller, 281, 283f, 381–383
deformed/undeformed propeller, 279, 280f
pre-pitched composite propeller, 283–285, 285f
SMAHC propeller, 295f
Optically excited lock-in/infrared thermography (OLT/IRT), 64–66, 65f
Orcaflex software, 404–405
Organic-based polymer matrix, 133
Osmosis, 23

P

PanMax, 431
Parallel feed strategy, 252–253
Passive control mechanisms, 349
Passive fire protection (PFP), 122–123
Peak of heat release rate (PHRR), 138
Peel ply, 243
Penetration process, 336, 342
Performance-based design, 120–122
Permeability, resin infusion, 231–232, 233f
Petrobras Technical specifications (PTS), 473, 475

pH effect, sea water conditioning techniques, 96
Phenol-formaldehyde resin (PF), 5
Pitch control, 349–350
Planar reinforcements, 18–19
Plastic Oceans, 25
Poly(acrylonitrile) (PAN), 14
Poly(lactic acid) (PLA), 47–48, 48f
Poly(propylene) (PP), 8–11
Polyamide 6, 41–43
Polyamide (PA), 12
Poly aryl ether ketones (PAEK), 12
Polybutylene terephthalate (PBT), 12
Polyester resins, 90
Polyesters, 12
Polyetheretherketone (PEEK), 31, 40–41, 40f, 41t
Polyethylene terephthalate (PET), 12
Polymer composite materials (PCMs), 115–116
advantages, 117–118, 119–120t
Polymeric resin, at elevated temperature, 133–135, 134f
Polymer sandwich composites
behavior, 142–143
failure mode, 144f
in-plane and out-of-plane deflection, 144f
Polypropylene repair, 46
Polyvinyl chloride (PVC) foams, 188–189, 304–305, 305t
axial modulus, 194f
buckling process, 192–193
closed-cell foams, 191f
compression test result, 192f
compressive properties, 192–198
damage evolution, 202f
densities, 195–196
energy-deflection curves, 199–200, 199f
HP250 specimen, 197f
impact properties, 198–201
during impact testing, 200f
load-deflection curves, 199–200, 199f
measurement, 192f, 196
microstructure, 191
moisture effects, 201–203
Poisson's ratio, 195
stitching, 196–198
stress and modulus, 193f
stress-strain curves, 194f, 198f

Polyvinyl chloride (PVC) foams *(Continued)*
  suppliers, 190–191
  tensile and compressive loading, 194f
  usage, 190–191
  z-pinning, 196–198
Pore pressures, composite bucket foundation, 336–341, 340f
Postfire mechanical properties, 143–145
PostPanmax, 431
Powdered fabrics, 34f
Power control system, 350–351, 356
Preforms, 21
Preimpregnated plies, 34f
Preimpregnated reinforcements (prepregs), 21, 21t
Pre-pitched propeller, 283–285, 284–285f
Pressure
  resin infusion, 232–234
  sea water conditioning techniques, 94–95
  surface, propeller, 364
Prestressed steel, 319–320, 326, 327f
Pretensioned rig, 424–426, 425f
Principle of Virtual Work (PVW), 463
Product forms, 33–34, 34f
Propeller. See also *specific propellers*
  component names, 364f
  composite (*see* Composite propeller)
  design of thickness, 281–282, 285–287, 286f
  material properties, 282t
  open water characteristics, 280f, 283f, 285f
  pre-pitched, 283–285, 285f
  SMAHC, 285–295, 286f
  surface model, 280f
PSF-2 program
  composite propeller, 369–370, 370f
  lifting surface theory, 371–372
Pulsed thermography (PT) technique, 69–70
Pultrusion, 399
PVC foams. See Polyvinyl chloride (PVC) foams
Pyrolysis reaction, 133–136
  matrix behavior at elevated temperature, 133–135
  reinforcing fiber at high temperature, 135–136

Q

Quintiere's fire growth model, 147–148

R

Realisation and Demonstration of Advanced Material Solutions for Sustainable and Efficient Ships (RAMSSES), 86
Recovery stress, SMA, 277, 287–292
  at different temperatures, 277–278
  lamina behavior, 277
Recyclamine, 5
Recycling
  and environmental impact, 46–48
  thermoset plastics, 58
Reinforcement forms, 18–21, 18–19t
Reinforcing fiber, at high temperature, 135–136
Release film, 243
Repair, thermoplastic matrix composites, 45–46
Representative Volume Element (RVE) technique, 481
Research community, challenges, 92–93
Resin galleries, 243–244
Resin infusion
  arrangement for infusion, 245
  back-up systems, 247–248, 265
  challenges, 230–231
  connectors, 248
  consumables, 242–244
  core fit, 262–263
  core materials, 241–242
  cost, 230
  cross sectional area, 235
  definition, 228–229
  delivery and management, 258–260
  dry layup phase, 264
  dry patches, resolving, 256–257
  environment, 229
  equipment, 246–247
  fiber placement, 261–262, 262f
  flow brake, 256f
  flow management, 254–256
  gel time, 265
  gravity, 257–258
  hand mixing, 258
  hand tools, 249
  heated tooling, 246
  inclined surfaces, 253–254
  infusion setup, 251–253
  in-process monitoring, 265
  leak detectors, 248–249

Index

leaks in vacuum bag, 265–266
machine mixing, 259–260
management, 260
materials selection and characterization, 236f, 237–244
performance, 229
permeability, 231–232
physics, 231–236
postinfusion management, 266
prediction, strategy, and setup, 250–258
preinfusion checks, 264
pressure, 232–234
process control, 230
productivity, 229
quality control, 229
raw materials and ambient conditions, 264
reinforcement, 240–241
resin traps, 248
risk, 230
selection, 237–240
shrinkage, 239–240
stages, 230–231
3D resin flow, 254
tooling, 244–246
training/skill set, 230
vacuum, 265
  gauges, 248
  receiver, 247
  ring main, 247
variables
  measuring, 263
  understanding, 263–264
viscosity, 235–236, 265
waste, 229
wet phase, 261
Resin selection, 90
Resin transfer moulding (RTM), 20–21
Reynold's averaged Navier-Stokes (RANS) equation, 273–274, 377–378
Reynolds number, 357
Reynolds stresses, 378
RIFLEX software, 405
Rig, 418
  aerodynamic shape platform, 419, 420f
  analysis technologies, 437–438
  angles, 423, 424f
  complexity, 420
  composites mechanical properties, 435–437
  configurations, 421, 422f
  design considerations, 422
    backstay arrangements, 430–431
    diamond jumper arrangements, 428
    headstay arrangements, 428–430, 430f
    mast section, 426–427, 427f
    number of spreaders, 427–428, 429f
    pretension, 424–426, 425f
    rigging angles, 423, 424f
    size and its effects, 431–432
    spreader sweep, 422–423, 423f
    stability, 426
  functions, 419
  lateral vs. longitudinal, 421
  loads, 438f, 439–442
  mast and spars, 444
  material selection, 432–437
  modern Ketch, 418f
  monohulls, 441, 445–447
  multihulls, 423, 441
  pitching motions, 439
  standing, 443–444
  statics and dynamics, 438–439
  stiffness, 419–420, 444–445
  strength, 420–421, 442–443
  structural members, 421
  tensile stiffness, 433, 434–435f
  weight savings, 432
  wing rigs, 447–448
Room corner test ISO 9705, 126–127
Rotating blade-fixed coordinate system, 374, 374f
Rule-of-thumb, 16–18

S

Sailing yacht rig, 418. *See also* Rig
  functions, 419
  physical environment, 421
Salty breezes, 468–469
Sandwich composites, 58
Sandwich structure, 21–22, 22t
  foam core sandwich structures, 189–190
  marine loading conditions, 190
  PVC foams
    compressive properties, 192–198
    impact properties, 198–201
    microstructure, 191
    moisture effects, 201–203

Sandwich structure *(Continued)*
  syntactic foams
    compressive properties, 206–209
    hollow particles and properties, 203–206
    impact properties, 209–214
    moisture effects, 214–216
    tailoring properties, 216–219
Sandwich-structured composites, 301, 305–306
Saturated specimens
  static testing, 99–101
  testing methodology, 98–99
Saturation level, temperature effect on, 108
Scaling approach, MHK turbine, 356–358
Schooner rig, 421
Screw propeller, 271–272 See also *specific propellers*
Seale construction, 453–454, 455*f*, 471
Sea water conditioning techniques, 93–98
  accelerated aging techniques, 103–106
  aging procedures, 97–98
  fatigue testing, 102
  flexural strength, 100*f*
  ILSS, 100*f*
  impact testing, 101–102
  manufacturing process effect, 97
  pH effect, 96
  pressure effect, 94–95
  specimen dimensions effect, 96
  specimen testing, 102–103
  static testing of conditioned specimens, 99–101
  temperature effect, 94
  testing environment, 95–96
  water composition effect, 95
Seawater environment
  acrylic composite, 44*f*
  carbon/PEEK, 41*t*
  thermoplastic matrix composites, 39
Seeman Composites Resin Infusion Molding Process (SCRIMP), 228
Self-weight, rig, 442
Semicrystalline polymers, 32
Semi-Implicit Method for Pressure Linked Equations (SIMPLE) algorithm, 274, 381
Semiproducts, 33–34, 34*f*
Semisubmersible, 454
Sequential feed strategy, 251–252

Shape memory alloy (SMA), 272
  constitutive equation, 277
  recovery stress
    at different temperatures, 277–278
    in lamina, 277
SIMLA software, 405
Single burning item (SBI) test, 129–130
Single skin laminates, 141–142
Sinking process, CBF
  earth pressure, 337–341, 338–341*f*
  negative-pressure, 333–334, 336–337, 342
Sizing selection, durability testing, 92
Sloop style rig, 421
SMA hybrid composite (SMAHC) propeller, 272. See also Propeller
  design of thickness, 285–287, 286*f*
  open water characteristics, 295*f*
  positioning inside propeller blade, 287–292
  Tsai-Hill index, 286–287, 287*t*
  twist and deformation in blade, 288–291*t*
    at different operating conditions, 292–295, 293*t*
Smoke growth rate index (SMOGRAta), 130–131
Smoke production, and toxicity, 140–141
Smoke production rate (SPR), 140–141
Socket terminations, 467, 468*f*
Softwoods, 304
Soil pressure distribution, 328, 328*f*
Soil properties, 324*t*
Specific extinction area (SEA), 140–141
Specimen dimensions effect, 96
Split-Hopkinson pressure bar, 196
Spreader sweep, 422–423, 423*f*
Standard fire, 126
Standing rig, 432–433, 443
  carbon, 433
  fiber-based, 432
  solutions, 443–444
Static testing, saturated specimens, 99–101
Steady fluid-structural coupling, 378–379
Steel cables
  final constructing cost for structures, 456*t*
  literature, 456
  mechanical properties, 455*t*
  requirements, 473
  Tension-Leg Platforms, 472
Steel plates, 321–323, 333
Stiffness

Index

in-plane tensile, 435, 436f
laminates, 165f
offshore offloading hoses, 410–412, 411f
rig, 419–420, 444–445
Stitched fabrics, 19–20
Stress-strain relationship, 275, 400
Structural health monitoring (SHM), 69, 69f
Structural stability, 129–130
Suction surface, propeller, 364
Superyachts, 418–419, 431, 437f
Sweep, spreader, 422–423, 423f
Syntactic foams, 188–189
   compressive properties, 206–209
   hollow particles and properties, 203–206
   impact properties, 209–214
   microstructures, 203, 204f
   moisture effects, 214–216
   tailoring properties, 216–219
   VE220-30 type, 212–213, 213f, 215f

T

Tangent intersection method, 332–333
Temperature effect
   diffusivity, 107
   durability testing, 88
   saturation level, 108
   sea water conditioning techniques, 94
Tensile stress, composite cables, 484–485
Tensile-tensile fatigue tests, 471–473, 474f
Tensile tests, 466–471, 472f, 477t
Tension controllers, 18–19
Tension-Leg Platforms (TLPs), 454, 469–472
Testing environment, 95–96
Tetraglycidyl 4,4'-diaminodiphenylmethane (TGDDM), 5
Thermo-gravimetric analysis (TGA), 149
Thermoplastic matrix composites, 31–32
   advantages and drawbacks, 33t
   autoclave, 35–36
   automated tape placement, 36–38
   compression moulding, 36
   heating, 34
   high-performance, 45
   influence of marine environment, 39–43
   infusion, 38
   manufacturing options, 33–39
   material options, 32–33
   morphology, 39
   process cycle, 35f
   recycling and environmental impact, 46–48
   repair, 45–46
   seawater immersion, 44f
   for underwater applications, 37f
   underwater structures, 44–45
Thermoplastic polymers
   performance characteristics, 13t
   poly(propylene), 8–11
   polyamide, 12
   polyesters, 12
Thermoset composite
   advantages and drawbacks, 33t
   material options, 32–33
Thermosetting resins
   bismaleimide resins, 8
   epoxide resin, 5
   phenol-formaldehyde resin, 5
   properties of, 9–11t
   unsaturated polyester resin, 5–7
   vinyl ester resin, 8
Thin-wing theory, 371–372
Three-dimensional solid model representation, 203, 204f
Three-dimensional woven fabrics, 20–21
Tidal turbine blades, 87
Tidal turbines, 346–347, 349, 351–352
Tiltrotor aircraft, 272
TLPs. See Tension-Leg Platforms (TLPs)
Tokyo Rope, 464, 467–468
Total heat release (THR), 130, 138, 153f
Total smoke production (TSP), 130–131, 140–141
Total smoke production rate (SPRtotal), 130–131
Traditional materials, 86
Trailing edge, 364
Transient fluid-structural coupling, 380–381
Tsai-Hill failure criterion, 279, 281, 286–287, 287t, 402
Tsai-Wu failure criterion, 168–169, 169f, 402
Tubing, 243
Turbulence, MHK turbine, 353
Turbulent flow, 273–274

U

Ultimate Limit State (ULS), 164–166
Underwater structures, 44–45

Unprotected glass fibers, 91
Unsaturated polyester resin (UPE), 5–7, 7f
USS Zumwalt (DDG 1000), 187

V

Vacuum, 265
  gauges, 248, 249f
  pumps, 246–247
  receiver, 247
  ring main, 247
Vacuum bag, 242
  hand layup, 189
  infusion, 92
  leaks in, 265–266
  processes, 21
Vessel structural computational design, 309–312
Vinylester cohesive material model, 75t
Vinyl ester resin (VE), 8
Vinyl esters, 90
Vinylsilane, 16–18

Viscosity, resin infusion, 235–236, 238, 265
Von-Mises stress, 382
Vortex-lattice method (VLM), 371–373

W

Warp-insertion weft-knit (WIWK) fabric, 20
Warrington construction, 453–454, 455f
Water composition effect, 95
Water diffusion, 39
Weibull bundle strength, 142
Weight savings, in rig, 432
Wind Blade Using Cost-Effective Advanced Lightweight Design (WALiD), 47
Wing rigs, 447–448
Wire helix, 398–400
Wire rope clamp terminations, 467, 468f
Woven fabrics, 19

Y

Yacht rig sailing. *See* Sailing yacht rig

Printed in the United States
By Bookmasters